Graduate Texts in Ma

Graduate Texts in Mathematics

(continued after index)

Steven Roman

Advanced Linear Algebra

Third Edition

Steven Roman
8 Night Star
Irvine, CA 92603
USA
sroman@romanpress.com

ISBN-13: 978-1-4419-2498-8 e-ISBN-13: 978-0-387-72831-5

Mathematics Subject Classification (2000): 15-01

Printed on acid-free paper.

9 8 7 6 5 4 3 2 1

springer.com

To Donna
and to
Rashelle, Carol and Dan

Preface to the Third Edition

Let me begin by thanking the readers of the second edition for their many helpful comments and suggestions, with special thanks to Joe Kidd and Nam Trang. For the third edition, I have corrected all known errors, polished and refined some arguments (such as the discussion of reflexivity, the rational canonical form, best approximations and the definitions of tensor products) and upgraded some proofs that were originally done only for finite-dimensional/rank cases. I have also moved some of the material on projection operators to an earlier position in the text.

A few new theorems have been added in this edition, including the spectral mapping theorem and a theorem to the effect that $\dim(V) \leq \dim(V^*)$, with equality if and only if V is finite-dimensional.

I have also added a new chapter on associative algebras that includes the well-known characterizations of the finite-dimensional division algebras over the real field (a theorem of Frobenius) and over a finite field (Wedderburn's theorem). The reference section has been enlarged considerably, with over a hundred references to books on linear algebra.

Steven Roman *Irvine, California, May 2007*

Preface to the Second Edition

Let me begin by thanking the readers of the first edition for their many helpful comments and suggestions. The second edition represents a major change from the first edition. Indeed, one might say that it is a totally new book, with the exception of the general range of topics covered.

The text has been completely rewritten. I hope that an additional 12 years and roughly 20 books worth of experience has enabled me to improve the quality of my exposition. Also, the exercise sets have been completely rewritten.

The second edition contains two new chapters: a chapter on convexity, separation and positive solutions to linear systems (Chapter 15) and a chapter on the QR decomposition, singular values and pseudoinverses (Chapter 17). The treatments of tensor products and the umbral calculus have been greatly expanded and I have included discussions of determinants (in the chapter on tensor products), the complexification of a real vector space, Schur's theorem and Geršgorin disks.

Steven Roman *Irvine, California February 2005*

Preface to the First Edition

This book is a thorough introduction to linear algebra, for the graduate or advanced undergraduate student. Prerequisites are limited to a knowledge of the basic properties of matrices and determinants. However, since we cover the basics of vector spaces and linear transformations rather rapidly, a prior course in linear algebra (even at the sophomore level), along with a certain measure of "mathematical maturity," is highly desirable.

Chapter 0 contains a summary of certain topics in modern algebra that are required for the sequel. *This chapter should be skimmed quickly and then used primarily as a reference.* Chapters 1–3 contain a discussion of the basic properties of vector spaces and linear transformations.

Chapter 4 is devoted to a discussion of modules, emphasizing a comparison between the properties of modules and those of vector spaces. Chapter 5 provides more on modules. The main goals of this chapter are to prove that any two bases of a free module have the same cardinality and to introduce Noetherian modules. However, the instructor may simply skim over this chapter, omitting all proofs. Chapter 6 is devoted to the theory of modules over a principal ideal domain, establishing the cyclic decomposition theorem for finitely generated modules. This theorem is the key to the structure theorems for finite-dimensional linear operators, discussed in Chapters 7 and 8.

Chapter 9 is devoted to real and complex inner product spaces. The emphasis here is on the finite-dimensional case, in order to arrive as quickly as possible at the finite-dimensional spectral theorem for normal operators, in Chapter 10. However, we have endeavored to state as many results as is convenient for vector spaces of arbitrary dimension.

The second part of the book consists of a collection of independent topics, with the one exception that Chapter 13 requires Chapter 12. Chapter 11 is on metric vector spaces, where we describe the structure of symplectic and orthogonal geometries over various base fields. Chapter 12 contains enough material on metric spaces to allow a unified treatment of topological issues for the basic

Hilbert space theory of Chapter 13. The rather lengthy proof that every metric space can be embedded in its completion may be omitted.

Chapter 14 contains a brief introduction to tensor products. In order to motivate the universal property of tensor products, without getting too involved in categorical terminology, we first treat both free vector spaces and the familiar direct sum, in a universal way. Chapter 15 (Chapter 16 in the second edition) is on affine geometry, emphasizing algebraic, rather than geometric, concepts.

The final chapter provides an introduction to a relatively new subject, called the umbral calculus. This is an algebraic theory used to study certain types of polynomial functions that play an important role in applied mathematics. We give only a brief introduction to the subject – emphasizing the algebraic aspects, rather than the applications. This is the first time that this subject has appeared in a true textbook.

One final comment. Unless otherwise mentioned, omission of a proof in the text is a tacit suggestion that the reader attempt to supply one.

Steven Roman *Irvine, California*

Contents

Preliminaries

In this chapter, we briefly discuss some topics that are needed for the sequel. This chapter should be skimmed quickly and used primarily as a reference.

Part 1 Preliminaries

Multisets

The following simple concept is much more useful than its infrequent appearance would indicate.

Definition *Let S be a nonempty set. A* **multiset** *M with* **underlying set** *S is a set of ordered pairs*

$$M = \{(s_i, n_i) \mid s_i \in S, n_i \in \mathbb{Z}^+, s_i \neq s_j \textit{ for } i \neq j\}$$

where $\mathbb{Z}^+ = \{1, 2, \ldots\}$. The number n_i is referred to as the **multiplicity** *of the elements s_i in M. If the underlying set of a multiset is finite, we say that the multiset is* **finite**. *The* **size** *of a finite multiset M is the sum of the multiplicities of all of its elements.*□

For example, $M = \{(a,2), (b,3), (c,1)\}$ is a multiset with underlying set $S = \{a, b, c\}$. The element a has multiplicity 2. One often writes out the elements of a multiset according to multiplicities, as in $M = \{a, a, b, b, b, c\}$.

Of course, two mutlisets are equal if their underlying sets are equal and if the multiplicity of each element in the common underlying set is the same in both multisets.

Matrices

The set of $m \times n$ matrices with entries in a field F is denoted by $\mathcal{M}_{m,n}(F)$ or by $\mathcal{M}_{m,n}$ when the field does not require mention. The set $\mathcal{M}_{n,n}(\mathcal{F})$ is denoted by $\mathcal{M}_n(F)$ or $\mathcal{M}_n$. If $A \in \mathcal{M}$, the (i,j)th entry of A will be denoted by $A_{i,j}$. The identity matrix of size $n \times n$ is denoted by I_n. The elements of the base field F are called **scalars**. We expect that the reader is familiar with the basic properties of matrices, including matrix addition and multiplication.

The **main diagonal** of an $m \times n$ matrix A is the sequence of entries

$$A_{1,1}, A_{2,2}, \ldots, A_{k,k}$$

where $k = \min\{m, n\}$.

Definition *The* **transpose** *of $A \in \mathcal{M}_{m,n}$ is the matrix A^t defined by*

$$(A^t)_{i,j} = A_{j,i}$$

A matrix A is **symmetric** *if $A = A^t$ and* **skew-symmetric** *if $A^t = -A$.*□

Theorem 0.1 (*Properties of the transpose*) *Let A, $B \in \mathcal{M}_{m,n}$. Then*
1) $(A^t)^t = A$
2) $(A + B)^t = A^t + B^t$
3) $(rA)^t = rA^t$ *for all $r \in F$*
4) $(AB)^t = B^t A^t$ *provided that the product AB is defined*
5) $\det(A^t) = \det(A)$.□

Partitioning and Matrix Multiplication

Let M be a matrix of size $m \times n$. If $B \subseteq \{1, \ldots, m\}$ and $C \subseteq \{1, \ldots, n\}$, then the **submatrix** $M[B, C]$ is the matrix obtained from M by keeping only the rows with index in B and the columns with index in C. Thus, all other rows and columns are discarded and $M[B, C]$ has size $|B| \times |C|$.

Suppose that $M \in \mathcal{M}_{m,n}$ and $N \in \mathcal{M}_{n,k}$. Let

1) $\mathcal{P} = \{B_1, \ldots, B_p\}$ be a partition of $\{1, \ldots, m\}$
2) $\mathcal{Q} = \{C_1, \ldots, C_q\}$ be a partition of $\{1, \ldots, n\}$
3) $\mathcal{R} = \{D_1, \ldots, D_r\}$ be a partition of $\{1, \ldots, k\}$

(Partitions are defined formally later in this chapter.) Then it is a very useful fact that matrix multiplication can be performed at the block level as well as at the entry level. In particular, we have

$$[MN][B_i, D_j] = \sum_{C_h \in \mathcal{Q}} M[B_i, C_h]N[C_h, D_j]$$

When the partitions in question contain only single-element blocks, this is precisely the usual formula for matrix multiplication

$$[MN]_{i,j} = \sum_{h=1}^{m} M_{i,h}N_{h,j}$$

Block Matrices

It will be convenient to introduce the notational device of a block matrix. If $B_{i,j}$ are matrices of the appropriate sizes, then by the **block matrix**

$$M = \begin{bmatrix} B_{1,1} & B_{1,2} & \cdots & B_{1,n} \\ \vdots & \vdots & & \vdots \\ B_{m,1} & B_{m,2} & \cdots & B_{m,n} \end{bmatrix}_{\text{block}}$$

we mean the matrix whose upper left *submatrix* is $B_{1,1}$, and so on. Thus, the $B_{i,j}$'s are *submatrices* of M and not entries. A square matrix of the form

$$M = \begin{bmatrix} B_1 & 0 & \cdots & 0 \\ 0 & \ddots & \ddots & \vdots \\ \vdots & \ddots & \ddots & 0 \\ 0 & \cdots & 0 & B_n \end{bmatrix}_{\text{block}}$$

where each B_i is square and 0 is a zero submatrix, is said to be a **block diagonal matrix**.

Elementary Row Operations

Recall that there are three types of elementary row operations. Type 1 operations consist of multiplying a row of A by a nonzero scalar. Type 2 operations consist of interchanging two rows of A. Type 3 operations consist of adding a scalar multiple of one row of A to another row of A.

If we perform an elementary operation of type k to an identity matrix I_n, the result is called an **elementary matrix** of type k. It is easy to see that all elementary matrices are invertible.

In order to perform an elementary row operation on $A \in \mathcal{M}_{m,n}$ we can perform that operation on the identity I_m, to obtain an elementary matrix E and then take the product EA. Note that multiplying on the right by E has the effect of performing column operations.

Definition *A matrix R is said to be in* **reduced row echelon form** *if*

1) *All rows consisting only of 0's appear at the bottom of the matrix.*
2) *In any nonzero row, the first nonzero entry is a 1. This entry is called a* **leading entry**.
3) *For any two consecutive rows, the leading entry of the lower row is to the right of the leading entry of the upper row.*
4) *Any column that contains a leading entry has 0's in all other positions.* □

Here are the basic facts concerning reduced row echelon form.

Theorem 0.2 *Matrices* $A, B \in \mathcal{M}_{m,n}$ *are* **row equivalent**, *denoted by* $A \sim B$, *if either one can be obtained from the other by a series of elementary row operations.*

1) *Row equivalence is an equivalence relation. That is,*
   a) $A \sim A$
   b) $A \sim B \Rightarrow B \sim A$
   c) $A \sim B$, $B \sim C \Rightarrow A \sim C$.
2) *A matrix A is row equivalent to one and only one matrix R that is in reduced row echelon form. The matrix R is called the* **reduced row echelon form** *of A. Furthermore,*

$$R = E_1 \cdots E_k A$$

   where E_i are the elementary matrices required to reduce A to reduced row echelon form.
3) *A is invertible if and only if its reduced row echelon form is an identity matrix. Hence, a matrix is invertible if and only if it is the product of elementary matrices.*□

The following definition is probably well known to the reader.

Definition *A square matrix is* **upper triangular** *if all of its entries below the main diagonal are* 0. *Similarly, a square matrix is* **lower triangular** *if all of its entries above the main diagonal are* 0. *A square matrix is* **diagonal** *if all of its entries off the main diagonal are* 0.□

Determinants

We assume that the reader is familiar with the following basic properties of determinants.

Theorem 0.3 *Let* $A \in \mathcal{M}_{n,n}(F)$. *Then* $\det(A)$ *is an element of* F. *Furthermore,*

1) *For any* $B \in \mathcal{M}_n(F)$,

$$\det(AB) = \det(A)\det(B)$$

2) *A is nonsingular (invertible) if and only if* $\det(A) \neq 0$.
3) *The determinant of an upper triangular or lower triangular matrix is the product of the entries on its main diagonal.*
4) *If a square matrix M has the block diagonal form*

$$M = \begin{bmatrix} B_1 & 0 & \cdots & 0 \\ 0 & \ddots & \ddots & \vdots \\ \vdots & \ddots & \ddots & 0 \\ 0 & \cdots & 0 & B_n \end{bmatrix}_{\text{block}}$$

   then $\det(M) = \prod \det(B_i)$.□

Polynomials

The set of all polynomials in the variable x with coefficients from a field F is denoted by $F[x]$. If $p(x) \in F[x]$, we say that $p(x)$ is a polynomial **over** F. If

$$p(x) = a_0 + a_1 x + \cdots + a_n x^n$$

is a polynomial with $a_n \neq 0$, then a_n is called the **leading coefficient** of $p(x)$ and the **degree** of $p(x)$ is n, written $\deg p(x) = n$. For convenience, the degree of the zero polynomial is $-\infty$. A polynomial is **monic** if its leading coefficient is 1.

Theorem 0.4 (Division algorithm) *Let $f(x), g(x) \in F[x]$ where $\deg g(x) > 0$. Then there exist unique polynomials $q(x), r(x) \in F[x]$ for which*

$$f(x) = q(x)g(x) + r(x)$$

where $r(x) = 0$ or $0 \leq \deg r(x) < \deg g(x)$.□

If $p(x)$ **divides** $q(x)$, that is, if there exists a polynomial $f(x)$ for which

$$q(x) = f(x)p(x)$$

then we write $p(x) \mid q(x)$. A nonzero polynomial $p(x) \in F[x]$ is said to **split** over F if $p(x)$ can be written as a product of linear factors

$$p(x) = (x - r_1) \cdots (x - r_n)$$

where $r_i \in F$.

Theorem 0.5 *Let $f(x), g(x) \in F[x]$. The* **greatest common divisor** *of $f(x)$ and $g(x)$, denoted by $\gcd(f(x), g(x))$, is the unique monic polynomial $p(x)$ over F for which*

1) $p(x) \mid f(x)$ and $p(x) \mid g(x)$
2) if $r(x) \mid f(x)$ and $r(x) \mid g(x)$ then $r(x) \mid p(x)$.

Furthermore, there exist polynomials $a(x)$ and $b(x)$ over F for which

$$\gcd(f(x), g(x)) = a(x)f(x) + b(x)g(x)$$ □

Definition *The polynomials $f(x), g(x) \in F[x]$ are* **relatively prime** *if $\gcd(f(x), g(x)) = 1$. In particular, $f(x)$ and $g(x)$ are relatively prime if and only if there exist polynomials $a(x)$ and $b(x)$ over F for which*

$$a(x)f(x) + b(x)g(x) = 1$$ □

Definition *A nonconstant polynomial $f(x) \in F[x]$ is* **irreducible** *if whenever $f(x) = p(x)q(x)$, then one of $p(x)$ and $q(x)$ must be constant.*□

The following two theorems support the view that irreducible polynomials behave like prime numbers.

Theorem 0.6 *A nonconstant polynomial $f(x)$ is irreducible if and only if it has the property that whenever $f(x) \mid p(x)q(x)$, then either $f(x) \mid p(x)$ or $f(x) \mid q(x)$.* □

Theorem 0.7 *Every nonconstant polynomial in $F[x]$ can be written as a product of irreducible polynomials. Moreover, this expression is unique up to order of the factors and multiplication by a scalar.* □

Functions

To set our notation, we should make a few comments about functions.

Definition *Let $f: S \to T$ be a function from a set S to a set T.*

1) *The* **domain** *of f is the set S and the* **range** *of f is T.*
2) *The* **image** *of f is the set* $\mathrm{im}(f) = \{f(s) \mid s \in S\}$.
3) *f is* **injective (one-to-one)**, *or an* **injection**, *if $x \neq y \Rightarrow f(x) \neq f(y)$.*
4) *f is* **surjective (onto T)**, *or a* **surjection**, *if* $\mathrm{im}(f) = T$.
5) *f is* **bijective**, *or a* **bijection**, *if it is both injective and surjective.*
6) *Assuming that $0 \in T$, the* **support** *of f is*

$$\mathrm{supp}(f) = \{s \in S \mid f(s) \neq 0\}$$ □

If $f: S \to T$ is injective, then its inverse $f^{-1}: \mathrm{im}(f) \to S$ exists and is well-defined as a function on $\mathrm{im}(f)$.

It will be convenient to apply f to subsets of S and T. In particular, if $X \subseteq S$ and if $Y \subseteq T$, we set

$$f(X) = \{f(x) \mid x \in X\}$$

and

$$f^{-1}(Y) = \{s \in S \mid f(s) \in Y\}$$

Note that the latter is defined even if f is not injective.

Let $f: S \to T$. If $A \subseteq S$, the **restriction** of f to A is the function $f|_A: A \to T$ defined by

$$f|_A(a) = f(a)$$

for all $a \in A$. Clearly, the restriction of an injective map is injective.

In the other direction, if $f: S \to T$ and if $S \subseteq U$, then an **extension** of f to U is a function $\overline{f}: U \to T$ for which $\overline{f}|_S = f$.

Equivalence Relations

The concept of an equivalence relation plays a major role in the study of matrices and linear transformations.

Definition *Let S be a nonempty set. A binary relation $\sim$ on S is called an* **equivalence relation** *on S if it satisfies the following conditions:*

1) **(Reflexivity)**
$$a \sim a$$
for all $a \in S$.
2) **(Symmetry)**
$$a \sim b \Rightarrow b \sim a$$
for all $a, b \in S$.
3) **(Transitivity)**
$$a \sim b, b \sim c \Rightarrow a \sim c$$
for all $a, b, c \in S$.□

Definition *Let $\sim$ be an equivalence relation on S. For $a \in S$, the set of all elements equivalent to a is denoted by*
$$[a] = \{b \in S \mid b \sim a\}$$
and called the **equivalence class** *of a.*□

Theorem 0.8 *Let $\sim$ be an equivalence relation on S. Then*
1) $b \in [a] \Leftrightarrow a \in [b] \Leftrightarrow [a] = [b]$
2) *For any $a, b \in S$, we have either $[a] = [b]$ or $[a] \cap [b] = \emptyset$.*□

Definition *A* **partition** *of a nonempty set S is a collection $\{A_1, \ldots, A_n\}$ of* nonempty *subsets of S, called the* **blocks** *of the partition, for which*
1) $A_i \cap A_j = \emptyset$ *for all $i \neq j$*
2) $S = A_1 \cup \cdots \cup A_n$.□

The following theorem sheds considerable light on the concept of an equivalence relation.

Theorem 0.9
1) *Let $\sim$ be an equivalence relation on S. Then the set of* distinct *equivalence classes with respect to $\sim$ are the blocks of a partition of S.*
2) *Conversely, if $\mathcal{P}$ is a partition of S, the binary relation $\sim$ defined by*
$$a \sim b \text{ if } a \text{ and } b \text{ lie in the same block of } \mathcal{P}$$

is an equivalence relation on S, whose equivalence classes are the blocks of $\mathcal{P}$.

This establishes a one-to-one correspondence between equivalence relations on S and partitions of S.$\square$

The most important problem related to equivalence relations is that of finding an efficient way to determine when two elements are equivalent. Unfortunately, in most cases, the definition does not provide an efficient test for equivalence and so we are led to the following concepts.

Definition *Let $\sim$ be an equivalence relation on S. A function $f\colon S \to T$, where T is any set, is called an* **invariant** *of $\sim$ if it is constant on the equivalence classes of $\sim$, that is,*

$$a \sim b \Rightarrow f(a) = f(b)$$

and a **complete invariant** *if it is constant and distinct on the equivalence classes of $\sim$, that is,*

$$a \sim b \Leftrightarrow f(a) = f(b)$$

A collection $\{f_1, \ldots, f_n\}$ of invariants is called a **complete system of invariants** *if*

$$a \sim b \Leftrightarrow f_i(a) = f_i(b) \text{ for all } i = 1, \ldots, n \qquad \square$$

Definition *Let $\sim$ be an equivalence relation on S. A subset $C \subseteq S$ is said to be a set of* **canonical forms** *(or just a* **canonical form***) for $\sim$ if for every $s \in S$, there is* exactly one *$c \in C$ such that $c \sim s$. Put another way, each equivalence class under $\sim$ contains* exactly one *member of C.*$\square$

Example 0.1 Define a binary relation $\sim$ on $F[x]$ by letting $p(x) \sim q(x)$ if and only if $p(x) = aq(x)$ for some nonzero constant $a \in F$. This is easily seen to be an equivalence relation. The function that assigns to each polynomial its degree is an invariant, since

$$p(x) \sim q(x) \Rightarrow \deg(p(x)) = \deg(q(x))$$

However, it is not a complete invariant, since there are inequivalent polynomials with the same degree. The set of all monic polynomials is a set of canonical forms for this equivalence relation.$\square$

Example 0.2 We have remarked that row equivalence is an equivalence relation on $\mathcal{M}_{m,n}(F)$. Moreover, the subset of reduced row echelon form matrices is a set of canonical forms for row equivalence, since every matrix is row equivalent to a unique matrix in reduced row echelon form.$\square$

Example 0.3 Two matrices $A, B \in \mathcal{M}_n(F)$ are row equivalent if and only if there is an invertible matrix P such that $A = PB$. Similarly, A and B are **column equivalent**, that is, A can be reduced to B using elementary column operations, if and only if there exists an invertible matrix Q such that $A = BQ$.

Two matrices A and B are said to be **equivalent** if there exist invertible matrices P and Q for which

$$A = PBQ$$

Put another way, A and B are equivalent if A can be reduced to B by performing a series of elementary row and/or column operations. (The use of the term equivalent is unfortunate, since it applies to all equivalence relations, not just this one. However, the terminology is standard, so we use it here.)

It is not hard to see that an $m \times n$ matrix R that is in both reduced row echelon form and reduced column echelon form must have the block form

$$J_k = \begin{bmatrix} I_k & 0_{k,n-k} \\ 0_{m-k,k} & 0_{m-k,n-k} \end{bmatrix}_{\text{block}}$$

We leave it to the reader to show that every matrix A in $\mathcal{M}_n$ is equivalent to exactly one matrix of the form J_k and so the set of these matrices is a set of canonical forms for equivalence. Moreover, the function f defined by $f(A) = k$, where $A \sim J_k$, is a complete invariant for equivalence.

Since the rank of J_k is k and since neither row nor column operations affect the rank, we deduce that the rank of A is k. Hence, rank is a complete invariant for equivalence. In other words, two matrices are equivalent if and only if they have the same rank.□

Example 0.4 Two matrices $A, B \in \mathcal{M}_n(F)$ are said to be **similar** if there exists an invertible matrix P such that

$$A = PBP^{-1}$$

Similarity is easily seen to be an equivalence relation on $\mathcal{M}_n$. As we will learn, two matrices are similar if and only if they represent the same linear operators on a given n-dimensional vector space V. Hence, similarity is extremely important for studying the structure of linear operators. One of the main goals of this book is to develop canonical forms for similarity.

We leave it to the reader to show that the determinant function and the trace function are invariants for similarity. However, these two invariants do not, in general, form a complete system of invariants.□

Example 0.5 Two matrices $A, B \in \mathcal{M}_n(F)$ are said to be **congruent** if there exists an invertible matrix P for which

$$A = PBP^t$$

where P^t is the transpose of P. This relation is easily seen to be an equivalence relation and we will devote some effort to finding canonical forms for congruence. For some base fields F (such as $\mathbb{R}$, $\mathbb{C}$ or a finite field), this is relatively easy to do, but for other base fields (such as $\mathbb{Q}$), it is extremely difficult.□

Zorn's Lemma

In order to show that any vector space has a basis, we require a result known as *Zorn's lemma.* To state this lemma, we need some preliminary definitions.

Definition *A* **partially ordered set** *is a pair $(P, \leq)$ where P is a nonempty set and $\leq$ is a binary relation called a* **partial order**, *read "less than or equal to," with the following properties:*

1) **(Reflexivity)** *For all $a \in P$,*

$$a \leq a$$

2) **(Antisymmetry)** *For all $a, b \in P$,*

$$a \leq b \text{ and } b \leq a \text{ implies } a = b$$

3) **(Transitivity)** *For all $a, b, c \in P$,*

$$a \leq b \text{ and } b \leq c \text{ implies } a \leq c$$

Partially ordered sets are also called **posets.**□

It is customary to use a phrase such as "Let P be a partially ordered set" when the partial order is understood. Here are some key terms related to partially ordered sets.

Definition Let P be a partially ordered set.

1) *The* **maximum (largest, top)** *element of P, should it exist, is an element $M \in P$ with the property that all elements of P are less than or equal to M, that is,*

$$p \in P \Rightarrow p \leq M$$

Similarly, the **mimimum (least, smallest, bottom)** *element of P, should it exist, is an element $N \in P$ with the property that all elements of P are greater than or equal to N, that is,*

$$p \in P \Rightarrow N \leq p$$

2) *A* **maximal element** *is an element $m \in P$ with the property that there is no larger element in P, that is,*

$$p \in P, m \leq p \Rightarrow m = p$$

Similarly, a **minimal element** *is an element* $n \in P$ *with the property that there is no smaller element in* P*, that is,*

$$p \in P, p \leq n \Rightarrow p = n$$

3) *Let* $a, b \in P$*. Then* $u \in P$ *is an* **upper bound** *for* a *and* b *if*

$$a \leq u \text{ and } b \leq u$$

The unique smallest upper bound for a *and* b*, if it exists, is called the* **least upper bound** *of* a *and* b *and is denoted by* $\text{lub}\{a, b\}$*.*

4) *Let* $a, b \in P$*. Then* $\ell \in P$ *is a* **lower bound** *for* a *and* b *if*

$$\ell \leq a \text{ and } \ell \leq b$$

The unique largest lower bound for a *and* b*, if it exists, is called the* **greatest lower bound** *of* a *and* b *and is denoted by* $\text{glb}\{a, b\}$*.*□

Let S be a subset of a partially ordered set P. We say that an element $u \in P$ is an **upper bound** for S if $s \leq u$ for all $s \in S$. Lower bounds are defined similarly.

Note that in a partially ordered set, it is possible that not all elements are comparable. In other words, it is possible to have $x, y \in P$ with the property that $x \not\leq y$ and $y \not\leq x$.

Definition *A partially ordered set in which every pair of elements is comparable is called a* **totally ordered set**, *or a* **linearly ordered set**. *Any totally ordered subset of a partially ordered set* P *is called a* **chain** *in* P*.*□

Example 0.6

1) The set $\mathbb{R}$ of real numbers, with the usual binary relation $\leq$, is a partially ordered set. It is also a totally ordered set. It has no maximal elements.
2) The set $\mathbb{N} = \{0, 1, \dots\}$ of natural numbers, together with the binary relation of divides, is a partially ordered set. It is customary to write $n \mid m$ to indicate that n divides m. The subset S of $\mathbb{N}$ consisting of all powers of 2 is a totally ordered subset of $\mathbb{N}$, that is, it is a chain in $\mathbb{N}$. The set $P = \{2, 4, 8, 3, 9, 27\}$ is a partially ordered set under $\mid$. It has two maximal elements, namely 8 and 27. The subset $Q = \{2, 3, 5, 7, 11\}$ is a partially ordered set in which every element is both maximal and minimal!
3) Let S be any set and let $\mathcal{P}(S)$ be the power set of S, that is, the set of all subsets of S. Then $\mathcal{P}(S)$, together with the subset relation $\subseteq$, is a partially ordered set.□

Now we can state Zorn's lemma, which gives a condition under which a partially ordered set has a maximal element.

Theorem 0.10 (Zorn's lemma) *If P is a partially ordered set in which every chain has an upper bound, then P has a maximal element.* □

We will use Zorn's lemma to prove that every vector space has a basis. Zorn's lemma is equivalent to the famous axiom of choice. As such, it is not subject to proof from the other axioms of ordinary (ZF) set theory. Zorn's lemma has many important equivalancies, one of which is the *well-ordering principle*. A **well ordering** on a nonempty set X is a total order on X with the property that every nonempty subset of X has a least element.

Theorem 0.11 (Well-ordering principle) *Every nonempty set has a well ordering.* □

Cardinality

Two sets S and T have the same **cardinality**, written

$$|S| = |T|$$

if there is a bijective function (a one-to-one correspondence) between the sets. The reader is probably aware of the fact that

$$|\mathbb{Z}| = |\mathbb{N}| \text{ and } |\mathbb{Q}| = |\mathbb{N}|$$

where $\mathbb{N}$ denotes the natural numbers, $\mathbb{Z}$ the integers and $\mathbb{Q}$ the rational numbers.

If S is in one-to-one correspondence with a *subset* of T, we write $|S| \leq |T|$. If S is in one-to-one correspondence with a *proper* subset of T but not all of T, then we write $|S| < |T|$. The second condition is necessary, since, for instance, $\mathbb{N}$ is in one-to-one correspondence with a proper subset of $\mathbb{Z}$ and yet $\mathbb{N}$ is also in one-to-one correspondence with $\mathbb{Z}$ itself. Hence, $|\mathbb{N}| = |\mathbb{Z}|$.

This is not the place to enter into a detailed discussion of cardinal numbers. The intention here is that the cardinality of a set, whatever that is, represents the "size" of the set. It is actually easier to talk about two sets having the same, or different, size (cardinality) than it is to explicitly define the size (cardinality) of a given set.

Be that as it may, we associate to each set S a cardinal number, denoted by $|S|$ or $\operatorname{card}(S)$, that is intended to measure the size of the set. Actually, cardinal numbers are just very special types of sets. However, we can simply think of them as vague amorphous objects that measure the size of sets.

Definition

1) *A set is* **finite** *if it can be put in one-to-one correspondence with a set of the form $\mathbb{Z}_n = \{0, 1, \ldots, n-1\}$, for some nonnegative integer n. A set that is* *not finite is* **infinite**. *The* **cardinal number** (*or* **cardinality**) *of a finite set is just the number of elements in the set.*

2) *The* **cardinal number** *of the set* $\mathbb{N}$ *of natural numbers is* $\aleph_0$ *(read "aleph nought"), where* $\aleph$ *is the first letter of the Hebrew alphabet. Hence,*

$$|\mathbb{N}| = |\mathbb{Z}| = |\mathbb{Q}| = \aleph_0$$

3) *Any set with cardinality* $\aleph_0$ *is called a* **countably infinite** *set and any finite or countably infinite set is called a* **countable** *set. An infinite set that is not countable is said to be* **uncountable**. □

Since it can be shown that $|\mathbb{R}| > |\mathbb{N}|$, the real numbers are uncountable.

If S and T are *finite* sets, then it is well known that

$$|S| \le |T| \text{ and } |T| \le |S| \Rightarrow |S| = |T|$$

The first part of the next theorem tells us that this is also true for infinite sets.

The reader will no doubt recall that the **power set** $\mathcal{P}(S)$ of a set S is the set of all subsets of S. For finite sets, the power set of S is always bigger than the set itself. In fact,

$$|S| = n \Rightarrow |\mathcal{P}(S)| = 2^n$$

The second part of the next theorem says that the power set of any set S is bigger (has larger cardinality) than S itself. On the other hand, the third part of this theorem says that, for infinite sets S, the set of all *finite* subsets of S is the same size as S.

Theorem 0.12

1) **(Schröder–Bernstein theorem)** *For any sets* S *and* T,

$$|S| \le |T| \text{ and } |T| \le |S| \Rightarrow |S| = |T|$$

2) **(Cantor's theorem)** *If* $\mathcal{P}(S)$ *denotes the power set of* S, *then*

$$|S| < |\mathcal{P}(S)|$$

3) *If* $\mathcal{P}_0(S)$ *denotes the set of all finite subsets of* S *and if* S *is an infinite set, then*

$$|S| = |\mathcal{P}_0(S)|$$

Proof. We prove only parts 1) and 2). Let $f: S \to T$ be an injective function from S into T and let $g: T \to S$ be an injective function from T into S. We want to use these functions to create a bijective function from S to T. For this purpose, we make the following definitions. The **descendants** of an element $s \in S$ are the elements obtained by repeated alternate applications of the functions f and g, namely

$$f(s), g(f(s)), f(g(f(s))), \ldots$$

If t is a descendant of s, then s is an **ancestor** of t. Descendants and ancestors of elements of T are defined similarly.

Now, by tracing an element's ancestry to its beginning, we find that there are three possibilities: the element may originate in S, or in T, or it may have no point of origin. Accordingly, we can write S as the union of three disjoint sets

$$\begin{aligned}\mathcal{S}_S &= \{s \in S \mid s \text{ originates in } S\}\\ \mathcal{S}_T &= \{s \in S \mid s \text{ originates in } T\}\\ \mathcal{S}_\infty &= \{s \in S \mid s \text{ has no originator}\}\end{aligned}$$

Similarly, T is the disjoint union of $\mathcal{T}_S$, $\mathcal{T}_T$ and $\mathcal{T}_\infty$.

Now, the restriction

$$f|_{\mathcal{S}_S}: \mathcal{S}_S \to \mathcal{T}_S$$

is a bijection. To see this, note that if $t \in \mathcal{T}_S$, then t originated in S and therefore must have the form $f(s)$ for some $s \in S$. But t and its ancestor s have the same point of origin and so $t \in \mathcal{T}_S$ implies $s \in \mathcal{S}_S$. Thus, $f|_{\mathcal{S}_S}$ is surjective and hence bijective. We leave it to the reader to show that the functions

$$(g|_{\mathcal{T}_T})^{-1}: \mathcal{S}_T \to \mathcal{T}_T \text{ and } f|_{\mathcal{S}_\infty}: \mathcal{S}_\infty \to \mathcal{T}_\infty$$

are also bijections. Putting these three bijections together gives a bijection between S and T. Hence, $|S| = |T|$, as desired.

We now prove Cantor's theorem. The map $\iota: S \to \mathcal{P}(S)$ defined by $\iota(s) = \{s\}$ is an injection from S to $\mathcal{P}(S)$ and so $|S| \le |\mathcal{P}(S)|$. To complete the proof we must show that no injective map $f: S \to \mathcal{P}(S)$ can be surjective. To this end, let

$$X = \{s \in S \mid s \notin f(s)\} \in \mathcal{P}(S)$$

We claim that X is not in $\operatorname{im}(f)$. For suppose that $X = f(x)$ for some $x \in S$. Then if $x \in X$, we have by the definition of X that $x \notin X$. On the other hand, if $x \notin X$, we have again by the definition of X that $x \in X$. This contradiction implies that $X \notin \operatorname{im}(f)$ and so f is not surjective.□

Cardinal Arithmetic

Now let us define addition, multiplication and exponentiation of cardinal numbers. If S and T are sets, the **cartesian product** $S \times T$ is the set of all ordered pairs

$$S \times T = \{(s,t) \mid s \in S, t \in T\}$$

The set of all functions from T to S is denoted by S^T.

Definition *Let κ and λ denote cardinal numbers. Let S and T be disjoint sets for which $|S| = \kappa$ and $|T| = \lambda$.*
1) The **sum** *$\kappa + \lambda$ is the cardinal number of $S \cup T$.*
2) The **product** *$\kappa\lambda$ is the cardinal number of $S \times T$.*
3) The **power** *κ^λ is the cardinal number of S^T.* □

We will not go into the details of why these definitions make sense. (For instance, they seem to depend on the sets S and T, but in fact they do not.) It can be shown, using these definitions, that cardinal addition and multiplication are associative and commutative and that multiplication distributes over addition.

Theorem 0.13 *Let κ, λ and μ be cardinal numbers. Then the following properties hold:*
1) **(Associativity)**

$$\kappa + (\lambda + \mu) = (\kappa + \lambda) + \mu \text{ and } \kappa(\lambda\mu) = (\kappa\lambda)\mu$$

2) **(Commutativity)**

$$\kappa + \lambda = \lambda + \kappa \text{ and } \kappa\lambda = \lambda\kappa$$

3) **(Distributivity)**

$$\kappa(\lambda + \mu) = \kappa\lambda + \kappa\mu$$

4) (Properties of Exponents)
a) $\kappa^{\lambda+\mu} = \kappa^\lambda \kappa^\mu$
b) $(\kappa^\lambda)^\mu = \kappa^{\lambda\mu}$
c) $(\kappa\lambda)^\mu = \kappa^\mu \lambda^\mu$ □

On the other hand, the arithmetic of cardinal numbers can seem a bit strange, as the next theorem shows.

Theorem 0.14 *Let κ and λ be cardinal numbers, at least one of which is infinite. Then*

$$\kappa + \lambda = \kappa\lambda = \max\{\kappa, \lambda\}$$ □

It is not hard to see that there is a one-to-one correspondence between the power set $\mathcal{P}(S)$ of a set S and the set of all functions from S to $\{0, 1\}$. This leads to the following theorem.

Theorem 0.15 *For any cardinal κ*
1) If $|S| = \kappa$, then $|\mathcal{P}(S)| = 2^\kappa$
2) $\kappa < 2^\kappa$ □

We have already observed that $|\mathbb{N}| = \aleph_0$. It can be shown that $\aleph_0$ is the smallest infinite cardinal, that is,

$$\kappa < \aleph_0 \Rightarrow \kappa \text{ is a natural number}$$

It can also be shown that the set $\mathbb{R}$ of real numbers is in one-to-one correspondence with the power set $\mathcal{P}(\mathbb{N})$ of the natural numbers. Therefore,

$$|\mathbb{R}| = 2^{\aleph_0}$$

The set of all points on the real line is sometimes called the **continuum** and so $2^{\aleph_0}$ is sometimes called the **power of the continuum** and denoted by c.

Theorem 0.14 shows that cardinal addition and multiplication have a kind of "absorption" quality, which makes it hard to produce larger cardinals from smaller ones. The next theorem demonstrates this more dramatically.

Theorem 0.16

1) *Addition applied a countable number of times or multiplication applied a finite number of times to the cardinal number $\aleph_0$, does not yield anything more than $\aleph_0$. Specifically, for any nonzero $n \in \mathbb{N}$, we have*

$$\aleph_0 \cdot \aleph_0 = \aleph_0 \text{ and } \aleph_0^n = \aleph_0$$

2) *Addition and multiplication applied a countable number of times to the cardinal number $2^{\aleph_0}$ does not yield more than $2^{\aleph_0}$. Specifically, we have*

$$\aleph_0 \cdot 2^{\aleph_0} = 2^{\aleph_0} \text{ and } (2^{\aleph_0})^{\aleph_0} = 2^{\aleph_0} \qquad \square$$

Using this theorem, we can establish other relationships, such as

$$2^{\aleph_0} \leq (\aleph_0)^{\aleph_0} \leq (2^{\aleph_0})^{\aleph_0} = 2^{\aleph_0}$$

which, by the Schröder–Bernstein theorem, implies that

$$(\aleph_0)^{\aleph_0} = 2^{\aleph_0}$$

We mention that the problem of evaluating κ^λ in general is a very difficult one and would take us far beyond the scope of this book.

We will have use for the following reasonable-sounding result, whose proof is omitted.

Theorem 0.17 *Let $\{A_k \mid k \in K\}$ be a collection of sets, indexed by the set K, with $|K| = \kappa$. If $|A_k| \leq \lambda$ for all $k \in K$, then*

$$\left| \bigcup_{k \in K} A_k \right| \leq \lambda\kappa \qquad \square$$

Let us conclude by describing the cardinality of some famous sets.

Theorem 0.18

1) *The following sets have cardinality* $\aleph_0$.
   a) *The rational numbers* $\mathbb{Q}$.
   b) *The set of all finite subsets of* $\mathbb{N}$.
   c) *The union of a countable number of countable sets.*
   d) *The set* $\mathbb{Z}^n$ *of all ordered* n*-tuples of integers.*
2) *The following sets have cardinality* $2^{\aleph_0}$.
   a) *The set of all points in* $\mathbb{R}^n$.
   b) *The set of all infinite sequences of natural numbers.*
   c) *The set of all infinite sequences of real numbers.*
   d) *The set of all finite subsets of* $\mathbb{R}$.
   e) *The set of all irrational numbers.* □

Part 2 Algebraic Structures

We now turn to a discussion of some of the many algebraic structures that play a role in the study of linear algebra.

Groups

Definition *A* **group** *is a nonempty set* G, *together with a binary operation denoted by* $*$, *that satisfies the following properties:*

1) (**Associativity**) *For all* $a, b, c \in G$,

$$(a*b)*c = a*(b*c)$$

2) (**Identity**) *There exists an element* $e \in G$ *for which*

$$e*a = a*e = a$$

   for all $a \in G$.
3) (**Inverses**) *For each* $a \in G$, *there is an element* $a^{-1} \in G$ *for which*

$$a*a^{-1} = a^{-1}*a = e$$

□

Definition *A group* G *is* **abelian**, *or* **commutative**, *if*

$$a*b = b*a$$

for all $a, b \in G$. *When a group is abelian, it is customary to denote the operation* $*$ *by* $+$, *thus writing* $a*b$ *as* $a + b$. *It is also customary to refer to the identity as the* **zero element** *and to denote the inverse* a^{-1} *by* $-a$, *referred to as the* **negative** *of* a. □

Example 0.7 The set $\mathcal{F}$ of all bijective functions from a set S to S is a group under composition of functions. However, in general, it is not abelian. □

Example 0.8 The set $\mathcal{M}_{m,n}(F)$ is an abelian group under addition of matrices. The identity is the zero matrix $0_{m,n}$ of size $m \times n$. The set $\mathcal{M}_n(F)$ is not a group under multiplication of matrices, since not all matrices have multiplicative inverses. However, the set of invertible matrices of size $n \times n$ is a (nonabelian) group under multiplication.□

A group G is **finite** if it contains only a finite number of elements. The cardinality of a finite group G is called its **order** and is denoted by $o(G)$ or simply $|G|$. Thus, for example, $\mathbb{Z}_n = \{0, 1, \ldots, n-1\}$ is a finite group under addition modulo n, but $\mathcal{M}_{m,n}(\mathbb{R})$ is not finite.

Definition *A* **subgroup** *of a group G is a nonempty subset S of G that is a group in its own right, using the same operations as defined on G.*□

Cyclic Groups

If a is a formal symbol, we can define a group G to be the set of all integral powers of a:

$$G = \{a^i \mid i \in \mathbb{Z}\}$$

where the product is defined by the formal rules of exponents:

$$a^i a^j = a^{i+j}$$

This group is denoted by $\langle a \rangle$ and called the **cyclic group generated by** a. The identity of $\langle a \rangle$ is $1 = a^0$. In general, a group G is **cyclic** if it has the form $G = \langle a \rangle$ for some $a \in G$.

We can also create a finite group $C_n(a)$ of arbitrary positive order n by declaring that $a^n = 1$. Thus,

$$C_n(a) = \{1 = a^0, a, a^2, \ldots, a^{n-1}\}$$

where the product is defined by the formal rules of exponents, followed by reduction modulo n:

$$a^i a^j = a^{(i+j) \bmod n}$$

This defines a group of order n, called a **cyclic group of order** n. The inverse of a^k is $a^{(-k) \bmod n}$.

Rings

Definition *A* **ring** *is a nonempty set R, together with two binary operations, called* **addition** *(denoted by $+$) and* **multiplication** *(denoted by juxtaposition), for which the following hold:*

1) *R is an abelian group under addition*

2) **(Associativity)** *For all $a, b, c \in R$,*

$$(ab)c = a(bc)$$

3) **(Distributivity)** *For all* $a, b, c \in R$,

$$(a+b)c = ac + bc \text{ and } c(a+b) = ca + cb$$

A ring R *is said to be* **commutative** *if* $ab = ba$ *for all* $a, b \in R$. *If a ring* R *contains an element* e *with the property that*

$$ae = ea = a$$

for all $a \in R$, *we say that* R *is a* **ring with identity**. *The identity* e *is usually denoted by* 1.□

A **field** F is a commutative ring with identity in which each nonzero element has a multiplicative inverse, that is, if $a \in F$ is nonzero, then there is a $b \in F$ for which $ab = 1$.

Example 0.9 The set $\mathbb{Z}_n = \{0, 1, \ldots, n-1\}$ is a commutative ring under addition and multiplication modulo n

$$a \oplus b = (a+b) \bmod n, \quad a \odot b = ab \bmod n$$

The element $1 \in \mathbb{Z}_n$ is the identity.□

Example 0.10 The set E of even integers is a commutative ring under the usual operations on $\mathbb{Z}$, but it has no identity.□

Example 0.11 The set $\mathcal{M}_n(F)$ is a noncommutative ring under matrix addition and multiplication. The identity matrix I_n is the identity for $\mathcal{M}_n(F)$.□

Example 0.12 Let F be a field. The set $F[x]$ of all polynomials in a single variable x, with coefficients in F, is a commutative ring under the usual operations of polynomial addition and multiplication. What is the identity for $F[x]$? Similarly, the set $F[x_1, \ldots, x_n]$ of polynomials in n variables is a commutative ring under the usual addition and multiplication of polynomials.□

Definition *If* R *and* S *are rings, then a function* $\sigma: R \to S$ *is a* **ring homomorphism** *if*

$$\begin{aligned} \sigma(a+b) &= \sigma a + \sigma b \\ \sigma(ab) &= \sigma(a)\sigma(b) \\ \sigma 1 &= 1 \end{aligned}$$

for all $a, b \in R$.□

Definition *A* **subring** *of a ring* R *is a subset* S *of* R *that is a ring in its own right, using the same operations as defined on* R *and having the same multiplicative identity as* R.□

The condition that a subring S have the same multiplicative identity as R is required. For example, the set S of all 2×2 matrices of the form

$$A_a = \begin{bmatrix} a & 0 \\ 0 & 0 \end{bmatrix}$$

for $a \in F$ is a ring under addition and multiplication of matrices (isomorphic to F). The multiplicative identity in S is the matrix A_1, which is not the identity I_2 of $\mathcal{M}_{2,2}(F)$. Hence, S is a ring under the same operations as $\mathcal{M}_{2,2}(F)$ but it is not a subring of $\mathcal{M}_{2,2}(F)$.

Applying the definition is not generally the easiest way to show that a subset of a ring is a subring. The following characterization is usually easier to apply.

Theorem 0.19 *A nonempty subset S of a ring R is a subring if and only if*

1) *The multiplicative identity 1_R of R is in S*
2) *S is closed under subtraction, that is,*

$$a, b \in S \Rightarrow a - b \in S$$

3) *S is closed under multiplication, that is,*

$$a, b \in S \Rightarrow ab \in S$$

□

Ideals

Rings have another important substructure besides subrings.

Definition *Let R be a ring. A nonempty subset $\mathcal{I}$ of R is called an* **ideal** *if*

1) *$\mathcal{I}$ is a subgroup of the abelian group R, that is, $\mathcal{I}$ is closed under subtraction:*

$$a, b \in \mathcal{I} \Rightarrow a - b \in \mathcal{I}$$

2) *$\mathcal{I}$ is closed under multiplication by* any *ring element, that is,*

$$a \in \mathcal{I}, r \in R \Rightarrow ar \in \mathcal{I} \text{ and } ra \in \mathcal{I}$$

□

Note that if an ideal $\mathcal{I}$ contains the unit element 1, then $\mathcal{I} = R$.

Example 0.13 Let $p(x)$ be a polynomial in $F[x]$. The set of all multiples of $p(x)$,

$$\langle p(x) \rangle = \{q(x)p(x) \mid q(x) \in F[x]\}$$

is an ideal in $F[x]$, called the *ideal generated by $p(x)$*.□

Definition *Let S be a subset of a ring R with identity. The set*

$$\langle S \rangle = \{r_1 s_1 + \cdots + r_n s_n \mid r_i \in R, s_i \in S, n \geq 1\}$$

of all finite linear combinations of elements of S*, with coefficients in* R*, is an ideal in* R*, called the* **ideal generated by** S*. It is the smallest (in the sense of set inclusion) ideal of* R *containing* S*. If* $S = \{s_1, \ldots, s_n\}$ *is a finite set, we write*

$$\langle s_1, \ldots, s_n \rangle = \{r_1 s_1 + \cdots + r_n s_n \mid r_i \in R, s_i \in S\} \qquad \square$$

Note that in the previous definition, we require that R have an identity. This is to ensure that $S \subseteq \langle S \rangle$.

Theorem 0.20 *Let* R *be a ring.*

1) *The intersection of any collection* $\{\mathcal{I}_k \mid k \in K\}$ *of ideals is an ideal.*
2) *If* $\mathcal{I}_1 \subseteq \mathcal{I}_2 \subseteq \cdots$ *is an ascending sequence of ideals, each one contained in the next, then the union* $\bigcup \mathcal{I}_k$ *is also an ideal.*
3) *More generally, if*

$$\mathcal{C} = \{\mathcal{I}_i \mid i \in I\}$$

is a chain of ideals in R*, then the union* $\mathcal{J} = \bigcup_{i \in I} \mathcal{I}_i$ *is also an ideal in* R*.*

Proof. To prove 1), let $\mathcal{J} = \bigcap \mathcal{I}_k$. Then if $a, b \in \mathcal{J}$, we have $a, b \in \mathcal{I}_k$ for all $k \in K$. Hence, $a - b \in \mathcal{I}_k$ for all $k \in K$ and so $a - b \in \mathcal{J}$. Hence, $\mathcal{J}$ is closed under subtraction. Also, if $r \in R$, then $ra \in \mathcal{I}_k$ for all $k \in K$ and so $ra \in \mathcal{J}$. Of course, part 2) is a special case of part 3). To prove 3), if $a, b \in \mathcal{J}$, then $a \in \mathcal{I}_i$ and $b \in \mathcal{I}_j$ for some $i, j \in I$. Since one of $\mathcal{I}_i$ and $\mathcal{I}_j$ is contained in the other, we may assume that $\mathcal{I}_i \subseteq \mathcal{I}_j$. It follows that $a, b \in \mathcal{I}_j$ and so $a - b \in \mathcal{I}_j \subseteq \mathcal{J}$ and if $r \in R$, then $ra \in \mathcal{I}_j \subseteq \mathcal{J}$. Thus $\mathcal{J}$ is an ideal.$\square$

Note that in general, the union of ideals is not an ideal. However, as we have just proved, the union of any *chain* of ideals is an ideal.

Quotient Rings and Maximal Ideals

Let S be a subset of a commutative ring R with identity. Let $\equiv$ be the binary relation on R defined by

$$a \equiv b \Leftrightarrow a - b \in S$$

It is easy to see that $\equiv$ is an equivalence relation. When $a \equiv b$, we say that a and b are **congruent modulo** S. The term "mod" is used as a colloquialism for modulo and $a \equiv b$ is often written

$$a \equiv b \bmod S$$

As shorthand, we write $a \equiv b$.

To see what the equivalence classes look like, observe that

$$\begin{aligned}[a] &= \{r \in R \mid r \equiv a\} \\ &= \{r \in R \mid r - a \in S\} \\ &= \{r \in R \mid r = a + s \text{ for some } s \in S\} \\ &= \{a + s \mid s \in S\} \\ &= a + S\end{aligned}$$

The set

$$a + S = \{a + s \mid s \in S\}$$

is called a **coset** of S in R. The element a is called a **coset representative** for $a + S$.

Thus, the equivalence classes for congruence mod S are the cosets $a + S$ of S in R. The set of all cosets is denoted by

$$R/S = \{a + S \mid a \in R\}$$

This is read "R mod S." We would like to place a ring structure on R/S. Indeed, if S is a subgroup of the abelian group R, then R/S is easily seen to be an abelian group as well under coset addition defined by

$$(a + S) + (b + S) = (a + b) + S$$

In order for the product

$$(a + S)(b + S) = ab + S$$

to be well-defined, we must have

$$b + S = b' + S \Rightarrow ab + S = ab' + S$$

or, equivalently,

$$b - b' \in S \Rightarrow a(b - b') \in S$$

But $b - b'$ may be any element of S and a may be any element of R and so this condition implies that S must be an ideal. Conversely, if S is an ideal, then coset multiplication is well defined.

Theorem 0.21 *Let R be a commutative ring with identity. Then the quotient $R/\mathcal{I}$ is a ring under coset addition and multiplication if and only if $\mathcal{I}$ is an ideal of R. In this case, $R/\mathcal{I}$ is called the* **quotient ring** *of R* **modulo** *$\mathcal{I}$, where addition and multiplication are defined by*

$$\begin{aligned}(a + S) + (b + S) &= (a + b) + S \\ (a + S)(b + S) &= ab + S\end{aligned}$$

□

Definition *An ideal $\mathcal{I}$ in a ring R is a* **maximal ideal** *if $\mathcal{I} \neq R$ and if whenever $\mathcal{J}$ is an ideal satisfying $\mathcal{I} \subseteq \mathcal{J} \subseteq R$, then either $\mathcal{J} = \mathcal{I}$ or $\mathcal{J} = R$.*$\square$

Here is one reason why maximal ideals are important.

Theorem 0.22 *Let R be a commutative ring with identity. Then the quotient ring $R/\mathcal{I}$ is a field if and only if $\mathcal{I}$ is a maximal ideal.*
Proof. First, note that for any ideal $\mathcal{I}$ of R, the ideals of $R/\mathcal{I}$ are precisely the quotients $\mathcal{J}/\mathcal{I}$ where $\mathcal{J}$ is an ideal for which $\mathcal{I} \subseteq \mathcal{J} \subseteq R$. It is clear that $\mathcal{J}/\mathcal{I}$ is an ideal of $R/\mathcal{I}$. Conversely, if $\mathcal{K}'$ is an ideal of $R/\mathcal{I}$, then let

$$\mathcal{K} = \{r \in R \mid r + \mathcal{I} \in \mathcal{K}'\}$$

It is easy to see that $\mathcal{K}$ is an ideal of R for which $\mathcal{I} \subseteq \mathcal{K} \subseteq R$.

Next, observe that a commutative ring S with identity is a field if and only if S has no nonzero proper ideals. For if S is a field and $\mathcal{I}$ is an ideal of S containing a nonzero element r, then $1 = r^{-1}r \in \mathcal{I}$ and so $\mathcal{I} = S$. Conversely, if S has no nonzero proper ideals and $0 \neq s \in S$, then the ideal $\langle s \rangle$ must be S and so there is an $r \in S$ for which $rs = 1$. Hence, S is a field.

Putting these two facts together proves the theorem.$\square$

The following result says that maximal ideals always exist.

Theorem 0.23 *Any nonzero commutative ring R with identity contains a maximal ideal.*
Proof. Since R is not the zero ring, the ideal $\{0\}$ is a proper ideal of R. Hence, the set $\mathcal{S}$ of all proper ideals of R is nonempty. If

$$\mathcal{C} = \{\mathcal{I}_i \mid i \in I\}$$

is a chain of proper ideals in R, then the union $\mathcal{J} = \bigcup_{i \in I} \mathcal{I}_i$ is also an ideal. Furthermore, if $\mathcal{J} = R$ is not proper, then $1 \in \mathcal{J}$ and so $1 \in \mathcal{I}_i$, for some $i \in I$, which implies that $\mathcal{I}_i = R$ is not proper. Hence, $\mathcal{J} \in \mathcal{S}$. Thus, any chain in $\mathcal{S}$ has an upper bound in $\mathcal{S}$ and so Zorn's lemma implies that $\mathcal{S}$ has a maximal element. This shows that R has a maximal ideal.$\square$

Integral Domains

Definition *Let R be a ring. A nonzero element $r \in R$ is called a* **zero divisor** *if there exists a nonzero $s \in R$ for which $rs = 0$. A commutative ring R with identity is called an* **integral domain** *if it contains no zero divisors.*$\square$

Example 0.14 If n is not a prime number, then the ring $\mathbb{Z}_n$ has zero divisors and so is not an integral domain. To see this, observe that if n is not prime, then $n = ab$ in $\mathbb{Z}$, where $a, b \geq 2$. But in $\mathbb{Z}_n$, we have

$$a \odot b = ab \bmod n = 0$$

and so a and b are both zero divisors. As we will see later, if n is a prime, then $\mathbb{Z}_n$ is a field (which is an integral domain, of course).□

Example 0.15 The ring $F[x]$ is an integral domain, since $p(x)q(x) = 0$ implies that $p(x) = 0$ or $q(x) = 0$.□

If R is a ring and $rx = ry$ where $r, x, y \in R$, then we cannot in general cancel the r's and conclude that $x = y$. For instance, in $\mathbb{Z}_4$, we have $2 \cdot 3 = 2 \cdot 1$, but canceling the 2's gives $3 = 1$. However, it is precisely the integral domains in which we can cancel. The simple proof is left to the reader.

Theorem 0.24 *Let R be a commutative ring with identity. Then R is an integral domain if and only if the cancellation law*

$$rx = ry, r \neq 0 \Rightarrow x = y$$

holds.□

The Field of Quotients of an Integral Domain

Any integral domain R can be embedded in a field. The **quotient field** (or **field of quotients**) of R is a field that is constructed from R just as the field of rational numbers is constructed from the ring of integers. In particular, we set

$$R^+ = \{(p, q) \mid p, q \in R, q \neq 0\}$$

where $(p, q) = (p', q')$ if and only if $pq' = p'q$. Addition and multiplication of fractions is defined by

$$(p, q) + (r, s) = (ps + qr, qs)$$

and

$$(p, q) \cdot (r, s) = (pr, qs)$$

It is customary to write (p, q) in the form p/q. Note that if R has zero divisors, then these definitions do not make sense, because qs may be 0 even if q and s are not. This is why we require that R be an integral domain.

Principal Ideal Domains

Definition *Let R be a ring with identity and let $a \in R$. The* **principal ideal** *generated by a is the ideal*

$$\langle a \rangle = \{ra \mid r \in R\}$$

An integral domain *R in which every ideal is a principal ideal is called a* **principal ideal domain.**□

Theorem 0.25 *The integers form a principal ideal domain. In fact, any ideal $\mathcal{I}$ in $\mathbb{Z}$ is generated by the smallest positive integer a that is contained in $\mathcal{I}$.*$\square$

Theorem 0.26 *The ring $F[x]$ is a principal ideal domain. In fact, any ideal $\mathcal{I}$ is generated by the unique monic polynomial of smallest degree contained in $\mathcal{I}$. Moreover, for polynomials $p_1(x),\ldots,p_n(x)$,*

$$\langle p_1(x),\ldots,p_n(x)\rangle = \langle \gcd\{p_1(x),\ldots,p_n(x)\}\rangle$$

Proof. Let $\mathcal{I}$ be an ideal in $F[x]$ and let $m(x)$ be a monic polynomial of smallest degree in $\mathcal{I}$. First, we observe that there is only one such polynomial in $\mathcal{I}$. For if $n(x) \in \mathcal{I}$ is monic and $\deg(n(x)) = \deg(m(x))$, then

$$b(x) = m(x) - n(x) \in \mathcal{I}$$

and since $\deg(b(x)) < \deg(m(x))$, we must have $b(x) = 0$ and so $n(x) = m(x)$.

We show that $\mathcal{I} = \langle m(x)\rangle$. Since $m(x) \in \mathcal{I}$, we have $\langle m(x)\rangle \subseteq \mathcal{I}$. To establish the reverse inclusion, if $p(x) \in \mathcal{I}$, then dividing $p(x)$ by $m(x)$ gives

$$p(x) = q(x)m(x) + r(x)$$

where $r(x) = 0$ or $0 \le \deg r(x) < \deg m(x)$. But since $\mathcal{I}$ is an ideal,

$$r(x) = p(x) - q(x)m(x) \in \mathcal{I}$$

and so $0 \le \deg r(x) < \deg m(x)$ is impossible. Hence, $r(x) = 0$ and

$$p(x) = q(x)m(x) \in \langle m(x)\rangle$$

This shows that $\mathcal{I} \subseteq \langle m(x)\rangle$ and so $\mathcal{I} = \langle m(x)\rangle$.

To prove the second statement, let $\mathcal{I} = \langle p_1(x),\ldots,p_n(x)\rangle$. Then, by what we have just shown,

$$\mathcal{I} = \langle p_1(x),\ldots,p_n(x)\rangle = \langle m(x)\rangle$$

where $m(x)$ is the unique monic polynomial $m(x)$ in $\mathcal{I}$ of smallest degree. In particular, since $p_i(x) \in \langle m(x)\rangle$, we have $m(x) \mid p_i(x)$ for each $i = 1,\ldots,n$. In other words, $m(x)$ is a common divisor of the $p_i(x)$'s.

Moreover, if $q(x) \mid p_i(x)$ for all i, then $p_i(x) \in \langle q(x)\rangle$ for all i, which implies that

$$m(x) \in \langle m(x)\rangle = \langle p_1(x),\ldots,p_n(x)\rangle \subseteq \langle q(x)\rangle$$

and so $q(x) \mid m(x)$. This shows that $m(x)$ is the *greatest* common divisor of the $p_i(x)$'s and completes the proof.$\square$

Example 0.16 The ring $R = F[x,y]$ of polynomials in two variables x and y is not a principal ideal domain. To see this, observe that the set $\mathcal{I}$ of all polynomials with zero constant term is an ideal in R. Now, suppose that $\mathcal{I}$ is the principal ideal $\mathcal{I} = \langle p(x,y) \rangle$. Since $x, y \in \mathcal{I}$, there exist polynomials $a(x,y)$ and $b(x,y)$ for which

$$x = a(x,y)p(x,y) \text{ and } y = b(x,y)p(x,y) \tag{0.1}$$

But $p(x,y)$ cannot be a constant, for then we would have $\mathcal{I} = R$. Hence, $\deg(p(x,y)) \geq 1$ and so $a(x,y)$ and $b(x,y)$ must both be constants, which implies that (0.1) cannot hold.□

Theorem 0.27 *Any principal ideal domain R satisfies the* **ascending chain condition**, *that is, R cannot have a strictly increasing sequence of ideals*

$$\mathcal{I}_1 \subset \mathcal{I}_2 \subset \cdots$$

where each ideal is properly contained in the next one.

Proof. Suppose to the contrary that there is such an increasing sequence of ideals. Consider the ideal

$$U = \bigcup \mathcal{I}_i$$

which must have the form $U = \langle a \rangle$ for some $a \in U$. Since $a \in \mathcal{I}_k$ for some k, we have $\mathcal{I}_k = \mathcal{I}_j$ for all $j \geq k$, contradicting the fact that the inclusions are proper.□

Prime and Irreducible Elements

We can define the notion of a prime element in any integral domain. For $r, s \in R$, we say that r **divides** s (written $r \mid s$) if there exists an $x \in R$ for which $s = xr$.

Definition *Let R be an integral domain.*

1) *An invertible element of R is called a* **unit**. *Thus, $u \in R$ is a unit if $uv = 1$ for some $v \in R$.*
2) *Two elements $a, b \in R$ are said to be* **associates** *if there exists a unit u for which $a = ub$. We denote this by writing $a \sim b$.*
3) *A nonzero nonunit $p \in R$ is said to be* **prime** *if*

$$p \mid ab \Rightarrow p \mid a \text{ or } p \mid b$$

4) *A nonzero nonunit $r \in R$ is said to be* **irreducible** *if*

$$r = ab \Rightarrow a \text{ or } b \text{ is a unit}$$

□

Note that if p is prime or irreducible, then so is up for any unit u.

The property of being associate is clearly an equivalence relation.

Definition *We will refer to the equivalence classes under the relation of being associate as the* **associate classes** *of* R.□

Theorem 0.28 *Let* R *be a ring.*
1) *An element* $u \in R$ *is a unit if and only if* $\langle u \rangle = R$.
2) $r \sim s$ *if and only if* $\langle r \rangle = \langle s \rangle$.
3) r *divides* s *if and only if* $\langle s \rangle \subseteq \langle r \rangle$.
4) r **properly divides** s, *that is,* $s = xr$ *where* x *is not a unit, if and only if* $\langle s \rangle \subset \langle r \rangle$.□

In the case of the integers, an integer is prime if and only if it is irreducible. In any integral domain, prime elements are irreducible, but the converse need not hold. (In the ring $\mathbb{Z}[\sqrt{-5}] = \{a + b\sqrt{-5} \mid a, b \in \mathbb{Z}\}$ the irreducible element 2 divides the product $(1 + \sqrt{-5})(1 - \sqrt{-5}) = 6$ but does not divide either factor.)

However, in principal ideal domains, the two concepts are equivalent.

Theorem 0.29 *Let* R *be a principal ideal domain.*
1) *An* $r \in R$ *is irreducible if and only if the ideal* $\langle r \rangle$ *is maximal.*
2) *An element in* R *is prime if and only if it is irreducible.*
3) *The elements* $a, b \in R$ *are* **relatively prime**, *that is, have no common nonunit factors, if and only if there exist* $r, s \in R$ *for which*

$$ra + sb = 1$$

This is denoted by writing $(a, b) = 1$.

Proof. To prove 1), suppose that r is irreducible and that $\langle r \rangle \subseteq \langle a \rangle \subseteq R$. Then $r \in \langle a \rangle$ and so $r = xa$ for some $x \in R$. The irreducibility of r implies that a or x is a unit. If a is a unit, then $\langle a \rangle = R$ and if x is a unit, then $\langle a \rangle = \langle xa \rangle = \langle r \rangle$. This shows that $\langle r \rangle$ is maximal. (We have $\langle r \rangle \neq R$, since r is not a unit.) Conversely, suppose that r is not irreducible, that is, $r = ab$ where neither a nor b is a unit. Then $\langle r \rangle \subseteq \langle a \rangle \subseteq R$. But if $\langle a \rangle = \langle r \rangle$, then $r \sim a$, which implies that b is a unit. Hence $\langle r \rangle \neq \langle a \rangle$. Also, if $\langle a \rangle = R$, then a must be a unit. So we conclude that $\langle r \rangle$ is not maximal, as desired.

To prove 2), assume first that p is prime and $p = ab$. Then $p \mid a$ or $p \mid b$. We may assume that $p \mid a$. Therefore, $a = xp = xab$. Canceling a's gives $1 = xb$ and so b is a unit. Hence, p is irreducible. (Note that this argument applies in any integral domain.)

Conversely, suppose that r is irreducible and let $r \mid ab$. We wish to prove that $r \mid a$ or $r \mid b$. The ideal $\langle r \rangle$ is maximal and so $\langle r, a \rangle = \langle r \rangle$ or $\langle r, a \rangle = R$. In the former case, $r \mid a$ and we are done. In the latter case, we have

$$1 = xa + yr$$

for some $x, y \in R$. Thus,

$$b = xab + yrb$$

and since r divides both terms on the right, we have $r \mid b$.

To prove 3), it is clear that if $ra + sb = 1$, then a and b are relatively prime. For the converse, consider the ideal $\langle a, b \rangle$, which must be principal, say $\langle a, b \rangle = \langle x \rangle$. Then $x \mid a$ and $x \mid b$ and so x must be a unit, which implies that $\langle a, b \rangle = R$. Hence, there exist $r, s \in R$ for which $ra + sb = 1$.□

Unique Factorization Domains

Definition *An integral domain R is said to be a* **unique factorization domain** *if it has the following factorization properties:*

1) *Every nonzero nonunit element $r \in R$ can be written as a product of a finite number of irreducible elements $r = p_1 \cdots p_n$.*
2) *The factorization into irreducible elements is unique in the sense that if $r = p_1 \cdots p_n$ and $r = q_1 \cdots q_m$ are two such factorizations, then $m = n$ and after a suitable reindexing of the factors, $p_i \sim q_i$.*□

Unique factorization is clearly a desirable property. Fortunately, principal ideal domains have this property.

Theorem 0.30 *Every principal ideal domain R is a unique factorization domain.*

Proof. Let $r \in R$ be a nonzero nonunit. If r is irreducible, then we are done. If not, then $r = r_1 r_2$, where neither factor is a unit. If r_1 and r_2 are irreducible, we are done. If not, suppose that r_2 is not irreducible. Then $r_2 = r_3 r_4$, where neither r_3 nor r_4 is a unit. Continuing in this way, we obtain a factorization of the form (after renumbering if necessary)

$$r = r_1 r_2 = r_1(r_3 r_4) = (r_1 r_3)(r_5 r_6) = (r_1 r_3 r_5)(r_7 r_8) = \cdots$$

Each step is a factorization of r into a product of nonunits. However, this process must stop after a finite number of steps, for otherwise it will produce an infinite sequence $s_1, s_2, \ldots$ of nonunits of R for which s_{i+1} properly divides s_i. But this gives the ascending chain of ideals

$$\langle s_1 \rangle \subset \langle s_2 \rangle \subset \langle s_3 \rangle \subset \langle s_4 \rangle \subset \cdots$$

where the inclusions are proper. But this contradicts the fact that a principal ideal domain satisfies the ascending chain condition. Thus, we conclude that every nonzero nonunit has a factorization into irreducible elements.

As to uniqueness, if $r = p_1 \cdots p_n$ and $r = q_1 \cdots q_m$ are two such factorizations, then because R is an integral domain, we may equate them and cancel like factors, so let us assume this has been done. Thus, $p_i \neq q_j$ for all i, j. If there are no factors on either side, we are done. If exactly one side has no factors left, then we have expressed 1 as a product of irreducible elements, which is not possible since irreducible elements are nonunits.

Suppose that both sides have factors left, that is,

$$p_1 \cdots p_n = q_1 \cdots q_m$$

where $p_i \neq q_j$. Then $q_m \mid p_1 \cdots p_n$, which implies that $q_m \mid p_i$ for some i. We can assume by reindexing if necessary that $p_n = a_n q_m$. Since p_n is irreducible a_n must be a unit. Replacing p_n by $a_n q_m$ and canceling q_m gives

$$a_n p_1 \cdots p_{n-1} = q_1 \cdots q_{m-1}$$

This process can be repeated until we run out of q's or p's. If we run out of q's first, then we have an equation of the form $up_1 \cdots p_k = 1$ where u is a unit, which is not possible since the p_i's are not units. By the same reasoning, we cannot run out of q's first and so $n = m$ and the p's and q's can be paired off as associates.□

Fields

For the record, let us give the definition of a field (a concept that we have been using).

Definition *A* **field** *is a set F, containing at least two elements, together with two binary operations, called* **addition** *(denoted by $+$) and* **multiplication** *(denoted by juxtaposition), for which the following hold:*

1) *F is an abelian group under addition.*
2) *The set F^* of all* nonzero *elements in F is an abelian group under multiplication.*
3) **(Distributivity)** *For all $a, b, c \in F$,*

$$(a+b)c = ac + bc \text{ and } c(a+b) = ca + cb \qquad \square$$

We require that F have at least two elements to avoid the pathological case in which $0 = 1$.

Example 0.17 The sets $\mathbb{Q}$, $\mathbb{R}$ and $\mathbb{C}$, of all rational, real and complex numbers, respectively, are fields, under the usual operations of addition and multiplication of numbers.□

Example 0.18 The ring $\mathbb{Z}_n$ is a field if and only if n is a prime number. We have already seen that $\mathbb{Z}_n$ is not a field if n is not prime, since a field is also an integral domain. Now suppose that $n = p$ is a prime.

We have seen that $\mathbb{Z}_p$ is an integral domain and so it remains to show that every nonzero element in $\mathbb{Z}_p$ has a multiplicative inverse. Let $0 \neq a \in \mathbb{Z}_p$. Since $a < p$, we know that a and p are relatively prime. It follows that there exist integers u and v for which

$$ua + vp = 1$$

Hence,

$$ua \equiv (1 - vp) \equiv 1 \bmod p$$

and so $u \odot a = 1$ in $\mathbb{Z}_p$, that is, u is the multiplicative inverse of a.□

The previous example shows that not all fields are infinite sets. In fact, finite fields play an extremely important role in many areas of abstract and applied mathematics.

A field F is said to be **algebraically closed** if every nonconstant polynomial over F has a root in F. This is equivalent to saying that every nonconstant polynomial splits over F. For example, the complex field $\mathbb{C}$ is algebraically closed but the real field $\mathbb{R}$ is not. We mention without proof that every field F is contained in an algebraically closed field $\overline{F}$, called the **algebraic closure** of F. For example, the algebraic closure of the real field is the complex field.

The Characteristic of a Ring

Let R be a ring with identity. If n is a positive integer, then by $n \cdot r$, we simply mean

$$n \cdot r = \underbrace{r + \cdots + r}_{n \text{ terms}}$$

Now, it may happen that there is a positive integer n for which

$$n \cdot 1 = 0$$

For instance, in $\mathbb{Z}_n$, we have $n \cdot 1 = n = 0$. On the other hand, in $\mathbb{Z}$, the equation $n \cdot 1 = 0$ implies $n = 0$ and so no such positive integer exists.

Notice that in any *finite* ring, there must exist such a positive integer n, since the members of the infinite sequence of numbers

$$1 \cdot 1, 2 \cdot 1, 3 \cdot 1, \ldots$$

cannot be distinct and so $i \cdot 1 = j \cdot 1$ for some $i < j$, whence $(j - i) \cdot 1 = 0$.

Definition *Let R be a ring with identity. The smallest positive integer c for which $c \cdot 1 = 0$ is called the* **characteristic** *of R. If no such number c exists, we say that R has characteristic 0. The characteristic of R is denoted by* $\mathrm{char}(R)$.□

If $\mathrm{char}(R) = c$, then for any $r \in R$, we have

$$c \cdot r = \underbrace{r + \cdots + r}_{c \text{ terms}} = (\underbrace{1 + \cdots + 1}_{c \text{ terms}})r = 0 \cdot r = 0$$

Theorem 0.31 *Any finite ring has nonzero characteristic. Any finite integral domain has prime characteristic.*

Proof. We have already seen that a finite ring has nonzero characteristic. Let F be a finite integral domain and suppose that $\text{char}(F) = c > 0$. If $c = pq$, where $p, q < c$, then $pq \cdot 1 = 0$. Hence, $(p \cdot 1)(q \cdot 1) = 0$, implying that $p \cdot 1 = 0$ or $q \cdot 1 = 0$. In either case, we have a contradiction to the fact that c is the smallest positive integer such that $c \cdot 1 = 0$. Hence, c must be prime.□

Notice that in any field F of characteristic 2, we have $2a = 0$ for all $a \in F$. Thus, in F,

$$a = -a \text{ for all } a \in F$$

This property takes a bit of getting used to and makes fields of characteristic 2 quite exceptional. (As it happens, there are many important uses for fields of characteristic 2.) It can be shown that all finite fields have size equal to a positive integral power p^n of a prime p and for each prime power p^n, there is a finite field of size p^n. In fact, up to isomorphism, there is exactly one finite field of size p^n.

Algebras

The final algebraic structure of which we will have use is a combination of a vector space and a ring. (We have not yet officially defined vector spaces, but we will do so before needing the following definition, which is placed here for easy reference.)

Definition *An* **algebra** $\mathcal{A}$ *over a field* F *is a nonempty set* $\mathcal{A}$*, together with three operations, called* **addition** *(denoted by* $+$*),* **multiplication** *(denoted by juxtaposition) and* **scalar multiplication** *(also denoted by juxtaposition), for which the following properties hold:*

1) $\mathcal{A}$ *is a vector space over* F *under addition and scalar multiplication.*
2) $\mathcal{A}$ *is a ring under addition and multiplication.*
3) If $r \in F$ *and* $a, b \in \mathcal{A}$*, then*

$$r(ab) = (ra)b = a(rb)$$

□

Thus, an algebra is a vector space in which we can take the product of vectors, or a ring in which we can multiply each element by a scalar (subject, of course, to additional requirements as given in the definition).

Part I—Basic Linear Algebra

Chapter 1
Vector Spaces

Vector Spaces

Let us begin with the definition of one of our principal objects of study.

Definition *Let F be a field, whose elements are referred to as* **scalars**. *A* **vector space** *over F is a nonempty set V, whose elements are referred to as* **vectors**, *together with two operations. The first operation, called* **addition** *and denoted by $+$, assigns to each pair (u,v) of vectors in V a vector $u+v$ in V. The second operation, called* **scalar multiplication** *and denoted by juxtaposition, assigns to each pair $(r,u) \in F \times V$ a vector ru in V. Furthermore, the following properties must be satisfied:*

1) **(Associativity of addition)** *For all vectors $u,v,w \in V$,*

$$u+(v+w)=(u+v)+w$$

2) **(Commutativity of addition)** *For all vectors $u,v \in V$,*

$$u+v=v+u$$

3) **(Existence of a zero)** *There is a vector $0 \in V$ with the property that*

$$0+u=u+0=u$$

for all vectors $u \in V$.

4) **(Existence of additive inverses)** *For each vector $u \in V$, there is a vector in V, denoted by $-u$, with the property that*

$$u+(-u)=(-u)+u=0$$

5) **(Properties of scalar multiplication)** *For all scalars $a, b \in F$ and for all vectors $u, v \in V$,*

$$\begin{aligned} a(u+v) &= au+av \\ (a+b)u &= au+bu \\ (ab)u &= a(bu) \\ 1u &= u \end{aligned}$$ □

Note that the first four properties in the definition of vector space can be summarized by saying that V is an abelian group under addition.

A vector space over a field F is sometimes called an ***F*-space**. A vector space over the real field is called a **real vector space** and a vector space over the complex field is called a **complex vector space**.

Definition *Let S be a nonempty subset of a vector space V. A* **linear combination** *of vectors in S is an expression of the form*

$$a_1v_1 + \cdots + a_nv_n$$

where $v_1, \ldots, v_n \in S$ and $a_1, \ldots, a_n \in F$. The scalars a_i are called the **coefficients** *of the linear combination. A linear combination is* **trivial** *if every coefficient a_i is zero. Otherwise, it is* **nontrivial**.□

Examples of Vector Spaces

Here are a few examples of vector spaces.

Example 1.1

1) Let F be a field. The set F^F of all functions from F to F is a vector space over F, under the operations of ordinary addition and scalar multiplication of functions:

$$(f+g)(x) = f(x) + g(x)$$

and

$$(af)(x) = a(f(x))$$

2) The set $\mathcal{M}_{m,n}(F)$ of all $m \times n$ matrices with entries in a field F is a vector space over F, under the operations of matrix addition and scalar multiplication.
3) The set F^n of all ordered n-tuples whose components lie in a field F, is a vector space over F, with addition and scalar multiplication defined componentwise:

$$(a_1, \ldots, a_n) + (b_1, \ldots, b_n) = (a_1 + b_1, \ldots, a_n + b_n)$$

and

$$c(a_1, \ldots, a_n) = (ca_1, \ldots, ca_n)$$

When convenient, we will also write the elements of F^n in column form. When F is a finite field F_q with q elements, we write $V(n, q)$ for F_q^n.

4) Many sequence spaces are vector spaces. The set $\mathrm{Seq}(F)$ of all infinite sequences with members from a field F is a vector space under the componentwise operations

$$(s_n) + (t_n) = (s_n + t_n)$$

and

$$a(s_n) = (as_n)$$

In a similar way, the set c_0 of all sequences of complex numbers that converge to 0 is a vector space, as is the set ℓ^∞ of all bounded complex sequences. Also, if p is a positive integer, then the set ℓ^p of all complex sequences (s_n) for which

$$\sum_{n=1}^{\infty} |s_n|^p < \infty$$

is a vector space under componentwise operations. To see that addition is a binary operation on ℓ^p, one verifies **Minkowski's inequality**

$$\left(\sum_{n=1}^{\infty} |s_n + t_n|^p\right)^{1/p} \leq \left(\sum_{n=1}^{\infty} |s_n|^p\right)^{1/p} + \left(\sum_{n=1}^{\infty} |t_n|^p\right)^{1/p}$$

which we will not do here.□

Subspaces

Most algebraic structures contain substructures, and vector spaces are no exception.

Definition *A* **subspace** *of a vector space V is a subset S of V that is a vector space in its own right under the operations obtained by restricting the operations of V to S. We use the notation $S \leq V$ to indicate that S is a subspace of V and $S < V$ to indicate that S is a* **proper subspace** *of V, that is, $S \leq V$ but $S \neq V$. The* **zero subspace** *of V is $\{0\}$.*□

Since many of the properties of addition and scalar multiplication hold a fortiori in a nonempty subset S, we can establish that S is a subspace merely by checking that S is closed under the operations of V.

Theorem 1.1 *A nonempty subset S of a vector space V is a subspace of V if and only if S is closed under addition and scalar multiplication or, equivalently,*

S is closed under linear combinations, that is,

$$a, b \in F, u, v \in S \Rightarrow au + bv \in S$$

□

Example 1.2 Consider the vector space $V(n,2)$ of all binary n-tuples, that is, n-tuples of 0's and 1's. The **weight** $\mathcal{W}(v)$ of a vector $v \in V(n,2)$ is the number of nonzero coordinates in v. For instance, $\mathcal{W}(101010) = 3$. Let E_n be the set of all vectors in V of even weight. Then E_n is a subspace of $V(n,2)$.

To see this, note that

$$\mathcal{W}(u+v) = \mathcal{W}(u) + \mathcal{W}(v) - 2\mathcal{W}(u \cap v)$$

where $u \cap v$ is the vector in $V(n,2)$ whose ith component is the product of the ith components of u and v, that is,

$$(u \cap v)_i = u_i \cdot v_i$$

Hence, if $\mathcal{W}(u)$ and $\mathcal{W}(v)$ are both even, so is $\mathcal{W}(u+v)$. Finally, scalar multiplication over F_2 is trivial and so E_n is a subspace of $V(n,2)$, known as the **even weight subspace** of $V(n,2)$.□

Example 1.3 Any subspace of the vector space $V(n,q)$ is called a **linear code**. Linear codes are among the most important and most studied types of codes, because their structure allows for efficient encoding and decoding of information.□

The Lattice of Subspaces

The set $\mathcal{S}(V)$ of all subspaces of a vector space V is partially ordered by set inclusion. The zero subspace $\{0\}$ is the smallest element in $\mathcal{S}(V)$ and the entire space V is the largest element.

If $S, T \in \mathcal{S}(V)$, then $S \cap T$ is the largest subspace of V that is contained in both S and T. In terms of set inclusion, $S \cap T$ is the *greatest lower bound* of S and T:

$$S \cap T = \text{glb}\{S, T\}$$

Similarly, if $\{S_i \mid i \in K\}$ is any collection of subspaces of V, then their intersection is the greatest lower bound of the subspaces:

$$\bigcap_{i \in K} S_i = \text{glb}\{S_i \mid i \in K\}$$

On the other hand, if $S, T \in \mathcal{S}(V)$ (and F is infinite), then $S \cup T \in \mathcal{S}(V)$ if and only if $S \subseteq T$ or $T \subseteq S$. Thus, the union of two subspaces is never a subspace in any "interesting" case. We also have the following.

Theorem 1.2 *A nontrivial vector space V over an infinite field F is not the union of a finite number of proper subspaces.*
Proof. Suppose that $V = S_1 \cup \cdots \cup S_n$, where we may assume that

$$S_1 \not\subseteq S_2 \cup \cdots \cup S_n$$

Let $w \in S_1 \setminus (S_2 \cup \cdots \cup S_n)$ and let $v \notin S_1$. Consider the infinite set

$$A = \{rw + v \mid r \in F\}$$

which is the "line" through v, parallel to w. We want to show that each S_i contains at most one vector from the infinite set A, which is contrary to the fact that $V = S_1 \cup \cdots \cup S_n$. This will prove the theorem.

If $rw + v \in S_1$ for $r \neq 0$, then $w \in S_1$ implies $v \in S_1$, contrary to assumption. Next, suppose that $r_1 w + v \in S_i$ and $r_2 w + v \in S_i$, for $i \geq 2$, where $r_1 \neq r_2$. Then

$$S_i \ni (r_1 w + v) - (r_2 w + v) = (r_1 - r_2)w$$

and so $w \in S_i$, which is also contrary to assumption.□

To determine the smallest subspace of V containing the subspaces S and T, we make the following definition.

Definition *Let S and T be subspaces of V. The* **sum** *$S + T$ is defined by*

$$S + T = \{u + v \mid u \in S, v \in T\}$$

More generally, the **sum** *of any collection $\{S_i \mid i \in K\}$ of subspaces is the set of all finite sums of vectors from the union $\bigcup S_i$:*

$$\sum_{i \in K} S_i = \left\{ s_1 + \cdots + s_n \,\middle|\, s_j \in \bigcup_{i \in K} S_i \right\}$$ □

It is not hard to show that the sum of any collection of subspaces of V is a subspace of V and that the sum is the least upper bound under set inclusion:

$$S + T = \text{lub}\{S, T\}$$

More generally,

$$\sum_{i \in K} S_i = \text{lub}\{S_i \mid i \in K\}$$

If a partially ordered set P has the property that every pair of elements has a least upper bound and greatest lower bound, then P is called a **lattice**. If P has a smallest element and a largest element and has the property that every collection of elements has a least upper bound and greatest lower bound, then P

is called a **complete lattice**. The least upper bound of a collection is also called the **join** of the collection and the greatest lower bound is called the **meet**.

Theorem 1.3 *The set $\mathcal{S}(V)$ of all subspaces of a vector space V is a complete lattice under set inclusion, with smallest element $\{0\}$, largest element V, meet*

$$\operatorname{glb}\{S_i \mid i \in K\} = \bigcap_{i \in K} S_i$$

and join

$$\operatorname{lub}\{S_i \mid i \in K\} = \sum_{i \in K} S_i$$

□

Direct Sums

As we will see, there are many ways to construct new vector spaces from old ones.

External Direct Sums

Definition *Let $V_1, \ldots, V_n$ be vector spaces over a field F. The* **external direct sum** *of $V_1, \ldots, V_n$, denoted by*

$$V = V_1 \boxplus \cdots \boxplus V_n$$

is the vector space V whose elements are ordered n-tuples:

$$V = \{(v_1, \ldots, v_n) \mid v_i \in V_i, i = 1, \ldots, n\}$$

with componentwise operations

$$(u_1, \ldots, u_n) + (v_1, \ldots, v_n) = (u_1 + v_1, \ldots, u_n + v_n)$$

and

$$r(v_1, \ldots, v_n) = (rv_1, \ldots, rv_n)$$

for all $r \in F$.□

Example 1.4 The vector space F^n is the external direct sum of n copies of F, that is,

$$F^n = F \boxplus \cdots \boxplus F$$

where there are n summands on the right-hand side.□

This construction can be generalized to any collection of vector spaces by generalizing the idea that an ordered n-tuple $(v_1, \ldots, v_n)$ is just a function $f: \{1, \ldots, n\} \to \bigcup V_i$ from the *index set* $\{1, \ldots, n\}$ to the union of the spaces with the property that $f(i) \in V_i$.

Definition *Let $\mathcal{F} = \{V_i \mid i \in K\}$ be any family of vector spaces over F. The* **direct product** *of $\mathcal{F}$ is the vector space*

$$\prod_{i \in K} V_i = \left\{ f\colon K \to \bigcup_{i \in K} V_i \,\middle|\, f(i) \in V_i \right\}$$

thought of as a subspace of the vector space of all functions from K to $\bigcup V_i$. □

It will prove more useful to restrict the set of functions to those with finite support.

Definition *Let $\mathcal{F} = \{V_i \mid i \in K\}$ be a family of vector spaces over F. The* **support** *of a function $f\colon K \to \bigcup V_i$ is the set*

$$\operatorname{supp}(f) = \{i \in K \mid f(i) \neq 0\}$$

Thus, a function f has **finite support** *if $f(i) = 0$ for all but a finite number of $i \in K$. The* **external direct sum** *of the family $\mathcal{F}$ is the vector space*

$$\bigoplus_{i \in K}^{\text{ext}} V_i = \left\{ f\colon K \to \bigcup_{i \in K} V_i \,\middle|\, f(i) \in V_i,\ f \text{ has finite support} \right\}$$

thought of as a subspace of the vector space of all functions from K to $\bigcup V_i$. □

An important special case occurs when $V_i = V$ for all $i \in K$. If we let V^K denote the set of all functions from K to V and $(V^K)_0$ denote the set of all functions in V^K that have finite support, then

$$\prod_{i \in K} V = V^K \quad \text{and} \quad \bigoplus_{i \in K}^{\text{ext}} V = (V^K)_0$$

Note that the direct product and the external direct sum are the same for a *finite* family of vector spaces.

Internal Direct Sums

An internal version of the direct sum construction is often more relevant.

Definition *A vector space V is the* **(internal) direct sum** *of a family $\mathcal{F} = \{S_i \mid i \in I\}$ of subspaces of V, written*

$$V = \bigoplus \mathcal{F} \quad \text{or} \quad V = \bigoplus_{i \in I} S_i$$

if the following hold:

1) **(Join of the family)** *V is the sum (join) of the family $\mathcal{F}$:*

$$V = \sum_{i \in I} S_i$$

2) **(Independence of the family)** *For each $i \in I$,*

$$S_i \cap \left(\sum_{j \neq i} S_j \right) = \{0\}$$

In this case, each S_i is called a **direct summand** *of V. If $\mathcal{F} = \{S_1, \ldots, S_n\}$ is a finite family, the direct sum is often written*

$$V = S_1 \oplus \cdots \oplus S_n$$

Finally, if $V = S \oplus T$, then T is called a **complement** *of S in V.*□

Note that the condition in part 2) of the previous definition is *stronger* than saying simply that the members of $\mathcal{F}$ are pairwise disjoint:

$$S_i \cap S_j = \emptyset$$

for all $i \neq j \in I$.

A word of caution is in order here: If S and T are subspaces of V, then we may always say that the sum $S + T$ exists. However, to say that the direct sum of S and T exists or to write $S \oplus T$ is to imply that $S \cap T = \{0\}$. Thus, while the sum of two subspaces always exists, the *direct* sum of two subspaces does not always exist. Similar statements apply to families of subspaces of V.

The reader will be asked in a later chapter to show that the concepts of internal and external direct sum are essentially equivalent (isomorphic). For this reason, the term "direct sum" is often used without qualification.

Once we have discussed the concept of a basis, the following theorem can be easily proved.

Theorem 1.4 *Any subspace of a vector space has a complement, that is, if S is a subspace of V, then there exists a subspace T for which $V = S \oplus T$.*□

It should be emphasized that a subspace generally has many complements (although they are isomorphic). The reader can easily find examples of this in $\mathbb{R}^2$.

We can characterize the uniqueness part of the definition of direct sum in other useful ways. First a remark. If S and T are distinct subspaces of V and if $x, y \in S \cap T$, then the sum $x + y$ can be thought of as a sum of vectors from the same subspace (say S) or from different subspaces—one from S and one from T. When we say that a vector v cannot be written as a sum of vectors from the distinct subspaces S and T, we mean that v cannot be written as a sum $x+y$ where x and y *can be interpreted* as coming from different subspaces, even if they can also be interpreted as coming from the same subspace. Thus, if $x,y \in S \cap T$, then $v = x+y$ *does* express v as a sum of vectors from distinct subspaces.

Theorem 1.5 *Let $\mathcal{F} = \{S_i \mid i \in I\}$ be a family of distinct subspaces of V. The following are equivalent:*

1) **(Independence of the family)** *For each $i \in I$,*

$$S_i \cap \left(\sum_{j \neq i} S_j\right) = \{0\}$$

2) **(Uniqueness of expression for 0)** *The zero vector* 0 *cannot be written as a sum of nonzero vectors from distinct subspaces of $\mathcal{F}$.*

3) **(Uniqueness of expression)** *Every nonzero $v \in V$ has a unique, except for order of terms, expression as a sum*

$$v = s_1 + \cdots + s_n$$

of nonzero vectors from distinct subspaces in $\mathcal{F}$.

Hence, a sum

$$V = \sum_{i \in I} S_i$$

is direct if and only if any one of 1)–3) holds.

Proof. Suppose that 2) fails, that is,

$$0 = s_{j_1} + \cdots + s_{j_n}$$

where the nonzero s_{j_i}'s are from distinct subspaces S_{j_i}. Then $n > 1$ and so

$$-s_{j_1} = s_{j_2} + \cdots + s_{j_n}$$

which violates 1). Hence, 1) implies 2). If 2) holds and

$$v = s_1 + \cdots + s_n \quad \text{and} \quad v = t_1 + \cdots + t_m$$

where the terms are nonzero and the s_i's belong to distinct subspaces in $\mathcal{F}$ and similarily for the t_i's, then

$$0 = s_1 + \cdots + s_n - t_1 - \cdots - t_m$$

By collecting terms from the same subspaces, we may write

$$0 = (s_{i_1} - t_{i_1}) + \cdots + (s_{i_k} - t_{i_k}) + s_{i_{k+1}} + \cdots + s_{i_n} - t_{i_{k+1}} - \cdots - t_{i_m}$$

Then 2) implies that $n = m = k$ and $s_{i_u} = t_{i_u}$ for all $u = 1, \ldots, k$. Hence, 2) implies 3).

Finally, suppose that 3) holds. If

$$0 \neq v \in S_i \cap \left(\sum_{j \neq i} S_j \right)$$

then $v = s_i \in S_i$ and

$$s_i = s_{j_1} + \cdots + s_{j_n}$$

where $s_{j_k} \in S_{j_k}$ are nonzero. But this violates 3).□

Example 1.5 Any matrix $A \in \mathcal{M}_n$ can be written in the form

$$A = \frac{1}{2}(A + A^t) + \frac{1}{2}(A - A^t) = B + C \tag{1.1}$$

where A^t is the transpose of A. It is easy to verify that B is symmetric and C is skew-symmetric and so (1.1) is a decomposition of A as the sum of a symmetric matrix and a skew-symmetric matrix.

Since the sets Sym and SkewSym of all symmetric and skew-symmetric matrices in $\mathcal{M}_n$ are subspaces of $\mathcal{M}_n$, we have

$$\mathcal{M}_n = \text{Sym} + \text{SkewSym}$$

Furthermore, if $S + T = S' + T'$, where S and S' are symmetric and T and T' are skew-symmetric, then the matrix

$$U = S - S' = T' - T$$

is both symmetric and skew-symmetric. Hence, provided that $\text{char}(F) \neq 2$, we must have $U = 0$ and so $S = S'$ and $T = T'$. Thus,

$$\mathcal{M}_n = \text{Sym} \oplus \text{SkewSym}$$ □

Spanning Sets and Linear Independence

A set of vectors *spans* a vector space if every vector can be written as a linear combination of some of the vectors in that set. Here is the formal definition.

Definition *The* **subspace spanned** (*or* **subspace generated**) *by a nonempty set S of vectors in V is the set of all linear combinations of vectors from S:*

$$\langle S \rangle = \text{span}(S) = \{r_1 v_1 + \cdots + r_n v_n \mid r_i \in F, v_i \in S\}$$

When $S = \{v_1, \ldots, v_n\}$ *is a finite set, we use the notation* $\langle v_1, \ldots, v_n \rangle$ *or* $\operatorname{span}(v_1, \ldots, v_n)$. *A set* S *of vectors in* V *is said to* **span** V, *or* **generate** V, *if* $V = \operatorname{span}(S)$. $\square$

It is clear that any superset of a spanning set is also a spanning set. Note also that all vector spaces have spanning sets, since V spans itself.

Linear Independence

Linear independence is a fundamental concept.

Definition *Let* V *be a vector space. A nonempty set* S *of vectors in* V *is* **linearly independent** *if for any distinct vectors* $s_1, \ldots, s_n$ *in* S,

$$a_1 s_1 + \cdots + a_n s_n = 0 \quad \Rightarrow \quad a_i = 0 \ \text{ for all } i$$

In words, S *is linearly independent if the only linear combination of vectors from* S *that is equal to* 0 *is the trivial linear combination, all of whose coefficients are* 0. *If* S *is not linearly independent, it is said to be* **linearly dependent**. $\square$

It is immediate that a linearly independent set of vectors cannot contain the zero vector, since then $1 \cdot 0 = 0$ violates the condition of linear independence.

Another way to phrase the definition of linear independence is to say that S is linearly independent if the zero vector has an "as unique as possible" expression as a linear combination of vectors from S. We can never prevent the zero vector from being written in the form $0 = 0s_1 + \cdots + 0s_n$, but we can prevent 0 from being written in any other way as a linear combination of the vectors in S.

For the introspective reader, the expression $0 = s_1 + (-1s_1)$ has two interpretations. One is $0 = as_1 + bs_1$ where $a = 1$ and $b = -1$, but this does not involve distinct vectors so is not relevant to the question of linear independence. The other interpretation is $0 = s_1 + t_1$ where $t_1 = -s_1 \neq s_1$ (assuming that $s_1 \neq 0$). Thus, if S is linearly independent, then S cannot contain both s_1 and $-s_1$.

Definition *Let* S *be a nonempty set of vectors in* V. *To say that a nonzero vector* $v \in V$ *is an* **essentially unique** *linear combination of the vectors in* S *is to say that, up to order of terms, there is one and only one way to express* v *as a linear combination*

$$v = a_1 s_1 + \cdots + a_n s_n$$

where the s_i*'s are distinct vectors in* S *and the coefficients* a_i *are nonzero. More explicitly,* $v \neq 0$ *is an essentially unique linear combination of the vectors in* S *if* $v \in \langle S \rangle$ *and if whenever*

$$v = a_1 s_1 + \cdots + a_n s_n \quad \textit{and} \quad v = b_1 t_1 + \cdots + b_m t_m$$

where the s_i*'s are distinct, the* t_i*'s are distinct and all coefficients are nonzero, then* $m = n$ *and after a reindexing of the* $b_i t_i$*'s if necessary, we have* $a_i = b_i$ *and* $s_i = t_i$ *for all* $i = 1, \ldots, n$*. (Note that this is stronger than saying that* $a_i s_i = b_i t_i$*.)*□

We may characterize linear independence as follows.

Theorem 1.6 *Let* $S \neq \{0\}$ *be a nonempty set of vectors in* V*. The following are equivalent:*

1) *S is linearly independent.*
2) *Every nonzero vector* $v \in \operatorname{span}(S)$ *is an essentially unique linear combination of the vectors in* S*.*
3) *No vector in* S *is a linear combination of other vectors in* S*.*

Proof. Suppose that 1) holds and that

$$0 \neq v = a_1 s_1 + \cdots + a_n s_n = b_1 t_1 + \cdots + b_m t_m$$

where the s_i's are distinct, the t_i's are distinct and the coefficients are nonzero. By subtracting and grouping s's and t's that are equal, we can write

$$\begin{aligned} 0 &= (a_{i_1} - b_{i_1}) s_{i_1} + \cdots + (a_{i_k} - b_{i_k}) s_{i_k} \\ &\quad + a_{i_{k+1}} s_{i_{k+1}} + \cdots + a_{i_n} s_{i_n} \\ &\quad - b_{i_{k+1}} t_{i_{k+1}} - \cdots - b_{i_m} t_{i_m} \end{aligned}$$

and so 1) implies that $n = m = k$ and $a_{i_u} = b_{i_u}$ and $s_{i_u} = t_{i_u}$ for all $i = 1, \ldots, k$. Thus, 1) implies 2).

If 2) holds and $s \in S$ can be written as

$$s = a_1 s_1 + \cdots + a_n s_n$$

where $s_i \in S$ are different from s, then we may collect like terms on the right and then remove all terms with 0 coefficient. The resulting expression violates 2). Hence, 2) implies 3). If 3) holds and

$$a_1 s_1 + \cdots + a_n s_n = 0$$

where the s_i's are distinct and $a_1 \neq 0$, then $n > 1$ and we may write

$$s_1 = -\frac{1}{a_1}(a_2 s_2 + \cdots + a_n s_n)$$

which violates 3).□

The following key theorem relates the notions of spanning set and linear independence.

Theorem 1.7 *Let S be a set of vectors in V. The following are equivalent:*

1) *S is linearly independent and spans V.*
2) *Every nonzero vector $v \in V$ is an essentially unique linear combination of vectors in S.*
3) *S is a minimal spanning set, that is, S spans V but any proper subset of S does not span V.*
4) *S is a maximal linearly independent set, that is, S is linearly independent, but any proper superset of S is not linearly independent.*

A set of vectors in V that satisfies any (and hence all) of these conditions is called a **basis** *for V.*

Proof. We have seen that 1) and 2) are equivalent. Now suppose 1) holds. Then S is a spanning set. If some proper subset S' of S also spanned V, then any vector in $S - S'$ would be a linear combination of the vectors in S', contradicting the fact that the vectors in S are linearly independent. Hence 1) implies 3).

Conversely, if S is a minimal spanning set, then it must be linearly independent. For if not, some vector $s \in S$ would be a linear combination of the other vectors in S and so $S - \{s\}$ would be a proper spanning subset of S, which is not possible. Hence 3) implies 1).

Suppose again that 1) holds. If S were not maximal, there would be a vector $v \in V - S$ for which the set $S \cup \{v\}$ is linearly independent. But then v is not in the span of S, contradicting the fact that S is a spanning set. Hence, S is a maximal linearly independent set and so 1) implies 4).

Conversely, if S is a maximal linearly independent set, then S must span V, for if not, we could find a vector $v \in V - S$ that is not a linear combination of the vectors in S. Hence, $S \cup \{v\}$ would be a linearly independent proper superset of S, which is a contradiction. Thus, 4) implies 1).□

Theorem 1.8 *A finite set $S = \{v_1, \dots, v_n\}$ of vectors in V is a basis for V if and only if*

$$V = \langle v_1 \rangle \oplus \cdots \oplus \langle v_n \rangle$$ □

Example 1.6 The ith **standard vector** in F^n is the vector e_i that has 0's in all coordinate positions except the ith, where it has a 1. Thus,

$$e_1 = (1, 0, \dots, 0), \quad e_2 = (0, 1, \dots, 0) \quad , \dots , \quad e_n = (0, \dots, 0, 1)$$

The set $\{e_1, \dots, e_n\}$ is called the **standard basis** for F^n.□

The proof that every nontrivial vector space has a basis is a classic example of the use of Zorn's lemma.

Theorem 1.9 *Let V be a nonzero vector space. Let I be a linearly independent set in V and let S be a spanning set in V containing I. Then there is a basis $\mathcal{B}$ for V for which $I \subseteq \mathcal{B} \subseteq S$. In particular,*

1) *Any vector space, except the zero space $\{0\}$, has a basis.*
2) *Any linearly independent set in V is contained in a basis.*
3) *Any spanning set in V contains a basis.*

Proof. Consider the collection $\mathcal{A}$ of all linearly independent subsets of V containing I and contained in S. This collection is not empty, since $I \in \mathcal{A}$. Now, if

$$\mathcal{C} = \{I_k \mid k \in K\}$$

is a chain in $\mathcal{A}$, then the union

$$U = \bigcup_{k \in K} I_i$$

is linearly independent and satisfies $I \subseteq U \subseteq S$, that is, $U \in \mathcal{A}$. Hence, every chain in $\mathcal{A}$ has an upper bound in $\mathcal{A}$ and according to Zorn's lemma, $\mathcal{A}$ must contain a maximal element $\mathcal{B}$, which is linearly independent.

Now, $\mathcal{B}$ is a basis for the vector space $\langle S \rangle = V$, for if any $s \in S$ is not a linear combination of the elements of $\mathcal{B}$, then $\mathcal{B} \cup \{s\} \subseteq S$ is linearly independent, contradicting the maximality of $\mathcal{B}$. Hence $S \subseteq \langle \mathcal{B} \rangle$ and so $V = \langle S \rangle \subseteq \langle \mathcal{B} \rangle$.$\square$

The reader can now show, using Theorem 1.9, that any subspace of a vector space has a complement.

The Dimension of a Vector Space

The next result, with its classical elegant proof, says that if a vector space V has a *finite* spanning set S, then the size of any linearly independent set cannot exceed the size of S.

Theorem 1.10 *Let V be a vector space and assume that the vectors $v_1, \ldots, v_n$ are linearly independent and the vectors $s_1, \ldots, s_m$ span V. Then $n \le m$.*

Proof. First, we list the two sets of vectors: the spanning set followed by the linearly independent set:

$$s_1, \ldots, s_m; v_1, \ldots, v_n$$

Then we move the first vector v_1 to the front of the first list:

$$v_1, s_1, \ldots, s_m; v_2, \ldots, v_n$$

Since $s_1, \ldots, s_m$ span V, v_1 is a linear combination of the s_i's. This implies that we may remove one of the s_i's, which by reindexing if necessary can be s_1, from the first list and still have a spanning set

$$v_1, s_2, \ldots, s_m; v_2, \ldots, v_n$$

Note that the first set of vectors still spans V and the second set is still linearly independent.

Now we repeat the process, moving v_2 from the second list to the first list

$$v_1, v_2, s_2, \ldots, s_m; v_3, \ldots, v_n$$

As before, the vectors in the first list are linearly dependent, since they spanned V before the inclusion of v_2. However, since the v_i's are linearly independent, any nontrivial linear combination of the vectors in the first list that equals 0 must involve at least one of the s_i's. Hence, we may remove that vector, which again by reindexing if necessary may be taken to be s_2 and still have a spanning set

$$v_1, v_2, s_3, \ldots, s_m; v_3, \ldots, v_n$$

Once again, the first set of vectors spans V and the second set is still linearly independent.

Now, if $m < n$, then this process will eventually exhaust the s_i's and lead to the list

$$v_1, v_2, \ldots, v_m; v_{m+1}, \ldots, v_n$$

where $v_1, v_2, \ldots, v_m$ span V, which is clearly not possible since v_n is not in the span of $v_1, v_2, \ldots, v_m$. Hence, $n \le m$.□

Corollary 1.11 *If V has a* finite *spanning set, then any two bases of V have the same size.*□

Now let us prove the analogue of Corollary 1.11 for arbitrary vector spaces.

Theorem 1.12 *If V is a vector space, then any two bases for V have the same cardinality.*
Proof. We may assume that all bases for V are infinite sets, for if any basis is finite, then V has a finite spanning set and so Corollary 1.11 applies.

Let $\mathcal{B} = \{b_i \mid i \in I\}$ be a basis for V and let $\mathcal{C}$ be another basis for V. Then any vector $c \in \mathcal{C}$ can be written as a finite linear combination of the vectors in $\mathcal{B}$, where all of the coefficients are nonzero, say

$$c = \sum_{i \in U_c} r_i b_i$$

But because $\mathcal{C}$ is a basis, we must have

$$\bigcup_{c \in \mathcal{C}} U_c = I$$

for if the vectors in $\mathcal{C}$ can be expressed as finite linear combinations of the vectors in a *proper* subset $\mathcal{B}'$ of $\mathcal{B}$, then $\mathcal{B}'$ spans V, which is not the case.

Since $|U_c| < \aleph_0$ for all $c \in \mathcal{C}$, Theorem 0.17 implies that

$$|\mathcal{B}| = |I| \leq \aleph_0 |\mathcal{C}| = |\mathcal{C}|$$

But we may also reverse the roles of $\mathcal{B}$ and $\mathcal{C}$, to conclude that $|\mathcal{C}| \leq |\mathcal{B}|$ and so the Schröder–Bernstein theorem implies that $|\mathcal{B}| = |\mathcal{C}|$.□

Theorem 1.12 allows us to make the following definition.

Definition *A vector space V is* **finite-dimensional** *if it is the zero space $\{0\}$, or if it has a finite basis. All other vector spaces are* **infinite-dimensional**. *The* **dimension** *of the zero space is 0 and the* **dimension** *of any nonzero vector space V is the cardinality of any basis for V. If a vector space V has a basis of cardinality κ, we say that V is* **κ-dimensional** *and write* $\dim(V) = \kappa$.□

It is easy to see that if S is a subspace of V, then $\dim(S) \leq \dim(V)$. If in addition, $\dim(S) = \dim(V) < \infty$, then $S = V$.

Theorem 1.13 *Let V be a vector space.*
1) If $\mathcal{B}$ is a basis for V and if $\mathcal{B} = \mathcal{B}_1 \cup \mathcal{B}_2$ and $\mathcal{B}_1 \cap \mathcal{B}_2 = \emptyset$, then

$$V = \langle \mathcal{B}_1 \rangle \oplus \langle \mathcal{B}_2 \rangle$$

2) Let $V = S \oplus T$. If $\mathcal{B}_1$ is a basis for S and $\mathcal{B}_2$ is a basis for T, then $\mathcal{B}_1 \cap \mathcal{B}_2 = \emptyset$ and $\mathcal{B} = \mathcal{B}_1 \cup \mathcal{B}_2$ is a basis for V.□

Theorem 1.14 *Let S and T be subspaces of a vector space V. Then*

$$\dim(S) + \dim(T) = \dim(S + T) + \dim(S \cap T)$$

In particular, if T is any complement of S in V, then

$$\dim(S) + \dim(T) = \dim(V)$$

that is,

$$\dim(S \oplus T) = \dim(S) + \dim(T)$$

Proof. Suppose that $\mathcal{B} = \{b_i \mid i \in I\}$ is a basis for $S \cap T$. Extend this to a basis $\mathcal{A} \cup \mathcal{B}$ for S where $\mathcal{A} = \{a_j \mid j \in J\}$ is disjoint from $\mathcal{B}$. Also, extend $\mathcal{B}$ to a basis $\mathcal{B} \cup \mathcal{C}$ for T where $\mathcal{C} = \{c_k \mid k \in K\}$ is disjoint from $\mathcal{B}$. We claim that $\mathcal{A} \cup \mathcal{B} \cup \mathcal{C}$ is a basis for $S + T$. It is clear that $\langle \mathcal{A} \cup \mathcal{B} \cup \mathcal{C} \rangle = S + T$.

To see that $\mathcal{A} \cup \mathcal{B} \cup \mathcal{C}$ is linearly independent, suppose to the contrary that

$$\alpha_1 v_1 + \cdots + \alpha_n v_n = 0$$

where $v_i \in \mathcal{A} \cup \mathcal{B} \cup \mathcal{C}$ and $\alpha_i \neq 0$ for all i. There must be vectors v_i in this expression from both $\mathcal{A}$ and $\mathcal{C}$, since $\mathcal{A} \cup \mathcal{B}$ and $\mathcal{B} \cup \mathcal{C}$ are linearly independent. Isolating the terms involving the vectors from $\mathcal{A}$ on one side of the equality shows that there is a nonzero vector in $x \in \langle \mathcal{A} \rangle \cap \langle \mathcal{B} \cup \mathcal{C} \rangle$. But then $x \in S \cap T$ and so $x \in \langle \mathcal{A} \rangle \cap \langle \mathcal{B} \rangle$, which implies that $x = 0$, a contradiction. Hence, $\mathcal{A} \cup \mathcal{B} \cup \mathcal{C}$ is linearly independent and a basis for $S + T$.

Now,

$$\begin{aligned}\dim(S) + \dim(T) &= |\mathcal{A} \cup \mathcal{B}| + |\mathcal{B} \cup \mathcal{C}| \\ &= |\mathcal{A}| + |\mathcal{B}| + |\mathcal{B}| + |\mathcal{C}| \\ &= |\mathcal{A}| + |\mathcal{B}| + |\mathcal{C}| + \dim(S \cap T) \\ &= \dim(S + T) + \dim(S \cap T)\end{aligned}$$

as desired.□

It is worth emphasizing that while the equation

$$\dim(S) + \dim(T) = \dim(S + T) + \dim(S \cap T)$$

holds for all vector spaces, we cannot write

$$\dim(S + T) = \dim(S) + \dim(T) - \dim(S \cap T)$$

unless $S + T$ is finite-dimensional.

Ordered Bases and Coordinate Matrices

It will be convenient to consider bases that have an order imposed on their members.

Definition *Let V be a vector space of dimension n. An* **ordered basis** *for V is an ordered n-tuple $(v_1, \ldots, v_n)$ of vectors for which the set $\{v_1, \ldots, v_n\}$ is a basis for V.*□

If $\mathcal{B} = (v_1, \ldots, v_n)$ is an ordered basis for V, then for each $v \in V$ there is a unique ordered n-tuple $(r_1, \ldots, r_n)$ of scalars for which

$$v = r_1 v_1 + \cdots + r_n v_n$$

Accordingly, we can define the **coordinate map** $\phi_{\mathcal{B}}: V \to F^n$ by

$$\phi_{\mathcal{B}}(v) = [v]_{\mathcal{B}} = \begin{bmatrix} r_1 \\ \vdots \\ r_n \end{bmatrix} \tag{1.3}$$

where the column matrix $[v]_{\mathcal{B}}$ is known as the **coordinate matrix** of v with respect to the ordered basis $\mathcal{B}$. Clearly, knowing $[v]_{\mathcal{B}}$ is equivalent to knowing v (assuming knowledge of $\mathcal{B}$).

Furthermore, it is easy to see that the coordinate map $\phi_{\mathcal{B}}$ is bijective and preserves the vector space operations, that is,

$$\phi_{\mathcal{B}}(r_1 v_1 + \cdots + r_n v_n) = r_1 \phi_{\mathcal{B}}(v_1) + \cdots + r_n \phi_{\mathcal{B}}(v_n)$$

or equivalently

$$[r_1 v_1 + \cdots + r_n v_n]_{\mathcal{B}} = r_1 [v_1]_{\mathcal{B}} + \cdots + r_n [v_n]_{\mathcal{B}}$$

Functions from one vector space to another that preserve the vector space operations are called *linear transformations* and form the objects of study in the next chapter.

The Row and Column Spaces of a Matrix

Let A be an $m \times n$ matrix over F. The rows of A span a subspace of F^n known as the **row space** of A and the columns of A span a subspace of F^m known as the **column space** of A. The dimensions of these spaces are called the **row rank** and **column rank**, respectively. We denote the row space and row rank by $\mathrm{rs}(A)$ and $\mathrm{rrk}(A)$ and the column space and column rank by $\mathrm{cs}(A)$ and $\mathrm{crk}(A)$.

It is a remarkable and useful fact that the row rank of a matrix is always equal to its column rank, despite the fact that if $m \neq n$, the row space and column space are not even in the same vector space!

Our proof of this fact hinges on the following simple observation about matrices.

Lemma 1.15 *Let A be an $m \times n$ matrix. Then elementary column operations do not affect the row rank of A. Similarly, elementary row operations do not affect the column rank of A.*

Proof. The second statement follows from the first by taking transposes. As to the first, the row space of A is

$$\mathrm{rs}(A) = \langle e_1 A, \ldots, e_n A \rangle$$

where e_i are the standard basis vectors in F^m. Performing an elementary column operation on A is equivalent to multiplying A on the right by an elementary matrix E. Hence the row space of AE is

$$\mathrm{rs}(AE) = \langle e_1 AE, \ldots, e_n AE \rangle$$

and since E is invertible,

$$\text{rrk}(A) = \dim(\text{rs}(A)) = \dim(\text{rs}(AE)) = \text{rrk}(AE)$$

as desired.□

Theorem 1.16 *If $A \in \mathcal{M}_{m,n}$, then* $\text{rrk}(A) = \text{crk}(A)$. *This number is called the* **rank** *of A and is denoted by* $\text{rk}(A)$.

Proof. According to the previous lemma, we may reduce A to reduced column echelon form without affecting the row rank. But this reduction does not affect the column rank either. Then we may further reduce A to reduced row echelon form without affecting either rank. The resulting matrix M has the same row and column ranks as A. But M is a matrix with 1's followed by 0's on the main diagonal (entries $M_{1,1}, M_{2,2}, \ldots$) and 0's elsewhere. Hence,

$$\text{rrk}(A) = \text{rrk}(M) = \text{crk}(M) = \text{crk}(A)$$

as desired.□

The Complexification of a Real Vector Space

If W is a complex vector space (that is, a vector space over $\mathbb{C}$), then we can think of W as a real vector space simply by restricting all scalars to the field $\mathbb{R}$. Let us denote this real vector space by $W_{\mathbb{R}}$ and call it the **real version** of W.

On the other hand, to each real vector space V, we can associate a complex vector space $V^{\mathbb{C}}$. This "complexification" process will play a useful role when we discuss the structure of linear operators on a real vector space. (Throughout our discussion V will denote a real vector space.)

Definition *If V is a real vector space, then the set $V^{\mathbb{C}} = V \times V$ of ordered pairs, with componentwise addition*

$$(u, v) + (x, y) = (u + x, v + y)$$

and scalar multiplication over $\mathbb{C}$ defined by

$$(a + bi)(u, v) = (au - bv, av + bu)$$

for $a, b \in \mathbb{R}$ is a complex vector space, called the **complexification** *of V.*□

It is convenient to introduce a notation for vectors in $V^{\mathbb{C}}$ that resembles the notation for complex numbers. In particular, we denote $(u, v) \in V^{\mathbb{C}}$ by $u + vi$ and so

$$V^{\mathbb{C}} = \{u + vi \mid u, v \in V\}$$

Addition now looks like ordinary addition of complex numbers,

$$(u + vi) + (x + yi) = (u + x) + (v + y)i$$

and scalar multiplication looks like ordinary multiplication of complex numbers,

$$(a+bi)(u+vi) = (au - bv) + (av + bu)i$$

Thus, for example, we immediately have for $a, b \in \mathbb{R}$,

$$\begin{aligned} a(u+vi) &= au + avi \\ bi(u+vi) &= -bv + bui \\ (a+bi)u &= au + bui \\ (a+bi)vi &= -bv + avi \end{aligned}$$

The **real part** of $z = u + vi$ is $u \in V$ and the **imaginary part** of z is $v \in V$. The essence of the fact that $z = u + vi \in V^{\mathbb{C}}$ is really an ordered pair is that z is 0 if and only if its real and imaginary parts are both 0.

We can define the **complexification map** cpx: $V \to V^{\mathbb{C}}$ by

$$\text{cpx}(v) = v + 0i$$

Let us refer to $v + 0i$ as the **complexification**, or **complex version** of $v \in V$. Note that this map is a group homomorphism, that is,

$$\text{cpx}(0) = 0 + 0i \quad \text{and} \quad \text{cpx}(u \pm v) = \text{cpx}(u) \pm \text{cpx}(v)$$

and it is injective:

$$\text{cpx}(u) = \text{cpx}(v) \Leftrightarrow u = v$$

Also, it preserves multiplication by *real* scalars:

$$\text{cpx}(au) = au + 0i = a(u + 0i) = a\text{cpx}(u)$$

for $a \in \mathbb{R}$. However, the complexification map is not surjective, since it gives only "real" vectors in $V^{\mathbb{C}}$.

The complexification map is an injective linear transformation (defined in the next chapter) from the real vector space V to the real version $(V^{\mathbb{C}})_{\mathbb{R}}$ of the complexification $V^{\mathbb{C}}$, that is, to the complex vector space $V^{\mathbb{C}}$ provided that scalars are restricted to real numbers. In this way, we see that $V^{\mathbb{C}}$ contains an embedded copy of V.

The Dimension of $V^{\mathbb{C}}$

The vector-space dimensions of V and $V^{\mathbb{C}}$ are the same. This should not necessarily come as a surprise because although $V^{\mathbb{C}}$ may seem "bigger" than V, the field of scalars is also "bigger."

Theorem 1.17 *If $\mathcal{B} = \{v_j \mid j \in I\}$ is a basis for V over $\mathbb{R}$, then the* **complexification** *of $\mathcal{B}$,*

$$\text{cpx}(\mathcal{B}) = \{v_j + 0i \mid v_j \in \mathcal{B}\}$$

is a basis for the vector space $V^{\mathbb{C}}$ over $\mathbb{C}$. Hence,

$$\dim(V^{\mathbb{C}}) = \dim(V)$$

Proof. To see that $\text{cpx}(\mathcal{B})$ spans $V^{\mathbb{C}}$ over $\mathbb{C}$, let $x + iy \in V^{\mathbb{C}}$. Then $x, y \in V$ and so there exist real numbers a_i and b_i (some of which may be 0) for which

$$\begin{aligned} x + yi &= \sum_{j=1}^{J} a_j v_j + \left[\sum_{j=1}^{J} b_j v_j\right] i \\ &= \sum_{j=1}^{J} (a_j v_j + b_j v_j i) \\ &= \sum_{j=1}^{J} (a_j + b_j i)(v_j + 0i) \end{aligned}$$

To see that $\text{cpx}(\mathcal{B})$ is linearly independent, if

$$\sum_{j=1}^{J} (a_j + b_j i)(v_j + 0i) = 0 + 0i$$

then the previous computations show that

$$\sum_{j=1}^{J} a_j v_j = 0 \text{ and } \sum_{j=1}^{J} b_j v_j = 0$$

The independence of $\mathcal{B}$ then implies that $a_i = 0$ and $b_i = 0$ for all i.□

If $v \in V$ and $\mathcal{B} = \{v_i \mid i \in I\}$ is a basis for V, then we may write

$$v = \sum_{i=1}^{n} a_i v_i$$

for $a_i \in \mathbb{R}$. Since the coefficients are real, we have

$$v + 0i = \sum_{i=1}^{n} a_i (v_i + 0i)$$

and so the coordinate matrices are equal:

$$[v + 0i]_{\text{cpx}(\mathcal{B})} = [v]_{\mathcal{B}}$$

Exercises

1. Let V be a vector space over F. Prove that $0v = 0$ and $r0 = 0$ for all $v \in V$ and $r \in F$. Describe the different 0's in these equations. Prove that if $rv = 0$, then $r = 0$ or $v = 0$. Prove that $rv = v$ implies that $v = 0$ or $r = 1$.

2. Prove Theorem 1.3.
3. a) Find an abelian group V and a field F for which V is a vector space over F in at least two different ways, that is, there are two different definitions of scalar multiplication making V a vector space over F.
   b) Find a vector space V over F and a subset S of V that is (1) a subspace of V and (2) a vector space using operations that differ from those of V.
4. Suppose that V is a vector space with basis $\mathcal{B} = \{b_i \mid i \in I\}$ and S is a subspace of V. Let $\{B_1, \ldots, B_k\}$ be a partition of $\mathcal{B}$. Then is it true that

$$S = \bigoplus_{i=1}^{k} (S \cap \langle B_i \rangle)$$

   What if $S \cap \langle B_i \rangle \neq \{0\}$ for all i?
5. Prove Theorem 1.8.
6. Let $S, T, U \in \mathcal{S}(V)$. Show that if $U \subseteq S$, then

$$S \cap (T + U) = (S \cap T) + U$$

   This is called the **modular law** for the lattice $\mathcal{S}(V)$.
7. For what vector spaces does the distributive law of subspaces

$$S \cap (T + U) = (S \cap T) + (S \cap U)$$

   hold?
8. A vector $v = (a_1, \ldots, a_n) \in \mathbb{R}^n$ is called **strongly positive** if $a_i > 0$ for all $i = 1, \ldots, n$.
   a) Suppose that v is strongly positive. Show that any vector that is "close enough" to v is also strongly positive. (Formulate carefully what "close enough" should mean.)
   b) Prove that if a subspace S of $\mathbb{R}^n$ contains a strongly positive vector, then S has a basis of strongly positive vectors.
9. Let M be an $m \times n$ matrix whose rows are linearly independent. Suppose that the k columns $c_{i_1}, \ldots, c_{i_k}$ of M span the column space of M. Let C be the matrix obtained from M by deleting all columns except $c_{i_1}, \ldots, c_{i_k}$. Show that the rows of C are also linearly independent.
10. Prove that the first two statements in Theorem 1.7 are equivalent.
11. Show that if S is a subspace of a vector space V, then $\dim(S) \leq \dim(V)$. Furthermore, if $\dim(S) = \dim(V) < \infty$ then $S = V$. Give an example to show that the finiteness is required in the second statement.
12. Let $\dim(V) < \infty$ and suppose that $V = U \oplus S_1 = U \oplus S_2$. What can you say about the relationship between S_1 and S_2? What can you say if $S_1 \subseteq S_2$?
13. What is the relationship between $S \oplus T$ and $T \oplus S$? Is the direct sum operation commutative? Formulate and prove a similar statement concerning associativity. Is there an "identity" for direct sum? What about "negatives"?

14. Let V be a finite-dimensional vector space over an infinite field F. Prove that if $S_1, \ldots, S_k$ are subspaces of V of equal dimension, then there is a subspace T of V for which $V = S_i \oplus T$ for all $i = 1, \ldots, k$. In other words, T is a common complement of the subspaces S_i.
15. Prove that the vector space $\mathcal{C}$ of all continuous functions from $\mathbb{R}$ to $\mathbb{R}$ is infinite-dimensional.
16. Show that Theorem 1.2 need not hold if the base field F is finite.
17. Let S be a subspace of V. The set $v + S = \{v + s \mid s \in S\}$ is called an **affine subspace** of V.
    a) Under what conditions is an affine subspace of V a subspace of V?
    b) Show that any two affine subspaces of the form $v + S$ and $w + S$ are either equal or disjoint.
18. If V and W are vector spaces over F for which $|V| = |W|$, then does it follow that $\dim(V) = \dim(W)$?
19. Let V be an n-dimensional real vector space and suppose that S is a subspace of V with $\dim(S) = n - 1$. Define an equivalence relation $\equiv$ on the set $V \setminus S$ by $v \equiv w$ if the "line segment"

$$L(v, w) = \{rv + (1 - r)w \mid 0 \leq r \leq 1\}$$

    has the property that $L(v, w) \cap S = \emptyset$. Prove that $\equiv$ is an equivalence relation and that it has exactly two equivalence classes.
20. Let F be a field. A **subfield** of F is a subset K of F that is a field in its own right using the same operations as defined on F.
    a) Show that F is a vector space over any subfield K of F.
    b) Suppose that F is an m-dimensional vector space over a subfield K of F. If V is an n-dimensional vector space over F, show that V is also a vector space over K. What is the dimension of V as a vector space over K?
21. Let F be a finite field of size q and let V be an n-dimensional vector space over F. The purpose of this exercise is to show that the number of subspaces of V of dimension k is

$$\binom{n}{k}_q = \frac{(q^n - 1)\cdots(q - 1)}{(q^k - 1)\cdots(q - 1)(q^{n-k} - 1)\cdots(q - 1)}$$

    The expressions $\binom{n}{k}_q$ are called **Gaussian coefficients** and have properties similar to those of the binomial coefficients. Let $S(n, k)$ be the number of k-dimensional subspaces of V.
    a) Let $N(n, k)$ be the number of k-tuples of linearly independent vectors $(v_1, \ldots, v_k)$ in V. Show that

$$N(n, k) = (q^n - 1)(q^n - q)\cdots(q^n - q^{k-1})$$

    b) Now, each of the k-tuples in a) can be obtained by first choosing a subspace of V of dimension k and then selecting the vectors from this subspace. Show that for any k-dimensional subspace of V, the number of k-tuples of independent vectors in this subspace is

$$(q^k - 1)(q^k - q)\cdots(q^k - q^{k-1})$$

c) Show that

$$N(n,k) = S(n,k)(q^k - 1)(q^k - q)\cdots(q^k - q^{k-1})$$

How does this complete the proof?

22. Prove that any subspace S of $\mathbb{R}^n$ is a closed set or, equivalently, that its set complement $S^c = \mathbb{R}^n \setminus S$ is open, that is, for any $x \in S^c$ there is an open ball $B(x,\epsilon)$ centered at x with radius $\epsilon > 0$ for which $B(x,\epsilon) \subseteq S^c$.
23. Let $\mathcal{B} = \{b_1, \ldots, b_n\}$ and $\mathcal{C} = \{c_1, \ldots, c_n\}$ be bases for a vector space V. Let $1 \le m \le n-1$. Show that there is a permutation σ of $\{1, \ldots, n\}$ such that

$$b_1, \ldots, b_m, c_{\sigma(m+1)}, \ldots, c_{\sigma(n)}$$

and

$$c_{\sigma(1)}, \ldots, c_{\sigma(m)}, b_{m+1}, \ldots, b_n$$

are both bases for V. *Hint*: You may use the fact that if M is an invertible $n \times n$ matrix and if $1 \le k \le n$, then it is possible to reorder the rows so that the upper left $k \times k$ submatrix and the lower right $(n-k) \times (n-k)$ submatrix are both invertible. (This follows, for example, from the general Laplace expansion theorem for determinants.)
24. Let V be an n-dimensional vector space over an infinite field F and suppose that $S_1, \ldots, S_k$ are subspaces of V with $\dim(S_i) \le m < n$. Prove that there is a subspace T of V of dimension $n-m$ for which $T \cap S_i = \{0\}$ for all i.
25. What is the dimension of the complexification $V^{\mathbb{C}}$ thought of as a real vector space?
26. (When is a subspace of a complex vector space a complexification?) Let V be a real vector space with complexification $V^{\mathbb{C}}$ and let U be a subspace of $V^{\mathbb{C}}$. Prove that there is a subspace S of V for which

$$U = S^{\mathbb{C}} = \{s + ti \mid s, t \in S\}$$

if and only if U is closed under complex conjugation $\chi\colon V^{\mathbb{C}} \to V^{\mathbb{C}}$ defined by $\chi(u + iv) = u - iv$.

Chapter 2
Linear Transformations

Linear Transformations

Loosely speaking, a linear transformation is a function from one vector space to another that *preserves* the vector space operations. Let us be more precise.

Definition *Let V and W be vector spaces over a field F. A function $\tau: V \to W$ is a* **linear transformation** *if*

$$\tau(ru + sv) = r\tau(u) + s\tau(v)$$

for all scalars $r, s \in F$ and vectors $u, v \in V$. The set of all linear transformations from V to W is denoted by $\mathcal{L}(V, W)$.

1) *A linear transformation from V to V is called a* **linear operator** *on V. The set of all linear operators on V is denoted by $\mathcal{L}(V)$. A linear operator on a real vector space is called a* **real operator** *and a linear operator on a complex vector space is called a* **complex operator**.
2) *A linear transformation from V to the base field F (thought of as a vector space over itself) is called a* **linear functional** *on V. The set of all linear functionals on V is denoted by V^* and called the* **dual space** *of V.* □

We should mention that some authors use the term linear operator for any linear transformation from V to W. Also, the application of a linear transformation τ on a vector v is denoted by $\tau(v)$ or by τv, parentheses being used when necessary, as in $\tau(u+v)$, or to improve readability, as in $\mu(\tau u)$ rather than $\mu(\tau(u))$.

Definition *The following terms are also employed:*

1) **homomorphism** *for linear transformation*
2) **endomorphism** *for linear operator*
3) **monomorphism** (*or* **embedding**) *for injective linear transformation*
4) **epimorphism** *for surjective linear transformation*
5) **isomorphism** *for bijective linear transformation.*

6) **automorphism** *for bijective linear operator.*□

Example 2.1

1) The derivative $D: V \to V$ is a linear operator on the vector space V of all infinitely differentiable functions on $\mathbb{R}$.
2) The integral operator $\tau: F[x] \to F[x]$ defined by
$$\tau f = \int_0^x f(t)dt$$
is a linear operator on $F[x]$.
3) Let A be an $m \times n$ matrix over F. The function $\tau_A: F^n \to F^m$ defined by $\tau_A v = Av$, where all vectors are written as column vectors, is a linear transformation from F^n to F^m. This function is just multiplication by A.
4) The coordinate map $\phi: V \to F^n$ of an n-dimensional vector space is a linear transformation from V to F^n.□

The set $\mathcal{L}(V,W)$ is a vector space in its own right and $\mathcal{L}(V)$ has the structure of an algebra, as defined in Chapter 0.

Theorem 2.1

1) The set $\mathcal{L}(V,W)$ is a vector space under ordinary addition of functions and scalar multiplication of functions by elements of F.
2) If $\sigma \in \mathcal{L}(U,V)$ and $\tau \in \mathcal{L}(V,W)$, then the composition $\tau\sigma$ is in $\mathcal{L}(U,W)$.
3) If $\tau \in \mathcal{L}(V,W)$ is bijective then $\tau^{-1} \in \mathcal{L}(W,V)$.
4) The vector space $\mathcal{L}(V)$ is an algebra, where multiplication is composition of functions. The identity map $\iota \in \mathcal{L}(V)$ is the multiplicative identity and the zero map $0 \in \mathcal{L}(V)$ is the additive identity.

Proof. We prove only part 3). Let $\tau: V \to W$ be a bijective linear transformation. Then $\tau^{-1}: W \to V$ is a well-defined function and since any two vectors w_1 and w_2 in W have the form $w_1 = \tau v_1$ and $w_2 = \tau v_2$, we have
$$\begin{aligned}\tau^{-1}(aw_1 + bw_2) &= \tau^{-1}(a\tau v_1 + b\tau v_2)\\ &= \tau^{-1}(\tau(av_1 + bv_2))\\ &= av_1 + bv_2\\ &= a\tau^{-1}(w_1) + b\tau^{-1}(w_2)\end{aligned}$$
which shows that τ^{-1} is linear.□

One of the easiest ways to define a linear transformation is to give its values on a basis. The following theorem says that we may assign these values arbitrarily and obtain a unique linear transformation by linear extension to the entire domain.

Theorem 2.2 *Let V and W be vector spaces and let $\mathcal{B} = \{v_i \mid i \in I\}$ be a basis for V. Then we can define a linear transformation $\tau \in \mathcal{L}(V,W)$ by* *specifying the values of* τv_i arbitrarily *for all* $v_i \in \mathcal{B}$ *and extending* τ *to* V *by linearity, that is,*

$$\tau(a_1v_1 + \cdots + a_nv_n) = a_1\tau v_1 + \cdots + a_n\tau v_n$$

This process defines a unique linear transformation, that is, if $\tau, \sigma \in \mathcal{L}(V, W)$ *satisfy* $\tau v_i = \sigma v_i$ *for all* $v_i \in \mathcal{B}$ *then* $\tau = \sigma$.

Proof. The crucial point is that the extension by linearity is well-defined, since each vector in V has an essentially unique representation as a linear combination of a finite number of vectors in $\mathcal{B}$. We leave the details to the reader.□

Note that if $\tau \in \mathcal{L}(V, W)$ and if S is a subspace of V, then the restriction $\tau|_S$ of τ to S is a linear transformation from S to W.

The Kernel and Image of a Linear Transformation

There are two very important vector spaces associated with a linear transformation τ from V to W.

Definition *Let* $\tau \in \mathcal{L}(V, W)$. *The subspace*

$$\ker(\tau) = \{v \in V \mid \tau v = 0\}$$

is called the **kernel** *of* τ *and the subspace*

$$\operatorname{im}(\tau) = \{\tau v \mid v \in V\}$$

is called the **image** *of* τ. *The dimension of* $\ker(\tau)$ *is called the* **nullity** *of* τ *and is denoted by* $\operatorname{null}(\tau)$. *The dimension of* $\operatorname{im}(\tau)$ *is called the* **rank** *of* τ *and is denoted by* $\operatorname{rk}(\tau)$.□

It is routine to show that $\ker(\tau)$ is a subspace of V and $\operatorname{im}(\tau)$ is a subspace of W. Moreover, we have the following.

Theorem 2.3 *Let* $\tau \in \mathcal{L}(V, W)$. *Then*

1) τ *is surjective if and only if* $\operatorname{im}(\tau) = W$

2) τ *is injective if and only if* $\ker(\tau) = \{0\}$

Proof. The first statement is merely a restatement of the definition of surjectivity. To see the validity of the second statement, observe that

$$\tau u = \tau v \Leftrightarrow \tau(u - v) = 0 \Leftrightarrow u - v \in \ker(\tau)$$

Hence, if $\ker(\tau) = \{0\}$, then $\tau u = \tau v \Leftrightarrow u = v$, which shows that τ is injective. Conversely, if τ is injective and $u \in \ker(\tau)$, then $\tau u = \tau 0$ and so $u = 0$. This shows that $\ker(\tau) = \{0\}$.□

Isomorphisms

Definition *A bijective linear transformation* $\tau: V \to W$ *is called an* **isomorphism** *from* V *to* W*. When an isomorphism from* V *to* W *exists, we say that* V *and* W *are* **isomorphic** *and write* $V \approx W$*.*□

Example 2.2 Let $\dim(V) = n$. For any ordered basis $\mathcal{B}$ of V, the coordinate map $\phi_{\mathcal{B}}: V \to F^n$ that sends each vector $v \in V$ to its coordinate matrix $[v]_{\mathcal{B}} \in F^n$ is an isomorphism. Hence, any n-dimensional vector space over F is isomorphic to F^n.□

Isomorphic vector spaces share many properties, as the next theorem shows. If $\tau \in \mathcal{L}(V, W)$ and $S \subseteq V$ we write

$$\tau S = \{\tau s \mid s \in S\}$$

Theorem 2.4 *Let* $\tau \in \mathcal{L}(V, W)$ *be an isomorphism. Let* $S \subseteq V$*. Then*
1) S *spans* V *if and only if* τS *spans* W*.*
2) S *is linearly independent in* V *if and only if* τS *is linearly independent in* W*.*
3) S *is a basis for* V *if and only if* τS *is a basis for* W*.*□

An isomorphism can be characterized as a linear transformation $\tau: V \to W$ that maps a basis for V to a basis for W.

Theorem 2.5 *A linear transformation* $\tau \in \mathcal{L}(V, W)$ *is an isomorphism if and only if there is a basis* $\mathcal{B}$ *for* V *for which* $\tau\mathcal{B}$ *is a basis for* W*. In this case,* τ *maps any basis of* V *to a basis of* W*.*□

The following theorem says that, up to isomorphism, there is only one vector space of any given dimension over a given field.

Theorem 2.6 *Let* V *and* W *be vector spaces over* F*. Then* $V \approx W$ *if and only if* $\dim(V) = \dim(W)$.□

In Example 2.2, we saw that any n-dimensional vector space is isomorphic to F^n. Now suppose that B is a set of cardinality κ and let $(F^B)_0$ be the vector space of all functions from B to F with finite support. We leave it to the reader to show that the functions $\delta_b \in (F^B)_0$ defined for all $b \in B$ by

$$\delta_b(x) = \begin{cases} 1 & \text{if } x = b \\ 0 & \text{if } x \neq b \end{cases}$$

form a basis for $(F^B)_0$, called the **standard basis**. Hence, $\dim((F^B)_0) = |B|$.

It follows that for any cardinal number κ, there is a vector space of dimension κ. Also, any vector space of dimension κ is isomorphic to $(F^B)_0$.

Theorem 2.7 *If n is a natural number, then any n-dimensional vector space over F is isomorphic to F^n. If κ is any cardinal number and if B is a set of cardinality κ, then any κ-dimensional vector space over F is isomorphic to the vector space $(F^B)_0$ of all functions from B to F with finite support.*□

The Rank Plus Nullity Theorem

Let $\tau \in \mathcal{L}(V,W)$. Since any subspace of V has a complement, we can write

$$V = \ker(\tau) \oplus \ker(\tau)^c$$

where $\ker(\tau)^c$ is a complement of $\ker(\tau)$ in V. It follows that

$$\dim(V) = \dim(\ker(\tau)) + \dim(\ker(\tau)^c)$$

Now, the restriction of τ to $\ker(\tau)^c$,

$$\tau^c : \ker(\tau)^c \to W$$

is injective, since

$$\ker(\tau^c) = \ker(\tau) \cap \ker(\tau)^c = \{0\}$$

Also, $\operatorname{im}(\tau^c) \subseteq \operatorname{im}(\tau)$. For the reverse inclusion, if $\tau v \in \operatorname{im}(\tau)$, then since $v = u + w$ for $u \in \ker(\tau)$ and $w \in \ker(\tau)^c$, we have

$$\tau v = \tau u + \tau w = \tau w = \tau^c w \in \operatorname{im}(\tau^c)$$

Thus $\operatorname{im}(\tau^c) = \operatorname{im}(\tau)$. It follows that

$$\ker(\tau)^c \approx \operatorname{im}(\tau)$$

From this, we deduce the following theorem.

Theorem 2.8 *Let $\tau \in \mathcal{L}(V,W)$.*
1) *Any complement of* $\ker(\tau)$ *is isomorphic to* $\operatorname{im}(\tau)$
2) **(The rank plus nullity theorem)**

$$\dim(\ker(\tau)) + \dim(\operatorname{im}(\tau)) = \dim(V)$$

or, in other notation,

$$\operatorname{rk}(\tau) + \operatorname{null}(\tau) = \dim(V)$$ □

Theorem 2.8 has an important corollary.

Corollary 2.9 *Let $\tau \in \mathcal{L}(V,W)$, where $\dim(V) = \dim(W) < \infty$. Then τ is injective if and only if it is surjective.*□

Note that this result fails if the vector spaces are not finite-dimensional. The reader is encouraged to find an example to support this statement.

Linear Transformations from F^n to F^m

Recall that for any $m \times n$ matrix A over F the multiplication map

$$\tau_A(v) = Av$$

is a linear transformation. In fact, any linear transformation $\tau \in \mathcal{L}(F^n, F^m)$ has this form, that is, τ is just multiplication by a matrix, for we have

$$\big(\tau e_1 \mid \cdots \mid \tau e_n\big)e_i = \big(\tau e_1 \mid \cdots \mid \tau e_n\big)^{(i)} = \tau e_i$$

and so $\tau = \tau_A$, where

$$A = \big(\tau e_1 \mid \cdots \mid \tau e_n\big)$$

Theorem 2.10

1) If A is an $m \times n$ matrix over F then $\tau_A \in \mathcal{L}(F^n, F^m)$.

2) If $\tau \in \mathcal{L}(F^n, F^m)$ then $\tau = \tau_A$, where

$$A = (\tau e_1 \mid \cdots \mid \tau e_n)$$

The matrix A is called the **matrix** *of τ.* □

Example 2.3 Consider the linear transformation $\tau: F^3 \to F^3$ defined by

$$\tau(x, y, z) = (x - 2y, z, x + y + z)$$

Then we have, in column form,

$$\tau\begin{bmatrix} x \\ y \\ z \end{bmatrix} = \begin{bmatrix} x-2y \\ z \\ x+y+z \end{bmatrix} = \begin{bmatrix} 1 & -2 & 0 \\ 0 & 0 & 1 \\ 1 & 1 & 1 \end{bmatrix}\begin{bmatrix} x \\ y \\ z \end{bmatrix}$$

and so the standard matrix of τ is

$$A = \begin{bmatrix} 1 & -2 & 0 \\ 0 & 0 & 1 \\ 1 & 1 & 1 \end{bmatrix}$$ □

If $A \in \mathcal{M}_{m,n}$, then since the image of τ_A is the column space of A, we have

$$\dim(\ker(\tau_A)) + \operatorname{rk}(A) = \dim(F^n)$$

This gives the following useful result.

Theorem 2.11 *Let A be an $m \times n$ matrix over F.*

1) $\tau_A: F^n \to F^m$ is injective if and only if $\operatorname{rk}(A) = n$.

2) $\tau_A: F^n \to F^m$ is surjective if and only if $\operatorname{rk}(A) = m$. □

Change of Basis Matrices

Suppose that $\mathcal{B} = (b_1, \ldots, b_n)$ and $\mathcal{C} = (c_1, \ldots, c_n)$ are ordered bases for a vector space V. It is natural to ask how the coordinate matrices $[v]_{\mathcal{B}}$ and $[v]_{\mathcal{C}}$ are related. Referring to Figure 2.1,

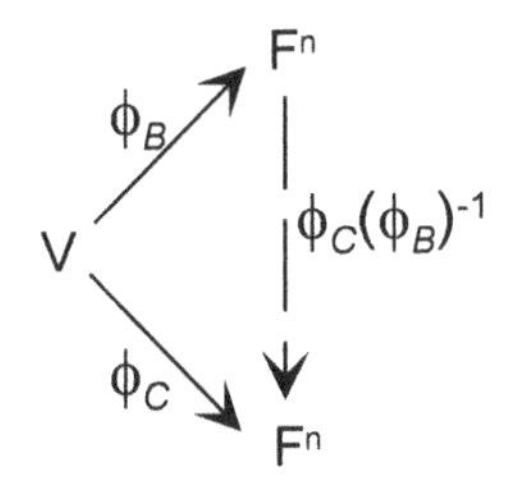

Figure 2.1

the map that takes $[v]_{\mathcal{B}}$ to $[v]_{\mathcal{C}}$ is $\phi_{\mathcal{B},\mathcal{C}} = \phi_{\mathcal{C}}\phi_{\mathcal{B}}^{-1}$ and is called the **change of basis operator** (or **change of coordinates operator**). Since $\phi_{\mathcal{B},\mathcal{C}}$ is an operator on F^n, it has the form τ_A, where

$$\begin{aligned} A &= (\phi_{\mathcal{B},\mathcal{C}}(e_1) \mid \cdots \mid \phi_{\mathcal{B},\mathcal{C}}(e_n)) \\ &= (\phi_{\mathcal{C}}\phi_{\mathcal{B}}^{-1}([b_1]_{\mathcal{B}}) \mid \cdots \mid \phi_{\mathcal{C}}\phi_{\mathcal{B}}^{-1}([b_n]_{\mathcal{B}})) \\ &= ([b_1]_{\mathcal{C}} \mid \cdots \mid [b_n]_{\mathcal{C}}) \end{aligned}$$

We denote A by $M_{\mathcal{B},\mathcal{C}}$ and call it the **change of basis matrix** from $\mathcal{B}$ to $\mathcal{C}$.

Theorem 2.12 *Let $\mathcal{B} = (b_1, \ldots, b_n)$ and $\mathcal{C}$ be ordered bases for a vector space V. Then the change of basis operator $\phi_{\mathcal{B},\mathcal{C}} = \phi_{\mathcal{C}}\phi_{\mathcal{B}}^{-1}$ is an automorphism of F^n, whose standard matrix is*

$$M_{\mathcal{B},\mathcal{C}} = ([b_1]_{\mathcal{C}} \mid \cdots \mid [b_n]_{\mathcal{C}})$$

Hence

$$[v]_{\mathcal{C}} = M_{\mathcal{B},\mathcal{C}}[v]_{\mathcal{B}}$$

and $M_{\mathcal{C},\mathcal{B}} = M_{\mathcal{B},\mathcal{C}}^{-1}$. □

Consider the equation

$$A = M_{\mathcal{B},\mathcal{C}}$$

or equivalently,

$$A = ([b_1]_{\mathcal{C}} \mid \cdots \mid [b_n]_{\mathcal{C}})$$

Then given any two of A (an invertible $n \times n$ matrix), $\mathcal{B}$ (an ordered basis for F^n) and $\mathcal{C}$ (an ordered basis for F^n), the third component is uniquely determined by this equation. This is clear if $\mathcal{B}$ and $\mathcal{C}$ are given or if A and $\mathcal{C}$ are given. If A and $\mathcal{B}$ are given, then there is a unique $\mathcal{C}$ for which $A^{-1} = M_{\mathcal{C},\mathcal{B}}$ and so there is a unique $\mathcal{C}$ for which $A = M_{\mathcal{B},\mathcal{C}}$.

Theorem 2.13 *If we are given any two of the following:*
1) an invertible $n \times n$ matrix A
2) an ordered basis $\mathcal{B}$ for F^n
3) an ordered basis $\mathcal{C}$ for F^n.
then the third is uniquely determined by the equation

$$A = M_{\mathcal{B},\mathcal{C}}$$

□

The Matrix of a Linear Transformation

Let $\tau: V \to W$ be a linear transformation, where $\dim(V) = n$ and $\dim(W) = m$ and let $\mathcal{B} = (b_1, \ldots, b_n)$ be an ordered basis for V and $\mathcal{C}$ an ordered basis for W. Then the map

$$\theta: [v]_{\mathcal{B}} \to [\tau v]_{\mathcal{C}}$$

is a *representation* of τ as a linear transformation from F^n to F^m, in the sense that knowing θ (along with $\mathcal{B}$ and $\mathcal{C}$, of course) is equivalent to knowing τ. Of course, this representation depends on the choice of ordered bases $\mathcal{B}$ and $\mathcal{C}$.

Since θ is a linear transformation from F^n to F^m, it is just multiplication by an $m \times n$ matrix A, that is,

$$[\tau v]_{\mathcal{C}} = A[v]_{\mathcal{B}}$$

Indeed, since $[b_i]_{\mathcal{B}} = e_i$, we get the columns of A as follows:

$$A^{(i)} = Ae_i = A[v_i]_{\mathcal{B}} = [\tau b_i]_{\mathcal{C}}$$

Theorem 2.14 *Let $\tau \in \mathcal{L}(V,W)$ and let $\mathcal{B} = (b_1, \ldots, b_n)$ and $\mathcal{C}$ be ordered bases for V and W, respectively. Then τ can be represented with respect to $\mathcal{B}$ and $\mathcal{C}$ as matrix multiplication, that is,*

$$[\tau v]_{\mathcal{C}} = [\tau]_{\mathcal{B},\mathcal{C}}[v]_{\mathcal{B}}$$

where

$$[\tau]_{\mathcal{B},\mathcal{C}} = ([\tau b_1]_{\mathcal{C}} \mid \cdots \mid [\tau b_n]_{\mathcal{C}})$$

is called the **matrix of τ with respect to the bases** *$\mathcal{B}$ and $\mathcal{C}$. When $V = W$ and $\mathcal{B} = \mathcal{C}$, we denote $[\tau]_{\mathcal{B},\mathcal{B}}$ by $[\tau]_{\mathcal{B}}$ and so*

$$[\tau v]_{\mathcal{B}} = [\tau]_{\mathcal{B}}[v]_{\mathcal{B}}$$

□

Example 2.4 Let $D: \mathcal{P}_2 \to \mathcal{P}_2$ be the derivative operator, defined on the vector space of all polynomials of degree at most 2. Let $\mathcal{B} = \mathcal{C} = (1, x, x^2)$. Then

$$[D(1)]_{\mathcal{C}}=[0]_{\mathcal{C}}=\begin{bmatrix}0\\0\\0\end{bmatrix},[D(x)]_{\mathcal{C}}=[1]_{\mathcal{C}}=\begin{bmatrix}1\\0\\0\end{bmatrix},[D(x^2)]_{\mathcal{C}}=[2x]_{\mathcal{C}}=\begin{bmatrix}0\\2\\0\end{bmatrix}$$

and so

$$[D]_{\mathcal{B}}=\begin{bmatrix}0&1&0\\0&0&2\\0&0&0\end{bmatrix}$$

Hence, for example, if $p(x)=5+x+2x^2$, then

$$[Dp(x)]_{\mathcal{C}}=[D]_{\mathcal{B}}\,[p(x)]_{\mathcal{B}}=\begin{bmatrix}0&1&0\\0&0&2\\0&0&0\end{bmatrix}\begin{bmatrix}5\\1\\2\end{bmatrix}=\begin{bmatrix}1\\4\\0\end{bmatrix}$$

and so $Dp(x)=1+4x$.□

The following result shows that we may work equally well with linear transformations or with the matrices that represent them (with respect to fixed ordered bases $\mathcal{B}$ and $\mathcal{C}$). This applies not only to addition and scalar multiplication, but also to matrix multiplication.

Theorem 2.15 *Let V and W be finite-dimensional vector spaces over F, with ordered bases $\mathcal{B}=(b_1,\ldots,b_n)$ and $\mathcal{C}=(c_1,\ldots,c_m)$, respectively.*

1) The map $\mu\colon\mathcal{L}(V,W)\to\mathcal{M}_{m,n}(F)$ defined by

$$\mu(\tau)=[\tau]_{\mathcal{B},\mathcal{C}}$$

is an isomorphism and so $\mathcal{L}(V,W)\approx\mathcal{M}_{m,n}(F)$. Hence,

$$\dim(\mathcal{L}(V,W))=\dim(\mathcal{M}_{m,n}(F))=m\times n$$

2) If $\sigma\in\mathcal{L}(U,V)$ and $\tau\in\mathcal{L}(V,W)$ and if $\mathcal{B}$, $\mathcal{C}$ and $\mathcal{D}$ are ordered bases for U, V and W, respectively, then

$$[\tau\sigma]_{\mathcal{B},\mathcal{D}}=[\tau]_{\mathcal{C},\mathcal{D}}[\sigma]_{\mathcal{B},\mathcal{C}}$$

Thus, the matrix of the product (composition) $\tau\sigma$ is the product of the matrices of τ and σ. In fact, this is the primary motivation for the definition of matrix multiplication.

Proof. To see that μ is linear, observe that for all i,

$$\begin{aligned}[s\sigma+t\tau]_{\mathcal{B},\mathcal{C}}[b_i]_{\mathcal{B}}&=[(s\sigma+t\tau)(b_i)]_{\mathcal{C}}\\&=[s\sigma(b_i)+t\tau(b_i)]_{\mathcal{C}}\\&=s[\sigma(b_i)]_{\mathcal{C}}+t[\tau(b_i)]_{\mathcal{C}}\\&=s[\sigma]_{\mathcal{B},\mathcal{C}}[b_i]_{\mathcal{B}}+t[\tau]_{\mathcal{B},\mathcal{C}}[b_i]_{\mathcal{B}}\\&=(s[\sigma]_{\mathcal{B},\mathcal{C}}+t[\tau]_{\mathcal{B},\mathcal{C}})[b_i]_{\mathcal{B}}\end{aligned}$$

and since $[b_i]_{\mathcal{B}} = e_i$ is a standard basis vector, we conclude that

$$[s\sigma + t\tau]_{\mathcal{B},\mathcal{C}} = s[\sigma]_{\mathcal{B},\mathcal{C}} + t[\tau]_{\mathcal{B},\mathcal{C}}$$

and so μ is linear. If $A \in \mathcal{M}_{m,n}$, we define τ by the condition $[\tau b_i]_{\mathcal{C}} = A^{(i)}$, whence $\mu(\tau) = A$ and μ is surjective. Also, $\ker(\mu) = \{0\}$ since $[\tau]_{\mathcal{B}} = 0$ implies that $\tau = 0$. Hence, the map μ is an isomorphism. To prove part 2), we have

$$[\tau\sigma]_{\mathcal{B},\mathcal{D}}[v]_{\mathcal{B}} = [\tau(\sigma v)]_{\mathcal{D}} = [\tau]_{\mathcal{C},\mathcal{D}}[\sigma v]_{\mathcal{C}} = [\tau]_{\mathcal{C},\mathcal{D}}[\sigma]_{\mathcal{B},\mathcal{C}}[v]_{\mathcal{B}} \qquad \square$$

Change of Bases for Linear Transformations

Since the matrix $[\tau]_{\mathcal{B},\mathcal{C}}$ that represents τ depends on the ordered bases $\mathcal{B}$ and $\mathcal{C}$, it is natural to wonder how to choose these bases in order to make this matrix as simple as possible. For instance, can we always choose the bases so that τ is represented by a diagonal matrix?

As we will see in Chapter 7, the answer to this question is no. In that chapter, we will take up the general question of how best to represent a linear operator by a matrix. For now, let us take the first step and describe the relationship between the matrices $[\tau]_{\mathcal{B},\mathcal{C}}$ and $[\tau]_{\mathcal{B}',\mathcal{C}'}$ of τ with respect to two different pairs $(\mathcal{B},\mathcal{C})$ and $(\mathcal{B}',\mathcal{C}')$ of ordered bases. Multiplication by $[\tau]_{\mathcal{B}',\mathcal{C}'}$ sends $[v]_{\mathcal{B}'}$ to $[\tau v]_{\mathcal{C}'}$. This can be reproduced by first switching from $\mathcal{B}'$ to $\mathcal{B}$, then applying $[\tau]_{\mathcal{B},\mathcal{C}}$ and finally switching from $\mathcal{C}$ to $\mathcal{C}'$, that is,

$$[\tau]_{\mathcal{B}',\mathcal{C}'} = M_{\mathcal{C},\mathcal{C}'}[\tau]_{\mathcal{B},\mathcal{C}}M_{\mathcal{B}',\mathcal{B}} = M_{\mathcal{C},\mathcal{C}'}[\tau]_{\mathcal{B},\mathcal{C}}M^{-1}_{\mathcal{B},\mathcal{B}'}$$

Theorem 2.16 *Let $\tau \in \mathcal{L}(V,W)$ and let $(\mathcal{B},\mathcal{C})$ and $(\mathcal{B}',\mathcal{C}')$ be pairs of ordered bases of V and W, respectively. Then*

$$[\tau]_{\mathcal{B}',\mathcal{C}'} = M_{\mathcal{C},\mathcal{C}'}[\tau]_{\mathcal{B},\mathcal{C}}M_{\mathcal{B}',\mathcal{B}} \qquad (2.1)\square$$

When $\tau \in \mathcal{L}(V)$ is a linear operator on V, it is generally more convenient to represent τ by matrices of the form $[\tau]_{\mathcal{B}}$, where the ordered bases used to represent vectors in the domain and image are the same. When $\mathcal{B} = \mathcal{C}$, Theorem 2.16 takes the following important form.

Corollary 2.17 *Let $\tau \in \mathcal{L}(V)$ and let $\mathcal{B}$ and $\mathcal{C}$ be ordered bases for V. Then the matrix of τ with respect to $\mathcal{C}$ can be expressed in terms of the matrix of τ with respect to $\mathcal{B}$ as follows:*

$$[\tau]_{\mathcal{C}} = M_{\mathcal{B},\mathcal{C}}[\tau]_{\mathcal{B}}M^{-1}_{\mathcal{B},\mathcal{C}} \qquad (2.2)\square$$

Equivalence of Matrices

Since the change of basis matrices are precisely the invertible matrices, (2.1) has the form

$$[\tau]_{\mathcal{B}',\mathcal{C}'} = P[\tau]_{\mathcal{B},\mathcal{C}}Q^{-1}$$

where P and Q are invertible matrices. This motivates the following definition.

Definition *Two matrices A and B are* **equivalent** *if there exist invertible matrices P and Q for which*

$$B = PAQ^{-1}$$ □

We have remarked that B is equivalent to A if and only if B can be obtained from A by a series of elementary row and column operations. Performing the row operations is equivalent to multiplying the matrix A on the left by P and performing the column operations is equivalent to multiplying A on the right by Q^{-1}.

In terms of (2.1), we see that performing row operations (premultiplying by P) is equivalent to changing the basis used to represent vectors in the image and performing column operations (postmultiplying by Q^{-1}) is equivalent to changing the basis used to represent vectors in the domain.

According to Theorem 2.16, if A and B are matrices that represent τ with respect to possibly different ordered bases, then A and B are equivalent. The converse of this also holds.

Theorem 2.18 *Let V and W be vector spaces with $\dim(V) = n$ and $\dim(W) = m$. Then two $m \times n$ matrices A and B are equivalent if and only if they represent the same linear transformation $\tau \in \mathcal{L}(V, W)$, but possibly with respect to different ordered bases. In this case, A and B represent exactly the same set of linear transformations in $\mathcal{L}(V, W)$.*

Proof. If A and B represent τ, that is, if

$$A = [\tau]_{\mathcal{B},\mathcal{C}} \quad \text{and} \quad B = [\tau]_{\mathcal{B}',\mathcal{C}'}$$

for ordered bases $\mathcal{B}, \mathcal{C}, \mathcal{B}'$ and $\mathcal{C}'$, then Theorem 2.16 shows that A and B are equivalent. Now suppose that A and B are equivalent, say

$$B = PAQ^{-1}$$

where P and Q are invertible. Suppose also that A represents a linear transformation $\tau \in \mathcal{L}(V, W)$ for some ordered bases $\mathcal{B}$ and $\mathcal{C}$, that is,

$$A = [\tau]_{\mathcal{B},\mathcal{C}}$$

Theorem 2.9 implies that there is a unique ordered basis $\mathcal{B}'$ for V for which $Q = M_{\mathcal{B},\mathcal{B}'}$ and a unique ordered basis $\mathcal{C}'$ for W for which $P = M_{\mathcal{C},\mathcal{C}'}$. Hence

$$B = M_{\mathcal{C},\mathcal{C}'}[\tau]_{\mathcal{B},\mathcal{C}}M_{\mathcal{B}',\mathcal{B}} = [\tau]_{\mathcal{B}',\mathcal{C}'}$$

Hence, B also represents τ. By symmetry, we see that A and B represent the same set of linear transformations. This completes the proof.$\square$

We remarked in Example 0.3 that every matrix is equivalent to exactly one matrix of the block form

$$J_k = \begin{bmatrix} I_k & 0_{k,n-k} \\ 0_{m-k,k} & 0_{m-k,n-k} \end{bmatrix}_{\text{block}}$$

Hence, the set of these matrices is a set of canonical forms for equivalence. Moreover, the rank is a complete invariant for equivalence. In other words, two matrices are equivalent if and only if they have the same rank.

Similarity of Matrices

When a linear operator $\tau \in \mathcal{L}(V)$ is represented by a matrix of the form $[\tau]_{\mathcal{B}}$, equation (2.2) has the form

$$[\tau]_{\mathcal{B}'} = P[\tau]_{\mathcal{B}}P^{-1}$$

where P is an invertible matrix. This motivates the following definition.

Definition *Two matrices A and B are* **similar**, *denoted by $A \sim B$, if there exists an invertible matrix P for which*

$$B = PAP^{-1}$$

The equivalence classes associated with similarity are called **similarity classes**.$\square$

The analog of Theorem 2.18 for square matrices is the following.

Theorem 2.19 *Let V be a vector space of dimension n. Then two $n \times n$ matrices A and B are similar if and only if they represent the same linear operator $\tau \in \mathcal{L}(V)$, but possibly with respect to different ordered bases. In this case, A and B represent exactly the same set of linear operators in $\mathcal{L}(V)$.*

Proof. If A and B represent $\tau \in \mathcal{L}(V)$, that is, if

$$A = [\tau]_{\mathcal{B}} \quad \text{and} \quad B = [\tau]_{\mathcal{C}}$$

for ordered bases $\mathcal{B}$ and $\mathcal{C}$, then Corollary 2.17 shows that A and B are similar. Now suppose that A and B are similar, say

$$B = PAP^{-1}$$

Suppose also that A represents a linear operator $\tau \in \mathcal{L}(V)$ for some ordered basis $\mathcal{B}$, that is,

$$A = [\tau]_{\mathcal{B}}$$

Theorem 2.9 implies that there is a unique ordered basis $\mathcal{C}$ for V for which

$P = M_{\mathcal{B},\mathcal{C}}$. Hence

$$B = M_{\mathcal{B},\mathcal{C}}[\tau]_{\mathcal{B}}M_{\mathcal{B},\mathcal{C}}^{-1} = [\tau]_{\mathcal{C}}$$

Hence, B also represents τ. By symmetry, we see that A and B represent the same set of linear operators. This completes the proof.□

We will devote much effort in Chapter 7 to finding a canonical form for similarity.

Similarity of Operators

We can also define similarity of operators.

Definition *Two linear operators $\tau, \sigma \in \mathcal{L}(V)$ are* **similar**, *denoted by $\tau \sim \sigma$, if there exists an automorphism $\phi \in \mathcal{L}(V)$ for which*

$$\sigma = \phi\tau\phi^{-1}$$

The equivalence classes associated with similarity are called **similarity classes.** □

Note that if $\mathcal{B} = (b_1, \ldots, b_n)$ and $\mathcal{C} = (c_1, \ldots, c_n)$ are ordered bases for V, then

$$M_{\mathcal{C},\mathcal{B}} = ([c_1]_{\mathcal{B}} \mid \cdots \mid [c_n]_{\mathcal{B}})$$

Now, the map defined by $\phi(b_i) = c_i$ is an automorphism of V and

$$M_{\mathcal{C},\mathcal{B}} = ([\phi(b_1)]_{\mathcal{B}} \mid \cdots \mid [\phi(b_n)]_{\mathcal{B}}) = [\phi]_{\mathcal{B}}$$

Conversely, if $\phi: V \to V$ is an automorphism and $\mathcal{B} = (b_1, \ldots, b_n)$ is an ordered basis for V, then $\mathcal{C} = (c_1 = \phi(b_1), \ldots, c_n = \phi(b_n))$ is also a basis:

$$[\phi]_{\mathcal{B}} = ([\phi(b_1)]_{\mathcal{B}} \mid \cdots \mid [\phi(b_n)]_{\mathcal{B}}) = M_{\mathcal{C},\mathcal{B}}$$

The analog of Theorem 2.19 for linear operators is the following.

Theorem 2.20 *Let V be a vector space of dimension n. Then two linear operators τ and σ on V are similar if and only if there is a matrix $A \in \mathcal{M}_n$ that represents both operators, but with respect to possibly different ordered bases. In this case, τ and σ are represented by exactly the same set of matrices in $\mathcal{M}_n$.*
Proof. If τ and σ are represented by $A \in \mathcal{M}_n$, that is, if

$$[\tau]_{\mathcal{B}} = A = [\sigma]_{\mathcal{C}}$$

for ordered bases $\mathcal{B}$ and $\mathcal{C}$, then

$$[\sigma]_{\mathcal{C}} = [\tau]_{\mathcal{B}} = M_{\mathcal{C},\mathcal{B}}[\tau]_{\mathcal{C}}M_{\mathcal{B},\mathcal{C}}$$

As remarked above, if $\phi: V \to V$ is defined by $\phi(c_i) = b_i$, then

$$[\phi]_{\mathcal{C}} = M_{\mathcal{B},\mathcal{C}}$$

and so

$$[\sigma]_{\mathcal{C}} = [\phi]_{\mathcal{C}}^{-1}[\tau]_{\mathcal{C}}[\phi]_{\mathcal{C}} = [\phi^{-1}\tau\phi]_{\mathcal{C}}$$

from which it follows that σ and τ are similar. Conversely, suppose that τ and σ are similar, say

$$\sigma = \phi\tau\phi^{-1}$$

where ϕ is an automorphism of V. Suppose also that τ is represented by the matrix $A \in \mathcal{M}_n$, that is,

$$A = [\tau]_{\mathcal{B}}$$

for some ordered basis $\mathcal{B}$. Then $[\phi]_{\mathcal{B}} = M_{\mathcal{C},\mathcal{B}}$ and so

$$[\sigma]_{\mathcal{B}} = [\phi\tau\phi^{-1}]_{\mathcal{B}} = [\phi]_{\mathcal{B}}[\tau]_{\mathcal{B}}[\phi]_{\mathcal{B}}^{-1} = M_{\mathcal{C},\mathcal{B}}[\tau]_{\mathcal{B}}M_{\mathcal{C},\mathcal{B}}^{-1}$$

It follows that

$$A = [\tau]_{\mathcal{B}} = M_{\mathcal{B},\mathcal{C}}[\sigma]_{\mathcal{B}}M_{\mathcal{B},\mathcal{C}}^{-1} = [\sigma]_{\mathcal{C}}$$

and so A also represents σ. By symmetry, we see that τ and σ are represented by the same set of matrices. This completes the proof.□

We can summarize the sitiation with respect to similarity in Figure 2.2. Each similarity class $\mathcal{S}$ in $\mathcal{L}(V)$ corresponds to a similarity class $\mathcal{T}$ in $\mathcal{M}_n(F)$: $\mathcal{T}$ is the set of all matrices that represent any $\tau \in \mathcal{S}$ and $\mathcal{S}$ is the set of all operators in $\mathcal{L}(V)$ that are represented by any $M \in \mathcal{T}$.

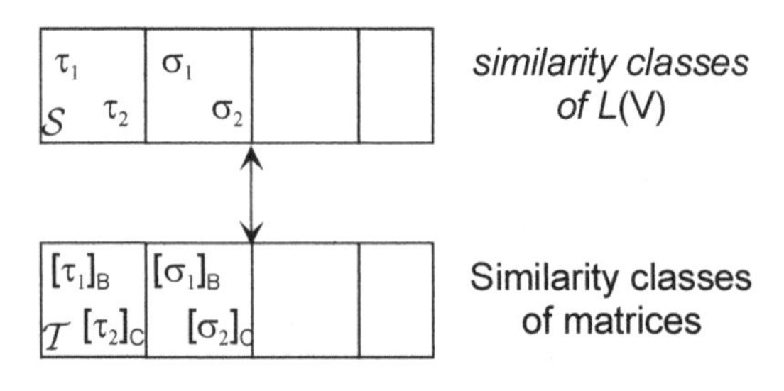

Figure 2.2

Invariant Subspaces and Reducing Pairs

The restriction of a linear operator $\tau \in \mathcal{L}(V)$ to a subspace S of V is not necessarily a linear operator on S. This prompts the following definition.

Definition *Let* $\tau \in \mathcal{L}(V)$. *A subspace* S *of* V *is said to be* **invariant under** τ *or* τ-**invariant** *if* $\tau S \subseteq S$, *that is, if* $\tau s \in S$ *for all* $s \in S$. *Put another way,* S *is invariant under* τ *if the restriction* $\tau|_S$ *is a linear operator on* S.$\square$

If

$$V = S \oplus T$$

then the fact that S is τ-invariant does not imply that the complement T is also τ-invariant. (The reader may wish to supply a simple example with $V = \mathbb{R}^2$.)

Definition *Let* $\tau \in \mathcal{L}(V)$. *If* $V = S \oplus T$ *and if both* S *and* T *are* τ*-invariant, we say that the pair* (S,T) **reduces** τ.$\square$

A reducing pair can be used to decompose a linear operator into a direct sum as follows.

Definition *Let* $\tau \in \mathcal{L}(V)$. *If* (S,T) *reduces* τ *we write*

$$\tau = \tau|_S \oplus \tau|_T$$

and call τ *the* **direct sum** *of* $\tau|_S$ *and* $\tau|_T$. *Thus, the expression*

$$\rho = \sigma \oplus \tau$$

means that there exist subspaces S *and* T *of* V *for which* (S,T) *reduces* ρ *and*

$$\sigma = \rho|_S \text{ and } \tau = \rho|_T \qquad \square$$

The concept of the direct sum of linear operators will play a key role in the study of the structure of a linear operator.

Projection Operators

We will have several uses for a special type of linear operator that is related to direct sums.

Definition *Let* $V = S \oplus T$. *The linear operator* $\rho_{S,T}: V \to V$ *defined by*

$$\rho_{S,T}(s+t) = s$$

where $s \in S$ *and* $t \in T$ *is called* **projection** *onto* S **along** T.$\square$

Whenever we say that the operator $\rho_{S,T}$ is a projection, it is with the understanding that $V = S \oplus T$. The following theorem describes a few basic properties of projection operators. We leave proof as an exercise.

Theorem 2.21 *Let* V *be a vector space and let* $\rho \in \mathcal{L}(V)$.

1) *If* $V = S \oplus T$ *then*

$$\rho_{S,T} + \rho_{T,S} = \iota$$

2) *If* $\rho = \rho_{S,T}$ *then*

$$\operatorname{im}(\rho) = S \quad \textit{and} \quad \ker(\rho) = T$$

and so

$$V = \operatorname{im}(\rho) \oplus \ker(\rho)$$

In other words, ρ *is projection onto its image along its kernel. Moreover,*

$$v \in \operatorname{im}(\rho) \quad \Leftrightarrow \quad \rho v = v$$

3) *If* $\sigma \in \mathcal{L}(V)$ *has the property that*

$$V = \operatorname{im}(\sigma) \oplus \ker(\sigma) \quad \textit{and} \quad \sigma|_{\operatorname{im}(\sigma)} = \iota$$

then σ *is projection onto* $\operatorname{im}(\sigma)$ *along* $\ker(\sigma)$.□

Projection operators are easy to characterize.

Definition *A linear operator* $\tau \in \mathcal{L}(V)$ *is* **idempotent** *if* $\tau^2 = \tau$.□

Theorem 2.22 *A linear operator* $\rho \in \mathcal{L}(V)$ *is a projection if and only if it is idempotent.*

Proof. If $\rho = \rho_{S,T}$, then for any $s \in S$ and $t \in T$,

$$\rho^2(s+t) = \rho s = s = \rho(s+t)$$

and so $\rho^2 = \rho$. Conversely, suppose that ρ is idempotent. If $v \in \operatorname{im}(\rho) \cap \ker(\rho)$, then $v = \rho x$ and so

$$0 = \rho v = \rho^2 x = \rho x = v$$

Hence $\operatorname{im}(\rho) \cap \ker(\rho) = \{0\}$. Also, if $v \in V$, then

$$v = (v - \rho v) + \rho v \in \ker(\rho) \oplus \operatorname{im}(\rho)$$

and so $V = \ker(\rho) \oplus \operatorname{im}(\rho)$. Finally, $\rho(\rho x) = \rho^2 x = \rho x$ and so $\rho|_{\operatorname{im}(\rho)} = \iota$. Hence, ρ is projection onto $\operatorname{im}(\rho)$ along $\ker(\rho)$.□

Projections and Invariance

Projections can be used to characterize invariant subspaces. Let $\tau \in \mathcal{L}(V)$ and let S be a subspace of V. Let $\rho = \rho_{S,T}$ for any complement T of S. The key is that the elements of S can be characterized as those vectors fixed by ρ, that is,

$s \in S$ if and only if $\rho s = s$. Hence, the following are equivalent:

$$\begin{aligned}\tau S &\subseteq S\\ \tau s &\in S \text{ for all } s \in S\\ \rho(\tau s) &= \tau s \text{ for all } s \in S\\ \rho(\tau\rho s) &= \tau\rho s \text{ for all } s \in S\end{aligned}$$

Thus, S is τ-invariant if and only if $\rho\tau\rho = \tau\rho$ for all vectors $s \in S$. But this is also true for all vectors in T, since both sides are equal to 0 on T. This proves the following theorem.

Theorem 2.23 *Let $\tau \in \mathcal{L}(V)$. Then a subspace S of V is τ-invariant if and only if there is a projection $\rho = \rho_{S,T}$ for which*

$$\rho\tau\rho = \tau\rho$$

in which case this holds for all projections of the form $\rho = \rho_{S,T}$. $\square$

We also have the following relationship between projections and reducing pairs.

Theorem 2.24 *Let $V = S \oplus T$. Then (S,T) reduces $\tau \in \mathcal{L}(V)$ if and only if τ commutes with $\rho_{S,T}$.*

Proof. Theorem 2.23 implies that S and T are τ-invariant if and only if

$$\rho_{S,T}\tau\rho_{S,T} = \rho_{S,T}\tau \quad \text{and} \quad (\iota - \rho_{S,T})\tau(\iota - \rho_{S,T}) = (\iota - \rho_{S,T})\tau$$

and a little algebra shows that this is equivalent to

$$\rho_{S,T}\tau\rho_{S,T} = \rho_{S,T}\tau \quad \text{and} \quad \rho_{S,T}\tau = \tau\rho_{S,T}$$

which is equivalent to $\rho_{S,T}\tau = \tau\rho_{S,T}$. $\square$

Orthogonal Projections and Resolutions of the Identity

Observe that if ρ is a projection, then

$$\rho(\iota - \rho) = (\iota - \rho)\rho = 0$$

Definition *Two projections $\rho, \sigma \in \mathcal{L}(V)$ are* **orthogonal**, *written $\rho \perp \sigma$, if*

$$\rho\sigma = \sigma\rho = 0$$

$\square$

Note that $\rho \perp \sigma$ if and only if

$$\operatorname{im}(\rho) \subseteq \ker(\sigma) \quad \text{and} \quad \operatorname{im}(\sigma) \subseteq \ker(\rho)$$

The following example shows that it is not enough to have $\rho\sigma = 0$ in the definition of orthogonality. In fact, it is possible for $\rho\sigma = 0$ and yet $\sigma\rho$ is not even a projection.

Example 2.5 Let $V = F^2$ and consider the X- and Y-axes and the diagonal:

$$X = \{(x,0) \mid x \in F\}$$
$$Y = \{(0,y) \mid y \in F\}$$
$$D = \{(x,x) \mid x \in F\}$$

Then

$$\rho_{D,X}\rho_{D,Y} = \rho_{D,Y} \neq \rho_{D,X} = \rho_{D,Y}\rho_{D,X}$$

From this we deduce that if ρ and σ are projections, it may happen that both products $\rho\sigma$ and $\sigma\rho$ are projections, but that they are not equal. We leave it to the reader to show that $\rho_{Y,X}\rho_{X,D} = 0$ (which is a projection), but that $\rho_{X,D}\rho_{Y,X}$ is not a projection.□

Since a projection ρ is idempotent, we can write the identity operator ι as s sum of two orthogonal projections:

$$\rho + (\iota - \rho) = \iota, \quad \rho \perp (\iota - \rho)$$

Let us generalize this to more than two projections.

Definition *A* **resolution of the identity** *on V is a sum of the form*

$$\rho_1 + \cdots + \rho_k = \iota$$

where the ρ_i's are pairwise orthogonal projections, that is, $\rho_i \perp \rho_j$ for $i \neq j$.□

There is a connection between the resolutions of the identity on V and direct sum decompositions of V. In general terms, if

$$\sigma_1 + \cdots + \sigma_k = \iota$$

for any linear operators $\sigma_i \in \mathcal{L}(V)$, then for all $v \in V$,

$$v = \sigma_1 v + \cdots + \sigma_k v \in \operatorname{im}(\sigma_1) + \cdots + \operatorname{im}(\sigma_k)$$

and so

$$V = \operatorname{im}(\sigma_1) + \cdots + \operatorname{im}(\sigma_k)$$

However, the sum need not be direct.

Theorem 2.25 *Let V be a vector space. Resolutions of the identity on V correspond to direct sum decompositions of V as follows:*

1) If $\rho_1 + \cdots + \rho_k = \iota$ is a resolution of the identity, then

$$V = \operatorname{im}(\rho_1) \oplus \cdots \oplus \operatorname{im}(\rho_k)$$

and ρ_i *is projection onto* $\operatorname{im}(\rho_i)$ *along*

$$\ker(\rho_i) = \bigoplus_{j \neq i} \operatorname{im}(\rho_j)$$

2) *Conversely, if*

$$V = S_1 \oplus \cdots \oplus S_k$$

and if ρ_i *is projection onto* S_i *along the direct sum* $\bigoplus_{j \neq i} S_j$,*, then* $\rho_1 + \cdots + \rho_k = \iota$ *is a resolution of the identity.*

Proof. To prove 1), if $\rho_1 + \cdots + \rho_k = \iota$ is a resolution of the identity, then

$$V = \operatorname{im}(\rho_1) + \cdots + \operatorname{im}(\rho_k)$$

Moreover, if

$$\rho_1 x_1 + \cdots + \rho_n x_n = 0$$

then applying ρ_i gives $\rho_i x_i = 0$ and so the sum is direct. As to the kernel of ρ_i, we have

$$\operatorname{im}(\rho_i) \oplus \ker(\rho_i) = V = \operatorname{im}(\rho_i) \oplus \left(\bigoplus_{j \neq i} \operatorname{im}(\rho_j) \right)$$

and since $\rho_i \rho_j = 0$, it follows that

$$\bigoplus_{j \neq i} \operatorname{im}(\rho_j) \subseteq \ker(\rho_i)$$

and so equality must hold. For part 2), suppose that

$$V = S_1 \oplus \cdots \oplus S_k$$

and ρ_i is projection onto S_i along $\bigoplus_{j \neq i} S_j$. If $i \neq j$, then

$$\operatorname{im}(\rho_i) = S_i \subseteq \ker(\rho_j)$$

and so $\rho_i \perp \rho_j$. Also, if $v = s_1 + \cdots + s_k$ for $s_i \in S_i$, then

$$v = s_1 + \cdots + s_k = \rho_1 v + \cdots + \rho_k v = (\rho_1 + \cdots + \rho_k) v$$

and so $\iota = \rho_1 + \cdots + \rho_k$ is a resolution of the identity.□

The Algebra of Projections

If ρ and σ are projections, it does not necessarily follow that $\rho + \sigma$, $\rho - \sigma$ or $\rho\sigma$ is a projection. For example, the sum $\rho + \sigma$ is a projection if and only if

$$(\rho + \sigma)^2 = \rho + \sigma$$

which is equivalent to

$$\rho\sigma = -\sigma\rho$$

Of course, this holds if $\rho\sigma = \sigma\rho = 0$, that is, if $\rho \perp \sigma$. But the converse is also true, provided that $\operatorname{char}(F) \neq 2$. To see this, we simply evaluate $\rho\sigma\rho$ in two ways:

$$(\rho\sigma)\rho = -(\sigma\rho)\rho = -\sigma\rho$$

and

$$\rho(\sigma\rho) = -\rho(\rho\sigma) = -\rho\sigma$$

Hence, $\sigma\rho = \rho\sigma = -\sigma\rho$ and so $\sigma\rho = 0$. It follows that $\rho\sigma = -\sigma\rho = 0$ and so $\rho \perp \sigma$. Thus, for $\operatorname{char}(F) \neq 2$, we have $\rho + \sigma$ is a projection if and only if $\rho \perp \sigma$.

Now suppose that $\rho + \sigma$ is a projection. For the kernel of $\rho + \sigma$, note that

$$(\rho + \sigma)v = 0 \quad \Rightarrow \quad \rho(\rho + \sigma)v = 0 \quad \Rightarrow \quad \rho v = 0$$

and similarly, $\sigma v = 0$. Hence, $\ker(\rho + \sigma) \subseteq \ker(\rho) \cap \ker(\sigma)$. But the reverse inclusion is obvious and so

$$\ker(\rho + \sigma) = \ker(\rho) \cap \ker(\sigma)$$

As to the image of $\rho + \sigma$, we have

$$v \in \operatorname{im}(\rho + \sigma) \quad \Rightarrow \quad v = (\rho + \sigma)v = \rho v + \sigma v \in \operatorname{im}(\rho) + \operatorname{im}(\sigma)$$

and so $\operatorname{im}(\rho + \sigma) \subseteq \operatorname{im}(\rho) + \operatorname{im}(\sigma)$. For the reverse inclusion, if $v = \rho x + \sigma y$, then

$$(\rho + \sigma)v = (\rho + \sigma)(\rho x + \sigma y) = \rho x + \sigma y = v$$

and so $v \in \operatorname{im}(\rho + \sigma)$. Thus, $\operatorname{im}(\rho + \sigma) = \operatorname{im}(\rho) + \operatorname{im}(\sigma)$. Finally, $\rho\sigma = 0$ implies that $\operatorname{im}(\sigma) \subseteq \ker(\rho)$ and so the sum is direct and

$$\operatorname{im}(\rho + \sigma) = \operatorname{im}(\rho) \oplus \operatorname{im}(\sigma)$$

The following theorem also describes the situation for the difference and product. Proof in these cases is left for the exercises.

Theorem 2.26 *Let V be a vector space over a field F of characteristic $\neq 2$ and let ρ and σ be projections.*

1) The sum $\rho + \sigma$ is a projection if and only if $\rho \perp \sigma$, in which case

$$\operatorname{im}(\rho + \sigma) = \operatorname{im}(\rho) \oplus \operatorname{im}(\sigma) \quad \textit{and} \quad \ker(\rho + \sigma) = \ker(\rho) \cap \ker(\sigma)$$

2) The difference $\rho - \sigma$ is a projection if and only if

$$\rho\sigma = \sigma\rho = \sigma$$

in which case

$$\operatorname{im}(\rho-\sigma)=\operatorname{im}(\rho)\cap\ker(\sigma)\quad \textit{and}\quad \ker(\rho-\sigma)=\ker(\rho)\oplus\operatorname{im}(\sigma)$$

3) *If ρ and σ commute, then $\rho\sigma$ is a projection, in which case*

$$\operatorname{im}(\rho\sigma)=\operatorname{im}(\rho)\cap\operatorname{im}(\sigma)\quad \textit{and}\quad \ker(\rho\sigma)=\ker(\rho)+\ker(\sigma)$$

(Example 2.5 shows that the converse may be false.)□

Topological Vector Spaces

This section is for readers with some familiarity with point-set topology.

The Definition

A pair $(V,\mathcal{T})$ where V is a real vector space V and $\mathcal{T}$ is a topology on the set V is called a **topological vector space** if the operations of addition

$$\mathcal{A}: V\times V\to V,\quad \mathcal{A}(v,w)=v+w$$

and scalar multiplication

$$\mathcal{M}: \mathbb{R}\times V\to V,\quad \mathcal{M}(r,v)=rv$$

are continuous functions.

The Standard Topology on $\mathbb{R}^n$

The vector space $\mathbb{R}^n$ is a topological vector space under the **standard topology**, which is the topology for which the set of **open rectangles**

$$\mathcal{B}=\{I_1\times\cdots\times I_n \mid I_i\text{'s are open intervals in }\mathbb{R}\}$$

is a base, that is, a subset of $\mathbb{R}^n$ is open if and only if it is a union of open rectangles. The standard topology is also the topology induced by the Euclidean metric on $\mathbb{R}^n$, since an open rectangle is the union of Euclidean open balls and an open ball is the union of open rectangles.

The standard topology on $\mathbb{R}^n$ has the property that the addition function

$$\mathcal{A}: \mathbb{R}^n\times\mathbb{R}^n\to\mathbb{R}^n: (v,w)\to v+w$$

and the scalar multiplication function

$$\mathcal{M}: \mathbb{R}\times\mathbb{R}^n\to\mathbb{R}^n: (r,v)\to rv$$

are continuous and so $\mathbb{R}^n$ is a topological vector space under this topology. Also, the linear functionals $f:\mathbb{R}^n\to\mathbb{R}$ are continuous maps.

For example, to see that addition is continuous, if

$$(u_1,\ldots,u_n)+(v_1,\ldots,v_n)\in(a_1,b_1)\times\cdots\times(a_n,b_n)\in\mathcal{B}$$

then $u_i + v_i \in (a_i, b_i)$ and so there is an $\epsilon > 0$ for which

$$(u_i - \epsilon, u_i + \epsilon) + (v_i - \epsilon, v_i + \epsilon) \subseteq (a_i, b_i)$$

for all i. It follows that if

$$(u_1, \ldots, u_n) \in I := (u_1 - \epsilon, u_1 + \epsilon) \times \cdots \times (u_n - \epsilon, u_n + \epsilon) \in \mathcal{B}$$

and

$$(v_1, \ldots, v_n) \in J := (v_1 - \epsilon, v_1 + \epsilon) \times \cdots \times (v_n - \epsilon, v_n + \epsilon) \in \mathcal{B}$$

then

$$(u_1, \ldots, u_n) + (v_1, \ldots, v_n) \in \mathcal{A}(I, J) \subseteq (a_1, b_1) \times \cdots \times (a_n, b_n)$$

The Natural Topology on V

Now let V be a real vector space of dimension n and fix an ordered basis $\mathcal{B} = (v_1, \ldots, v_n)$ for V. We wish to show that there is precisely one topology $\mathcal{T}$ on V for which $(V, \mathcal{T})$ is a topological vector space and all linear functionals are continuous. This topology is called the **natural topology** on V.

Our plan is to show that if $(V, \mathcal{T})$ is a topological vector space and if all linear functionals on V are continuous, then the coordinate map $\phi_{\mathcal{B}}: V \approx \mathbb{R}_n$ is a homeomorphism. This implies that if $\mathcal{T}$ does exist, it must be unique. Then we use $\psi = \phi_{\mathcal{B}}^{-1}$ to move the standard topology from $\mathbb{R}^n$ to V, thus giving V a topology $\mathcal{T}$ for which $\phi_{\mathcal{B}}$ is a homeomorphism. Finally, we show that $(V, \mathcal{T})$ is a topological vector space and that all linear functionals on V are continuous.

The first step is to show that if $(V, \mathcal{T})$ is a topological vector space, then ψ is continuous. Since $\psi = \sum \psi_i$ where $\psi_i: \mathbb{R}^n \to V$ is defined by

$$\psi_i(a_1, \ldots, a_n) = a_i v_i$$

it is sufficient to show that these maps are continuous. (The sum of continuous maps is continuous.) Let O be an open set in $\mathcal{T}$. Then

$$\mathcal{M}^{-1}(O) = \{(r, x) \in \mathbb{R} \times V \mid rx \in O\}$$

is open in $\mathbb{R} \times V$. This implies that if $rx \in O$, then there is an open interval $I \subseteq \mathbb{R}$ containing r for which

$$Ix = \{sx \mid s \in I\} \subseteq O$$

We need to show that the set $\psi_i^{-1}(O)$ is open. But

$$\begin{aligned}\psi_i^{-1}(O) &= \{(a_1, \ldots, a_n) \in \mathbb{R}^n \mid a_i v_i \in O\} \\ &= \mathbb{R} \times \cdots \times \mathbb{R} \times \{a_i \in \mathbb{R} \mid a_i v_i \in O\} \times \mathbb{R} \times \cdots \times \mathbb{R}\end{aligned}$$

In words, an n-tuple $(a_1, \ldots, a_n)$ is in $\psi_i^{-1}(O)$ if the ith coordinate a_i times v_i is in O. But if $a_i v_i \in O$, then there is an open interval $I \subseteq \mathbb{R}$ for which $a_i \in I$ and $I v_i \subseteq O$. Hence, the entire open set

$$U = \mathbb{R} \times \cdots \times \mathbb{R} \times I \times \mathbb{R} \times \cdots \times \mathbb{R}$$

where the factor I is in the ith position is in $\psi_i^{-1}(O)$, that is,

$$(a_1, \ldots, a_n) \in U \subseteq \psi_i^{-1}(O)$$

Thus, $\psi_i^{-1}(O)$ is open and ψ_i, and therefore also ψ, is continuous.

Next we show that if every linear functional on V is continuous under a topology $\mathcal{T}$ on V, then the coordinate map ϕ is continuous. If $v \in V$ denote by $[v]_{\mathcal{B},i}$ the ith coordinate of $[v]_{\mathcal{B}}$. The map $\mu: V \to \mathbb{R}$ defined by $\mu v = [v]_{\mathcal{B},i}$ is a linear functional and so is continuous by assumption. Hence, for any open interval $I_i \in \mathbb{R}$ the set

$$A_i = \{v \in V \mid [v]_{\mathcal{B},i} \in I_i\}$$

is open. Now, if I_i are open intervals in $\mathbb{R}$, then

$$\phi^{-1}(I_1 \times \cdots \times I_n) = \{v \in V \mid [v]_{\mathcal{B}} \in I_1 \times \cdots \times I_n\} = \bigcap A_i$$

is open. Thus, ϕ is continuous.

We have shown that if a topology $\mathcal{T}$ has the property that $(V, \mathcal{T})$ is a topological vector space under which every linear functional is continuous, then ϕ and $\psi = \phi^{-1}$ are homeomorphisms. This means that if $\mathcal{T}$ exists, its open sets must be the images under ψ of the open sets in the standard topology of $\mathbb{R}^n$. It remains to prove that the topology $\mathcal{T}$ on V that makes ϕ a homeomorphism makes $(V, \mathcal{T})$ a topological vector space for which any linear functional f on V is continuous.

The addition map on V is a composition

$$\mathcal{A} = \phi^{-1} \circ \mathcal{A}' \circ (\phi \times \phi)$$

where $\mathcal{A}': \mathbb{R}^n \times \mathbb{R}^n \to \mathbb{R}^n$ is addition in $\mathbb{R}^n$ and since each of the maps on the right is continuous, so is $\mathcal{A}$.

Similarly, scalar multiplication in V is

$$\mathcal{M} = \phi^{-1} \circ \mathcal{M}' \circ (\iota \times \phi)$$

where $\mathcal{M}': \mathbb{R} \times \mathbb{R}^n \to \mathbb{R}^n$ is scalar multiplication in $\mathbb{R}^n$. Hence, $\mathcal{M}$ is continuous.

Now let f be a linear functional. Since ϕ is continuous if and only if $f \circ \phi^{-1}$ is continuous, we can confine attention to $V = \mathbb{R}^n$. In this case, if $e_1, \ldots, e_n$ is the standard basis for $\mathbb{R}^n$ and $|f(e_i)| \le M$ for all i, then for any

$x = (a_1, \ldots, a_n) \in \mathbb{R}^n$, we have

$$|f(x)| = \left|\sum a_i f(e_i)\right| \le \sum |a_i||f(e_i)| \le M \sum |a_i|$$

Now, if $|x| < \epsilon/Mn$, then $|a_i| < \epsilon/Mn$ and so $|f(x)| < \epsilon$, which implies that f is continuous at $x = 0$.

According to the Riesz representation theorem (Theorem 9.18) and the Cauchy–Schwarz inequality, we have

$$\|f(x)\| \le \|\mathcal{R}_f\| \|x\|$$

where $R_f \in \mathbb{R}^n$. Hence, $x_n \to 0$ implies $f(x_n) \to 0$ and so by linearity, $x_n \to x$ implies $f(x_n) \to x$ and so f is continuous at all x.

Theorem 2.27 *Let V be a real vector space of dimension n. There is a unique topology on V, called the* **natural topology**, *for which V is a topological vector space and for which all linear functionals on V are continuous. This topology is determined by the fact that the coordinate map $\phi: V \to \mathbb{R}^n$ is a homeomorphism, where $\mathbb{R}^n$ has the standard topology induced by the Euclidean metric.* □

Linear Operators on $V^{\mathbb{C}}$

A linear operator τ on a real vector space V can be extended to a linear operator $\tau^{\mathbb{C}}$ on the complexification $V^{\mathbb{C}}$ by defining

$$\tau^{\mathbb{C}}(u + vi) = \tau(u) + \tau(v)i$$

Here are the basic properties of this **complexification** of τ.

Theorem 2.28 *If $\tau, \sigma \in \mathcal{L}(V)$, then*
1) $(a\tau)^{\mathbb{C}} = a\tau^{\mathbb{C}}$, $a \in \mathbb{R}$
2) $(\tau + \sigma)^{\mathbb{C}} = \tau^{\mathbb{C}} + \sigma^{\mathbb{C}}$
3) $(\tau\sigma)^{\mathbb{C}} = \tau^{\mathbb{C}}\sigma^{\mathbb{C}}$
4) $[\tau v]^{\mathbb{C}} = \tau^{\mathbb{C}}(v^{\mathbb{C}})$. □

Let us recall that for any ordered basis $\mathcal{B}$ for V and any vector $v \in V$ we have

$$[v + 0i]_{\text{cpx}(\mathcal{B})} = [v]_{\mathcal{B}}$$

Now, if $\mathcal{B}$ is an ordered basis for V, then the ith column of $[\tau]_{\mathcal{B}}$ is

$$[\tau b_i]_{\mathcal{B}} = [\tau b_i + 0i]_{\text{cpx}(\mathcal{B})} = [\tau^{\mathbb{C}}(b_i + 0i)]_{\text{cpx}(\mathcal{B})}$$

which is the ith column of the coordinate matrix of $\tau^{\mathbb{C}}$ with respect to the basis $\text{cpx}(\mathcal{B})$. Thus we have the following theorem.

Theorem 2.29 *Let $\tau \in \mathcal{L}(V)$ where V is a real vector space. The matrix of $\tau^{\mathbb{C}}$ with respect to the ordered basis* $\operatorname{cpx}(\mathcal{B})$ *is equal to the matrix of τ with respect to the ordered basis $\mathcal{B}$:*

$$[\tau^{\mathbb{C}}]_{\operatorname{cpx}(\mathcal{B})} = [\tau]_{\mathcal{B}}$$

Hence, if a real matrix A represents a linear operator τ on V, then A also represents the complexification $\tau^{\mathbb{C}}$ of τ on $V^{\mathbb{C}}$.$\square$

Exercises

1. Let $A \in \mathcal{M}_{m,n}$ have rank k. Prove that there are matrices $X \in \mathcal{M}_{m,k}$ and $Y \in \mathcal{M}_{k,n}$, both of rank k, for which $A = XY$. Prove that A has rank 1 if and only if it has the form $A = x^t y$ where x and y are row matrices.
2. Prove Corollary 2.9 and find an example to show that the corollary does not hold without the finiteness condition.
3. Let $\tau \in \mathcal{L}(V, W)$. Prove that τ is an isomorphism if and only if it carries a basis for V to a basis for W.
4. If $\tau \in \mathcal{L}(V_1, W_1)$ and $\sigma \in \mathcal{L}(V_2, W_2)$ we define the external direct sum $\tau \boxplus \sigma \in \mathcal{L}(V_1 \boxplus V_2, W_1 \boxplus W_2)$ by
$$(\tau \boxplus \sigma)((v_1, v_2)) = (\tau v_1, \sigma v_2)$$
Show that $\tau \boxplus \sigma$ is a linear transformation.
5. Let $V = S \oplus T$. Prove that $S \oplus T \approx S \boxplus T$. Thus, internal and external direct sums are equivalent up to isomorphism.
6. Let $V = A + B$ and consider the external direct sum $E = A \boxplus B$. Define a map $\tau: A \boxplus B \to V$ by $\tau(v, w) = v + w$. Show that τ is linear. What is the kernel of τ? When is τ an isomorphism?
7. Let $\tau \in \mathcal{L}_F(V)$ where $\dim(V) = n < \infty$. Let $A \in \mathcal{M}_n(F)$. Suppose that there is an isomorphism $\sigma: V \approx F^n$ with the property that $\sigma(\tau v) = A(\sigma v)$. Prove that there is an ordered basis $\mathcal{B}$ for which $A = [\tau]_{\mathcal{B}}$.
8. Let $\mathcal{T}$ be a subset of $\mathcal{L}(V)$. A subspace S of V is $\mathcal{T}$**-invariant** if S is τ-invariant for every $\tau \in \mathcal{T}$. Also, V is $\mathcal{T}$**-irreducible** if the only $\mathcal{T}$-invariant subspaces of V are $\{0\}$ and V. Prove the following form of *Schur's lemma.* Suppose that $\mathcal{T}_V \subseteq \mathcal{L}(V)$ and $\mathcal{T}_W \subseteq \mathcal{L}(W)$ and V is $\mathcal{T}_V$-irreducible and W is $\mathcal{T}_W$-irreducible. Let $\alpha \in \mathcal{L}(V, W)$ satisfy $\alpha\mathcal{T}_V = \mathcal{T}_W\alpha$, that is, for any $\mu \in \mathcal{T}_V$ there is a $\lambda \in \mathcal{T}_W$ such that $\alpha\mu = \lambda\alpha$ and for any $\lambda \in \mathcal{T}_W$ there is a $\mu \in \mathcal{T}_V$ such that $\alpha\mu = \lambda\alpha$. Prove that $\alpha = 0$ or α is an isomorphism.
9. Let $\tau \in \mathcal{L}(V)$ where $\dim(V) < \infty$. If $\operatorname{rk}(\tau^2) = \operatorname{rk}(\tau)$ show that $\operatorname{im}(\tau) \cap \ker(\tau) = \{0\}$.
10. Let $\tau \in \mathcal{L}(U,V)$ and $\sigma \in \mathcal{L}(V, W)$. Show that
$$\operatorname{rk}(\sigma\tau) \le \min\{\operatorname{rk}(\tau), \operatorname{rk}(\sigma)\}$$
11. Let $\tau \in \mathcal{L}(U, V)$ and $\sigma \in \mathcal{L}(V, W)$. Show that
$$\operatorname{null}(\sigma\tau) \le \operatorname{null}(\tau) + \operatorname{null}(\sigma)$$

12. Let $\tau, \sigma \in \mathcal{L}(V)$ where τ is invertible. Show that
$$\mathrm{rk}(\tau\sigma) = \mathrm{rk}(\sigma\tau) = \mathrm{rk}(\sigma)$$
13. Let $\tau, \sigma \in \mathcal{L}(V, W)$. Show that
$$\mathrm{rk}(\tau + \sigma) \le \mathrm{rk}(\tau) + \mathrm{rk}(\sigma)$$
14. Let S be a subspace of V. Show that there is a $\tau \in \mathcal{L}(V)$ for which $\ker(\tau) = S$. Show also that there exists a $\sigma \in \mathcal{L}(V)$ for which $\mathrm{im}(\sigma) = S$.
15. Suppose that $\tau, \sigma \in \mathcal{L}(V)$.
    a) Show that $\sigma = \tau\mu$ for some $\mu \in \mathcal{L}(V)$ if and only if $\mathrm{im}(\sigma) \subseteq \mathrm{im}(\tau)$.
    b) Show that $\sigma = \mu\tau$ for some $\mu \in \mathcal{L}(V)$ if and only if $\ker(\tau) \subseteq \ker(\sigma)$.
16. Let $\dim(V) < \infty$ and suppose that $\tau \in \mathcal{L}(V)$ satisfies $\tau^2 = 0$. Show that $2\mathrm{rk}(\tau) \le \dim(V)$.
17. Let A be an $m \times n$ matrix over F. What is the relationship between the linear transformation $\tau_A: F^n \to F^m$ and the system of equations $AX = B$? Use your knowledge of linear transformations to state and prove various results concerning the system $AX = B$, especially when $B = 0$.
18. Let V have basis $\mathcal{B} = \{v_1, \ldots, v_n\}$ and assume that the base field F for V has characteristic 0. Suppose that for each $1 \le i, j \le n$ we define $\tau_{i,j} \in \mathcal{L}(V)$ by
$$\tau_{i,j}(v_k) = \begin{cases} v_k & \text{if } k \ne i \\ v_i + v_j & \text{if } k = i \end{cases}$$
    Prove that the $\tau_{i,j}$ are invertible and form a basis for $\mathcal{L}(V)$.
19. Let $\tau \in \mathcal{L}(V)$. If S is a τ-invariant subspace of V must there be a subspace T of V for which (S, T) reduces τ?
20. Find an example of a vector space V and a proper subspace S of V for which $V \approx S$.
21. Let $\dim(V) < \infty$. If τ, $\sigma \in \mathcal{L}(V)$ prove that $\sigma\tau = \iota$ implies that τ and σ are invertible and that $\sigma = p(\tau)$ for some polynomial $p(x) \in F[x]$.
22. Let $\tau \in \mathcal{L}(V)$. If $\tau\sigma = \sigma\tau$ for all $\sigma \in \mathcal{L}(V)$ show that $\tau = a\iota$, for some $a \in F$, where ι is the identity map.
23. Let V be a vector space over a field F of characteristic $\ne 2$ and let ρ and σ be projections. Prove the following:
    a) The difference $\rho - \sigma$ is a projection if and only if
$$\rho\sigma = \sigma\rho = \sigma$$
    in which case
$$\mathrm{im}(\rho - \sigma) = \mathrm{im}(\rho) \cap \ker(\sigma) \quad \text{and} \quad \ker(\rho - \sigma) = \ker(\rho) \oplus \mathrm{im}(\sigma)$$
    Hint: ρ is a projection if and only if $\iota - \rho$ is a projection and so $\rho - \sigma$ is a projection if and only if

$$\theta = \iota - (\rho - \sigma) = (\iota - \rho) + \sigma$$

is a projection.

b) If ρ and σ commute, then $\rho\sigma$ is a projection, in which case

$$\operatorname{im}(\rho\sigma) = \operatorname{im}(\rho) \cap \operatorname{im}(\sigma) \quad \text{and} \quad \ker(\rho\sigma) = \ker(\rho) + \ker(\sigma)$$

24. Let $f\colon \mathbb{R}^n \to \mathbb{R}$ be a continuous function with the property that

$$f(x+y) = f(x) + f(y)$$

Prove that f is a linear functional on $\mathbb{R}^n$.

25. Prove that any linear functional $f\colon \mathbb{R}^n \to \mathbb{R}$ is a continuous map.
26. Prove that any subspace S of $\mathbb{R}^n$ is a closed set or, equivalently, that $S^c = \mathbb{R}^n \setminus S$ is open, that is, for any $x \in S^c$ there is an open ball $B(x,\epsilon)$ centered at x with radius $\epsilon > 0$ for which $B(x,\epsilon) \subseteq S^c$.
27. Prove that any linear transformation $\tau\colon V \to W$ is continuous under the natural topologies of V and W.
28. Prove that any surjective linear transformation τ from V to W (both finite-dimensional topological vector spaces under the natural topology) is an open map, that is, τ maps open sets to open sets.
29. Prove that any subspace S of a finite-dimensional vector space V is a closed set or, equivalently, that S^c is open, that is, for any $x \in S^c$ there is an open ball $B(x,\epsilon)$ centered at x with radius $\epsilon > 0$ for which $B(x,\epsilon) \subseteq S^c$.
30. Let S be a subspace of V with $\dim(V) < \infty$.
    a) Show that the subspace topology on S inherited from V is the natural topology.
    b) Show that the natural topology on V/S is the topology for which the natural projection map $\pi\colon V \to V/S$ continuous and open.
31. If V is a real vector space, then $V^{\mathbb{C}}$ is a complex vector space. Thinking of $V^{\mathbb{C}}$ as a vector space $(V^{\mathbb{C}})_{\mathbb{R}}$ over $\mathbb{R}$, show that $(V^{\mathbb{C}})_{\mathbb{R}}$ is isomorphic to the external direct product $V \boxplus V$.
32. (When is a complex linear map a complexification?) Let V be a real vector space with complexification $V^{\mathbb{C}}$ and let $\sigma \in \mathcal{L}(V^{\mathbb{C}})$. Prove that σ is a complexification, that is, σ has the form $\tau^{\mathbb{C}}$ for some $\tau \in \mathcal{L}(V)$ if and only if σ commutes with the conjugate map $\chi\colon V^{\mathbb{C}} \to V^{\mathbb{C}}$ defined by $\chi(u+iv) = u - iv$.
33. Let W be a complex vector space.
    a) Consider replacing the scalar multiplication on W by the operation

    $$(z, w) \to \overline{z}w$$

    where $z \in \mathbb{C}$ and $w \in W$. Show that the resulting set with the addition defined for the vector space W and with this scalar multiplication is a complex vector space, which we denote by $\overline{W}$.
    b) Show, without using dimension arguments, that $(W_{\mathbb{R}})^{\mathbb{C}} \approx W \boxplus \overline{W}$.

Chapter 3
The Isomorphism Theorems

Quotient Spaces

Let S be a subspace of a vector space V. It is easy to see that the binary relation on V defined by

$$u \equiv v \quad \Leftrightarrow \quad u - v \in S$$

is an equivalence relation. When $u \equiv v$, we say that u and v are **congruent modulo** S. The term *mod* is used as a colloquialism for modulo and $u \equiv v$ is often written

$$u \equiv v \bmod S$$

When the subspace in question is clear, we will simply write $u \equiv v$.

To see what the equivalence classes look like, observe that

$$\begin{aligned} [v] &= \{u \in V \mid u \equiv v\} \\ &= \{u \in V \mid u - v \in S\} \\ &= \{u \in V \mid u = v + s \text{ for some } s \in S\} \\ &= \{v + s \mid s \in S\} \\ &= v + S \end{aligned}$$

The set

$$[v] = v + S = \{v + s \mid s \in S\}$$

is called a **coset** of S in V and v is called a **coset representative** for $v + S$. (Thus, any member of a coset is a coset representative.)

The set of all cosets of S in V is denoted by

$$V/S = \{v + S \mid v \in V\}$$

This is read "V mod S" and is called the **quotient space of** V **modulo** S. Of course, the term space is a hint that we intend to define vector space operations on V/S.

The natural choice for these vector space operations is

$$(u+S)+(v+S)=(u+v)+S$$

and

$$r(u+S)=(ru)+S$$

but we must check that these operations are well-defined, that is,

1) $u_1+S=u_2+S, v_1+S=v_2+S \Rightarrow (u_1+v_1)+S=(u_2+v_2)+S$
2) $u_1+S=u_2+S \Rightarrow ru_1+S=ru_2+S$

Equivalently, the equivalence relation $\equiv$ must be *consistent* with the vector space operations on V, that is,

3) $u_1 \equiv u_2, v_1 \equiv v_2 \Rightarrow (u_1+v_1) \equiv (u_2+v_2)$
4) $u_1 \equiv u_2 \Rightarrow ru_1 \equiv ru_2$

This senario is a recurring one in algebra. An equivalence relation on an algebraic structure, such as a group, ring, module or vector space is called a **congruence relation** if it preserves the algebraic operations. In the case of a vector space, these are conditions 3) and 4) above.

These conditions follow easily from the fact that S is a subspace, for if $u_1 \equiv u_2$ and $v_1 \equiv v_2$, then

$$\begin{aligned} u_1-u_2 \in S, v_1-v_2 \in S &\Rightarrow r(u_1-u_2)+s(v_1-v_2) \in S \\ &\Rightarrow (ru_1+sv_1)-(ru_2+sv_2) \in S \\ &\Rightarrow ru_1+sv_1 \equiv ru_2+sv_2 \end{aligned}$$

which verifies both conditions at once. We leave it to the reader to verify that V/S is indeed a vector space over F under these well-defined operations.

Actually, we are lucky here: For *any* subspace S of V, the quotient V/S is a vector space under the natural operations. In the case of groups, not all subgroups have this property. Indeed, it is precisely the *normal* subgroups N of G that have the property that the quotient G/N is a group. Also, for rings, it is precisely the *ideals* (not the subrings) that have the property that the quotient is a ring.

Let us summarize.

Theorem 3.1 *Let S be a subspace of V. The binary relation*

$$u \equiv v \quad \Leftrightarrow \quad u - v \in S$$

is an equivalence relation on V, whose equivalence classes are the **cosets**

$$v + S = \{v + s \mid s \in S\}$$

of S in V. The set V/S of all cosets of S in V, called the **quotient space** *of V modulo S, is a vector space under the well-defined operations*

$$r(u + S) = ru + S$$
$$(u + S) + (v + S) = (u + v) + S$$

The zero vector in V/S is the coset $0 + S = S$.□

The Natural Projection and the Correspondence Theorem

If S is a subspace of V, then we can define a map $\pi_S: V \to V/S$ by sending each vector to the coset containing it:

$$\pi_S(v) = v + S$$

This map is called the **canonical projection** or **natural projection** of V onto V/S, or simply **projection modulo** S. (Not to be confused with the projection operators $\rho_{S,T}$.) It is easily seen to be linear, for we have (writing π for π_S)

$$\pi(ru + sv) = (ru + sv) + S = r(u + S) + s(v + S) = r\pi(u) + s\pi(v)$$

The canonical projection is clearly surjective. To determine the kernel of π, note that

$$v \in \ker(\pi) \Leftrightarrow \pi(v) = 0 \Leftrightarrow v + S = S \Leftrightarrow v \in S$$

and so

$$\ker(\pi) = S$$

Theorem 3.2 *The canonical projection $\pi_S: V \to V/S$ defined by*

$$\pi_S(v) = v + S$$

is a surjective linear transformation with $\ker(\pi_S) = S$.□

If S is a subspace of V, then the subspaces of the quotient space V/S have the form T/S for some intermediate subspace T satisfying $S \subseteq T \subseteq V$. In fact, as shown in Figure 3.1, the projection map π_S provides a one-to-one correspondence between intermediate subspaces $S \subseteq T \subseteq V$ and subspaces of the quotient space V/S. The proof of the following theorem is left as an exercise.

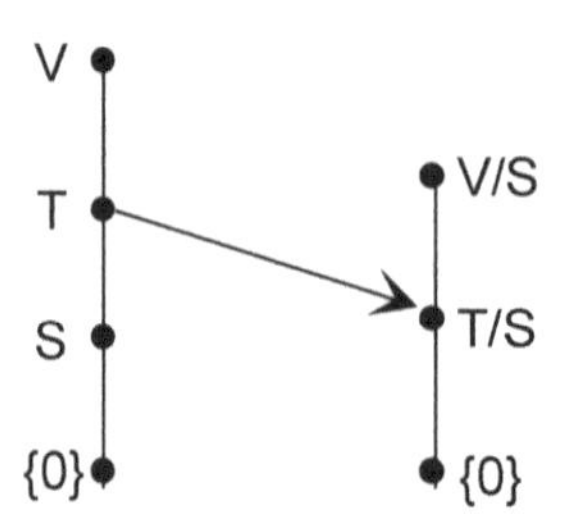

Figure 3.1: The correspondence theorem

Theorem 3.3 (The correspondence theorem) *Let S be a subspace of V. Then the function that assigns to each intermediate subspace $S \subseteq T \subseteq V$ the subspace T/S of V/S is an order-preserving (with respect to set inclusion) one-to-one correspondence between the set of all subspaces of V containing S and the set of all subspaces of V/S.*

Proof. We prove only that the correspondence is surjective. Let

$$X = \{u + S \mid u \in U\}$$

be a subspace of V/S and let T be the union of all cosets in X:

$$T = \bigcup_{u \in U} (u + S)$$

We show that $S \leq T \leq V$ and that $T/S = X$. If $x, y \in T$, then $x + S$ and $y + S$ are in X and since $X \leq V/S$, we have

$$rx + S, (x + y) + S \in X$$

which implies that $rx, x + y \in T$. Hence, T is a subspace of V containing S. Moreover, if $t + S \in T/S$, then $t \in T$ and so $t + S \in X$. Conversely, if $u + S \in X$, then $u \in T$ and therefore $u + S \in T/S$. Thus, $X = T/S$.□

The Universal Property of Quotients and the First Isomorphism Theorem

Let S be a subspace of V. The pair $(V/S, \pi_S)$ has a very special property, known as the *universal property*—a term that comes from the world of category theory.

Figure 3.2 shows a linear transformation $\tau \in \mathcal{L}(V, W)$, along with the canonical projection π_S from V to the quotient space V/S.

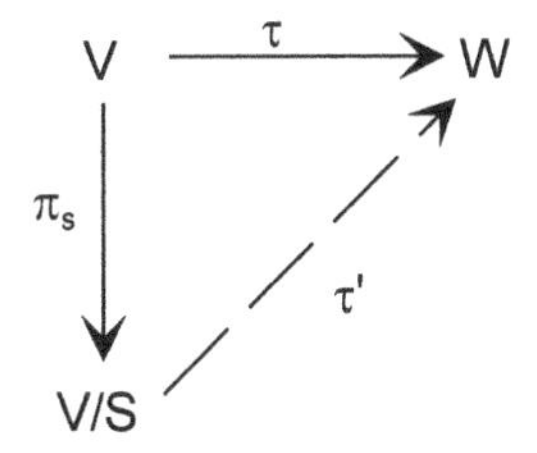

Figure 3.2: The universal property

The universal property states that if $\ker(\tau) \supseteq S$, then there is a unique $\tau'\colon V/S \to W$ for which

$$\tau' \circ \pi_S = \tau$$

Another way to say this is that any such $\tau \in \mathcal{L}(V,W)$ can be *factored through* the canonical projection π_S.

Theorem 3.4 *Let S be a subspace of V and let $\tau \in \mathcal{L}(V,W)$ satisfy $S \subseteq \ker(\tau)$. Then, as pictured in Figure 3.2, there is a unique linear transformation $\tau'\colon V/S \to W$ with the property that*

$$\tau' \circ \pi_S = \tau$$

Moreover, $\ker(\tau') = \ker(\tau)/S$ *and* $\operatorname{im}(\tau') = \operatorname{im}(\tau)$.

Proof. We have no other choice but to define τ' by the condition $\tau' \circ \pi_S = \tau$, that is,

$$\tau'(v+S) = \tau v$$

This function is well-defined if and only if

$$v+S = u+S \Rightarrow \tau'(v+S) = \tau'(u+S)$$

which is equivalent to each of the following statements:

$$\begin{aligned} v+S = u+S &\Rightarrow \tau v = \tau u \\ v-u \in S &\Rightarrow \tau(v-u) = 0 \\ x \in S &\Rightarrow \tau x = 0 \\ S &\subseteq \ker(\tau) \end{aligned}$$

Thus, $\tau'\colon V/S \to W$ is well-defined. Also,

$$\operatorname{im}(\tau') = \{\tau'(v+S) \mid v \in V\} = \{\tau v \mid v \in V\} = \operatorname{im}(\tau)$$

and

$$\begin{aligned}\ker(\tau') &= \{v+S \mid \tau'(v+S)=0\}\\ &= \{v+S \mid \tau v = 0\}\\ &= \{v+S \mid v\in \ker(\tau)\}\\ &= \ker(\tau)/S\end{aligned}$$

The uniqueness of τ' is evident.□

Theorem 3.4 has a very important corollary, which is often called the *first isomorphism theorem* and is obtained by taking $S = \ker(\tau)$.

Theorem 3.5 (The first isomorphism theorem) *Let $\tau: V \to W$ be a linear transformation. Then the linear transformation $\tau': V/\ker(\tau) \to W$ defined by*

$$\tau'(v + \ker(\tau)) = \tau v$$

is injective and

$$\frac{V}{\ker(\tau)} \approx \operatorname{im}(\tau)$$ □

According to Theorem 3.5, the image of any linear transformation on V is isomorphic to a quotient space of V. Conversely, any quotient space V/S of V is the image of a linear transformation on V: the canonical projection π_S. Thus, up to isomorphism, quotient spaces are equivalent to homomorphic images.

Quotient Spaces, Complements and Codimension

The first isomorphism theorem gives some insight into the relationship between complements and quotient spaces. Let S be a subspace of V and let T be a complement of S, that is,

$$V = S \oplus T$$

Applying the first isomorphism theorem to the projection operator $\rho_{T,S}: V \to T$ gives

$$T \approx V/S$$

Theorem 3.6 *Let S be a subspace of V. All complements of S in V are isomorphic to V/S and hence to each other.*□

The previous theorem can be rephrased by writing

$$A \oplus B = A \oplus C \Rightarrow B \approx C$$

On the other hand, quotients and complements do not behave as nicely with respect to isomorphisms as one might casually think. We leave it to the reader to show the following:

1) It is possible that
$$A \oplus B = C \oplus D$$
with $A \approx C$ but $B \not\approx D$. Hence, $A \approx C$ does *not* imply that a complement of A is isomorphic to a complement of C.
2) It is possible that $V \approx W$ and
$$V = S \oplus B \text{ and } W = S \oplus D$$
but $B \not\approx D$. Hence, $V \approx W$ does *not* imply that $V/S \approx W/S$. (However, according to the previous theorem, if V *equals* W then $B \approx D$.)

Corollary 3.7 *Let S be a subspace of a vector space V. Then*
$$\dim(V) = \dim(S) + \dim(V/S) \qquad \square$$

Definition *If S is a subspace of V, then* $\dim(V/S)$ *is called the* **codimension** *of S in V and is denoted by* $\operatorname{codim}(S)$ *or* $\operatorname{codim}_V(S)$.$\square$

Thus, the codimension of S in V is the dimension of any complement of S in V and when V is *finite-dimensional*, we have
$$\operatorname{codim}_V(S) = \dim(V) - \dim(S)$$
(This makes no sense, in general, if V is not finite-dimensional, since infinite cardinal numbers cannot be subtracted.)

Additional Isomorphism Theorems

There are other isomorphism theorems that are direct consequences of the first isomorphism theorem. As we have seen, if $V = S \oplus T$ then $V/T \approx S$. This can be written
$$\frac{S \oplus T}{T} \approx \frac{S}{S \cap T}$$
This applies to nondirect sums as well.

Theorem 3.7 (The second isomorphism theorem) *Let V be a vector space and let S and T be subspaces of V. Then*
$$\frac{S + T}{T} \approx \frac{S}{S \cap T}$$

Proof. Let $\tau\colon (S+T) \to S/(S \cap T)$ be defined by
$$\tau(s+t) = s + (S \cap T)$$
We leave it to the reader to show that τ is a well-defined surjective linear transformation, with kernel T. An application of the first isomorphism theorem then completes the proof.$\square$

The following theorem demonstrates one way in which the expression V/S behaves like a fraction.

Theorem 3.8 (The third isomorphism theorem) *Let V be a vector space and suppose that $S \subseteq T \subseteq V$ are subspaces of V. Then*

$$\frac{V/S}{T/S} \approx \frac{V}{T}$$

Proof. Let $\tau: V/S \to V/T$ be defined by $\tau(v+S) = v+T$. We leave it to the reader to show that τ is a well-defined surjective linear transformation whose kernel is T/S. The rest follows from the first isomorphism theorem.□

The following theorem demonstrates one way in which the expression V/S does not behave like a fraction.

Theorem 3.9 *Let V be a vector space and let S be a subspace of V. Suppose that $V = V_1 \oplus V_2$ and $S = S_1 \oplus S_2$ with $S_i \subseteq V_i$. Then*

$$\frac{V}{S} = \frac{V_1 \oplus V_2}{S_1 \oplus S_2} \approx \frac{V_1}{S_1} \boxplus \frac{V_2}{S_2}$$

Proof. Let $\tau: V \to (V_1/S_1) \boxplus (V_2/S_2)$ be defined by

$$\tau(v_1 + v_2) = (v_1 + S_1, v_2 + S_2)$$

This map is well-defined, since the sum $V = V_1 \oplus V_2$ is direct. We leave it to the reader to show that τ is a surjective linear transformation, whose kernel is $S_1 \oplus S_2$. The rest follows from the first isomorphism theorem.□

Linear Functionals

Linear transformations from V to the base field F (thought of as a vector space over itself) are extremely important.

Definition *Let V be a vector space over F. A linear transformation $f \in \mathcal{L}(V, F)$ whose values lie in the base field F is called a* **linear functional** *(or simply* **functional***) on V. (Some authors use the term* linear function.*) The vector space of all linear functionals on V is denoted by V^* and is called the* **algebraic dual space** *of V.*□

The adjective *algebraic* is needed here, since there is another type of dual space that is defined on general normed vector spaces, where continuity of linear transformations makes sense. We will discuss the so-called *continuous dual space* briefly in Chapter 13. However, until then, the term "dual space" will refer to the algebraic dual space.

To help distinguish linear functionals from other types of linear transformations, we will usually denote linear functionals by lowercase italic letters, such as f, g and h.

Example 3.1 The map $f: F[x] \to F$ defined by $f(p(x)) = p(0)$ is a linear functional, known as **evaluation at** 0.□

Example 3.2 Let $\mathcal{C}[a,b]$ denote the vector space of all continuous functions on $[a,b] \subseteq \mathbb{R}$. Let $f: \mathcal{C}[a,b] \to \mathbb{R}$ be defined by

$$f(\alpha(x)) = \int_a^b \alpha(x)\,dx$$

Then $f \in \mathcal{C}[a,b]^*$.□

For any $f \in V^*$, the rank plus nullity theorem is

$$\dim(\ker(f)) + \dim(\operatorname{im}(f)) = \dim(V)$$

But since $\operatorname{im}(f) \subseteq F$, we have either $\operatorname{im}(f) = \{0\}$, in which case f is the zero linear functional, or $\operatorname{im}(f) = F$, in which case f is surjective. In other words, a nonzero linear functional is surjective. Moreover, if $f \neq 0$, then

$$\operatorname{codim}(\ker(f)) = \dim\left(\frac{V}{\ker(f)}\right) = 1$$

and if $\dim(V) < \infty$, then

$$\dim(\ker(f)) = \dim(V) - 1$$

Thus, in dimensional terms, the kernel of a linear functional is a very "large" subspace of the domain V.

The following theorem will prove very useful.

Theorem 3.10

1) *For any nonzero vector $v \in V$, there exists a linear functional $f \in V^*$ for which $f(v) \neq 0$.*
2) *A vector $v \in V$ is zero if and only if $f(v) = 0$ for all $f \in V^*$.*
3) *Let $f \in V^*$. If $f(x) \neq 0$, then*

$$V = \langle x \rangle \oplus \ker(f)$$

4) *Two nonzero linear functionals $f, g \in V^*$ have the same kernel if and only if there is a nonzero scalar λ such that $f = \lambda g$.*

Proof. For part 3), if $0 \neq v \in \langle x \rangle \cap \ker(f)$, then $f(v) = 0$ and $v = ax$ for $0 \neq a \in F$, whence $f(x) = 0$, which is false. Hence, $\langle x \rangle \cap \ker(f) = \{0\}$ and the direct sum $S = \langle x \rangle \oplus \ker(f)$ exists. Also, for any $v \in V$ we have

$$v = \frac{f(v)}{f(x)}x + \left(v - \frac{f(v)}{f(x)}x\right) \in \langle x \rangle + \ker(f)$$

and so $V = \langle x \rangle \oplus \ker(f)$.

For part 4), if $f = \lambda g$ for $\lambda \neq 0$, then $\ker(f) = \ker(g)$. Conversely, if $K = \ker(f) = \ker(g)$, then for $x \notin K$ we have by part 3),

$$V = \langle x \rangle \oplus K$$

Of course, $f|_K = \lambda g|_K$ for any λ. Therefore, if $\lambda = f(x)/g(x)$, it follows that $\lambda g(x) = f(x)$ and hence $f = \lambda g$.$\square$

Dual Bases

Let V be a vector space with basis $\mathcal{B} = \{v_i \mid i \in I\}$. For each $i \in I$, we can define a linear functional $v_i^* \in V^*$ by the orthogonality condition

$$v_i^*(v_j) = \delta_{i,j}$$

where $\delta_{i,j}$ is the **Kronecker delta function**, defined by

$$\delta_{i,j} = \begin{cases} 1 & \text{if } i = j \\ 0 & \text{if } i \neq j \end{cases}$$

Then the set $\mathcal{B}^* = \{v_i^* \mid i \in I\}$ is linearly independent, since applying the equation

$$0 = a_{i_1}v_{i_1}^* + \cdots + a_{i_n}v_{i_n}^*$$

to the basis vector v_{i_k} gives

$$0 = \sum_{j=1}^{n} a_{i_j}v_{i_j}^*(v_{i_k}) = \sum_{j=1}^{n} a_{i_j}\delta_{i_j,i_k} = a_{i_k}$$

for all i_k.

Theorem 3.11 *Let V be a vector space with basis $\mathcal{B} = \{v_i \mid i \in I\}$.*
1) The set $\mathcal{B}^* = \{v_i^* \mid i \in I\}$ is linearly independent.
2) If V is finite-dimensional, then $\mathcal{B}^*$ is a basis for V^*, called the **dual basis** *of $\mathcal{B}$.*

Proof. For part 2), for any $f \in V^*$, we have

$$\sum_j f(v_j)v_j^*(v_i) = \sum_j f(v_j)\delta_{i,j} = f(v_i)$$

and so $f = \sum f(v_j)v_j^*$ is in the span of $\mathcal{B}^*$. Hence, $\mathcal{B}^*$ is a basis for V^*.$\square$

It follows from the previous theorem that if $\dim(V) < \infty$, then

$$\dim(V^*) = \dim(V)$$

since the dual vectors also form a basis for V^*. Our goal now is to show that the converse of this also holds. But first, let us consider an example.

Example 3.3 Let V be an infinite-dimensional vector space over the field $F = \mathbb{Z}_2 = \{0, 1\}$, with basis $\mathcal{B}$. Since the only coefficients in F are 0 and 1, a finite linear combination over F is just a finite sum. Hence, V is the set of all finite sums of vectors in $\mathcal{B}$ and so according to Theorem 0.12,

$$|V| \le |\mathcal{P}_0(\mathcal{B})| = |\mathcal{B}|$$

On the other hand, each linear functional $f \in V^*$ is uniquely defined by specifying its values on the basis $\mathcal{B}$. Since these values must be either 0 or 1, specifying a linear functional is equivalent to specifying the subset of $\mathcal{B}$ on which f takes the value 1. In other words, there is a one-to-one correspondence between linear functionals on V and all subsets of $\mathcal{B}$. Hence,

$$|V^*| = |\mathcal{P}(\mathcal{B})| > |\mathcal{B}| \ge |V|$$

This shows that V^* cannot be isomorphic to V, nor to any proper subset of V. Hence, $\dim(V^*) > \dim(V)$.□

We wish to show that the behavior in the previous example is typical, in particular, that

$$\dim(V) \le \dim(V^*)$$

with equality if and only if V is finite-dimensional. The proof uses the concept of the **prime subfield** of a field K, which is defined as the smallest subfield of the field K. Since $0, 1 \in K$, it follows that K contains a copy of the integers

$$0, 1, 2 = 1 + 1, 3 = 1 + 1 + 1, \ldots$$

If K has prime characteristic p, then $p = 0$ and so K contains the elements

$$\mathbb{Z}_p = \{0, 1, 2, , \ldots, p - 1\}$$

which form a subfield of K. Since any subfield F of K contains 0 and 1, we see that $\mathbb{Z}_p \subseteq F$ and so $\mathbb{Z}_p$ is the prime subfield of K. On the other hand, if K has characteristic 0, then K contains a "copy" of the integers $\mathbb{Z}$ and therefore also the rational numbers $\mathbb{Q}$, which is the prime subfield of K. Our main interest in the prime subfield is that in either case, the prime subfield is *countable*.

Theorem 3.12 *Let V be a vector space. Then*

$$\dim(V) \le \dim(V^*)$$

with equality if and only if V is finite-dimensional.

Proof. For any vector space V, we have

$$\dim(V) \le \dim(V^*)$$

since the dual vectors to a basis $\mathcal{B}$ for V are linearly independent in V^*. We have already seen that if V is finite-dimensional, then $\dim(V) = \dim(V^*)$. We wish to show that if V is infinite-dimensional, then $\dim(V) < \dim(V^*)$. (The author is indebted to Professor Richard Foote for suggesting this line of proof.)

If $\mathcal{B}$ is a basis for V and if K is the base field for V, then Theorem 2.7 implies that

$$V \approx (K^{\mathcal{B}})_0$$

where $(K^{\mathcal{B}})_0$ is the set of all functions with finite support from $\mathcal{B}$ to K and

$$V^* \approx K^{\mathcal{B}}$$

where $K^{\mathcal{B}}$ is the set of all functions from $\mathcal{B}$ to K. Thus, we can work with the vector spaces $(K^{\mathcal{B}})_0$ and $K^{\mathcal{B}}$.

The plan is to show that if F is a countable subfield of K and if $\mathcal{B}$ is infinite, then

$$\dim_K\big((K^{\mathcal{B}})_0\big) = \dim_F\big((F^{\mathcal{B}})_0\big) < \dim_F(F^{\mathcal{B}}) \le \dim_K\big(K^{\mathcal{B}}\big)$$

Since we may take F to be the prime subfield of K, this will prove the theorem. The first equality follows from the fact that the K-space $(K^{\mathcal{B}})_0$ and the F-space $(F^{\mathcal{B}})_0$ each have a basis consisting of the "standard" linear functionals $\{f_i \mid i \in \mathcal{B}\}$ defined by

$$f_i v_j = \delta_{i,j}$$

for all $v_i \in \mathcal{B}$, where $\delta_{i,j}$ is the Kronecker delta function.

For the final inequality, suppose that $\{f_i\} \subseteq F^{\mathcal{B}}$ is linearly independent over F and that

$$\sum_i \alpha_i f_i = 0$$

where $\alpha_i \in K$. If $\{\kappa_j\}$ is a basis for K over F, then $\alpha_i = \sum_j a_{i,j}\kappa_j$ for $a_{i,j} \in F$ and so

$$0 = \sum_i \alpha_i f_i = \sum_i \sum_j a_{i,j}\kappa_j f_i$$

Evaluating at any $v \in \mathcal{B}$ gives

$$0=\sum_i\sum_j a_{i,j}\kappa_j f_i(v)=\sum_j\left(\sum_i a_{i,j}f_i(v)\right)\kappa_j$$

and since the inner sums are in F and $\{\kappa_j\}$ is F-independent, the inner sums must be zero:

$$\sum_i a_{i,j}f_i(v)=0$$

Since this holds for all $v\in\mathcal{B}$, we have

$$\sum_i a_{i,j}f_i=0$$

which implies that $a_{i,j}=0$ for all i,j. Hence, $\{f_i\}$ is linearly independent over K. This proves that $\dim_F(F^{\mathcal{B}})\le\dim_K(K^{\mathcal{B}})$.

For the center inequality, it is clear that

$$\dim_F\big((F^{\mathcal{B}})_0\big)\le\dim_F(F^{\mathcal{B}})$$

We will show that the inequality must be strict by showing that the cardinality of $(F^{\mathcal{B}})_0$ is $|\mathcal{B}|$ whereas the cardinality of $F^{\mathcal{B}}$ is greater than $|\mathcal{B}|$. To this end, the set $(F^{\mathcal{B}})_0$ can be partitioned into blocks based on the support of the function. In particular, for each finite subset S of $\mathcal{B}$, if we let

$$A_S=\{f\in(F^{\mathcal{B}})_0\mid\operatorname{supp}(f)=S\}$$

then

$$(F^{\mathcal{B}})_0=\bigcup_{\substack{S\subseteq\mathcal{B}\\ S\text{ finite}}}A_S$$

where the union is disjoint. Moreover, if $|S|=n$, then

$$|A_S|\le|F|^n\le\aleph_0$$

and so

$$\big|(F^{\mathcal{B}})_0\big|=\sum_{\substack{S\subseteq\mathcal{B}\\ S\text{ finite}}}|A_S|\le|\mathcal{B}|\cdot\aleph_0=\max(|\mathcal{B}|,\aleph_0)=|\mathcal{B}|$$

But since the reverse inequality is easy to establish, we have

$$\big|(F^{\mathcal{B}})_0\big|=|\mathcal{B}|$$

As to the cardinality of $F^{\mathcal{B}}$, for each subset T of $\mathcal{B}$, there is a function $f_T\in F^{\mathcal{B}}$ that sends every element of T to 1 and every element of $\mathcal{B}\setminus T$ to 0. Clearly, each distinct subset T gives rise to a distinct function f_T and so Cantor's theorem implies that

$$|F^{\mathcal{B}}| \geq |2^{\mathcal{B}}| > |\mathcal{B}| = |(F^{\mathcal{B}})_0|$$

This shows that

$$\dim_F((F^{\mathcal{B}})_0) < \dim_F(F^{\mathcal{B}})$$

and completes the proof.□

Reflexivity

If V is a vector space, then so is the dual space V^* and so we may form the **double (algebraic) dual space** V^{**}, which consists of all linear functionals $\sigma: V^* \to F$. In other words, an element σ of V^{**} is a linear functional that assigns a scalar to each linear functional on V.

With this firmly in mind, there is one rather obvious way to obtain an element of V^{**}. Namely, if $v \in V$, consider the map $\overline{v}: V^* \to F$ defined by

$$\overline{v}(f) = f(v)$$

which sends the linear functional f to the scalar $f(v)$. The map $\overline{v}$ is called **evaluation at** v. To see that $\overline{v} \in V^{**}$, if $f, g \in V^*$ and $a, b \in F$, then

$$\overline{v}(af + bg) = (af + bg)(v) = af(v) + bg(v) = a\overline{v}(f) + b\overline{v}(g)$$

and so $\overline{v}$ is indeed linear.

We can now define a map $\tau: V \to V^{**}$ by

$$\tau v = \overline{v}$$

This is called the **canonical map** (or the **natural map**) from V to V^{**}. This map is injective and hence in the finite-dimensional case, it is also surjective.

Theorem 3.13 *The canonical map $\tau: V \to V^{**}$ defined by $\tau v = \overline{v}$, where $\overline{v}$ is evaluation at v, is a monomorphism. If V is finite-dimensional, then τ is an isomorphism.*

Proof. The map τ is linear since

$$\overline{au + bv}(f) = f(au + bv) = af(u) + bf(v) = (a\overline{u} + b\overline{v})(f)$$

for all $f \in V^*$. To determine the kernel of τ, observe that

$$\begin{aligned} \tau v = 0 &\Rightarrow \overline{v} = 0 \\ &\Rightarrow \overline{v}(f) = 0 \text{ for all } f \in V^* \\ &\Rightarrow f(v) = 0 \text{ for all } f \in V^* \\ &\Rightarrow v = 0 \end{aligned}$$

by Theorem 3.10 and so $\ker(\tau) = \{0\}$. In the finite-dimensional case, since

$$\dim(V^{**}) = \dim(V^*) = \dim(V)$$

it follows that τ is also surjective, hence an isomorphism.□

Note that if $\dim(V) < \infty$, then since the dimensions of V and V^{**} are the same, we deduce immediately that $V \approx V^{**}$. This is not the point of Theorem 3.13. The point is that the *natural map* $v \to \overline{v}$ is an isomorphism. Because of this, V is said to be **algebraically reflexive**. Theorem 3.13 and Theorem 3.12 together imply that a vector space is algebraically reflexive if and only if it is finite-dimensional.

If V is finite-dimensional, it is customary to identify the double dual space V^{**} with V and to think of the elements of V^{**} simply as vectors in V. Let us consider a specific example to show how algebraic reflexivity fails in the infinite-dimensional case.

Example 3.4 Let V be the vector space over $\mathbb{Z}_2$ with basis

$$e_k = (0, \ldots, 0, 1, 0, \ldots)$$

where the 1 is in the kth position. Thus, V is the set of all infinite binary sequences with a finite number of 1's. Define the **order** $o(v)$ of any $v \in V$ to be the largest coordinate of v with value 1. Then $o(v) < \infty$ for all $v \in V$.

Consider the dual vectors e_k^*, defined (as usual) by

$$e_k^*(e_j) = \delta_{k,j}$$

For any $v \in V$, the evaluation functional $\overline{v}$ has the property that

$$\overline{v}(e_k^*) = e_k^*(v) = 0 \text{ if } k > o(v)$$

However, since the dual vectors e_k^* are linearly independent, there is a linear functional $f \in V^{**}$ for which

$$f(e_k^*) = 1$$

for all $k \geq 1$. Hence, f does not have the form $\overline{v}$ for any $v \in V$. This shows that the canonical map is not surjective and so V is not algebraically reflexive.□

Annihilators

The functions $f \in V^*$ are defined on vectors in V, but we may also define f on subsets M of V by letting

$$f(M) = \{f(v) \mid v \in M\}$$

Definition *Let M be a nonempty subset of a vector space V. The* **annihilator** *M^0 of M is*

$$M^0 = \{f \in V^* \mid f(M) = \{0\}\}$$

□

The term annihilator is quite descriptive, since M^0 consists of all linear functionals that *annihilate* (send to 0) every vector in M. It is not hard to see that M^0 is a subspace of V^*, even when M is not a subspace of V.

The basic properties of annihilators are contained in the following theorem.

Theorem 3.14

1) **(Order-reversing)** *If M and N are nonempty subsets of V, then*

$$M \subseteq N \Rightarrow N^0 \subseteq M^0$$

2) If $\dim(V) < \infty$, then for any nonempty subset M of V the natural map

$$\tau\colon \text{span}(M) \approx M^{00}$$

is an isomorphism from $\text{span}(M)$ onto M^{00}. In particular, if S is a subspace of V, then $S^{00} \approx S$.

3) If S and T are subspaces of V, then

$$(S \cap T)^0 = S^0 + T^0 \textit{ and } (S + T)^0 = S^0 \cap T^0$$

Proof. We leave proof of part 1) for the reader. For part 2), since

$$M^{00} = (\text{span}(M))^{00}$$

it is sufficient to prove that $\tau\colon S \approx S^{00}$ is an isomorphism, where S is a subspace of V. Now, we know that τ is a monomorphism, so it remains to prove that $\tau S = S^{00}$. If $s \in S$, then $\tau s = \overline{s}$ has the property that for all $f \in S^0$,

$$\overline{s}(f) = fs = 0$$

and so $\tau s = \overline{s} \in S^{00}$, which implies that $\tau S \subseteq S^{00}$. Moreover, if $\overline{v} \in S^{00}$, then for all $f \in S^0$ we have

$$f(v) = \overline{v}(f) = 0$$

and so every linear functional that annihilates S also annihilates v. But if $v \notin S$, then there is a linear functional $g \in V^*$ for which $g(S) = \{0\}$ and $g(v) \neq 0$. (We leave proof of this as an exercise.) Hence, $v \in S$ and so $\overline{v} = \tau v \in \tau S$ and so $S^{00} \subseteq \tau S$.

For part 3), it is clear that f annihilates $S + T$ if and only if f annihilates both S and T. Hence, $(S + T)^0 = S^0 \cap T^0$. Also, if $f = g + h \in S^0 + T^0$ where $g \in S^0$ and $h \in T^0$, then $g, h \in (S \cap T)^0$ and so $f \in (S \cap T)^0$. Thus,

$$S^0 + T^0 \subseteq (S \cap T)^0$$

For the reverse inclusion, suppose that $f \in (S \cap T)^0$. Write

$$V = S' \oplus (S \cap T) \oplus T' \oplus U$$

where $S = S' \oplus (S \cap T)$ and $T = (S \cap T) \oplus T'$. Define $g \in V^*$ by

$$g|_{S'} = f, \quad g|_{S \cap T} = f|_{S \cap T} = 0, \quad g|_{T'} = 0, \quad g|_U = f$$

and define $h \in V^*$ by

$$h|_{S'} = 0, \quad h|_{S \cap T} = f|_{S \cap T} = 0, \quad h|_{T'} = f, \quad h|_U = 0$$

It follows that $g \in T^0$, $h \in S^0$ and $g + h = f$.□

Annihilators and Direct Sums

Consider a direct sum decomposition

$$V = S \oplus T$$

Then any linear functional $f \in T^*$ can be extended to a linear functional $\overline{f}$ on V by setting $\overline{f}(S) = 0$. Let us call this **extension by** 0. Clearly, $\overline{f} \in S^0$ and it is easy to see that the extension by 0 map $f \to \overline{f}$ is an isomorphism from T^* to S^0, whose inverse is the restriction to T.

Theorem 3.15 *Let $V = S \oplus T$.*

a) The extension by 0 map is an isomorphism from T^* to S^0 and so

$$T^* \approx S^0$$

b) If V is finite-dimensional, then

$$\dim(S^0) = \operatorname{codim}_V(S) = \dim(V) - \dim(S) \qquad \square$$

Example 3.5 Part b) of Theorem 3.15 may fail in the infinite-dimensional case, since it may easily happen that $S^0 \approx V^*$. As an example, let V be the vector space over $\mathbb{Z}_2$ with a countably infinite ordered basis $\mathcal{B} = (e_1, e_2, \ldots)$. Let $S = \langle e_1 \rangle$ and $T = \langle e_2, e_3, \ldots \rangle$. It is easy to see that $S^0 \approx T^* \approx V^*$ and that $\dim(V^*) > \dim(V)$.□

The annihilator provides a way to describe the dual space of a direct sum.

Theorem 3.16 *A linear functional on the direct sum $V = S \oplus T$ can be written as a sum of a linear functional that annihilates S and a linear functional that annihilates T, that is,*

$$(S \oplus T)^* = S^0 \oplus T^0$$

Proof. Clearly $S^0 \cap T^0 = \{0\}$, since any functional that annihilates both S and T must annihilate $S \oplus T = V$. Hence, the sum $S^0 + T^0$ is direct. The rest follows from Theorem 3.14, since

$$V^* = \{0\}^0 = (S \cap T)^0 = S^0 + T^0 = S^0 \oplus T^0$$

Alternatively, since $\rho_T + \rho_S = \iota$ is the identity map, if $f \in V^*$, then we can write

$$f = f \circ (\rho_T + \rho_S) = (f \circ \rho_T) + (f \circ \rho_S) \in S^0 \oplus T^0$$

and so $V^* = S^0 \oplus T^0$.□

Operator Adjoints

If $\tau \in \mathcal{L}(V, W)$, then we may define a map $\tau^\times : W^* \to V^*$ by

$$\tau^\times(f) = f \circ \tau = f\tau$$

for $f \in W^*$. (We will write composition as juxtaposition.) Thus, for any $v \in V$,

$$[\tau^\times(f)](v) = f(\tau v)$$

The map $\tau^\times$ is called the **operator adjoint** of τ and can be described by the phrase "apply τ first."

Theorem 3.17 (Properties of the Operator Adjoint)

1) For $\tau, \sigma \in \mathcal{L}(V, W)$ and $a, b \in F$,

$$(a\tau + b\sigma)^\times = a\tau^\times + b\sigma^\times$$

2) For $\sigma \in \mathcal{L}(V, W)$ and $\tau \in \mathcal{L}(W, U)$,

$$(\tau\sigma)^\times = \sigma^\times \tau^\times$$

3) For any invertible $\tau \in \mathcal{L}(V)$,

$$(\tau^{-1})^\times = (\tau^\times)^{-1}$$

Proof. Proof of part 1) is left for the reader. For part 2), we have for all $f \in U^*$,

$$(\tau\sigma)^\times(f) = f(\tau\sigma) = \sigma^\times(f\tau) = \sigma^\times(\tau^\times(f)) = (\sigma^\times\tau^\times)(f)$$

Part 3) follows from part 2) and

$$\tau^\times(\tau^{-1})^\times = (\tau^{-1}\tau)^\times = \iota^\times = \iota$$

and in the same way, $(\tau^{-1})^\times \tau^\times = \iota$. Hence $(\tau^{-1})^\times = (\tau^\times)^{-1}$.□

If $\tau \in \mathcal{L}(V, W)$, then $\tau^\times \in \mathcal{L}(W^*, V^*)$ and so $\tau^{\times\times} \in \mathcal{L}(V^{**}, W^{**})$. Of course, $\tau^{\times\times}$ is not equal to τ. However, in the finite-dimensional case, if we use the natural maps to identify V^{**} with V and W^{**} with W, then we can think of $\tau^{\times\times}$

as being in $\mathcal{L}(V,W)$. Using these identifications, we do have equality in the finite-dimensional case.

Theorem 3.18 *Let V and W be finite-dimensional and let $\tau \in \mathcal{L}(V,W)$. If we identify V^{**} with V and W^{**} with W using the natural maps, then $\tau^{\times\times}$ is identified with τ.*

Proof. For any $x \in V$ let the corresponding element of V^{**} be denoted by $\overline{x}$ and similarly for W. Then before making any identifications, we have for $v \in V$,

$$\tau^{\times\times}(\overline{v})(f) = \overline{v}[\tau^{\times}(f)] = \overline{v}(f\tau) = f(\tau v) = \overline{\tau v}(f)$$

for all $f \in W^*$ and so

$$\tau^{\times\times}(\overline{v}) = \overline{\tau v} \in W^{**}$$

Therefore, using the canonical identifications for both V^{**} and W^{**} we have

$$\tau^{\times\times}(v) = \tau v$$

for all $v \in V$.□

The next result describes the kernel and image of the operator adjoint.

Theorem 3.19 *Let $\tau \in \mathcal{L}(V,W)$. Then*

1) $\ker(\tau^{\times}) = \operatorname{im}(\tau)^0$

2) $\operatorname{im}(\tau^{\times}) = \ker(\tau)^0$

Proof. For part 1),

$$\begin{aligned}\ker(\tau^{\times}) &= \{f \in W^* \mid \tau^{\times}(f) = 0\}\\ &= \{f \in W^* \mid f(\tau V) = \{0\}\}\\ &= \{f \in W^* \mid f(\operatorname{im}(\tau)) = \{0\}\}\\ &= \operatorname{im}(\tau)^0\end{aligned}$$

For part 2), if $f = g\tau = \tau^{\times}g \in \operatorname{im}(\tau^{\times})$, then $\ker(\tau) \subseteq \ker(f)$ and so $f \in \ker(\tau)^0$.

For the reverse inclusion, let $f \in \ker(\tau)^0 \subseteq V^*$. We wish to show that $f = \tau^{\times}g = g\tau$ for some $g \in W^*$. On $K = \ker(\tau)$, there is no problem since f and $\tau^{\times}g = g\tau$ agree on K for any $g \in W^*$. Let S be a complement of $\ker(\tau)$. Then τ maps a basis $\mathcal{B} = \{b_i \mid i \in I\}$ for S to a linearly independent set

$$\tau\mathcal{B} = \{\tau b_i \mid i \in I\}$$

in W and so we can define $g \in W^*$ on $\tau\mathcal{B}$ by setting

$$g(\tau b_i) = fb_i$$

and extending to all of W. Then $f = g\tau = \tau^{\times}g$ on $\mathcal{B}$ and therefore on S. Thus, $f = \tau^{\times}g \in \operatorname{im}(\tau^{\times})$.□

Corollary 3.20 *Let $\tau \in \mathcal{L}(V,W)$, where V and W are finite-dimensional. Then* $\operatorname{rk}(\tau) = \operatorname{rk}(\tau^\times)$.$\square$

In the finite-dimensional case, τ and $\tau^\times$ can both be represented by matrices. Let

$$\mathcal{B} = (b_1, \ldots, b_n) \text{ and } \mathcal{C} = (c_1, \ldots, c_m)$$

be ordered bases for V and W, respectively, and let

$$\mathcal{B}^* = (b_1^*, \ldots, b_n^*) \text{ and } \mathcal{C}^* = (c_1^*, \ldots, c_m^*)$$

be the corresponding dual bases. Then

$$([\tau]_{\mathcal{B},\mathcal{C}})_{i,j} = ([\tau b_j]_{\mathcal{C}})_i = c_i^*[\tau b_j]$$

and

$$([\tau^\times]_{\mathcal{C}^*,\mathcal{B}^*})_{i,j} = ([\tau^\times(c_j^*)]_{\mathcal{B}^*})_i = b_i^{**}[\tau^\times(c_j^*)] = \tau^\times(c_j^*)(b_i) = c_j^*(\tau b_i)$$

Comparing the last two expressions we see that they are the same except that the roles of i and j are reversed. Hence, the matrices in question are transposes.

Theorem 3.21 *Let $\tau \in \mathcal{L}(V,W)$, where V and W are finite-dimensional. If $\mathcal{B}$ and $\mathcal{C}$ are ordered bases for V and W, respectively, and $\mathcal{B}^*$ and $\mathcal{C}^*$ are the corresponding dual bases, then*

$$[\tau^\times]_{\mathcal{C}^*,\mathcal{B}^*} = ([\tau]_{\mathcal{B},\mathcal{C}})^t$$

In words, the matrices of τ and its operator adjoint $\tau^\times$ are transposes of one another.$\square$

Exercises

1. If V is infinite-dimensional and S is an infinite-dimensional subspace, must the dimension of V/S be finite? Explain.
2. Prove the correspondence theorem.
3. Prove the first isomorphism theorem.
4. Complete the proof of Theorem 3.9.
5. Let S be a subspace of V. Starting with a basis $\{s_1, \ldots, s_k\}$ for S, how would you find a basis for V/S?
6. Use the first isomorphism theorem to prove the rank-plus-nullity theorem
$$\operatorname{rk}(\tau) + \operatorname{null}(\tau) = \dim(V)$$
for $\tau \in \mathcal{L}(V,W)$ and $\dim(V) < \infty$.
7. Let $\tau \in \mathcal{L}(V)$ and suppose that S is a subspace of V. Define a map $\tau' : V/S \to V/S$ by

$$\tau'(v+S)=\tau v+S$$

When is τ' well-defined? If τ' is well-defined, is it a linear transformation? What are $\operatorname{im}(\tau')$ and $\ker(\tau')$?

8. Show that for any nonzero vector $v\in V$, there exists a linear functional $f\in V^*$ for which $f(v)\neq 0$.
9. Show that a vector $v\in V$ is zero if and only if $f(v)=0$ for all $f\in V^*$.
10. Let S be a proper subspace of a finite-dimensional vector space V and let $v\in V\setminus S$. Show that there is a linear functional $f\in V^*$ for which $f(v)=1$ and $f(s)=0$ for all $s\in S$.
11. Find a vector space V and decompositions

$$V=A\oplus B=C\oplus D$$

with $A\approx C$ but $B\not\approx D$. Hence, $A\approx C$ does *not* imply that $A^c\approx C^c$.

12. Find isomorphic vectors spaces V and W with

$$V=S\oplus B \text{ and } W=S\oplus D$$

but $B\not\approx D$. Hence, $V\approx W$ does *not* imply that $V/S\approx W/S$.

13. Let V be a vector space with

$$V=S_1\oplus T_1=S_2\oplus T_2$$

Prove that if S_1 and S_2 have finite codimension in V, then so does $S_1\cap S_2$ and

$$\operatorname{codim}(S_1\cap S_2)\le\dim(T_1)+\dim(T_2)$$

14. Let V be a vector space with

$$V=S_1\oplus T_1=S_2\oplus T_2$$

Suppose that S_1 and S_2 have finite codimension. Hence, by the previous exercise, so does $S_1\cap S_2$. Find a direct sum decomposition $V=W\oplus X$ for which (1) W has finite codimension, (2) $W\subseteq S_1\cap S_2$ and (3) $X\supseteq T_1+T_2$.

15. Let $\mathcal{B}$ be a basis for an infinite-dimensional vector space V and define, for all $b\in\mathcal{B}$, the map $b'\in V^*$ by $b'(c)=1$ if $c=b$ and 0 otherwise, for all $c\in\mathcal{B}$. Does $\{b'\mid b\in\mathcal{B}\}$ form a basis for V^*? What do you conclude about the concept of a dual basis?
16. Prove that if S and T are subspaces of V, then $(S\oplus T)^*\approx S^*\boxplus T^*$.
17. Prove that $0^\times=0$ and $\iota^\times=\iota$ where 0 is the zero linear operator and ι is the identity.
18. Let S be a subspace of V. Prove that $(V/S)^*\approx S^0$.
19. Verify that
    a) $(\tau+\sigma)^\times=\tau^\times+\sigma^\times$ for $\tau,\sigma\in\mathcal{L}(V,W)$.
    b) $(r\tau)^\times=r\tau^\times$ for any $r\in F$ and $\tau\in\mathcal{L}(V,W)$
20. Let $\tau\in\mathcal{L}(V,W)$, where V and W are finite-dimensional. Prove that $\operatorname{rk}(\tau)=\operatorname{rk}(\tau^\times)$.

Chapter 4
Modules I: Basic Properties

Motivation

Let V be a vector space over a field F and let $\tau \in \mathcal{L}(V)$. Then for any polynomial $p(x) \in F[x]$, the operator $p(\tau)$ is well-defined. For instance, if $p(x) = 1 + 2x + x^3$, then

$$p(\tau) = \iota + 2\tau + \tau^3$$

where ι is the identity operator and τ^3 is the threefold composition $\tau \circ \tau \circ \tau$.

Thus, using the operator τ we can define the product of a polynomial $p(x) \in F[x]$ and a vector $v \in V$ by

$$p(x)v = p(\tau)(v) \tag{4.1}$$

This product satisfies the usual properties of scalar multiplication, namely, for all $r(x),\ s(x) \in F[x]$ and $u, v \in V$,

$$\begin{aligned} r(x)(u+v) &= r(x)u + r(x)v \\ (r(x)+s(x))u &= r(x)u + s(x)u \\ [r(x)s(x)]u &= r(x)[s(x)u] \\ 1u &= u \end{aligned}$$

Thus, for a fixed $\tau \in \mathcal{L}(V)$, we can think of V as being endowed with the operations of addition and multiplication of an element of V by a *polynomial* in $F[x]$. However, since $F[x]$ is not a field, these two operations do not make V into a vector space. Nevertheless, the situation in which the scalars form a ring but not a field is extremely important, not only in this context but in many others.

Modules

Definition *Let R be a commutative ring with identity, whose elements are called* **scalars***. An R-***module** *(or a* **module over** *R) is a nonempty set M,* *together with two operations. The first operation, called* **addition** *and denoted by* $+$, *assigns to each pair* $(u,v) \in M \times M$, *an element* $u+v \in M$. *The second operation, denoted by juxtaposition, assigns to each pair* $(r,v) \in R \times M$, *an element* $rv \in M$. *Furthermore, the following properties must hold:*

1) M *is an abelian group under addition.*
2) *For all* $r,s \in R$ *and* $u,v \in M$

$$\begin{aligned} r(u+v) &= ru+rv \\ (r+s)u &= ru+su \\ (rs)u &= r(su) \\ 1u &= u \end{aligned}$$

The ring R *is called the* **base ring** *of* M.□

Note that vector spaces are just special types of modules: a vector space is a module over a field.

When we turn in a later chapter to the study of the structure of a linear transformation $\tau \in \mathcal{L}(V)$, we will think of V as having the structure of a vector space over F as well as a module over $F[x]$ and we will use the notation V_τ. Put another way, V_τ is an abelian group under addition, with two scalar multiplications—one whose scalars are elements of F and one whose scalars are polynomials over F. This viewpoint will be of tremendous benefit for the study of τ. For now, we concentrate only on modules.

Example 4.1

1) If R is a ring, the set R^n of all ordered n-tuples whose components lie in R is an R-module, with addition and scalar multiplication defined componentwise (just as in F^n),

$$(a_1,\ldots,a_n)+(b_1,\ldots,b_n)=(a_1+b_1,\ldots,a_n+b_n)$$

and

$$r(a_1,\ldots,a_n)=(ra_1,\ldots,ra_n)$$

for a_i, b_i, $r \in R$. For example, $\mathbb{Z}^n$ is the $\mathbb{Z}$-module of all ordered n-tuples of integers.

2) If R is a ring, the set $\mathcal{M}_{m,n}(R)$ of all matrices of size $m \times n$ is an R-module, under the usual operations of matrix addition and scalar multiplication over R. Since R is a ring, we can also take the product of matrices in $\mathcal{M}_{m,n}(R)$. One important example is $R=F[x]$, whence $\mathcal{M}_{m,n}(F[x])$ is the $F[x]$-module of all $m \times n$ matrices whose entries are polynomials.
3) Any commutative ring R with identity is a module over itself, that is, R is an R-module. In this case, scalar multiplication is just multiplication by

elements of R, that is, scalar multiplication is the ring multiplication. The defining properties of a ring imply that the defining properties of the R-module R are satisfied. We shall use this example many times in the sequel.□

Importance of the Base Ring

Our definition of a module requires that the ring R of scalars be commutative. Modules over noncommutative rings can exhibit quite a bit more unusual behavior than modules over commutative rings. Indeed, as one would expect, the general behavior of R-modules improves as we impose more structure on the base ring R. If we impose the very strict structure of a field, the result is the very well behaved vector space.

To illustrate, we will give an example of a module over a *noncommutative* ring that has a basis of size n for every integer $n > 0$! As another example, if the base ring is an integral domain, then whenever $v_1, \ldots, v_n$ are linearly independent over R so are $rv_1, \ldots, rv_n$ for any nonzero $r \in R$. This can fail when R is not an integral domain.

We will also consider the property on the base ring R that all of its ideals are finitely generated. In this case, any finitely generated R-module M has the property that all of its submodules are also finitely generated. This property of R-modules fails if R does not have the stated property.

When R is a principal ideal domain (such as $\mathbb{Z}$ or $F[x]$), each of its ideals is generated by a single element. In this case, the R-modules are "reasonably" well behaved. For instance, in general, a module may have a basis and yet possess a submodule that has no basis. However, if R is a principal ideal domain, this cannot happen.

Nevertheless, even when R is a principal ideal domain, R-modules are less well behaved than vector spaces. For example, there are modules over a principal ideal domain that do not have any linearly independent elements. Of course, such modules cannot have a basis.

Submodules

Many of the basic concepts that we defined for vector spaces can also be defined for modules, although their properties are often quite different. We begin with submodules.

Definition *A* **submodule** *of an R-module M is a nonempty subset S of M that is an R-module in its own right, under the operations obtained by restricting the operations of M to S. We write $S \leq M$ to denote the fact that S is a submodule of M.*□

Theorem 4.1 *A nonempty subset S of an R-module M is a submodule if and only if it is closed under the taking of linear combinations, that is,*

$$r, s \in R, u, v \in S \Rightarrow ru + sv \in S$$ □

Theorem 4.2 *If S and T are submodules of M, then $S \cap T$ and $S + T$ are also submodules of M.*□

We have remarked that a commutative ring R with identity is a module over itself. As we will see, this type of module provides some good examples of non-vector-space-like behavior.

When we think of a ring R as an R-module rather than as a ring, multiplication is treated as *scalar* multiplication. This has some important implications. In particular, if S is a submodule of R, then it is closed under scalar multiplication, which means that it is closed under multiplication by *all* elements of the ring R. In other words, S is an ideal of the ring R. Conversely, if $\mathcal{I}$ is an ideal of the ring R, then $\mathcal{I}$ is also a submodule of the module R. Hence, *the submodules of the R-module R are precisely the ideals of the ring R.*

Spanning Sets

The concept of spanning set carries over to modules as well.

Definition *The* **submodule spanned** (*or* **generated**) *by a subset S of a module M is the set of all* **linear combinations** *of elements of S:*

$$\langle\langle S \rangle\rangle = \{r_1 v_1 + \cdots + r_n v_n \mid r_i \in R, v_i \in S, n \geq 1\}$$

A subset $S \subseteq M$ is said to **span** *M or* **generate** *M if $M = \langle\langle S \rangle\rangle$.*□

We use a double angle bracket notation for the submodule generated by a set because when we study the F-vector space/$F[x]$-module V_τ, we will need to make a distinction between the subspace $\langle v \rangle = Fv$ generated by $v \in V$ and the submodule $\langle\langle v \rangle\rangle = F[x]v$ generated by v.

One very important point to note is that if a nontrivial linear combination of the elements $v_1, \ldots, v_n$ in an R-module M is 0,

$$r_1 v_1 + \cdots + r_n v_n = 0$$

where not all of the coefficients are 0, then we *cannot* conclude, as we could in a vector space, that one of the elements v_i is a linear combination of the others. After all, this involves dividing by one of the coefficients, which may not be possible in a ring. For instance, for the $\mathbb{Z}$-module $\mathbb{Z} \times \mathbb{Z}$ we have

$$2(3,6) - 3(2,4) = (0,0)$$

but neither $(3,6)$ nor $(2,4)$ is an integer multiple of the other.

The following simple submodules play a special role in the theory.

Definition *Let M be an R-module. A submodule of the form*

$$\langle\langle v \rangle\rangle = Rv = \{rv \mid r \in R\}$$

for $v \in M$ is called the **cyclic submodule** *generated by v.* □

Of course, any finite-dimensional vector space is the direct sum of cyclic submodules, that is, one-dimensional subspaces. One of our main goals is to show that a finitely generated module over a principal ideal domain has this property as well.

Definition *An R-module M is said to be* **finitely generated** *if it contains a finite set that generates M. More specifically, M is* **n-generated** *if it has a generating set of size n (although it may have a smaller generating set as well).* □

Of course, a vector space is finitely generated if and only if it has a finite basis, that is, if and only if it is finite-dimensional. For modules, life is more complicated. The following is an example of a finitely generated module that has a submodule that is not finitely generated.

Example 4.2 Let R be the ring $F[x_1, x_2, \ldots]$ of all polynomials in infinitely many variables over a field F. It will be convenient to use X to denote $x_1, x_2, \ldots$ and write a polynomial in R in the form $p(X)$. (Each polynomial in R, being a finite sum, involves only finitely many variables, however.) Then R is an R-module and as such, is finitely generated by the identity element $p(X) = 1$.

Now consider the submodule S of all polynomials with zero constant term. This module is generated by the variables themselves,

$$S = \langle\langle x_1, x_2, \ldots \rangle\rangle$$

However, S is not finitely generated. To see this, suppose that $G = \{p_1, \ldots, p_n\}$ is a finite generating set for S. Choose a variable x_k that does not appear in any of the polynomials in G. Then no linear combination of the polynomials in G can be equal to x_k. For if

$$x_k = \sum_{i=1}^{n} a_i(X) p_i(X)$$

then let $a_i(X) = x_k q_i(X) + r_i(X)$ where $r_i(X)$ does not involve x_k. This gives

$$\begin{aligned} x_k &= \sum_{i=1}^{n} [x_k q_i(X) + r_i(X)] p_i(X) \\ &= x_k \sum_{i=1}^{n} q_i(X) p_i(X) + \sum_{i=1}^{n} r_i(X) p_i(X) \end{aligned}$$

The last sum does not involve x_k and so it must equal 0. Hence, the first sum must equal 1, which is not possible since $p_i(X)$ has no constant term.□

Linear Independence

The concept of linear independence also carries over to modules.

Definition *A subset S of an R-module M is* **linearly independent** *if for any distinct $v_1, \ldots, v_n \in S$ and $r_1, \ldots, r_n \in R$, we have*

$$r_1 v_1 + \cdots + r_n v_n = 0 \Rightarrow r_i = 0 \text{ for all } i$$

A set S that is not linearly independent is **linearly dependent**.□

It is clear from the definition that any subset of a linearly independent set is linearly independent.

Recall that in a vector space, a set S of vectors is linearly dependent if and only if some vector in S is a linear combination of the other vectors in S. For arbitrary modules, this is not true.

Example 4.3 Consider $\mathbb{Z}$ as a $\mathbb{Z}$-module. The elements $2, 3 \in \mathbb{Z}$ are linearly dependent, since

$$3(2) - 2(3) = 0$$

but neither one is a linear combination (i.e., integer multiple) of the other.□

The problem in the previous example (as noted earlier) is that

$$r_1 v_1 + \cdots + r_n v_n = 0$$

implies that

$$r_1 v_1 = -r_2 v_2 - \cdots - r_n v_n$$

but in general, we cannot divide both sides by r_1, since it may not have a multiplicative inverse in the ring R.

Torsion Elements

In a vector space V over a field F, singleton sets $\{v\}$ where $v \neq 0$ are linearly independent. Put another way, $r \neq 0$ and $v \neq 0$ imply $rv \neq 0$. However, in a module, this need not be the case.

Example 4.4 The abelian group $\mathbb{Z}_n = \{0, 1, \ldots, n-1\}$ is a $\mathbb{Z}$-module, with scalar multiplication defined by $za = (z \cdot a) \bmod n$, for all $z \in \mathbb{Z}$ and $a \in \mathbb{Z}_n$. However, since $na = 0$ for all $a \in \mathbb{Z}_n$, no singleton set $\{a\}$ is linearly independent. Indeed, $\mathbb{Z}_n$ has no linearly independent sets.□

This example motivates the following definition.

Definition *Let M be an R-module. A nonzero element $v \in M$ for which $rv = 0$ for some nonzero $r \in R$ is called a* **torsion element** *of M. A module that has no nonzero torsion elements is said to be* **torsion-free**. *If all elements of M are torsion elements, then M is a* **torsion module**. *The set of all torsion elements of M, together with the zero element, is denoted by M_{tor}.*□

If M is a module over an *integral domain*, it is not hard to see that M_{tor} is a submodule of M and that M/M_{tor} is torsion-free. (We will define quotient modules shortly: they are defined in the same way as for vector spaces.)

Annihilators

Closely associated with the notion of a torsion element is that of an annihilator.

Definition *Let M be an R-module. The* **annihilator** *of an element $v \in M$ is*

$$\operatorname{ann}(v) = \{r \in R \mid rv = 0\}$$

and the **annihilator** *of a submodule N of M is*

$$\operatorname{ann}(N) = \{r \in R \mid rN = \{0\}\}$$

where $rN = \{rv \mid v \in N\}$. Annihilators are also called **order ideals**.□

It is easy to see that $\operatorname{ann}(v)$ and $\operatorname{ann}(N)$ are ideals of R. Clearly, $v \in M$ is a torsion element if and only if $\operatorname{ann}(v) \neq \{0\}$. Also, if A and B are submodules of M, then

$$A \leq B \quad \Rightarrow \quad \operatorname{ann}(B) \leq \operatorname{ann}(A)$$

(note the reversal of order).

Let $M = \langle\langle u_1, \ldots, u_n \rangle\rangle$ be a finitely generated module over an integral domain R and assume that each of the generators u_i is torsion, that is, for each i, there is a nonzero $a_i \in \operatorname{ann}(u_i)$. Then, the nonzero product $a = a_1 \cdots a_n$ annihilates each generator of M and therefore every element of M, that is, $a \in \operatorname{ann}(M)$. This shows that $\operatorname{ann}(M) \neq \{0\}$. On the other hand, this may fail if R is not an integral domain. Also, there are torsion modules whose annihilators are trivial. (We leave verification of these statements as an exercise.)

Free Modules

The definition of a basis for a module parallels that of a basis for a vector space.

Definition *Let M be an R-module. A subset $\mathcal{B}$ of M is a* **basis** *if $\mathcal{B}$ is linearly independent and spans M. An R-module M is said to be* **free** *if $M = \{0\}$ or if M has a basis. If $\mathcal{B}$ is a basis for M, we say that M is* **free on** *$\mathcal{B}$.*□

We have the following analog of part of Theorem 1.7.

Theorem 4.3 *A subset $\mathcal{B}$ of a module M is a basis if and only if every nonzero $v \in M$ is an essentially unique linear combination of the vectors in $\mathcal{B}$.*□

In a vector space, a set of vectors is a basis if and only if it is a minimal spanning set, or equivalently, a maximal linearly independent set. For modules, the following is the best we can do in general. We leave proof to the reader.

Theorem 4.4 *Let $\mathcal{B}$ be a basis for an R-module M. Then*
1) $\mathcal{B}$ is a minimal spanning set.
2) $\mathcal{B}$ is a maximal linearly independent set.□

The $\mathbb{Z}$-module $\mathbb{Z}_n$ has no basis since it has no linearly independent sets. But since the entire module is a spanning set, we deduce that a minimal spanning set need not be a basis. In the exercises, the reader is asked to give an example of a module M that has a finite basis, but with the property that not every spanning set in M contains a basis and not every linearly independent set in M is contained in a basis. It follows in this case that a maximal linearly independent set need not be a basis.

The next example shows that even free modules are not very much like vector spaces. It is an example of a free module that has a submodule that is not free.

Example 4.5 The set $\mathbb{Z} \times \mathbb{Z}$ is a free module over itself, using componentwise scalar multiplication

$$(n, m)(a, b) = (na, mb)$$

with basis $\{(1, 1)\}$. But the submodule $\mathbb{Z} \times \{0\}$ is not free since it has no linearly independent elements and hence no basis.□

Theorem 2.2 says that a linear transformation can be defined by specifying its values arbitrarily on a basis. The same is true for *free* modules.

Theorem 4.5 *Let M and N be R-modules where M is free with basis $\mathcal{B}=\{b_i \mid i \in I\}$. Then we can define a unique R-map $\tau: M \to N$ by specifying the values of τb_i arbitrarily for all $b_i \in \mathcal{B}$ and then extending τ to M by linearity, that is,*

$$\tau(a_1 v_1 + \cdots + a_n v_n) = a_1 \tau v_1 + \cdots + a_n \tau v_n$$ □

Homomorphisms

The term *linear transformation* is special to vector spaces. However, the concept applies to most algebraic structures.

Definition *Let M and N be R-modules. A function $\tau: M \to N$ is an* **R-homomorphism** *or* **R-map** *if it preserves the module operations, that is,*

$$\tau(ru + sv) = r\tau(u) + s\tau(v)$$

for all $r, s \in R$ and $u, v \in M$. The set of all R-homomorphisms from M to N is denoted by $\hom_R(M, N)$. *The following terms are also employed:*

1) *An R-***endomorphism** *is an R-homomorphism from M to itself.*
2) *An R-***monomorphism** *or R-***embedding** *is an injective R-homomorphism.*
3) *An R-***epimorphism** *is a surjective R-homomorphism.*
4) *An R-***isomorphism** *is a bijective R-homomorphism.*□

It is easy to see that $\hom_R(M, N)$ is itself an R-module under addition of functions and scalar multiplication defined by

$$(r\tau)(v) = r(\tau v) = \tau(rv)$$

Theorem 4.6 *Let $\tau \in \hom_R(M, N)$. The kernel and image of τ, defined as for linear transformations by*

$$\ker(\tau) = \{v \in M \mid \tau v = 0\}$$

and

$$\operatorname{im}(\tau) = \{\tau v \mid v \in M\}$$

are submodules of M and N, respectively. Moreover, τ is a monomorphism if and only if $\ker(\tau) = \{0\}$.□

If N is a submodule of the R-module M, then the map $j: N \to M$ defined by $j(v) = v$ is evidently an R-monomorphism, called **injection** of N into M.

Quotient Modules

The procedure for defining quotient modules is the same as that for defining quotient vector spaces. We summarize in the following theorem.

Theorem 4.7 *Let S be a submodule of an R-module M. The binary relation*

$$u \equiv v \Leftrightarrow u - v \in S$$

is an equivalence relation on M, whose equivalence classes are the **cosets**

$$v + S = \{v + s \mid s \in S\}$$

of S in M. The set M/S of all cosets of S in M, called the **quotient module** *of M* **modulo** *S, is an R-module under the well-defined operations*

$$\begin{aligned}(u + S) + (v + S) &= (u + v) + S \\ r(u + S) &= ru + S\end{aligned}$$

The zero element in M/S is the coset $0 + S = S$.$\square$

One question that immediately comes to mind is whether a quotient module of a free module must be free. As the next example shows, the answer is no.

Example 4.6 As a module over itself, $\mathbb{Z}$ is free on the set $\{1\}$. For any $n > 0$, the set $\mathbb{Z}n = \{zn \mid z \in \mathbb{Z}\}$ is a free cyclic submodule of $\mathbb{Z}$, but the quotient $\mathbb{Z}$-module $\mathbb{Z}/\mathbb{Z}n$ is isomorphic to $\mathbb{Z}_n$ via the map

$$\tau(u + \mathbb{Z}n) = u \bmod n$$

and since $\mathbb{Z}_n$ is not free as a $\mathbb{Z}$-module, neither is $\mathbb{Z}/\mathbb{Z}n$.$\square$

The Correspondence and Isomorphism Theorems

The correspondence and isomorphism theorems for vector spaces have analogs for modules.

Theorem 4.8 (The correspondence theorem) *Let S be a submodule of M. Then the function that assigns to each intermediate submodule $S \subseteq T \subseteq M$ the quotient submodule T/S of M/S is an order-preserving (with respect to set inclusion) one-to-one correspondence between submodules of M containing S and all submodules of M/S.*$\square$

Theorem 4.9 (The first isomorphism theorem) *Let $\tau: M \to N$ be an R-homomorphism. Then the map $\tau': M/\ker(\tau) \to N$ defined by*

$$\tau'(v + \ker(\tau)) = \tau v$$

is an R-embedding and so

$$\frac{M}{\ker(\tau)} \approx \operatorname{im}(\tau) \qquad \square$$

Theorem 4.10 (The second isomorphism theorem) *Let M be an R-module and let S and T be submodules of M. Then*

$$\frac{S+T}{T} \approx \frac{S}{S\cap T}$$ □

Theorem 4.11 (The third isomorphism theorem) *Let M be an R-module and suppose that $S \subseteq T$ are submodules of M. Then*

$$\frac{M/S}{T/S} \approx \frac{M}{T}$$ □

Direct Sums and Direct Summands

The definition of direct sum of a family of submodules is a direct analog of the definition for vector spaces.

Definition *The* **external direct sum** *of R-modules $M_1,\ldots,M_n$, denoted by*

$$M = M_1 \boxplus \cdots \boxplus M_n$$

is the r-module whose elements are ordered n-tuples

$$M = \{(v_1,\ldots,v_n) \mid v_i \in M_i, i = 1,\ldots,n\}$$

with componentwise operations

$$(u_1,\ldots,u_n) + (v_1,\ldots,v_n) = (u_1+v_1,\ldots,u_n+v_n)$$

and

$$r(v_1,\ldots,v_n) = (rv_1,\ldots,rv_n)$$

for $r \in R$.□

We leave it to the reader to formulate the definition of external direct sums and products for arbitrary families of modules, in direct analogy with the case of vector spaces.

Definition *An R-module M is the* **(internal) direct sum** *of a family $\mathcal{F} = \{S_i \mid i \in I\}$ of submodules of M, written*

$$M = \bigoplus \mathcal{F} \quad \text{or} \quad M = \bigoplus_{i\in I} S_i$$

if the following hold:

1) **(Join of the family)** *M is the sum (join) of the family $\mathcal{F}$:*

$$V = \sum_{i\in I} S_i$$

2) **(Independence of the family)** *For each* $i \in I$,

$$S_i \cap \left(\sum_{j \neq i} S_j \right) = \{0\}$$

In this case, each S_i *is called a* **direct summand** *of* M. *If* $\mathcal{F} = \{S_1, \ldots, S_n\}$ *is a finite family, the direct sum is often written*

$$M = S_1 \oplus \cdots \oplus S_n$$

Finally, if $M = S \oplus T$, *then* S *is said to be* **complemented** *and* T *is called a* **complement** *of* S *in* M.$\square$

As with vector spaces, we have the following useful characterization of direct sums.

Theorem 4.12 *Let* $\mathcal{F} = \{S_i \mid i \in I\}$ *be a family of distinct submodules of an* R-*module* M. *The following are equivalent:*

1) **(Independence of the family)** *For each* $i \in I$,

$$S_i \cap \left(\sum_{j \neq i} S_j \right) = \{0\}$$

2) **(Uniqueness of expression for 0)** *The zero element* 0 *cannot be written as a sum of nonzero elements from distinct submodules in* $\mathcal{F}$.
3) **(Uniqueness of expression)** *Every nonzero* $v \in M$ *has a unique, except for order of terms, expression as a sum*

$$v = s_1 + \cdots + s_n$$

of nonzero elements from distinct submodules in $\mathcal{F}$.

Hence, a sum

$$M = \sum_{i \in I} S_i$$

is direct if and only if any one of 1)–3) holds.$\square$

In the case of vector spaces, every subspace is a direct summand, that is, every subspace has a complement. However, as the next example shows, this is not true for modules.

Example 4.7 The set $\mathbb{Z}$ of integers is a $\mathbb{Z}$-module. Since the submodules of $\mathbb{Z}$ are precisely the ideals of the ring $\mathbb{Z}$ and since $\mathbb{Z}$ is a principal ideal domain, the submodules of $\mathbb{Z}$ are the sets

$$\langle\langle n \rangle\rangle = \mathbb{Z}n = \{zn \mid z \in \mathbb{Z}\}$$

Hence, any two nonzero proper submodules of $\mathbb{Z}$ have nonzero intersection, for if $n \neq m > 0$, then

$$\mathbb{Z}n \cap \mathbb{Z}m = \mathbb{Z}k$$

where $k = \operatorname{lcm}\{n, m\}$. It follows that the only complemented submodules of $\mathbb{Z}$ are $\mathbb{Z}$ and $\{0\}$.$\square$

In the case of vector spaces, there is an intimate connection between subspaces and quotient spaces, as we saw in Theorem 3.6. The problem we face in generalizing this to modules is that not all submodules are complemented. However, this is the only problem.

Theorem 4.13 *Let S be a complemented submodule of M. All complements of S are isomorphic to M/S and hence to each other.*

Proof. For any complement T of S, the first isomorphism theorem applied to the projection $\rho_{T,S}\colon M \to T$ gives $T \approx M/S$.$\square$

Direct Summands and Extensions of Isomorphisms

Direct summands play a role in questions relating to whether certain module homomorphisms $\sigma\colon N \to M_1$ can be extended from a submodule $N \leq M$ to the full module M. The discussion will be a bit simpler if we restrict attention to epimorphisms.

If $M = N \oplus H$, then a module epimorphism $\sigma\colon N \to M_1$ can be extended to an epimorphism $\overline{\sigma}\colon M \to M_1$ simply by sending the elements of H to zero, that is, by setting

$$\overline{\sigma}(n + h) = \sigma n$$

This is easily seen to be an R-map with

$$\ker(\overline{\sigma}) = \ker(\sigma) \oplus H$$

Moreover, if τ is another extension of σ with the same kernel as $\overline{\sigma}$, then τ and $\overline{\sigma}$ agree on H as well as on N, whence $\tau = \overline{\sigma}$. Thus, there is a *unique* extension of σ with kernel $\ker(\sigma) \oplus H$.

Now suppose that $\sigma\colon N \approx M_1$ is an isomorphism. If N is complemented, that is, if

$$G = N \oplus H$$

then we have seen that there is a *unique* extension $\overline{\sigma}$ of σ for which $\ker(\overline{\sigma}) = H$. Thus, the correspondence

$$H \mapsto \overline{\sigma}, \quad \text{where } \ker(\overline{\sigma}) = H$$

from complements of N to extensions of σ is an injection. To see that this correspondence is a bijection, if $\overline{\sigma}\colon M \to M_1$ is an extension of σ, then

$$M = N \oplus \ker(\overline{\sigma})$$

To see this, we have

$$N \cap \ker(\overline{\sigma}) = \ker(\sigma) = \{0\}$$

and if $a \in M$, then there is a $b \in N$ for which $\sigma b = \overline{\sigma}a$ and so

$$\overline{\sigma}(a - b) = \overline{\sigma}a - \sigma b = 0$$

Thus,

$$a = b + (a - b) \in N + \ker(\overline{\sigma})$$

which shows that $\ker(\overline{\sigma})$ is a complement of N.

Theorem 4.14 *Let M and M_1 be R-modules and let $N \le M$.*

1) *If $M = N \oplus H$, then any R-epimorphism $\sigma: N \to M_1$ has a unique extension $\overline{\sigma}: M \to M_1$ to an epimorphism with*

$$\ker(\overline{\sigma}) = \ker(\sigma) \oplus H$$

2) *Let $\sigma: N \approx M_1$ be an R-isomorphism. Then the correspondence*

$$H \mapsto \overline{\sigma}, \quad \textit{where } \ker(\overline{\sigma}) = H$$

is a bijection from complements of N onto the extensions of σ. Thus, an isomorphism $\sigma: N \approx M_1$ has an extension to M if and only if N is complemented.□

Definition *Let $N \le M$. When the identity map $\iota: N \approx N$ has an extension to $\sigma: M \to N$, the submodule N is called a* **retract** *of M and σ is called the* **retraction map**.□

Corollary 4.15 *A submodule $N \le M$ is a retract of M if and only if N has a complement in M.*□

Direct Summands and One-Sided Invertibility

Direct summands are also related to one-sided invertibility of R-maps.

Definition *Let $\tau: A \to B$ be a module homomorphism.*

1) *A* **left inverse** *of τ is a module homomorphism $\tau_L: B \to A$ for which $\tau_L \circ \tau = \iota$.*
2) *A* **right inverse** *of τ is a module homomorphism $\tau_R: B \to A$ for which $\tau \circ \tau_R = \iota$.*

Left and right inverses are called **one-sided inverses**. *An ordinary inverse is called a* **two-sided inverse**.□

Unlike a two-sided inverse, one-sided inverses need not be unique.

A left-invertible homomorphism σ must be injective, since

$$\sigma a = \sigma b \Rightarrow \sigma_L \circ \sigma a = \sigma_L \circ \sigma b \Rightarrow a = b$$

Also, a right-invertible homomorphism $\sigma: A \to B$ must be surjective, since if $b \in B$, then

$$b = \sigma[\sigma_R(b)] \in \operatorname{im}(\sigma)$$

For *set* functions, the converses of these statements hold: σ is left-invertible if and only if it is injective and σ is right-invertible if and only if it is surjective. However, this is not the case for R-maps.

Let $\sigma: M \to M_1$ be an injective R-map. Referring to Figure 4.1,

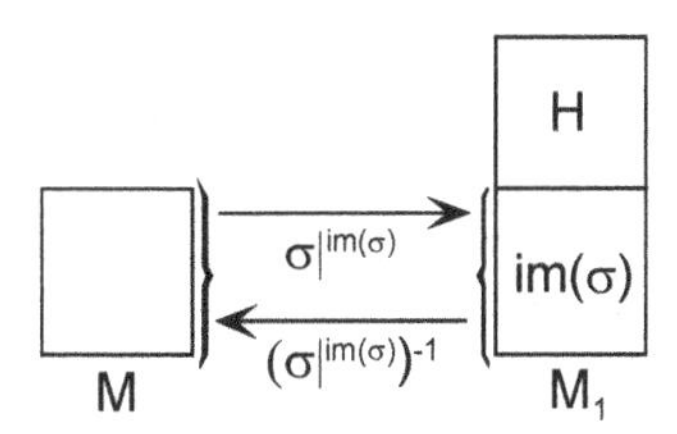

Figure 4.1

the map $\sigma|^{\operatorname{im}(\sigma)}: M \approx \operatorname{im}(\sigma)$ obtained from σ by restricting its range to $\operatorname{im}(\sigma)$ is an isomorphism and the left inverses σ_L of σ are precisely the extensions of $(\sigma|^{\operatorname{im}(\sigma)})^{-1}: \operatorname{im}(\sigma) \approx M$ to M_1. Hence, Theorem 4.14 says that the correspondence

$$H \mapsto \text{extension of } (\sigma|^{\operatorname{im}(\sigma)})^{-1} \text{ with kernel } H$$

is a bijection from the complements H of $\operatorname{im}(\sigma)$ onto the left inverses of σ.

Now let $\sigma: M \to M_1$ be a surjective R-map. Referring to Figure 4.2,

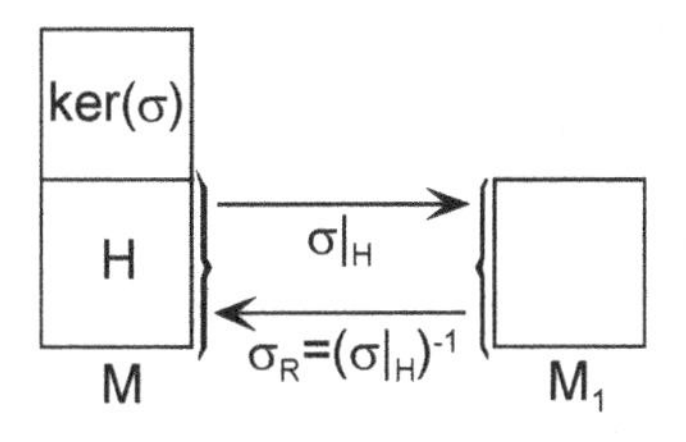

Figure 4.2

if $\ker(\sigma)$ is complemented, that is, if

$$M = \ker(\sigma) \oplus H$$

then $\sigma|_H: H \approx M_1$ is an isomorphism. Thus, a map $\tau: M_1 \to M$ is a right inverse of σ if and only if τ is a *range-extension* of $(\sigma|_H)^{-1}: M_1 \approx H$, the only difference being in the ranges of the two functions. Hence, $(\sigma|_H)^{-1}: M_1 \to M$ is the only right inverse of σ with image H. It follows that the correspondence

$$H \mapsto (\sigma|_H)^{-1}: M_1 \to M$$

is an injection from the complements H of $\ker(\sigma)$ to the right inverses of σ. Moreover, this map is a bijection, since if $\sigma_R: M_1 \to M$ is a right inverse of σ, then $\sigma_R: M_1 \approx \operatorname{im}(\sigma_R)$ and σ is an extension of $\sigma_R^{-1}: \operatorname{im}(\sigma_R) \approx M_1$, which implies that

$$M = \operatorname{im}(\sigma_R) \oplus \ker(\sigma)$$

Theorem 4.16 *Let M and M_1 be R-modules and let $\sigma: M \to M_1$ be an R-map.*

1) *Let $\sigma: M \hookrightarrow M_1$ be injective. The map*

$$H \mapsto \textit{extension of } (\sigma|^{\operatorname{im}(\sigma)})^{-1} \textit{ with kernel } H$$

is a bijection from the complements H of $\operatorname{im}(\sigma)$ onto the left inverses of σ. Thus, there is exactly one left inverse of σ for each complement of $\operatorname{im}(\sigma)$ and that complement is the kernel of the left inverse.

2) *Let $\sigma: M \to M_1$ be surjective. The map*

$$H \mapsto (\sigma|_H)^{-1}: M_1 \to M$$

is a bijection from the complements H of $\ker(\sigma)$ to the right inverses of σ. Thus, there is exactly one right inverse of σ for each complement H of $\ker(\sigma)$ and that complement is the image of the right inverse. Thus,

$$M = \ker(\sigma) \oplus H \approx \ker(\sigma) \boxplus \operatorname{im}(\sigma)$$

□

The last part of the previous theorem is worth further comment. Recall that if $\tau: V \to W$ is a linear transformation on vector spaces, then

$$V \approx \ker(\tau) \boxplus \operatorname{im}(\tau)$$

This holds for modules as well *provided that* $\ker(\tau)$ *is a direct summand.*

Modules Are Not as Nice as Vector Spaces

Here is a list of some of the properties of modules (over commutative rings with identity) that emphasize the differences between modules and vector spaces.

1) A submodule of a module need not have a complement.
2) A submodule of a finitely generated module need not be finitely generated.
3) There exist modules with no linearly independent elements and hence with no basis.
4) A minimal spanning set or maximal linearly independent set is not necessarily a basis.

5) There exist free modules with submodules that are not free.
6) There exist free modules with linearly independent sets that are not contained in a basis and spanning sets that do not contain a basis.

Recall also that a module over a *noncommutative* ring may have bases of different sizes. However, all bases for a free module over a commutative ring with identity have the same size, as we will prove in the next chapter.

Exercises

1. Give the details to show that any commutative ring with identity is a module over itself.
2. Let $S = \{v_1, \ldots, v_n\}$ be a subset of a module M. Prove that $N = \langle\langle S \rangle\rangle$ is the *smallest* submodule of M containing S. First you will need to formulate precisely what it means to be the smallest submodule of M containing S.
3. Let M be an R-module and let I be an ideal in R. Let IM be the set of all finite sums of the form
$$r_1 v_1 + \cdots + r_n v_n$$
where $r_i \in I$ and $v_i \in M$. Is IM a submodule of M?
4. Show that if S and T are submodules of M, then (with respect to set inclusion)
$$S \cap T = \text{glb}\{S, T\} \text{ and } S + T = \text{lub}\{S, T\}$$
5. Let $S_1 \subseteq S_2 \subseteq \cdots$ be an ascending sequence of submodules of an R-module M. Prove that the union $\bigcup S_i$ is a submodule of M.
6. Give an example of a module M that has a finite basis but with the property that not every spanning set in M contains a basis and not every linearly independent set in M is contained in a basis.
7. Show that, just as in the case of vector spaces, an R-homomorphism can be defined by assigning arbitrary values on the elements of a basis and extending by linearity.
8. Let $\tau \in \hom_R(M, N)$ be an R-isomorphism. If $\mathcal{B}$ is a basis for M, prove that $\tau\mathcal{B} = \{\tau b \mid b \in \mathcal{B}\}$ is a basis for N.
9. Let M be an R-module and let $\tau \in \hom_R(M, M)$ be an R-endomorphism. If τ is **idempotent**, that is, if $\tau^2 = \tau$, show that
$$M = \ker(\tau) \oplus \text{im}(\tau)$$
Does the converse hold?
10. Consider the ring $R = F[x, y]$ of polynomials in two variables. Show that the set M consisting of all polynomials in R that have zero constant term is an R-module. Show that M is not a free R-module.
11. Prove that if R is an integral domain, then all R-modules M have the following property: If $v_1, \ldots, v_n$ is linearly independent over R, then so is $rv_1, \ldots, rv_n$ for any nonzero $r \in R$.

12. Prove that if a nonzero commutative ring R with identity has the property that every finitely generated R-module is free then R is a field.
13. Let M and N be R-modules. If S is a submodule of M and T is a submodule of N show that

$$\frac{M \oplus N}{S \oplus T} \approx \frac{M}{S} \boxplus \frac{N}{T}$$

14. If R is a commutative ring with identity and $\mathcal{I}$ is an ideal of R, then $\mathcal{I}$ is an R-module. What is the maximum size of a linearly independent set in $\mathcal{I}$? Under what conditions is $\mathcal{I}$ free?
15. a) Show that for any module M over an integral domain the set M_{tor} of all torsion elements in a module M is a submodule of M.
    b) Find an example of a ring R with the property that for some R-module M the set M_{tor} is not a submodule.
    c) Show that for any module M over an integral domain, the quotient module M/M_{tor} is torsion-free.
16. a) Find a module M that is finitely generated by torsion elements but for which $\mathrm{ann}(M) = \{0\}$.
    b) Find a torsion module M for which $\mathrm{ann}(M) = \{0\}$.
17. Let N be an abelian group together with a scalar multiplication over a ring R that satisfies all of the properties of an R-module except that $1v$ does not necessarily equal v for all $v \in N$. Show that N can be written as a direct sum of an R-module N_0 and another "pseudo R-module" N_1.
18. Prove that $\hom_R(M, N)$ is an R-module under addition of functions and scalar multiplication defined by

$$(r\tau)(v) = r(\tau v) = \tau(rv)$$

19. Prove that any R-module M is isomorphic to the R-module $\hom_R(R, M)$.
20. Let R and S be commutative rings with identity and let $f\colon R \to S$ be a ring homomorphism. Show that any S-module is also an R-module under the scalar multiplication

$$rv = f(r)v$$

21. Prove that $\hom_{\mathbb{Z}}(\mathbb{Z}_n, \mathbb{Z}_m) \approx \mathbb{Z}_d$ where $d = \gcd(n, m)$.
22. Suppose that R is a commutative ring with identity. If $\mathcal{I}$ and $\mathcal{J}$ are ideals of R for which $R/\mathcal{I} \approx R/\mathcal{J}$ as R-modules, then prove that $\mathcal{I} = \mathcal{J}$. Is the result true if $R/\mathcal{I} \approx R/\mathcal{J}$ as rings?

Chapter 5
Modules II: Free and Noetherian Modules

The Rank of a Free Module

Since all bases for a vector space V have the same cardinality, the concept of vector space dimension is well-defined. A similar statement holds for free R-modules when the base ring is commutative (but not otherwise).

Theorem 5.1 *Let M be a free module over a commutative ring R with identity.*
1) Then any two bases of M have the same cardinality.
2) The cardinality of a spanning set is greater than or equal to that of a basis.

Proof. The plan is to find a vector space V with the property that, for any basis for M, there is a basis of the same cardinality for V. Then we can appeal to the corresponding result for vector spaces.

Let $\mathcal{I}$ be a maximal ideal of R, which exists by Theorem 0.23. Then $R/\mathcal{I}$ is a field. Our first thought might be that M is a vector space over $R/\mathcal{I}$, but that is not the case. In fact, scalar multiplication using the field $R/\mathcal{I}$,

$$(r+\mathcal{I})v = rv$$

is not even well-defined, since this would require that $\mathcal{I}M = \{0\}$. On the other hand, we can fix precisely this problem by factoring out the submodule

$$\mathcal{I}M = \{a_1v_1 + \cdots + a_nv_n \mid a_i \in \mathcal{I}, v_i \in M\}$$

Indeed, $M/\mathcal{I}M$ is a vector space over $R/\mathcal{I}$, with scalar multiplication defined by

$$(r+\mathcal{I})(u+\mathcal{I}M) = ru + \mathcal{I}M$$

To see that this is well-defined, we must show that the conditions

$$\begin{aligned} r+\mathcal{I} &= r'+\mathcal{I} \\ u+\mathcal{I}M &= u'+\mathcal{I}M \end{aligned}$$

imply

$$ru + \mathcal{I}M = r'u' + \mathcal{I}M$$

But this follows from the fact that

$$ru - r'u' = r(u - u') + (r - r')u' \in \mathcal{I}M$$

Hence, scalar multiplication is well-defined. We leave it to the reader to show that $M/\mathcal{I}M$ is a vector space over $R/\mathcal{I}$.

Consider now a set $\mathcal{B} = \{b_i \mid i \in I\} \subseteq M$ and the corresponding set

$$\mathcal{B} + \mathcal{I}M = \{b_i + \mathcal{I}M \mid i \in I\} \subseteq \frac{M}{\mathcal{I}M}$$

If $\mathcal{B}$ spans M over R, then $\mathcal{B} + \mathcal{I}M$ spans $M/\mathcal{I}M$ over $R/\mathcal{I}$. To see this, note that any $v \in M$ has the form $v = \Sigma r_i b_i$ for $r_i \in R$ and so

$$\begin{aligned} v + \mathcal{I}M &= \left(\sum_j r_{i_j} b_{i_j}\right) + \mathcal{I}M \\ &= \sum_j r_{i_j}(b_{i_j} + \mathcal{I}M) \\ &= \sum_j (r_{i_j} + \mathcal{I})(b_{i_j} + \mathcal{I}M) \end{aligned}$$

which shows that $\mathcal{B} + \mathcal{I}M$ spans $M/\mathcal{I}M$.

Now suppose that $\mathcal{B} = \{b_i \mid i \in I\}$ is a basis for M over R. We show that $\mathcal{B} + \mathcal{I}M$ is a basis for $M/\mathcal{I}M$ over $R/\mathcal{I}$. We have seen that $\mathcal{B} + \mathcal{I}M$ spans $M/\mathcal{I}M$. Also, if

$$\sum_j (r_{i_j} + \mathcal{I})(b_{i_j} + \mathcal{I}M) = \mathcal{I}M$$

then $\sum_j r_{i_j} b_{i_j} \in \mathcal{I}M$ and so

$$\sum_j r_{i_j} b_{i_j} = \sum_k a_{i_k} b_{i_k}$$

where $a_{i_k} \in \mathcal{I}$. From the linear independence of $\mathcal{B}$ we deduce that $r_{i_j} \in \mathcal{I}$ for all j and so $r_{i_j} + \mathcal{I} = \mathcal{I}$. Hence $\mathcal{B} + \mathcal{I}M$ is linearly independent and therefore a basis, as desired.

To see that $|\mathcal{B}| = |\mathcal{B} + \mathcal{I}M|$, note that if $b_i + \mathcal{I}M = b_k + \mathcal{I}M$, then

$$b_i - b_k = \sum_j a_{i_j} b_{i_j}$$

where $a_{i_j} \in \mathcal{I}$. If $b_i \neq b_k$, then the coefficient of b_i on the right must be equal to

1 and so $1 \in \mathcal{I}$, which is not possible since $\mathcal{I}$ is a maximal ideal. Hence, $b_i = b_k$.

Thus, if $\mathcal{B}$ is a basis for M over R, then

$$|\mathcal{B}| = |\mathcal{B} + \mathcal{I}M| = \dim_{R/\mathcal{I}}(M/\mathcal{I}M)$$

and so all bases for M over R have the same cardinality, which proves part 1).

Finally, if $\mathcal{B}$ spans M over R, then $\mathcal{B} + \mathcal{I}M$ spans $M/\mathcal{I}M$ and so

$$\dim_{R/\mathcal{I}}(M/\mathcal{I}M) \le |\mathcal{B} + \mathcal{I}M| \le |\mathcal{B}|$$

Thus, $\mathcal{B}$ has cardinality at least as great as that of any basis for M over R.□

The previous theorem allows us to define the *rank* of a free module. (The term *dimension* is not used for modules in general.)

Definition *Let R be a commutative ring with identity. The* **rank** $\operatorname{rk}(M)$ *of a nonzero free R-module M is the cardinality of any basis for M. The rank of the trivial module $\{0\}$ is* 0.□

Theorem 5.1 fails if the underlying ring of scalars is not commutative. The next example describes a module over a noncommutative ring that has the remarkable property of possessing a basis of size n for any positive integer n.

Example 5.1 Let V be a vector space over F with a countably infinite basis $\mathcal{B} = \{b_0, b_1, \ldots\}$. Let $\mathcal{L}(V)$ be the ring of linear operators on V. Observe that $\mathcal{L}(V)$ is not commutative, since composition of functions is not commutative.

The ring $\mathcal{L}(V)$ is an $\mathcal{L}(V)$-module and as such, the identity map ι forms a basis for $\mathcal{L}(V)$. However, we can also construct a basis for $\mathcal{L}(V)$ of any desired finite size n. To understand the idea, consider the case $n = 2$ and define the operators β_1 and β_2 by

$$\beta_1(b_{2k}) = b_k, \beta_1(b_{2k+1}) = 0$$

and

$$\beta_2(b_{2k}) = 0, \beta_2(b_{2k+1}) = b_k$$

These operators are linearly independent essentially because they are surjective and their supports are disjoint. In particular, if

$$f\beta_1 + g\beta_2 = 0$$

then

$$0 = (f\beta_1 + g\beta_2)(b_{2k}) = f(b_k)$$

and

$$0 = (f\beta_1 + g\beta_2)(b_{2k+1}) = g(b_k)$$

which shows that $f = 0$ and $g = 0$. Moreover, if $h \in \mathcal{L}(V)$, then we define f and g by

$$\begin{aligned} f(b_k) &= h(b_{2k}) \\ g(b_k) &= h(b_{2k+1}) \end{aligned}$$

from which it follows easily that

$$h = f\beta_1 + g\beta_2$$

which shows that $\{\beta_1, \beta_2\}$ is a basis for $\mathcal{L}(V)$.

More generally, we begin by partitioning $\mathcal{B}$ into n blocks. For each $s = 0, \ldots, n-1$, let

$$\mathcal{B}_s = \{b_i \mid i \equiv s \bmod n\}$$

Now we define elements $\beta_s \in \mathcal{L}(V)$ by

$$\beta_s(b_{kn+t}) = \delta_{t,s} b_k$$

where $0 \le t < n$ and where $\delta_{t,s}$ is the Kronecker delta function. These functions are surjective and have disjoint support. It follows that $\mathcal{C}_n = \{\beta_0, \ldots, \beta_{n-1}\}$ is linearly independent. For if

$$0 = \alpha_0\beta_0 + \cdots + \alpha_{n-1}\beta_{n-1}$$

where $\alpha_s \in \mathcal{L}(V)$, then, applying this to b_{kn+t} gives

$$0 = \alpha_t\beta_t(b_{kn+t}) = \alpha_t(b_k)$$

for all k. Hence, $\alpha_t = 0$.

Also, $\mathcal{C}_n$ spans $\mathcal{L}(V)$, for if $\tau \in \mathcal{L}(V)$, we define $\alpha_s \in \mathcal{L}(V)$ by

$$\alpha_s(b_k) = \tau(b_{kn+s})$$

to get

$$(\alpha_0\beta_0 + \cdots + \alpha_{n-1}\beta_{n-1})(b_{kn+t}) = \alpha_t\beta_t(b_{kn+t}) = \alpha_t(b_k) = \tau(b_{kn+t})$$

and so

$$\tau = \alpha_0\beta_0 + \cdots + \alpha_{n-1}\beta_{n-1}$$

Thus, $\mathcal{C}_n = \{\beta_0, \ldots, \beta_{n-1}\}$ is a basis for $\mathcal{L}(V)$ of size n.$\square$

Recall that if B is a basis for a vector space V over F, then V is isomorphic to the vector space $(F^B)_0$ of all functions from B to F that have finite support. A similar result holds for free R-modules. We begin with the fact that $(R^B)_0$ is a free R-module. The simple proof is left to the reader.

Theorem 5.2 *Let B be any set and let R be a commutative ring with identity. The set $(R^B)_0$ of all functions from B to R that have finite support is a free R-module of rank $|B|$ with basis $\mathcal{B} = \{\delta_b\}$ where*

$$\delta_b(x) = \begin{cases} 1 & \text{if } x = b \\ 0 & \text{if } x \neq b \end{cases}$$

This basis is referred to as the **standard basis** for $(R^B)_0$.□

Theorem 5.3 *Let M be an R-module. If B is a basis for M, then M is isomorphic to $(R^B)_0$.*

Proof. Consider the map $\tau: M \to (R^B)_0$ defined by setting

$$\tau b = \delta_b$$

where δ_b is defined in Theorem 5.2 and extending τ to M by linearity. Since τ maps a basis for M to a basis $\mathcal{B} = \{\delta_b\}$ for $(R^B)_0$, it follows that τ is an isomorphism from M to $(R^B)_0$.□

Theorem 5.4 *Two free R-modules (over a commutative ring) are isomorphic if and only if they have the same rank.*

Proof. If $M \approx N$, then any isomorphism τ from M to N maps a basis for M to a basis for N. Since τ is a bijection, we have $\text{rk}(M) = \text{rk}(N)$. Conversely, suppose that $\text{rk}(M) = \text{rk}(N)$. Let $\mathcal{B}$ be a basis for M and let $\mathcal{C}$ be a basis for N. Since $|\mathcal{B}| = |\mathcal{C}|$, there is a bijective map $\tau: \mathcal{B} \to \mathcal{C}$. This map can be extended by linearity to an isomorphism of M onto N and so $M \approx N$.□

We have seen that the cardinality of a (minimal) spanning set for a free module M is at least equal to $\text{rk}(M)$. Let us now speak about the cardinality of maximal linearly independent sets.

Theorem 5.5 *Let R be an integral domain and let M be a free R-module. Then all linearly independent sets have cardinality at most* $\text{rk}(M)$.

Proof. Since $M \approx (R^\kappa)_0$ we need only prove the result for $(R^\kappa)_0$. Let Q be the field of quotients of R. Then $(Q^\kappa)_0$ is a vector space. Now, if

$$\mathcal{B} = \{v_i \mid i \in I\} \subseteq (R^\kappa)_0 \subseteq (Q^\kappa)_0$$

is linearly independent over Q as a subset of $(Q^\kappa)_0$, then $\mathcal{B}$ is clearly linearly independent over R as a subset of $(R^\kappa)_0$. Conversely, suppose that $\mathcal{B}$ is linearly independent over R and

$$\frac{r_1}{s_1}v_{i_1} + \cdots + \frac{r_k}{s_k}v_{i_k} = 0$$

where $s_i \neq 0$ for all i and $r_j \neq 0$ for some j. Multiplying by $s = s_1 \cdots s_k \neq 0$ produces a nontrivial linear dependency over R,

$$\frac{s}{s_1} r_1 v_{i_1} + \cdots + \frac{s}{s_k} r_k v_{i_k} = 0$$

which implies that $r_i = 0$ for all i. Thus $\mathcal{B}$ is linearly dependent over R if and only if it is linearly dependent over Q. But in the vector space $(Q^\kappa)_0$, all sets of cardinality greater than κ are linearly dependent over Q and hence all subsets of $(R^\kappa)_0$ of cardinality greater than κ are linearly dependent over R.□

Free Modules and Epimorphisms

If $\sigma: M \to F$ is a module epimorphism where F is free on $\mathcal{B}$, then it is easy to define a right inverse for σ, since we can define an R-map $\sigma_R: F \to M$ by specifying its values arbitrarily on $\mathcal{B}$ and extending by linearity. Thus, we take $\sigma_R(b)$ to be any member of $\sigma^{-1}(b)$. Then Theorem 4.16 implies that $\ker(\sigma)$ is a direct summand of M and

$$M \approx \ker(\sigma) \boxplus F$$

This discussion applies to the canonical projection $\pi: M \to M/S$ provided that the quotient M/S is free.

Theorem 5.6 *Let R be a commutative ring with identity.*

1) *If $\sigma: M \to F$ is an R-epimorphism and F is free, then $\ker(\sigma)$ is complemented and*

$$M = \ker(\sigma) \oplus N \approx \ker(\sigma) \boxplus F$$

where $N \approx F$.

2) *If S is a submodule of M and if M/S is free, then S is complemented and*

$$M \approx S \boxplus \frac{M}{S}$$

If M, S and M/S are free, then

$$\operatorname{rk}(M) = \operatorname{rk}(S) + \operatorname{rk}\left(\frac{M}{S}\right)$$

and if the ranks are all finite, then

$$\operatorname{rk}\left(\frac{M}{S}\right) = \operatorname{rk}(M) - \operatorname{rk}(S)$$

□

Noetherian Modules

One of the most desirable properties of a finitely generated R-module M is that all of its submodules be finitely generated:

$$M \text{ finitely generated}, \quad S \le M \quad \Rightarrow \quad S \text{ finitely generated}$$

Example 4.2 shows that this is not always the case and leads us to search for conditions on the ring R that will guarantee this property for R-modules.

Definition *An R-module M is said to satisfy the* **ascending chain condition** *(abbreviated ACC) on submodules if every ascending sequence of submodules*

$$S_1 \subseteq S_2 \subseteq S_3 \subseteq \cdots$$

of M is eventually constant, that is, there exists an index k for which

$$S_k = S_{k+1} = S_{k+2} = \cdots$$

Modules with the ascending chain condition on submodules are also called **Noetherian modules** *(after Emmy Noether, one of the pioneers of module theory).* □

Since a ring R is a module over itself and since the submodules of the module R are precisely the ideals of the ring R, the preceding definition can be formulated for rings as follows.

Definition *A ring R is said to satisfy the* **ascending chain condition** *(abbreviated ACC) on ideals if any ascending sequence*

$$\mathcal{I}_1 \subseteq \mathcal{I}_2 \subseteq \mathcal{I}_3 \subseteq \cdots$$

of ideals of R is eventually constant, that is, there exists an index k for which

$$\mathcal{I}_k = \mathcal{I}_{k+1} = \mathcal{I}_{k+2} = \cdots$$

A ring that satisfies the ascending chain condition on ideals is called a **Noetherian ring.** □

The following theorem describes the relevance of this to the present discussion.

Theorem 5.7

1) *An R-module M is Noetherian if and only if every submodule of M is finitely generated.*
2) *In particular, a ring R is Noetherian if and only if every ideal of R is finitely generated.*

Proof. Suppose that all submodules of M are finitely generated and that M contains an infinite ascending sequence

$$S_1 \subseteq S_2 \subseteq S_3 \subseteq \cdots \tag{5.1}$$

of submodules. Then the union

$$S = \bigcup_j S_j$$

is easily seen to be a submodule of M. Hence, S is finitely generated, say $S = \langle\langle u_1, \ldots, u_n \rangle\rangle$. Since $u_i \in S$, there exists an index k_i such that $u_i \in S_{k_i}$. Therefore, if $k = \max\{k_1, \ldots, k_n\}$, we have

$$\{u_1, \ldots, u_n\} \subseteq S_k$$

and so

$$S = \langle\langle u_1, \ldots, u_n \rangle\rangle \subseteq S_k \subseteq S_{k+1} \subseteq S_{k+2} \subseteq \cdots \subseteq S$$

which shows that the chain (5.1) is eventually constant.

For the converse, suppose that M satisfies the ACC on submodules and let S be a submodule of M. Pick $u_1 \in S$ and consider the submodule $S_1 = \langle\langle u_1 \rangle\rangle \subseteq S$ generated by u_1. If $S_1 = S$, then S is finitely generated. If $S_1 \neq S$, then there is a $u_2 \in S - S_1$. Now let $S_2 = \langle\langle u_1, u_2 \rangle\rangle$. If $S_2 = S$, then S is finitely generated. If $S_2 \neq S$, then pick $u_3 \in S - S_2$ and consider the submodule $S_3 = \langle\langle u_1, u_2, u_3 \rangle\rangle$.

Continuing in this way, we get an ascending chain of submodules

$$\langle\langle u_1 \rangle\rangle \subseteq \langle\langle u_1, u_2 \rangle\rangle \subseteq \langle\langle u_1, u_2, u_3 \rangle\rangle \subseteq \cdots \subseteq S$$

If none of these submodules were equal to S, we would have an infinite ascending chain of submodules, each properly contained in the next, which contradicts the fact that M satisfies the ACC on submodules. Hence, $S = \langle\langle u_1, \ldots, u_n \rangle\rangle$ for some n and so S is finitely generated.□

Our goal is to find conditions under which all finitely generated R-modules are Noetherian. The very pleasing answer is that all finitely generated R-modules are Noetherian if and only if R is Noetherian as an R-module, or equivalently, as a ring.

Theorem 5.8 *Let R be a commutative ring with identity.*

1) *R is Noetherian if and only if every finitely generated R-module is Noetherian.*
2) *Let R be a principal ideal domain. If an R-module M is n-generated, then any submodule of M is also n-generated.*

Proof. For part 1), one direction is evident. Assume that R is Noetherian and let $M = \langle\langle u_1, \ldots, u_n \rangle\rangle$ be a finitely generated R-module. Consider the epimorphism $\tau: R^n \to M$ defined by

$$\tau(r_1, \ldots, r_n) = r_1 u_1 + \cdots + r_n u_n$$

Let S be a submodule of M. Then

$$\tau^{-1}(S) = \{u \in R^n \mid \tau u \in S\}$$

is a submodule of R^n and $\tau(\tau^{-1}S) = S$. If every submodule of R^n is finitely generated, then $\tau^{-1}(S)$ is finitely generated and so $\tau^{-1}(S) = \langle\langle v_1, \ldots, v_k \rangle\rangle$. Then S is finitely generated by $\{\tau v_1, \ldots, \tau v_k\}$. Thus, it is sufficient to prove the theorem for R^n, which we do by induction on n.

If $n = 1$, any submodule of R is an ideal of R, which is finitely generated by assumption. Assume that every submodule of R^k is finitely generated for all $1 \le k < n$ and let S be a submodule of R^n.

If $n > 1$, we can extract from S something that is isomorphic to an ideal of R and so will be finitely generated. In particular, let S_1 be the "last coordinates" in S, specifically, let

$$S_1 = \{(0, \ldots, 0, a_n) \mid (a_1, \ldots, a_{n-1}, a_n) \in S \text{ for some } a_1, \ldots, a_{n-1} \in R\}$$

The set S_1 is isomorphic to an ideal of R and is therefore finitely generated, say $S_1 = \langle\langle \mathcal{G}_1 \rangle\rangle$, where $\mathcal{G}_1 = \{g_1, \ldots, g_k\}$ is a finite subset of S_1.

Also, let

$$S_2 = \{v \in S \mid v = (a_1, \ldots, a_{n-1}, 0) \text{ for some } a_1, \ldots, a_{n-1} \in R\}$$

be the set of all elements of S that have last coordinate equal to 0. Note that S_2 is a submodule of R^n and is isomorphic to a submodule of R^{n-1}. Hence, the inductive hypothesis implies that S_2 is finitely generated, say $S_2 = \langle\langle \mathcal{G}_2 \rangle\rangle$, where $\mathcal{G}_2$ is a finite subset of S.

By definition of S_1, each $g_i \in \mathcal{G}_1$ has the form

$$g_i = (0, \ldots, 0, g_{i,n})$$

for $g_{i,n} \in R$ where there is a $\overline{g}_i \in S$ of the form

$$\overline{g}_i = (g_{i,1}, \ldots, g_{i,n-1}, g_{i,n})$$

Let $\overline{\mathcal{G}}_1 = \{\overline{g}_1, \ldots, \overline{g}_k\}$. We claim that S is generated by the finite set $\overline{\mathcal{G}}_1 \cup \mathcal{G}_2$.

To see this, let $v = (a_1, \ldots, a_n) \in S$. Then $(0, \ldots, 0, a_n) \in S_1$ and so

$$(0, \ldots, 0, a_n) = \sum_{i=1}^{k} r_i g_i$$

for $r_i \in R$. Consider now the sum

$$w = \sum_{i=1}^{k} r_i \overline{g}_i \in \langle\langle \overline{\mathcal{G}}_1 \rangle\rangle$$

The last coordinate of this sum is

$$\sum_{i=1}^{k} r_i g_{i,n} = a_n$$

and so the difference $v - w$ has last coordinate 0 and is thus in $S_2 = \langle\langle \mathcal{G}_2 \rangle\rangle$. Hence

$$v = (v - w) + w \in \langle\langle \overline{\mathcal{G}}_1 \rangle\rangle + \langle\langle \mathcal{G}_2 \rangle\rangle = \langle\langle \overline{\mathcal{G}}_1 \cup \mathcal{G}_2 \rangle\rangle$$

as desired.

For part 2), we leave it to the reader to review the proof and make the necessary changes. The key fact is that S_1 is isomorphic to an ideal of R, which is principal. Hence, S_1 is generated by a single element of M.□

The Hilbert Basis Theorem

Theorem 5.8 naturally leads us to ask which familiar rings are Noetherian. The following famous theorem describes one very important case.

Theorem 5.9 (Hilbert basis theorem) *If a ring R is Noetherian, then so is the polynomial ring $R[x]$.*
Proof. We wish to show that any ideal $\mathcal{I}$ in $R[x]$ is finitely generated. Let L denote the set of all leading coefficients of polynomials in $\mathcal{I}$, together with the 0 element of R. Then L is an ideal of R.

To see this, observe that if $\alpha \in L$ is the leading coefficient of $f(x) \in \mathcal{I}$ and if $r \in R$, then either $r\alpha = 0$ or else $r\alpha$ is the leading coefficient of $rf(x) \in \mathcal{I}$. In either case, $r\alpha \in L$. Similarly, suppose that $\beta \in L$ is the leading coefficient of $g(x) \in \mathcal{I}$. We may assume that $\deg f(x) = i$ and $\deg g(x) = j$, with $i \le j$. Then $h(x) = x^{j-i} f(x)$ is in $\mathcal{I}$, has leading coefficient α and has the same degree as $g(x)$. Hence, either $\alpha - \beta$ is 0 or $\alpha - \beta$ is the leading coefficient of $h(x) - g(x) \in \mathcal{I}$. In either case $\alpha - \beta \in L$.

Since L is an ideal of the Noetherian ring R, it must be finitely generated, say $L = \langle a_1, \ldots, a_m \rangle$. Since $a_i \in L$, there exist polynomials $f_i(x) \in \mathcal{I}$ with leading coefficient a_i. By multiplying each $f_i(x)$ by a suitable power of x, we may assume that

$$\deg f_i(x) = d = \max\{\deg f_i(x)\}$$

for all $i = 1, \ldots, m$.

Now for $k = 0, \ldots, d-1$ let L_k be the set of all leading coefficients of polynomials in $\mathcal{I}$ of degree k, together with the 0 element of R. A similar argument shows that L_k is an ideal of R and so L_k is also finitely generated. Hence, we can find polynomials $P_k = \{p_{k,1}(x), \ldots, p_{k,n_k}(x)\}$ in $\mathcal{I}$ whose leading coefficients constitute a generating set for L_k.

Consider now the finite set

$$P = \left(\bigcup_{k=0}^{d-1} P_k\right) \cup \{f_1(x), \ldots, f_m(x)\}$$

If $\mathcal{J}$ is the ideal generated by P, then $\mathcal{J} \subseteq \mathcal{I}$. An induction argument can be used to show that $\mathcal{J} = \mathcal{I}$. If $g(x) \in \mathcal{I}$ has degree 0, then it is a linear combination of the elements of P_0 (which are constants) and is thus in $\mathcal{J}$. Assume that any polynomial in $\mathcal{I}$ of degree less than k is in $\mathcal{J}$ and let $g(x) \in \mathcal{I}$ have degree k.

If $k < d$, then some linear combination $h(x)$ over R of the polynomials in P_k has the same leading coefficient as $g(x)$ and if $k \geq d$, then some linear combination $h(x)$ of the polynomials

$$\{x^{k-d} f_1(x), \ldots, x^{k-d} f_m(x)\} \subseteq \mathcal{J}$$

has the same leading coefficient as $g(x)$. In either case, there is a polynomial $h(x) \in \mathcal{J}$ that has the same leading coefficient as $g(x)$. Since $g(x) - h(x) \in \mathcal{I}$ has degree strictly smaller than that of $g(x)$ the induction hypothesis implies that

$$g(x) - h(x) \in \mathcal{J}$$

and so

$$g(x) = [g(x) - h(x)] + h(x) \in \mathcal{J}$$

This completes the induction and shows that $\mathcal{I} = \mathcal{J}$ is finitely generated.□

Exercises

1. If M is a free R-module and $\tau: M \to N$ is an epimorphism, then must N also be free?
2. Let $\mathcal{I}$ be an ideal of R. Prove that if $R/\mathcal{I}$ is a free R-module, then $\mathcal{I}$ is the zero ideal.
3. Prove that the union of an ascending chain of submodules is a submodule.
4. Let S be a submodule of an R-module M. Show that if M is finitely generated, so is the quotient module M/S.
5. Let S be a submodule of an R-module. Show that if both S and M/S are finitely generated, then so is M.
6. Show that an R-module M satisfies the ACC for submodules if and only if the following condition holds. Every nonempty collection $\mathcal{S}$ of submodules

of M has a maximal element. That is, for every nonempty collection $\mathcal{S}$ of submodules of M there is an $S \in \mathcal{S}$ with the property that $T \in \mathcal{S} \Rightarrow T \subseteq S$.

7. Let $\tau: M \to N$ be an R-homomorphism.
   a) Show that if M is finitely generated, then so is $\text{im}(\tau)$.
   b) Show that if $\ker(\tau)$ and $\text{im}(\tau)$ are finitely generated, then $M = \ker(\tau) + S$ where S is a finitely generated submodule of M. Hence, M is finitely generated.
8. If R is Noetherian and $\mathcal{I}$ is an ideal of R show that $R/\mathcal{I}$ is also Noetherian.
9. Prove that if R is Noetherian, then so is $R[x_1, \ldots, x_n]$.
10. Find an example of a commutative ring with identity that does not satisfy the ascending chain condition.
11. a) Prove that an R-module M is cyclic if and only if it is isomorphic to $R/\mathcal{I}$ where $\mathcal{I}$ is an ideal of R.
    b) Prove that an R-module M is **simple** ($M \neq \{0\}$ and M has no proper nonzero submodules) if and only if it is isomorphic to $R/\mathcal{I}$ where $\mathcal{I}$ is a maximal ideal of R.
    c) Prove that for any nonzero commutative ring R with identity, a simple R-module exists.
12. Prove that the condition that R be a principal ideal domain in part 2) of Theorem 5.8 is required.
13. Prove Theorem 5.8 in the following way.
    a) Show that if $T \subseteq S$ are submodules of M and if T and S/T are finitely generated, then so is S.
    b) The proof is again by induction. Assuming it is true for any module generated by n elements, let $M = \langle\langle v_1, \ldots, v_{n+1} \rangle\rangle$ and let $M' = \langle\langle v_1, \ldots, v_n \rangle\rangle$. Then let $T = S \cap M'$ in part a).
14. Prove that any R-module M is isomorphic to the quotient of a free module F. If M is finitely generated, then F can also be taken to be finitely generated.
15. Prove that if S and T are isomorphic submodules of a module M it does not necessarily follow that the quotient modules M/S and M/T are isomorphic. Prove also that if $S \oplus T_1 \approx S \oplus T_2$ as modules it does not necessarily follow that $T_1 \approx T_2$. Prove that these statements do hold if all modules are free and have finite rank.

Chapter 6
Modules over a Principal Ideal Domain

We remind the reader of a few of the basic properties of principal ideal domains.

Theorem 6.1 *Let R be a principal ideal domain.*
1) *An element $r \in R$ is irreducible if and only if the ideal $\langle r \rangle$ is maximal.*
2) *An element in R is prime if and only if it is irreducible.*
3) *R is a unique factorization domain.*
4) *R satisfies the ascending chain condition on ideals. Hence, so does any finitely generated R-module M. Moreover, if M is n-generated, then any submodule of M is n-generated.*

Annihilators and Orders

When R is a principal ideal domain, all annihilators are generated by a single element. This permits the following definition.

Definition *Let R be a principal ideal domain and let M be an R-module.*
1) *If N is a submodule of M, then any generator of* $\operatorname{ann}(N)$ *is called an* **order** *of N.*
2) *An* **order** *of an element $v \in M$ is an order of the submodule $\langle\langle v \rangle\rangle$.*□

For readers acquainted with group theory, we mention that the order of a module corresponds to the smallest exponent of a group, *not* to the order of the group.

Theorem 6.2 *Let R be a principal ideal domain and let M be an R-module.*
1) *If α is an order of $N \leq M$, then the orders of N are precisely the associates of α. We denote any order of N by $o(N)$ and, as is customary, refer to $o(N)$ as "the" order of N.*
2) *If $M = A \oplus B$, then*

$$o(M) = \operatorname{lcm}(o(A), o(B))$$

that is, the orders of M are precisely the least common multiples of the orders of A and B.

Proof. We leave proof of part 1) for the reader. For part 2), suppose that

$$o(M) = \delta, \quad o(A) = \alpha, \quad o(B) = \beta, \quad \lambda = \operatorname{lcm}(\alpha, \beta)$$

Then $\delta A = \{0\}$ and $\delta B = \{0\}$ imply that $\alpha \mid \delta$ and $\beta \mid \delta$ and so $\lambda \mid \delta$. On the other hand, λ annihilates both A and B and therefore also $M = A \oplus B$. Hence, $\delta \mid \lambda$ and so $\lambda \sim \delta$ is an order of M.□

Cyclic Modules

The simplest type of nonzero module is clearly a cyclic module. Despite their simplicity, cyclic modules will play a very important role in our study of linear operators on a finite-dimensional vector space and so we want to explore some of their basic properties, including their composition and decomposition.

Theorem 6.3 *Let R be a principal ideal domain.*

1) *If $\langle\langle v\rangle\rangle$ is a cyclic R-module with annihilator $\langle\alpha\rangle$, then the multiplication map $\tau\colon R \to \langle\langle v\rangle\rangle$ defined by $\tau r = rv$ is an R-epimorphism with kernel $\langle\alpha\rangle$. Hence the induced map*

$$\overline{\sigma}\colon \frac{R}{\langle\alpha\rangle} \to \langle\langle v\rangle\rangle$$

defined by

$$\overline{\sigma}(r + \langle\alpha\rangle) = rv$$

is an isomorphism. In other words, cyclic R-modules are isomorphic to quotient modules of the base ring R.

2) *Any submodule of a cyclic R-module is cyclic.*
3) *If $\langle\langle v\rangle\rangle$ is a cyclic submodule of M of order α, then for $\beta \in R$,*

$$o(\langle\langle \beta v\rangle\rangle) = \frac{\alpha}{\gcd(\beta, \alpha)}$$

Also,

$$\langle\langle \beta v\rangle\rangle = \langle\langle v\rangle\rangle \quad \Leftrightarrow \quad (o(v), \beta) = 1 \quad \Leftrightarrow \quad o(\beta v) = o(v)$$

Proof. We leave proof of part 1) as an exercise. For part 2), let $S \le \langle\langle v\rangle\rangle$. Then $I = \{r \in R \mid rv \in S\}$ is an ideal of R and so $I = \langle s\rangle$ for some $s \in R$. Thus,

$$S = Iv = Rsv = \langle\langle sv\rangle\rangle$$

For part 3), we have $r(\beta v) = 0$ if and only if $(r\beta)v = 0$, that is, if and only if $\alpha \mid r\beta$, which is equivalent to

$$\gamma := \frac{\alpha}{\gcd(\alpha, \beta)} \Bigg| r$$

Thus, $r \in \text{ann}(\beta v)$ if and only if $r \in \langle\gamma\rangle$ and so $\text{ann}(\beta v) = \langle\gamma\rangle$. For the second statement, if $(\alpha, \beta) = 1$ then there exist $a, b \in R$ for which $a\alpha + b\beta = 1$ and so

$$v = (a\alpha + b\beta)v = b\beta v \in \langle\langle\beta v\rangle\rangle \subseteq \langle\langle v\rangle\rangle$$

and so $\langle\langle\beta v\rangle\rangle = \langle\langle v\rangle\rangle$. Of course, if $\langle\langle\beta v\rangle\rangle = \langle\langle v\rangle\rangle$ then $o(\beta v) = \alpha$. Finally, if $o(\beta v) = \alpha$, then

$$\alpha = o(\beta v) = \frac{\alpha}{\gcd(\alpha, \beta)}$$

and so $(\alpha, \beta) = 1$.□

The Decomposition of Cyclic Modules

The following theorem shows how cyclic modules can be composed and decomposed.

Theorem 6.4 *Let M be an R-module.*

1) **(Composing cyclic modules)** *If $u_1, \ldots, u_n \in M$ have relatively prime orders, then*

$$o(u_1 + \cdots + u_n) = o(u_1) \cdots o(u_n)$$

and

$$\langle\langle u_1\rangle\rangle \oplus \cdots \oplus \langle\langle u_n\rangle\rangle = \langle\langle u_1 + \cdots + u_n\rangle\rangle$$

Consequently, if

$$M = A_1 + \cdots + A_n$$

where the submodules A_i have relatively prime orders, then the sum is direct.

2) **(Decomposing cyclic modules)** *If $o(v) = \alpha_1 \cdots \alpha_n$ where the α_i's are pairwise relatively prime, then v has the form*

$$v = u_1 + \cdots + u_n$$

where $o(u_i) = \alpha_i$ and so

$$\langle\langle v\rangle\rangle = \langle\langle u_1 + \cdots + u_n\rangle\rangle = \langle\langle u_1\rangle\rangle \oplus \cdots \oplus \langle\langle u_n\rangle\rangle$$

Proof. For part 1), let $\alpha_k = o(u_k)$, $\mu := \alpha_1 \cdots \alpha_n$ and $v := u_1 + \cdots + u_n$. Then since μ annihilates v, the order of v divides μ. If $o(v)$ is a proper divisor of μ, then for some index k, there is a prime $p \mid \alpha_k$ for which μ/p annihilates v. But μ/p annihilates each u_i for $i \neq k$. Thus,

$$0 = \frac{\mu}{p} v = \frac{\mu}{p} u_k = \frac{\alpha_k}{p} \left(\frac{\mu}{\alpha_k} \right) u_k$$

Since $o(u_k)$ and μ/α_k are relatively prime, the order of $(\mu/\alpha_k)u_k$ is equal to $o(u_k) = \alpha_k$, which contradicts the equation above. Hence, $o(v) = \mu$.

It is clear that $\langle\langle u_1 + \cdots + u_n \rangle\rangle \subseteq \langle\langle u_1 \rangle\rangle \oplus \cdots \oplus \langle\langle u_n \rangle\rangle$. For the reverse inclusion, since α_1 and μ/α_1 are relatively prime, there exist $r, s \in R$ for which

$$r\alpha_1 + s\frac{\mu}{\alpha_1} = 1$$

Hence

$$u_1 = \left(r\alpha_1 + s\frac{\mu}{\alpha_1} \right) u_1 = s\frac{\mu}{\alpha_1} u_1 = s\frac{\mu}{\alpha_1}(u_1 + \cdots + u_n) \in \langle\langle u_1 + \cdots + u_n \rangle\rangle$$

Similarly, $u_k \in \langle\langle u_1 + \cdots + u_n \rangle\rangle$ for all k and so we get the reverse inclusion.

Finally, to see that the sum above is direct, note that if

$$v_1 + \cdots + v_n = 0$$

where $v_i \in A_i$, then each v_i must be 0, for otherwise the order of the sum on the left would be different from 1.

For part 2), the scalars $\beta_k = \mu/\alpha_k$ are relatively prime and so there exist $a_i \in R$ for which

$$a_1\beta_1 + \cdots + a_n\beta_n = 1$$

Hence,

$$v = (a_1\beta_1 + \cdots + a_n\beta_n)v = a_1\beta_1 v + \cdots + a_n\beta_n v$$

Since $o(\beta_k v) = \mu/\gcd(\mu, \beta_k) = \alpha_k$ and since a_k and α_k are relatively prime, we have $o(a_k\beta_k v) = \alpha_k$. The second statement follows from part 1).□

Free Modules over a Principal Ideal Domain

We have seen that a submodule of a free module need not be free: The submodule $\mathbb{Z} \times \{0\}$ of the module $\mathbb{Z} \times \mathbb{Z}$ over itself is not free. However, if R is a principal ideal domain this cannot happen.

Theorem 6.5 *Let M be a free module over a principal ideal domain R. Then any submodule S of M is also free and* $\text{rk}(S) \le \text{rk}(M)$.

Proof. We will give the proof first for modules of finite rank and then generalize to modules of arbitrary rank. Since $M \approx R^n$ where $n = \text{rk}(M)$ is finite, we may in fact assume that $M = R^n$. For each $1 \le k \le n$, let

$$I_k = \{r \in R \mid (a_1, \ldots, a_{k-1}, r, 0, \ldots, 0) \in S \text{ for some } a_1, \ldots, a_{k-1} \in R\}$$

Then it is easy to see that I_k is an ideal of R and so $I_k = \langle r_k \rangle$ for some $r_k \in R$. Let

$$u_k = (a_1, \ldots, a_{k-1}, r_k, 0, \ldots, 0) \in S$$

We claim that

$$\mathcal{B} = \{u_k \mid k = 1, \ldots, n \text{ and } r_k \neq 0\}$$

is a basis for S. As to linear independence, suppose that

$$\mathcal{B} = \{u_{i_1}, \ldots, u_{i_m}\}$$

and that

$$a_{j_1} u_{j_1} + \cdots + a_{j_s} u_{j_s} = 0$$

Then comparing the j_sth coordinates gives $a_{j_s} r_{j_s} = 0$ and since $r_{j_s} \neq 0$, it follows that $a_{j_s} = 0$. In a similar way, all coefficients are 0 and so $\mathcal{B}$ is linearly independent.

To see that $\mathcal{B}$ spans S, we partition the elements $x \in S$ according to the largest coordinate index $i(x)$ with nonzero entry and induct on $i(x)$. If $i(x) = 0$, then $x = 0$, which is in the span of $\mathcal{B}$. Suppose that all $x \in S$ with $i(x) < k$ are in the span of $\mathcal{B}$ and let $i(x) = k$, that is,

$$x = (a_1, \ldots, a_k, 0, \ldots, 0)$$

where $a_k \neq 0$. Then $a_k \in I_k$ and so $r_k \neq 0$ and $a_k = c r_k$ for some $c \in R$. Hence, $i(x - c u_k) < k$ and so $y = x - c u_k \in \langle\langle \mathcal{B} \rangle\rangle$ and therefore $x \in \langle\langle \mathcal{B} \rangle\rangle$. Thus, $\mathcal{B}$ is a basis for S.

The previous proof can be generalized in a more or less direct way to modules of arbitrary rank. In this case, we may assume that $M = (R^\kappa)_0$ is the R-module of functions with finite support from κ to R, where κ is a cardinal number. We use the fact that κ is a well-ordered set, that is, κ is a totally ordered set in which any nonempty subset has a smallest element. If $\alpha \in \kappa$, the **closed interval** $[0, \alpha]$ is

$$[0, \alpha] = \{x \in \kappa \mid 0 \leq x \leq \alpha\}$$

Let $S \leq M$. For each $0 < \alpha \leq \kappa$, let

$$M_\alpha = \{f \in S \mid \text{supp}(f) \subseteq [0, \alpha]\}$$

Then the set

$$I_\alpha = \{f(\alpha) \mid f \in M_\alpha\}$$

is an ideal of R and so $I_\alpha = \langle f_\alpha(\alpha) \rangle$ for some $f_\alpha \in S$. We show that

$$\mathcal{B} = \{f_\alpha \mid 0 < \alpha \le \kappa, f_\alpha(\alpha) \ne 0\}$$

is a basis for S. First, suppose that

$$r_1 f_{\alpha_1} + \cdots + r_n f_{\alpha_n} = 0$$

where $\alpha_i < \alpha_j$ for $i < j$. Applying this to α_n gives

$$r_n f_{\alpha_n}(\alpha_n) = 0$$

and since R is an integral domain, $r_n = 0$. Similarly, $r_i = 0$ for all i and so $\mathcal{B}$ is linearly independent.

To show that $\mathcal{B}$ spans S, since any $f \in S$ has finite support, there is a largest index $\alpha_f = i(f)$ for which $f(\alpha_f) \ne 0$. Now, if $\langle\langle \mathcal{B} \rangle\rangle < S$, then since κ is well-ordered, we may choose a $g \in S \setminus \langle\langle \mathcal{B} \rangle\rangle$ for which $\alpha = \alpha_g = i(g)$ is as small as possible. Then $g \in M_\alpha$. Moreover, since $0 \ne g(\alpha) \in I_\alpha$, it follows that $f_\alpha(\alpha) \ne 0$ and $g(\alpha) = c f_\alpha(\alpha)$ for some $c \in R$. Then

$$\operatorname{supp}(g - c f_\alpha) \subseteq [0, \alpha]$$

and

$$(g - c f_\alpha)(\alpha) = g(\alpha) - c f_\alpha(\alpha) = 0$$

and so $i(g - c f_\alpha) < \alpha$, which implies that $g - c f_\alpha \in \langle\langle \mathcal{B} \rangle\rangle$. But then

$$g = (g - c f_\alpha) + c f_\alpha \in \langle\langle \mathcal{B} \rangle\rangle$$

a contradiction. Thus, $\mathcal{B}$ is a basis for S.$\square$

In a vector space of dimension n, any set of n linearly independent vectors is a basis. This fails for modules. For example, $\mathbb{Z}$ is a $\mathbb{Z}$-module of rank 1 but the independent set $\{2\}$ is not a basis. On the other hand, the fact that a spanning set of size n is a basis does hold for modules over a principal ideal domain, as we now show.

Theorem 6.6 *Let M be a free R-module of finite rank n, where R is a principal ideal domain. Let $S = \{s_1, \ldots, s_n\}$ be a spanning set for M. Then S is a basis for M.*
Proof. Let $\mathcal{B} = \{b_1, \ldots, b_n\}$ be a basis for M and define the map $\tau\colon M \to M$ by $\tau b_i = s_i$ and extending to a surjective R-homomorphism. Since M is free, Theorem 5.6 implies that

$$M \approx \ker(\tau) \boxplus \operatorname{im}(\tau) = \ker(\tau) \boxplus M$$

Since $\ker(\tau)$ is a submodule of the free module and since R is a principal ideal domain, we know that $\ker(\tau)$ is free of rank at most n. It follows that

$$\text{rk}(M) = \text{rk}(\ker(\tau)) + \text{rk}(M)$$

and so $\text{rk}(\ker(\tau)) = 0$, that is, $\ker(\tau) = \{0\}$, which implies that τ is an R-isomorphism and so S is a basis.□

In general, a basis for a submodule of a free module over a principal ideal domain cannot be extended to a basis for the entire module. For example, the set $\{2\}$ is a basis for the submodule $2\mathbb{Z}$ of the $\mathbb{Z}$-module $\mathbb{Z}$, but this set cannot be extended to a basis for $\mathbb{Z}$ itself. We state without proof the following result along these lines.

Theorem 6.7 *Let M be a free R-module of rank n, where R is a principal ideal domain. Let N be a submodule of M that is free of rank $k \le n$. Then there is a basis $\mathcal{B}$ for M that contains a subset $S = \{v_1, \ldots, v_k\}$ for which $\{r_1 v_1, \ldots, r_k v_k\}$ is a basis for N, for some nonzero elements $r_1, \ldots, r_k$ of R.*□

Torsion-Free and Free Modules

Let us explore the relationship between the concepts of torsion-free and free. It is not hard to see that any free module over an integral domain is torsion-free. The converse does not hold, unless we strengthen the hypotheses by requiring that the module be finitely generated.

Theorem 6.8 *A finitely generated module over a principal ideal domain is free if and only if it is torsion-free.*

Proof. We leave proof that a free module over an integral domain is torsion-free to the reader. Let $G = \{v_1, \ldots, v_n\}$ be a generating set for M. Consider first the case $n = 1$, whence $G = \{v\}$. Then G is a basis for M since singleton sets are linearly independent in a torsion-free module. Hence, M is free.

Now suppose that $G = \{u, v\}$ is a generating set with $u, v \neq 0$. If G is linearly independent, we are done. If not, then there exist nonzero $r, s \in R$ for which $ru = sv$. It follows that $sM = s\langle\langle u, v\rangle\rangle \subseteq \langle\langle u\rangle\rangle$ and so sM is a submodule of a free module and is therefore free by Theorem 6.5. But the map $\tau\colon M \to sM$ defined by $\tau v = sv$ is an isomorphism because M is torsion-free. Thus M is also free.

Now we can do the general case. Write

$$G = \{u_1, \ldots, u_k, v_1, \ldots, v_{n-k}\}$$

where $S = \{u_1, \ldots, u_k\}$ is a maximal linearly independent subset of G. (Note that S is nonempty because singleton sets are linearly independent.)

For each v_i, the set $\{u_1, \ldots, u_k, v_i\}$ is linearly dependent and so there exist $a_i \in R$ and $r_1, \ldots, r_k \in R$ for which

$$a_i v_i + r_1 u_1 + \cdots + r_k u_k = 0$$

If $a = a_1 \cdots a_{n-k}$, then

$$aM = a\langle\langle u_1, \ldots, u_k, v_1, \ldots, v_{n-k} \rangle\rangle \subseteq \langle\langle u_1, \ldots, u_k \rangle\rangle$$

and since the latter is a free module, so is aM, and therefore so is M.□

The Primary Cyclic Decomposition Theorem

The first step in the decomposition of a finitely generated module M over a principal ideal domain R is an easy one.

Theorem 6.9 *Any finitely generated module M over a principal ideal domain R is the direct sum of a finitely generated free R-module and a finitely generated torsion R-module*

$$M = M_{\text{free}} \oplus M_{\text{tor}}$$

The torsion part M_{tor} is unique, since it must be the set of all torsion elements of M, whereas the free part M_{free} is unique only up to isomorphism, that is, the rank of the free part is unique.

Proof. It is easy to see that the set M_{tor} of all torsion elements is a submodule of M and the quotient M/M_{tor} is torsion-free. Moreover, since M is finitely generated, so is M/M_{tor}. Hence, Theorem 6.8 implies that M/M_{tor} is free. Hence, Theorem 5.6 implies that

$$M = M_{\text{tor}} \oplus F$$

where $F \approx M/M_{\text{tor}}$ is free.

As to the uniqueness of the torsion part, suppose that $M \doteq T \oplus G$ where T is torsion and G is free. Then $T \subseteq M_{\text{tor}}$. But if $v = t + g \in M_{\text{tor}}$ for $t \in T$ and $g \in G$, then $g = v - t \in M_{\text{tor}}$ and so $g = 0$ and $v \in T$. Thus, $T = M_{\text{tor}}$.

For the free part, since $M = M_{\text{tor}} \oplus F = M_{\text{tor}} \oplus G$, the submodules F and G are both complements of M_{tor} and hence are isomorphic.□

Note that if $\{w_1, \ldots, w_m\}$ is a basis for M_{free} we can write

$$M = \langle\langle w_1 \rangle\rangle \oplus \cdots \oplus \langle\langle w_m \rangle\rangle \oplus M_{\text{tor}}$$

where each cyclic submodule $\langle\langle w_i \rangle\rangle$ has zero annihilator. This is a partial decomposition of M into a direct sum of cyclic submodules.

The Primary Decomposition

In view of Theorem 6.9, we turn our attention to the decomposition of finitely generated torsion modules M over a principal ideal domain. The first step is to decompose M into a direct sum of *primary* submodules, defined as follows.

Definition *Let p be a prime in R. A p-***primary** *(or just* **primary***) module is a module whose order is a power of p.*$\square$

Theorem 6.10 (The primary decomposition theorem) *Let M be a torsion module over a principal ideal domain R, with order*

$$\mu = p_1^{e_1} \cdots p_n^{e_n}$$

where the p_i's are distinct nonassociate primes in R.

1) M is the direct sum

$$M = M_{p_1} \oplus \cdots \oplus M_{p_n}$$

where

$$M_{p_i} = \frac{\mu}{p_i^{e_i}} M = \{v \in M \mid p_i^{e_i} v = 0\}$$

is a primary submodule of order $p_i^{e_i}$. This decomposition of M into primary submodules is called the **primary decomposition** *of M.*

2) The primary decomposition of M is unique up to order of the summands. That is, if

$$M = N_{q_1} \oplus \cdots \oplus N_{q_m}$$

where N_{q_i} is primary of order $q_i^{f_i}$ and $q_1, \ldots, q_m$ are distinct nonassociate primes, then $m = n$ and, after a possible reindexing, $N_{q_i} = M_{p_i}$. Hence, $f_i = e_i$ and $q_i \sim p_i$, for $i = 1, \ldots, n$.

3) Two R-modules M and N are isomorphic if and only if the summands in their primary decompositions are pairwise isomorphic, that is, if

$$M = M_{p_1} \oplus \cdots \oplus M_{p_n}$$

and

$$N = N_{q_1} \oplus \cdots \oplus N_{q_m}$$

are primary decompositions, then $m = n$ and, after a possible reindexing, $M_{p_i} \approx N_{q_i}$ for $i = 1, \ldots, n$.

Proof. Let us write $\mu_i = \mu / p_i^{e_i}$ and show first that

$$M_{p_i} = \mu_i M = \{\mu_i v \mid v \in M\}$$

Since $p_i^{e_i}(\mu_i M) = \mu M = \{0\}$, we have $\mu_i M \subseteq M_{p_i}$. On the other hand, since μ_i and $p_i^{e_i}$ are relatively prime, there exist $a, b \in R$ for which

$$a\mu_i + bp_i^{e_i} = 1$$

and so if $x \in M_{p_i}$ then

$$x = (a\mu_i + bp_i^{e_i})x = a\mu_i x \in \mu_i M$$

Hence $M_{p_i} = \mu_i M$.

For part 1), since $\gcd(\mu_1, \ldots, \mu_n) = 1$, there exist scalars a_i for which

$$a_1\mu_1 + \cdots + a_n\mu_n = 1$$

and so for any $x \in M$,

$$x = (a_1\mu_1 + \cdots + a_n\mu_n)x \in \sum_{i=1}^{n} \mu_i M$$

Moreover, since the $o(\mu_i M) \mid p_i^{e_i}$ and the $p_i^{e_i}$'s are pairwise relatively prime, it follows that the sum of the submodules $\mu_i M$ is direct, that is,

$$M = \mu_1 M \oplus \cdots \oplus \mu_n M = M_{p_1} \oplus \cdots \oplus M_{p_n}$$

As to the annihilators, it is clear that $\langle p_i^{e_i} \rangle \subseteq \operatorname{ann}(\mu_i M)$. For the reverse inclusion, if $r \in \operatorname{ann}(\mu_i M)$, then $r\mu_i \in \operatorname{ann}(M)$ and so $p_i^{e_i}\mu_i \mid r\mu_i$, that is, $p_i^{e_i} \mid r$ and so $r \in \langle p_i^{e_i} \rangle$. Thus $\operatorname{ann}(\mu_i M) = \langle p_i^{e_i} \rangle$.

As to uniqueness, we claim that $q = q_1^{f_1} \cdots q_m^{f_m}$ is an order of M. It is clear that q annihilates M and so $\mu \mid q$. On the other hand, N_{q_i} contains an element u_i of order $q_i^{f_i}$ and so the sum $v = u_1 + \cdots + u_m$ has order q, which implies that $q \mid \mu$. Hence, q and μ are associates.

Unique factorization in R now implies that $m = n$ and, after a suitable reindexing, that $f_i = e_i$ and q_i and p_i are associates. Hence, N_{q_i} is primary of order $p_i^{e_i}$. For convenience, we can write N_{q_i} as N_{p_i}. Hence,

$$N_{p_i} \subseteq \{v \in M \mid p_i^{e_i} v = 0\} = M_{p_i}$$

But if

$$N_{p_1} \oplus \cdots \oplus N_{p_n} = M_{p_1} \oplus \cdots \oplus M_{p_n}$$

and $N_{p_i} \subseteq M_{p_i}$ for all i, we must have $N_{p_i} = M_{p_i}$ for all i.

For part 3), if $m = n$ and $\sigma_i \colon M_{p_i} \approx N_{q_i}$, then the map $\sigma \colon M \to N$ defined by

$$\sigma(a_1 + \cdots + a_n) = \sigma_1(a_1) + \cdots + \sigma_n(a_n)$$

is an isomorphism and so $M \approx N$. Conversely, suppose that $\sigma \colon M \approx N$. Then M and N have the same annihilators and therefore the same order

$$\mu = p_1^{e_1} \cdots p_n^{e_n}$$

Hence, part 1) and part 2) imply that $m = n$ and after a suitable reindexing,

$q_i = p_i$. Moreover, since

$$a \in M_{p_i} \Leftrightarrow \mu_i a = 0 \Leftrightarrow \sigma(\mu_i a) = 0 \Leftrightarrow \mu_i \sigma a = 0 \Leftrightarrow \sigma a \in N_{p_i}$$

it follows that $\sigma\colon M_{p_i} \approx N_{p_i}$.□

The Cyclic Decomposition of a Primary Module

The next step in the decomposition process is to show that a primary module can be decomposed into a direct sum of cyclic submodules. While this decomposition is not unique (see the exercises), the set of annihilators is unique, as we will see. To establish this uniqueness, we use the following result.

Lemma 6.11 *Let M be a module over a principal ideal domain R and let $p \in R$ be a prime.*

1) *If $pM = \{0\}$, then M is a vector space over the field $R/\langle p\rangle$ with scalar multiplication defined by*

$$(r + \langle p\rangle)v = rv$$

for all $v \in M$.

2) *For any submodule S of M the set*

$$S^{(p)} = \{v \in S \mid pv = 0\}$$

is also a submodule of M and if $M = S \oplus T$, then

$$M^{(p)} = S^{(p)} \oplus T^{(p)}$$

Proof. For part 1), since p is prime, the ideal $\langle p\rangle$ is maximal and so $R/\langle p\rangle$ is a field. We leave the proof that M is a vector space over $R/\langle p\rangle$ to the reader. For part 2), it is straightforward to show that $S^{(p)}$ is a submodule of M. Since $S^{(p)} \subseteq S$ and $T^{(p)} \subseteq T$ we see that $S^{(p)} \cap T^{(p)} = \{0\}$. Also, if $v \in M^{(p)}$, then $pv = 0$. But $v = s + t$ for some $s \in S$ and $t \in T$ and so $0 = pv = ps + pt$. Since $ps \in S$ and $pt \in T$ we deduce that $ps = pt = 0$, whence $v \in S^{(p)} \oplus T^{(p)}$. Thus, $M^{(p)} \subseteq S^{(p)} \oplus T^{(p)}$. But the reverse inequality is manifest.□

Theorem 6.12 (The cyclic decomposition theorem of a primary module) *Let M be a primary finitely generated torsion module over a principal ideal domain R, with order p^e.*

1) *M is a direct sum*

$$M = \langle\langle v_1\rangle\rangle \oplus \cdots \oplus \langle\langle v_n\rangle\rangle \tag{6.1}$$

of cyclic submodules with annihilators $\operatorname{ann}(\langle\langle v_i\rangle\rangle) = \langle p^{e_i}\rangle$, which can be arranged in ascending order

$$\operatorname{ann}(\langle\langle v_1\rangle\rangle) \subseteq \cdots \subseteq \operatorname{ann}(\langle\langle v_n\rangle\rangle)$$

or equivalently,

$$e = e_1 \geq e_2 \geq \cdots \geq e_n$$

2) *As to uniqueness, suppose that* M *is also the direct sum*

$$M = \langle\langle u_1 \rangle\rangle \oplus \cdots \oplus \langle\langle u_m \rangle\rangle$$

of cyclic submodules with annihilators $\text{ann}(\langle\langle u_i \rangle\rangle) = \langle q^{f_i} \rangle$*, arranged in ascending order*

$$\text{ann}(\langle\langle u_1 \rangle\rangle) \subseteq \cdots \subseteq \text{ann}(\langle\langle u_m \rangle\rangle)$$

or equivalently

$$f_1 \geq f_2 \geq \cdots \geq f_m$$

Then the two chains of annihilators are identical, that is, $m = n$ *and*

$$\text{ann}(\langle\langle u_i \rangle\rangle) = \text{ann}(\langle\langle v_i \rangle\rangle)$$

for all i*. Thus,* $p \sim q$ *and* $f_i = e_i$ *for all* i*.*

3) *Two* p*-primary* R*-modules*

$$M = \langle\langle v_1 \rangle\rangle \oplus \cdots \oplus \langle\langle v_n \rangle\rangle$$

and

$$N = \langle\langle u_1 \rangle\rangle \oplus \cdots \oplus \langle\langle u_m \rangle\rangle$$

are isomorphic if and only if they have the same annihilator chains, that is, if and only if $m = n$ *and, after a possible reindexing,*

$$\text{ann}(\langle\langle u_i \rangle\rangle) = \text{ann}(\langle\langle v_i \rangle\rangle)$$

Proof. Let $v_1 \in M$ have order equal to the order of M, that is,

$$\text{ann}(v_1) = \text{ann}(M) = \langle p^e \rangle$$

Such an element must exist since $o(v_1) \leq p^e$ for all $v \in M$ and if this inequality is strict, then p^{e-1} will annihilate M.

If we show that $\langle\langle v_1 \rangle\rangle$ is complemented, that is, $M = \langle\langle v_1 \rangle\rangle \oplus S_1$ for some submodule S_1, then since S_1 is also a finitely generated primary torsion module over R, we can repeat the process to get

$$M = \langle\langle v_1 \rangle\rangle \oplus \langle\langle v_2 \rangle\rangle \oplus S_2$$

where $\text{ann}(v_i) = \langle p^{e_i} \rangle$. We can continue this decomposition:

$$M = \langle\langle v_1 \rangle\rangle \oplus \langle\langle v_2 \rangle\rangle \oplus \cdots \oplus \langle\langle v_n \rangle\rangle \oplus S_n$$

as long as $S_n \neq \{0\}$. But the ascending sequence of submodules

$$\langle\langle v_1 \rangle\rangle \subseteq \langle\langle v_1 \rangle\rangle \oplus \langle\langle v_2 \rangle\rangle \subseteq \cdots$$

must terminate since M is Noetherian and so there is an integer n for which eventually $S_n = \{0\}$, giving (6.1).

Let $v = v_1$. The direct sum $M_1 = \langle\langle v \rangle\rangle \oplus \{0\}$ clearly exists. Suppose that the direct sum

$$M_k = \langle\langle v \rangle\rangle \oplus S_k$$

exists. We claim that if $M_k < M$, then it is possible to find a submodule S_{k+1} for which $S_k < S_{k+1}$ and for which the direct sum $M_{k+1} = \langle\langle v \rangle\rangle \oplus S_{k+1}$ also exists. This process must also stop after a finite number of steps, giving $M = \langle\langle v \rangle\rangle \oplus S$ as desired.

If $M_k < M$ and $u \in M \setminus M_k$ let

$$S_{k+1} = \langle\langle S_k, u - \alpha v \rangle\rangle$$

for $\alpha \in R$. Then $S_k < S_{k+1}$ since $u \notin M_k$. We wish to show that for some $\alpha \in R$, the direct sum

$$\langle\langle v \rangle\rangle \oplus S_{k+1}$$

exists, that is,

$$x \in \langle\langle v \rangle\rangle \cap \langle\langle S_k, u - \alpha v \rangle\rangle \Rightarrow x = 0$$

Now, there exist scalars a and b for which

$$x = av = s + b(u - \alpha v)$$

for $s \in S_k$ and so if we find a scalar α for which

$$b(u - \alpha v) \in S_k \tag{6.2}$$

then $\langle\langle v \rangle\rangle \cap S_k = \{0\}$ implies that $x = 0$ and the proof of existence will be complete.

Solving for bu gives

$$bu = (a + \alpha b)v - s \in \langle\langle v \rangle\rangle \oplus S_k = M_k$$

so let us consider the ideal of all such scalars:

$$\mathcal{I} = \{r \in R \mid ru \in M_k\}$$

Since $p^e \in \mathcal{I}$ and $\mathcal{I}$ is principal, we have

$$\mathcal{I} = \langle p^f \rangle$$

for some $f \leq e$. Also, $f > 0$ since $u \notin M_k$ implies that $1 \notin \mathcal{I}$.

Since $b \in \mathcal{I}$, we have $b = \beta p^f$ and there exist $d \in R$ and $t \in S_k$ for which

$$p^f u = dv + t$$

Hence,

$$bu = \beta p^f u = \beta(dv + t) = \beta dv + \beta t$$

Now we need more information about d. Multiplying the expression for $p^f u$ by p^{e-f} gives

$$0 = p^e u = p^{e-f}(p^f u) = p^{e-f} dv + p^{e-f} t$$

and since $\langle\langle v \rangle\rangle \cap S_k = \{0\}$, it follows that $p^{e-f} dv = 0$. Hence, $p^e \mid p^{e-f} d$, that is, $p^f \mid d$ and so $d = \delta p^f$ for some $\delta \in R$. Now we can write

$$bu = \beta \delta p^f v + \beta t$$

and so

$$b(u - \delta v) = \beta t \in S_k$$

Thus, we take $\alpha = \delta$ to get (6.2) and that completes the proof of existence.

For uniqueness, note first that M has orders p^{e_1} and q^{f_1} and so p and q are associates and $e_1 = f_1$. Next we show that $n = m$. According to part 2) of Lemma 6.10,

$$M^{(p)} = \langle\langle v_1 \rangle\rangle^{(p)} \oplus \cdots \oplus \langle\langle v_n \rangle\rangle^{(p)}$$

and

$$M^{(p)} = \langle\langle u_1 \rangle\rangle^{(p)} \oplus \cdots \oplus \langle\langle u_m \rangle\rangle^{(p)}$$

where all summands are nonzero. Since $pM^{(p)} = \{0\}$, it follows from Lemma 6.10 that $M^{(p)}$ is a vector space over $R/\langle p \rangle$ and so each of the preceding decompositions expresses $M^{(p)}$ as a direct sum of one-dimensional vector subspaces. Hence, $m = \dim(M^{(p)}) = n$.

Finally, we show that the exponents e_i and f_i are equal using induction on e_1. If $e_1 = 1$, then $e_i = 1$ for all i and since $f_1 = e_1$, we also have $f_i = 1$ for all i. Suppose the result is true whenever $e_1 \leq k - 1$ and let $e_1 = k$. Write

$$(e_1, \ldots, e_n) = (e_1, \ldots, e_s, 1, \ldots, 1), e_s > 1$$

and

$$(f_1, \ldots, f_n) = (f_1, \ldots, f_t, 1, \ldots, 1), f_t > 1$$

Then

$$pM = p\langle\langle v_1\rangle\rangle \oplus \cdots \oplus p\langle\langle v_s\rangle\rangle$$

and

$$pM = p\langle\langle u_1\rangle\rangle \oplus \cdots \oplus p\langle\langle u_t\rangle\rangle$$

But $p\langle\langle v_1\rangle\rangle = \langle\langle pv_1\rangle\rangle$ is a cyclic submodule of M with annihilator $\langle p^{e_i-1}\rangle$ and so by the induction hypothesis

$$s = t \text{ and } e_1 = f_1, \ldots, e_s = f_s$$

which concludes the proof of uniqueness.

For part 3), suppose that $\sigma\colon M \approx N$ and M has annihilator chain

$$\text{ann}(\langle\langle v_1\rangle\rangle) \subseteq \cdots \subseteq \text{ann}(\langle\langle v_n\rangle\rangle)$$

and N has annihilator chain

$$\text{ann}(\langle\langle u_1\rangle\rangle) \subseteq \cdots \subseteq \text{ann}(\langle\langle u_m\rangle\rangle)$$

Then

$$N = \sigma M = \langle\langle \sigma v_1\rangle\rangle \oplus \cdots \oplus \langle\langle \sigma v_n\rangle\rangle$$

and so $m = n$ and after a suitable reindexing,

$$\text{ann}(\langle\langle v_i\rangle\rangle) = \text{ann}(\langle\langle \sigma v_i\rangle\rangle) = \text{ann}(\langle\langle u_i\rangle\rangle)$$

Conversely, suppose that

$$M = \langle\langle v_1\rangle\rangle \oplus \cdots \oplus \langle\langle v_n\rangle\rangle$$

and

$$N = \langle\langle u_1\rangle\rangle \oplus \cdots \oplus \langle\langle u_m\rangle\rangle$$

have the same annihilator chains, that is, $m = n$ and

$$\text{ann}(\langle\langle u_i\rangle\rangle) = \text{ann}(\langle\langle v_i\rangle\rangle)$$

Then

$$\langle\langle u_i\rangle\rangle \approx \frac{R}{\text{ann}(\langle\langle u_i\rangle\rangle)} = \frac{R}{\text{ann}(\langle\langle v_i\rangle\rangle)} \approx \langle\langle v_i\rangle\rangle \qquad \square$$

The Primary Cyclic Decomposition

Now we can combine the various decompositions.

Theorem 6.13 (The primary cyclic decomposition theorem) *Let M be a finitely generated torsion module over a principal ideal domain R.*

1) *If M has order*

$$\mu = p_1^{e_1} \cdots p_n^{e_n}$$

where the p_i's are distinct nonassociate primes in R, then M can be uniquely decomposed (up to the order of the summands) into the direct sum

$$M = M_{p_1} \oplus \cdots \oplus M_{p_n}$$

where

$$M_{p_i} = \frac{\mu}{p_i^{e_i}} M = \{v \in M \mid p_i^{e_i} v = 0\}$$

is a primary submodule with annihilator $\langle p_i^{e_i} \rangle$. Finally, each primary submodule M_{p_i} can be written as a direct sum of cyclic submodules, so that

$$M = \underbrace{[\langle\langle v_{1,1} \rangle\rangle \oplus \cdots \oplus \langle\langle v_{1,k_1} \rangle\rangle]}_{M_{p_1}} \oplus \cdots \oplus \underbrace{[\langle\langle v_{n,1} \rangle\rangle \oplus \cdots \oplus \langle\langle v_{n,k_n} \rangle\rangle]}_{M_{p_n}}$$

where $\operatorname{ann}(\langle\langle v_{i,j} \rangle\rangle) = \langle p_i^{e_{i,j}} \rangle$ and the terms in each cyclic decomposition can be arranged so that, for each i,

$$\operatorname{ann}(\langle\langle v_{i,1} \rangle\rangle) \subseteq \cdots \subseteq \operatorname{ann}(\langle\langle v_{i,k_i} \rangle\rangle)$$

or, equivalently,

$$e_i = e_{i,1} \geq e_{i,2} \geq \cdots \geq e_{i,k_i}$$

2) *As for uniqueness, suppose that*

$$M = \underbrace{[\langle\langle u_{1,1} \rangle\rangle \oplus \cdots \oplus \langle\langle u_{1,j_1} \rangle\rangle]}_{N_{q_1}} \oplus \cdots \oplus \underbrace{[\langle\langle u_{m,1} \rangle\rangle \oplus \cdots \oplus \langle\langle u_{m,j_m} \rangle\rangle]}_{N_{q_m}}$$

is also a primary cyclic decomposition of M. Then,

a) *The number of summands is the same in both decompositions; in fact, $m = n$ and after possible reindexing, $k_u = j_u$ for all u.*
b) *The primary submodules are the same; that is, after possible reindexing, $q_i \sim p_i$ and $N_{q_i} = M_{p_i}$*
c) *For each primary submodule pair $N_{q_i} = M_{p_i}$, the cyclic submodules have the same annihilator chains; that is, after possible reindexing,*

$$\operatorname{ann}(\langle\langle u_{i,j} \rangle\rangle) = \operatorname{ann}(\langle\langle v_{i,j} \rangle\rangle)$$

for all i, j.

In summary, the primary submodules and annihilator chains are uniquely determined by the module M.

3) *Two R-modules M and N are isomorphic if and only if they have the same annihilator chains.* □

Elementary Divisors

Since the chain of annihilators

$$\operatorname{ann}(\langle\langle v_{i,j}\rangle\rangle) = \langle p_i^{e_{i,j}}\rangle$$

is unique except for order, the multiset $\{p_i^{e_{i,j}}\}$ of generators is uniquely determined up to associate. The generators $p_i^{e_{i,j}}$ are called the **elementary divisors** of M. Note that for each prime p_i, the elementary divisor $p_i^{e_{i,j}}$ of largest exponent is precisely the factor of $o(M)$ associated to p_i.

Let us write $\operatorname{ElemDiv}(M)$ to denote the multiset of *all* elementary divisors of M. Thus, if $r \in \operatorname{ElemDiv}(M)$, then any associate of r is also in $\operatorname{ElemDiv}(M)$. We can now say that $\operatorname{ElemDiv}(M)$ is a complete invariant for isomorphism. Technically, the function $M \mapsto \operatorname{ElemDiv}(M)$ is the complete invariant, but this hair is not worth splitting. Also, we could work with a system of distinct representatives for the associate classes of the elementary divisors, but in general, there is no way to single out a special representative.

Theorem 6.14 *Let R be a principal ideal domain. The multiset* $\operatorname{ElemDiv}(M)$ *is a complete invariant for isomorphism of finitely generated torsion R-modules, that is,*

$$M \approx N \quad \Leftrightarrow \quad \operatorname{ElemDiv}(M) = \operatorname{ElemDiv}(N) \qquad \square$$

We have seen (Theorem 6.2) that if

$$M = A \oplus B$$

then

$$o(M) = \operatorname{lcm}(o(A), o(B))$$

Let us now compare the elementary divisors of M to those of A and B.

Theorem 6.15 *Let M be a finitely generated torsion module over a principal ideal domain and suppose that*

$$M = A \oplus B$$

1) The primary cyclic decomposition of M is the direct sum of the primary cyclic decompositons of A and B; that is, if

$$A = \bigoplus \langle\langle a_{i,j}\rangle\rangle \quad \text{and} \quad B = \bigoplus \langle\langle b_{i,j}\rangle\rangle$$

are the primary cyclic decompositions of A and B, respectively, then

$$M = \left(\bigoplus \langle\langle a_{i,j}\rangle\rangle\right) \oplus \left(\bigoplus \langle\langle b_{i,j}\rangle\rangle\right)$$

is the primary cyclic decomposition of M.

2) *The elementary divisors of M are*

$$\mathrm{ElemDiv}(M) = \mathrm{ElemDiv}(A) \cup \mathrm{ElemDiv}(B)$$

where the union is a multiset union; that is, we keep all duplicate members. $\square$

The Invariant Factor Decomposition

According to Theorem 6.4, if S and T are cyclic submodules with relatively prime orders, then $S \oplus T$ is a cyclic submodule whose order is the product of the orders of S and T. Accordingly, in the primary cyclic decomposition of M,

$$M = \underbrace{[\langle\langle v_{1,1}\rangle\rangle \oplus \cdots \oplus \langle\langle v_{1,k_1}\rangle\rangle]}_{M_{p_1}} \oplus \cdots \oplus \underbrace{[\langle\langle v_{n,1}\rangle\rangle \oplus \cdots \oplus \langle\langle v_{n,k_n}\rangle\rangle]}_{M_{p_n}}$$

with elementary divisors $p_i^{e_{i,j}}$ satisfying

$$e_i = e_{i,1} \geq e_{i,2} \geq \cdots \geq e_{i,k_i} \tag{6.3}$$

we can combine cyclic summands with relatively prime orders. One judicious way to do this is to take the leftmost (highest-order) cyclic submodules from each group to get

$$D_1 = \langle\langle v_{1,1}\rangle\rangle \oplus \cdots \oplus \langle\langle v_{n,1}\rangle\rangle$$

and repeat the process

$$\begin{aligned} D_2 &= \langle\langle v_{1,2}\rangle\rangle \oplus \cdots \oplus \langle\langle v_{n,2}\rangle\rangle \\ D_3 &= \langle\langle v_{1,3}\rangle\rangle \oplus \cdots \oplus \langle\langle v_{n,3}\rangle\rangle \\ &\vdots \end{aligned}$$

Of course, some summands may be missing here since different primary modules M_{p_i} do not necessarily have the same number of summands. In any case, the result of this regrouping and combining is a decomposition of the form

$$M = D_1 \oplus \cdots \oplus D_m$$

which is called an *invariant factor decomposition* of M.

For example, suppose that

$$M = [\langle\langle v_{1,1}\rangle\rangle \oplus \langle\langle v_{1,2}\rangle\rangle] \oplus [\langle\langle v_{2,1}\rangle\rangle] \oplus [\langle\langle v_{3,1}\rangle\rangle \oplus \langle\langle v_{3,2}\rangle\rangle \oplus \langle\langle v_{3,3}\rangle\rangle]$$

Then the resulting regrouping and combining gives

$$M = \underbrace{[\langle\langle v_{1,1}\rangle\rangle \oplus \langle\langle v_{2,1}\rangle\rangle \oplus \langle\langle v_{3,1}\rangle\rangle]}_{D_1} \oplus \underbrace{[\langle\langle v_{1,2}\rangle\rangle \oplus \langle\langle v_{3,2}\rangle\rangle]}_{D_2} \oplus \underbrace{[\langle\langle v_{3,3}\rangle\rangle]}_{D_3}$$

As to the orders of the summands, referring to (6.3), if D_i has order d_i, then since the highest powers of each prime p_i are taken for d_1, the second–highest for d_2 and so on, we conclude that

$$d_m \mid d_{m-1} \mid \cdots \mid d_2 \mid d_1 \tag{6.4}$$

or equivalently,

$$\operatorname{ann}(D_1) \subseteq \operatorname{ann}(D_2) \subseteq \cdots$$

The numbers d_i are called *invariant factors* of the decomposition.

For instance, in the example above suppose that the elementary divisors are

$$p_1^3, p_1^2, p_2, p_3^3, p_3^3, p_3$$

Then the invariant factors are

$$\begin{aligned} d_1 &= p_1^3 p_2 p_3^3 \\ d_2 &= p_1^2 p_3^3 \\ d_3 &= p_3 \end{aligned}$$

The process described above that passes from a sequence $p_i^{e_{i,j}}$ of elementary divisors in order (6.3) to a sequence of invariant factors in order (6.4) is reversible. The inverse process takes a sequence $d_1, \ldots, d_m$ satisfying (6.4), factors each d_i into a product of distinct nonassociate prime powers with the primes in the same order and then "peels off" like prime powers from the left. (The reader may wish to try it on the example above.)

This fact, together with Theorem 6.4, implies that primary cyclic decompositions and invariant factor decompositions are essentially equivalent. Therefore, since the multiset of elementary divisors of M is unique up to associate, the multiset of invariant factors of M is also unique up to associate. Furthermore, the multiset of invariant factors is a complete invariant for isomorphism.

Theorem 6.16 (The invariant factor decomposition theorem) *Let M be a finitely generated torsion module over a principal ideal domain R. Then*

$$M = D_1 \oplus \cdots \oplus D_m$$

where D_i is a cyclic submodule of M, with order d_i, where

$$d_m \mid d_{m-1} \mid \cdots \mid d_2 \mid d_1$$

This decomposition is called an **invariant factor decomposition** *of M and the scalars d_i are called the* **invariant factors** *of M.*

1) *The multiset of invariant factors is uniquely determined up to associate by the module M.*
2) *The multiset of invariant factors is a complete invariant for isomorphism.* □

The annihilators of an invariant factor decomposition are called the **invariant ideals** of M. The chain of invariant ideals is unique, as is the chain of annihilators in the primary cyclic decomposition. Note that d_1 is an order of M, that is,

$$\operatorname{ann}(M) = \langle d_1 \rangle$$

Note also that the product

$$\gamma = d_1 \cdots d_m$$

of the invariant factors of M has some nice properties. For example, γ is the product of all the elementary divisors of M. We will see in a later chapter that in the context of a linear operator τ on a vector space, γ is the characteristic polynomial of τ.

Characterizing Cyclic Modules

The primary cyclic decomposition can be used to characterize cyclic modules via their elementary divisors.

Theorem 6.17 *Let M be a finitely generated torsion module over a principal ideal domain, with order*

$$\mu = p_1^{e_1} \cdots p_n^{e_n}$$

The following are equivalent:

1) M is cyclic.

2) M is the direct sum

$$M = \langle\langle v_1 \rangle\rangle \oplus \cdots \oplus \langle\langle v_k \rangle\rangle$$

of primary cyclic submodules $\langle\langle v_i \rangle\rangle$ of order $p_i^{e_i}$.

3) The elementary divisors of M are precisely the prime power factors of μ:

$$\operatorname{ElemDiv}(M) = \{p_1^{e_1}, \ldots, p_n^{e_n}\}$$

Proof. Suppose that M is cyclic. Then the primary decomposition of M is a primary *cyclic* decomposition, since any submodule of a cyclic module is cyclic. Hence, 1) implies 2). Conversely, if 2) holds, then since the orders are relatively prime, Theorem 6.4 implies that M is cyclic. We leave the rest of the proof to the reader.□

Indecomposable Modules

The primary cyclic decomposition of M is a decomposition of M into a direct sum of submodules that cannot be further decomposed. In fact, this characterizes the primary cyclic decomposition of M. Before justifying these statements, we make the following definition.

Definition *A module M is* **indecomposable** *if it cannot be written as a direct sum of proper submodules.*□

We leave proof of the following as an exercise.

Theorem 6.18 *Let M be a finitely generated torsion module over a principal ideal domain. The following are equivalent:*
1) M is indecomposable
2) M is primary cyclic
3) M has only one elementary divisor:

$$\mathrm{ElemDiv}(M) = \{p^e\}$$
□

Thus, the primary cyclic decomposition of M is a decomposition of M into a direct sum of indecomposable modules. Conversely, if

$$M = A_1 \oplus \cdots \oplus A_m$$

is a decomposition of M into a direct sum of indecomposable submodules, then each submodule A_i is primary cyclic and so this is the primary cyclic decomposition of M.

Indecomposable Submodules of Prime Order

Readers acquainted with group theory know that any group of prime order is cyclic. However, as mentioned earlier, the order of a module corresponds to the smallest exponent of a group, not to the order of a group. Indeed, there are modules of prime order that are not cyclic. Nevertheless, cyclic modules of prime order are important.

Indeed, if M is a finitely generated torsion module over a principal ideal domain, with order μ, then each prime factor p of μ gives rise to a cyclic submodule W of M whose order is p and so W is also indecomposable. Unfortunately, W need not be complemented and so we cannot use it to decompose M. Nevertheless, the theorem is still useful, as we will see in a later chapter.

Theorem 6.19 *Let M be a finitely generated torsion module over a principal ideal domain, with order μ. If p is a prime divisor of μ, then M has a cyclic (equivalently, indecomposable) submodule W of prime order p.*
Proof. If $\mu = pq$, then there is a $v \in M$ for which $w = qv \neq 0$ but $pw = 0$. Then $W = \langle\langle w \rangle\rangle$ is annihilated by p and so $o(w) \mid p$. But p is prime and $o(w) \neq 1$ and so $o(w) = p$. Since W has prime order, Theorem 6.18 implies that W is cyclic if and only if it is indecomposable.□

Exercises

1. Show that any free module over an integral domain is torsion-free.
2. Let M be a finitely generated torsion module over a principal ideal domain. Prove that the following are equivalent:
   a) M is indecomposable
   b) M has only one elementary divisor (including multiplicity)

c) M is cyclic of prime power order.

3. Let R be a principal ideal domain and R^+ the field of quotients. Then R^+ is an R-module. Prove that any nonzero finitely generated submodule of R^+ is a free module of rank 1.
4. Let R be a principal ideal domain. Let M be a finitely generated torsion-free R-module. Suppose that N is a submodule of M for which N is a free R-module of rank 1 and M/N is a torsion module. Prove that M is a free R-module of rank 1.
5. Show that the primary cyclic decomposition of a torsion module over a principal ideal domain is not unique (even though the elementary divisors are).
6. Show that if M is a finitely generated R-module where R is a principal ideal domain, then the free summand in the decomposition $M = F \oplus M_{\text{tor}}$ need not be unique.
7. If $\langle\langle v \rangle\rangle$ is a cyclic R-module of order a show that the map $\tau: R \to \langle\langle v \rangle\rangle$ defined by $\tau r = rv$ is a surjective R-homomorphism with kernel $\langle a \rangle$ and so
$$\langle\langle v \rangle\rangle \approx \frac{R}{\langle a \rangle}$$
8. If R is an integral domain with the property that all submodules of cyclic R-modules are cyclic, show that R is a principal ideal domain.
9. Suppose that F is a finite field and let F^* be the set of all nonzero elements of F.
   a) Show that if $p(x) \in F[x]$ is a nonconstant polynomial over F and if $r \in F$ is a root of $p(x)$, then $x - r$ is a factor of $p(x)$.
   b) Prove that a nonconstant polynomial $p(x) \in F[x]$ of degree n can have at most n distinct roots in F.
   c) Use the invariant factor or primary cyclic decomposition of a finite $\mathbb{Z}$-module to prove that F^* is cyclic.
10. Let R be a principal ideal domain. Let $M = \langle\langle v \rangle\rangle$ be a cyclic R-module with order α. We have seen that any submodule of M is cyclic. Prove that for each $\beta \in R$ such that $\beta \mid \alpha$ there is a unique submodule of M of order β.
11. Suppose that M is a free module of finite rank over a principal ideal domain R. Let N be a submodule of M. If M/N is torsion, prove that $\text{rk}(N) = \text{rk}(M)$.
12. Let $F[x]$ be the ring of polynomials over a field F and let $F'[x]$ be the ring of all polynomials in $F[x]$ that have coefficient of x equal to 0. Then $F[x]$ is an $F'[x]$-module. Show that $F[x]$ is finitely generated and torsion-free but not free. Is $F'[x]$ a principal ideal domain?
13. Show that the rational numbers $\mathbb{Q}$ form a torsion-free $\mathbb{Z}$-module that is not free.

More on Complemented Submodules

14. Let R be a principal ideal domain and let M be a free R-module.

a) Prove that a submodule N of M is complemented if and only if M/N is free.
b) If M is also finitely generated, prove that N is complemented if and only if M/N is torsion-free.

15. Let M be a free module of finite rank over a principal ideal domain R.
 a) Prove that if N is a complemented submodule of M, then $\operatorname{rk}(N) = \operatorname{rk}(M)$ if and only if $N = M$.
 b) Show that this need not hold if N is not complemented.
 c) Prove that N is complemented if and only if any basis for N can be extended to a basis for M.
16. Let M and N be free modules of finite rank over a principal ideal domain R. Let $\tau\colon M \to N$ be an R-homomorphism.
 a) Prove that $\ker(\tau)$ is complemented.
 b) What about $\operatorname{im}(\tau)$?
 c) Prove that

$$\operatorname{rk}(M) = \operatorname{rk}(\ker(\tau)) + \operatorname{rk}(\operatorname{im}(\tau)) = \operatorname{rk}(\ker(\tau)) + \operatorname{rk}\left(\frac{M}{\ker(\tau)}\right)$$

 d) If τ is surjective, then τ is an isomorphism if and only if $\operatorname{rk}(M) = \operatorname{rk}(N)$.
 e) If L is a submodule of M and if M/L is free, then

$$\operatorname{rk}\left(\frac{M}{L}\right) = \operatorname{rk}(M) - \operatorname{rk}(L)$$

17. A submodule N of a module M is said to be **pure in** M if whenever $v \notin M \setminus N$, then $rv \notin N$ for all nonzero $r \in R$.
 a) Show that N is pure if and only if $v \in N$ and $v = rw$ for $r \in R$ implies $w \in N$.
 b) Show that N is pure if and only if M/N is torsion-free.
 c) If R is a principal ideal domain and M is finitely generated, prove that N is pure if and only if M/N is free.
 d) If L and N are pure submodules of M, then so are $L \cap N$ and $L \cup N$. What about $L + N$?
 e) If N is pure in M, then show that $L \cap N$ is pure in L for any submodule L of M.
18. Let M be a free module of finite rank over a principal ideal domain R. Let L and N be submodules of M with L complemented in M. Prove that

$$\operatorname{rk}(L + N) + \operatorname{rk}(L \cap N) = \operatorname{rk}(L) + \operatorname{rk}(N)$$

Chapter 7
The Structure of a Linear Operator

In this chapter, we study the structure of a linear operator on a finite-dimensional vector space, using the powerful module decomposition theorems of the previous chapter. *Unless otherwise noted, all vector spaces will be assumed to be finite-dimensional.*

Let V be a finite-dimensional vector space. Let us recall two earler theorems (Theorem 2.19 and Theorem 2.20).

Theorem 7.1 *Let V be a vector space of dimension n.*

1) *Two $n \times n$ matrices A and B are similar* (*written* $A \sim B$) *if and only if they represent the same linear operator $\tau \in \mathcal{L}(V)$, but possibly with respect to different ordered bases. In this case, the matrices A and B represent exactly the same set of linear operators in $\mathcal{L}(V)$.*
2) *Then two linear operators τ and σ on V are similar* (*written* $\tau \sim \sigma$) *if and only if there is a matrix $A \in \mathcal{M}_n$ that represents both operators, but with respect to possibly different ordered bases. In this case, τ and σ are represented by exactly the same set of matrices in $\mathcal{M}_n$.*□

Theorem 7.1 implies that the matrices that represent a given linear operator are precisely the matrices that lie in one similarity class. Hence, in order to uniquely represent all linear operators on V, we would like to find a set consisting of one simple representative of each similarity class, that is, a set of simple canonical forms for similarity.

One of the simplest types of matrix is the diagonal matrix. However, these are too simple, since some operators cannot be represented by a diagonal matrix. A less simple type of matrix is the upper triangular matrix. However, these are not simple enough: Every operator (over an algebraically closed field) can be represented by an upper triangular matrix but some operators can be represented by more than one upper triangular matrix.

This gives rise to two different directions for further study. First, we can search for a characterization of those linear operators that can be represented by diagonal matrices. Such operators are called *diagonalizable*. Second, we can search for a different type of "simple" matrix that does provide a set of canonical forms for similarity. We will pursue both of these directions.

The Module Associated with a Linear Operator

If $\tau \in \mathcal{L}(V)$, we will think of V not only as a vector space over a field F but also as a module over $F[x]$, with scalar multiplication defined by

$$p(x)v = p(\tau)(v)$$

We will write V_τ to indicate the dependence on τ. Thus, V_τ and V_σ are modules with the same ring of scalars $F[x]$, although with different scalar multiplication if $\tau \neq \sigma$.

Our plan is to interpret the concepts of the previous chapter for the module V_τ. First, if $\dim(V) = n$, then $\dim(\mathcal{L}(V)) = n^2$. This implies that V_τ is a torsion module. In fact, the $n^2 + 1$ vectors

$$\iota, \tau, \tau^2, \ldots, \tau^{n^2}$$

are linearly dependent in $\mathcal{L}(V)$, which implies that $p(\tau) = 0$ for some nonzero polynomial $p(x) \in F[x]$. Hence, $p(x) \in \operatorname{ann}(V_\tau)$ and so $\operatorname{ann}(V_\tau)$ is a nonzero principal ideal of $F[x]$.

Also, since V is finitely generated as a vector space, it is, a fortiori, finitely generated as an $F[x]$-module. Thus, V is a finitely generated torsion module over a principal ideal domain $F[x]$ and so we may apply the decomposition theorems of the previous chapter. In the first part of this chapter, we embark on a "translation project" to translate the powerful results of the previous chapter into the language of the modules V_τ.

Let us first characterize when two modules V_τ and V_σ are isomorphic.

Theorem 7.2 *If $\tau, \sigma \in \mathcal{L}(V)$, then*

$$V_\tau \approx V_\sigma \quad \Leftrightarrow \quad \tau \sim \sigma$$

In particular, $\phi: V_\tau \to V_\sigma$ is a module isomorphism if and only if ϕ is a vector space automorphism of V satisfying

$$\sigma = \phi\tau\phi^{-1}$$

Proof. Suppose that $\phi: V_\tau \to V_\sigma$ is a module isomorphism. Then for $v \in V$,

$$\phi(xv) = x(\phi v)$$

which is equivalent to

$$\phi(\tau v) = \sigma(\phi v)$$

and since ϕ is bijective, this is equivalent to

$$(\phi\tau\phi^{-1})v = \sigma v$$

that is, $\sigma = \phi\tau\phi^{-1}$. Since a module isomorphism from V_τ to V_σ is a vector space isomorphism as well, the result follows.

For the converse, suppose that ϕ is a vector space automorphism of V and $\sigma = \phi\tau\phi^{-1}$, that is, $\phi\tau = \sigma\phi$. Then

$$\phi(x^k v) = \phi(\tau^k v) = \sigma^k(\phi v) = x^k(\phi v)$$

and the F-linearity of ϕ implies that for any polynomial $p(x) \in F[x]$,

$$\phi(p(\tau)v) = p(\sigma)\phi v$$

Hence, ϕ is a module isomorphism from V_τ to V_σ.□

Submodules and Invariant Subspaces

There is a simple connection between the submodules of the $F[x]$-module V_τ and the subspaces of the vector space V. Recall that a subspace S of V is τ-**invariant** if $\tau S \subseteq S$.

Theorem 7.3 *A subset $S \subseteq V$ is a submodule of V_τ if and only if S is a τ-invariant subspace of V.*□

Orders and the Minimal Polynomial

We have seen that the annihilator of V_τ,

$$\text{ann}(V_\tau) = \{p(x) \in F[x] \mid p(x)V_\tau = \{0\}\}$$

is a nonzero principal ideal of $F[x]$, say

$$\text{ann}(V_\tau) = \langle m(x) \rangle$$

Since the elements of the base ring $F[x]$ of V_τ are polynomials, for the first time in our study of modules there is a logical choice among all scalars in a given associate class: Each associate class contains exactly one *monic* polynomial.

Definition *Let $\tau \in \mathcal{L}(V)$. The unique monic order of V_τ is called the* **minimal polynomial** *for τ and is denoted by $m_\tau(x)$ or* $\min(\tau)$. *Thus,*

$$\text{ann}(V_\tau) = \langle m_\tau(x) \rangle$$ □

In treatments of linear algebra that do not emphasize the role of the module V_τ, the minimal polynomial of a linear operator τ is simply defined as the unique *monic* polynomial $m_\tau(x)$ of *smallest degree* for which $m_\tau(\tau) = 0$. This definition is equivalent to our definition.

The concept of minimal polynomial is also defined for matrices. The **minimal polynomial** $m_A(x)$ of matrix $A \in \mathcal{M}_n(F)$ is defined as the minimal polynomial of the multiplication operator τ_A. Equivalently, $m_A(x)$ is the unique monic polynomial $p(x) \in F[x]$ of smallest degree for which $p(A) = 0$.

Theorem 7.4

1) *If $\tau \sim \sigma$ are similar linear operators on V, then $m_\tau(x) = m_\sigma(x)$. Thus, the minimal polynomial is an invariant under similarity of operators.*
2) *If $A \sim B$ are similar matrices, then $m_A(x) = m_B(x)$. Thus, the minimal polynomial is an invariant under similarity of matrices.*
3) *The minimal polynomial of $\tau \in \mathcal{L}(V)$ is the same as the minimal polynomial of any matrix that represents τ.*□

Cyclic Submodules and Cyclic Subspaces

Let us now look at the cyclic submodules of V_τ:

$$\langle\langle v \rangle\rangle = F[x]v = \{p(\tau)(v) \mid p(x) \in F[x]\}$$

which are τ-invariant subspaces of V. Let $m(x)$ be the minimal polynomial of $\tau|_{\langle\langle v \rangle\rangle}$ and suppose that $\deg(m(x)) = n$. If $p(x)v \in \langle\langle v \rangle\rangle$, then writing

$$p(x) = q(x)m(x) + r(x)$$

where $\deg r(x) < \deg m(x)$ gives

$$p(x)v = [q(x)m(x) + r(x)]v = r(x)v$$

and so

$$\langle\langle v \rangle\rangle = \{r(x)v \mid \deg r(x) < n\}$$

Hence, the set

$$\mathcal{B} = \{v, xv, \ldots, x^{n-1}v\} = \{v, \tau v, \ldots, \tau^{n-1}v\}$$

spans the *vector space* $\langle\langle v \rangle\rangle$. To see that $\mathcal{B}$ is a basis for $\langle\langle v \rangle\rangle$, note that any linear combination of the vectors in $\mathcal{B}$ has the form $r(x)v$ for $\deg(r(x)) < n$ and so is equal to 0 if and only if $r(x) = 0$. Thus, $\mathcal{B}$ is an ordered basis for $\langle\langle v \rangle\rangle$.

Definition *Let $\tau \in \mathcal{L}(V)$. A τ-invariant subspace S of V is τ-***cyclic** *if S has a basis of the form*

$$\mathcal{B} = \{v, \tau v, \ldots, \tau^{n-1}v\}$$

*for some $v \in V$ and $n > 0$. The basis $\mathcal{B}$ is called a τ-***cyclic basis** *for V.*□

Thus, a cyclic submodule $\langle\langle v\rangle\rangle$ of V_τ with order $m(x)$ of degree n is a τ-cyclic subspace of V of dimension n. The converse is also true, for if

$$\mathcal{B}=\{v,\tau v,\ldots,\tau^{n-1}v\}$$

is a basis for a τ-invariant subspace S of V, then S is a submodule of V_τ. Moreover, the minimal polynomial of $\tau|_S$ has degree n, since if

$$\tau^n v=-a_0v-a_1\tau v-\cdots-a_{n-1}\tau^{n-1}v$$

then $\tau|_S$ satisfies the polynomial

$$m(x)=a_0+a_1x+\cdots+a_{n-1}x^{n-1}+x^n$$

but none of smaller degree since $\mathcal{B}$ is linearly independent.

Theorem 7.5 *Let V be a finite-dimenional vector space and let $S\subseteq V$. The following are equivalent:*
1) S is a cyclic submodule of V_τ with order $m(x)$ of degree n
2) S is a τ-cyclic subspace of V of dimension n.□

We will have more to say about cyclic modules a bit later in the chapter.

Summary

The following table summarizes the connection between the module concepts and the vector space concepts that we have discussed so far.

$F[x]$**-Module** V_τ	F**-Vector Space** V
Scalar multiplication: $p(x)v$	Action of $p(\tau)$: $p(\tau)(v)$
Submodule of V_τ	τ-Invariant subspace of V
Annihilator: $\operatorname{ann}(V_\tau)=\{p(x)\mid p(x)V_\tau=\{0\}\}$	Annihilator: $\operatorname{ann}(V)=\{p(x)\mid p(\tau)(V)=\{0\}\}$
Monic order $m(x)$ of V_τ: $\operatorname{ann}(V_\tau)=\langle m(x)\rangle$	Minimal polynomial of τ: $m(x)$ has smallest deg with $m(\tau)=0$
Cyclic submodule of V_τ: $\langle\langle v\rangle\rangle=\{p(x)v\mid \deg p(x)<\deg m(x)\}$	τ-cyclic subspace of V: $\langle v,\tau v,\ldots,\tau^{m-1}(v)\rangle, m=\deg(p(x))$

The Primary Cyclic Decomposition of V_τ

We are now ready to translate the cyclic decomposition theorem into the language of V_τ.

Definition *Let $\tau\in\mathcal{L}(V)$.*
1) The **elementary divisors** *and* **invariant factors** *of τ are the* monic *elementary divisors and invariant factors, respectively, of the module V_τ. We denote the multiset of elementary divisors of τ by* ElemDiv(τ) *and the multiset of invariant factors of τ by* InvFact(τ).

2) *The* **elementary divisors** *and* **invariant factors** *of a matrix A are the elementary divisors and invariant factors, respectively, of the multiplication operator τ_A:*

$$\text{ElemDiv}(A) = \text{ElemDiv}(\tau_A) \quad \textit{and} \quad \text{InvFact}(A) = \text{InvFact}(\tau_A) \qquad \square$$

We emphasize that the elementary divisors and invariant factors of an operator or matrix are *monic* by definition. Thus, we no longer need to worry about uniqueness up to associate.

Theorem 7.6 (The primary cyclic decomposition theorem for V) *Let V be finite-dimensional and let $\tau \in \mathcal{L}(V)$ have minimal polynomial*

$$m_\tau(x) = p_1^{e_1}(x) \cdots p_n^{e_n}(x)$$

where the polynomials $p_i(x)$ are distinct monic primes.

1) **(Primary decomposition)** *The $F[x]$-module V_τ is the direct sum*

$$V_\tau = V_{p_1} \oplus \cdots \oplus V_{p_n}$$

where

$$V_{p_i} = \frac{m_\tau(x)}{p_i^{e_i}(x)} V = \{v \in V \mid p_i^{e_i}(\tau)(v) = 0\}$$

is a primary submodule of V_τ of order $p_i^{e_i}(x)$. In vector space terms, V_{p_i} is a τ-invariant subspace of V and the minimal polynomial of $\tau|_{V_{p_i}}$ is

$$\min(\tau|_{V_{p_i}}) = p_i^{e_i}(x)$$

2) **(Cyclic decomposition)** *Each primary summand V_{p_i} can be decomposed into a direct sum*

$$V_{p_i} = \langle\langle v_{i,1} \rangle\rangle \oplus \cdots \oplus \langle\langle v_{i,k_i} \rangle\rangle$$

of τ-cyclic submodules $\langle\langle v_{i,j} \rangle\rangle$ of order $p_i^{e_{i,j}}(x)$ with

$$e_i = e_{i,1} \geq e_{i,2} \geq \cdots \geq e_{i,k_i}$$

In vector space terms, $\langle\langle v_{i,j} \rangle\rangle$ is a τ-cyclic subspace of V_{p_i} and the minimal polynomial of $\tau|_{\langle\langle v_{i,j} \rangle\rangle}$ is

$$\min(\tau|_{\langle\langle v_{i,j} \rangle\rangle}) = p_i^{e_{i,j}}(x)$$

3) **(The complete decomposition)** *This yields the decomposition of V_τ into a direct sum of τ-cyclic subspaces*

$$V_\tau = (\langle\langle v_{1,1} \rangle\rangle \oplus \cdots \oplus \langle\langle v_{1,k_1} \rangle\rangle) \oplus \cdots \oplus (\langle\langle v_{n,1} \rangle\rangle \oplus \cdots \oplus \langle\langle v_{n,k_n} \rangle\rangle)$$

4) **(Elementary divisors and dimensions)** *The multiset of elementary divisors $\{p_i^{e_{i,j}}(x)\}$ is uniquely determined by τ. If $\deg(p_i^{e_{i,j}}(x)) = d_{i,j}$, then the τ-*

cyclic subspace $\langle\langle v_{i,j}\rangle\rangle$ *has* τ*-cyclic basis*

$$\mathcal{B}_{i,j} = \left(v_{i,j}, \tau v_{i,j}, \ldots, \tau^{d_{i,j}-1} v_{i,j}\right)$$

and $\dim(\langle\langle v_{i,j}\rangle\rangle) = \deg(p_i^{e_{i,j}})$. *Hence,*

$$\dim(V_{p_i}) = \sum_{j=1}^{k_i} \deg(p_i^{e_{i,j}})$$

We will call the basis

$$\mathcal{R} = \bigcup_{i,j} \mathcal{B}_{i,j}$$

for V *the* **elementary divisor basis** *for* V_τ.□

Recall that if $V = A \oplus B$ and if both A and B are τ-invariant subspaces of V, the pair (A, B) is said to **reduce** τ. In module language, the pair (A, B) reduces τ if A and B are submodules of V_τ and

$$V_\tau = A_\tau \oplus B_\tau$$

We can now translate Theorem 6.15 into the current context.

Theorem 7.7 *Let* $\tau \in \mathcal{L}(V)$ *and let*

$$V_\tau = A_\tau \oplus B_\tau$$

1) *The minimal polynomial of* τ *is*

$$m_\tau(x) = \operatorname{lcm}(m_{\tau|_A}(x), m_{\tau|_B}(x))$$

2) *The primary cyclic decomposition of* V_τ *is the direct sum of the primary cyclic decompositons of* A_τ *and* B_τ*; that is, if*

$$A_\tau = \bigoplus \langle\langle a_{i,j}\rangle\rangle \quad \text{and} \quad B_\tau = \bigoplus \langle\langle b_{i,j}\rangle\rangle$$

are the primary cyclic decompositions of A_τ *and* B_τ*, respectively, then*

$$V_\tau = \left(\bigoplus \langle\langle a_{i,j}\rangle\rangle\right) \oplus \left(\bigoplus \langle\langle b_{i,j}\rangle\rangle\right)$$

is the primary cyclic decomposition of V_τ.

3) *The elementary divisors of* τ *are*

$$\operatorname{ElemDiv}(\tau) = \operatorname{ElemDiv}(\tau|_A) \cup \operatorname{ElemDiv}(\tau|_B)$$

where the union is a multiset union; that is, we keep all duplicate members.□

The Characteristic Polynomial

To continue our translation project, we need a definition. Recall that in the characterization of cyclic modules in Theorem 6.17, we made reference to the product of the elementary divisors, one from each associate class. Now that we have singled out a special representative from each associate class, we can make a useful definition.

Definition *Let $\tau \in \mathcal{L}(V)$. The* **characteristic polynomial** *$c_\tau(x)$ of τ is the product of all of the elementary divisors of τ:*

$$c_\tau(x) = \prod_{i,j} p_i^{e_{i,j}}(x)$$

Hence,

$$\deg(c_\tau(x)) = \dim(V)$$

Similarly, the **characteristic polynomial** *$c_M(x)$ of a matrix M is the product of the elementary divisors of M.*□

The following theorem describes the relationship between the minimal and characteristic polynomials.

Theorem 7.8 *Let $\tau \in \mathcal{L}(V)$.*

1) **(The Cayley–Hamilton theorem)** *The minimal polynomial of τ divides the characteristic polynomial of τ:*

$$m_\tau(x) \mid c_\tau(x)$$

Equivalently, τ satisfies its own characteristic polynomial, that is,

$$c_\tau(\tau) = 0$$

2) The minimal polynomial

$$m_\tau(x) = p_1^{e_{1,1}}(x) \cdots p_n^{e_{n,1}}(x)$$

and characteristic polynomial

$$c_\tau(x) = \prod_{i,j} p_i^{e_{i,j}}(x)$$

of τ have the same set of prime factors $p_i(x)$ and hence the same set of roots (not counting multiplicity).□

We have seen that the multiset of elementary divisors forms a complete invariant for similarity. The reader should construct an example to show that the pair $(m_\tau(x), c_\tau(x))$ is *not* a complete invariant for similarity, that is, this pair of polynomials does not uniquely determine the multiset of elementary divisors of the operator τ.

In general, the minimal polynomial of a linear operator is hard to find. One of the virtues of the characteristic polynomial is that it is comparatively easy to find and we will discuss this in detail a bit later in the chapter.

Note that since $m_\tau(x) \mid c_\tau(x)$ and both polynomials are monic, it follows that

$$m_\tau(x) = c_\tau(x) \quad \Leftrightarrow \quad \deg(m_\tau(x)) = \deg(c_\tau(x))$$

Definition *A linear operator $\tau \in \mathcal{L}(V)$ is* **nonderogatory** *if its minimal polynomial is equal to its characteristic polynomial:*

$$m_\tau(x) = c_\tau(x)$$

or equivalently, if

$$\deg(m_\tau(x)) = \deg(c_\tau(x))$$

or if

$$\deg(m_\tau(x)) = \dim(V)$$

Similar statements hold for matrices. □

Cyclic and Indecomposable Modules

We have seen (Theorem 6.17) that cyclic submodules can be characterized by their elementary divisors. Let us translate this theorem into the language of V_τ (and add one more equivalence related to the characteristic polynomial).

Theorem 7.9 *Let $\tau \in \mathcal{L}(V)$ have minimal polynomial*

$$m_\tau(x) = p_1^{e_1}(x) \cdots p_n^{e_n}(x)$$

where $p_i(x)$ are distinct monic primes. The following are equivalent:

1) *V_τ is cyclic.*
2) *V_τ is the direct sum*

$$V_\tau = \langle\langle v_1 \rangle\rangle \oplus \cdots \oplus \langle\langle v_k \rangle\rangle$$

of τ-cyclic submodules $\langle\langle v_i \rangle\rangle$ of order $p_i^{e_i}(x)$.

3) *The elementary divisors of τ are*

$$\operatorname{ElemDiv}(\tau) = \{p_1^{e_1}(x), \ldots, p_n^{e_n}(x)\}$$

4) *τ is nonderogatory, that is,*

$$m_\tau(x) = c_\tau(x)$$ □

Indecomposable Modules

We have also seen (Theorem 6.19) that, in the language of V_τ, each prime factor $p(x)$ of the minimal polynomial $m_\tau(x)$ gives rise to a cyclic submodule W of V of prime order $p(x)$.

Theorem 7.10 *Let $\tau \in \mathcal{L}(V)$ and let $p(x)$ be a prime factor of $m_\tau(x)$. Then V_τ has a cyclic submodule W_τ of prime order $p(x)$.*□

For a module of prime order, we have the following.

Theorem 7.11 *For a module W_τ of prime order $m_\tau(x)$, the following are equivalent:*
1) W_τ *is cyclic*
2) W_τ *is indecomposable*
3) $c_\tau(x)$ *is irreducible*
4) τ *is nonderogatory, that is,* $c_\tau(x) = m_\tau(x)$
5) $\dim(W_\tau) = \deg(p(x))$.□

Our translation project is now complete and we can begin to look at issues that are specific to the modules V_τ.

Companion Matrices

We can also characterize the cyclic modules V_τ via the matrix representations of the operator τ, which is obviously something that we could not do for arbitrary modules. Let $V_\tau = \langle\langle v \rangle\rangle$ be a cyclic module, with order

$$m_\tau(x) = a_0 + a_1 x + \cdots + a_{n-1}x^{n-1} + x^n$$

and ordered τ-cyclic basis

$$\mathcal{B} = \left(v, \tau v, \ldots, \tau^{n-1}v\right)$$

Then

$$\tau(\tau^i v) = \tau^{i+1}v$$

for $0 \le i \le n-2$ and

$$\begin{aligned}\tau(\tau^{n-1}v) &= \tau^n v\\ &= -(a_0 + a_1\tau + \cdots + a_{n-1}\tau^{n-1})v\\ &= -a_0 v - a_1 \tau v - \cdots - a_{n-1}\tau^{n-1}v\end{aligned}$$

and so

$$[\tau]_{\mathcal{B}}=\begin{bmatrix}0&0&\cdots&0&-a_0\\1&0&\cdots&0&-a_1\\0&1&\ddots&&\vdots\\\vdots&\vdots&\ddots&0&-a_{n-2}\\0&0&\cdots&1&-a_{n-1}\end{bmatrix}$$

This matrix is known as the *companion matrix* for the polynomial $m_\tau(x)$.

Definition *The* **companion matrix** *of a monic polyomial*

$$p(x)=a_0+a_1x+\cdots+a_{n-1}x^{n-1}+x^n$$

is the matrix

$$C[p(x)]=\begin{bmatrix}0&0&\cdots&0&-a_0\\1&0&\cdots&0&-a_1\\0&1&\ddots&&\vdots\\\vdots&\vdots&\ddots&0&-a_{n-2}\\0&0&\cdots&1&-a_{n-1}\end{bmatrix}$$ □

Note that companion matrices are defined only for *monic* polynomials. Companion matrices are nonderogatory. Also, companion matrices are precisely the matrices that represent operators on τ-cyclic subspaces.

Theorem 7.12 *Let* $p(x)\in F[x]$.
1) A companion matrix $A=C[p(x)]$ *is nonderogatory; in fact,*

$$c_A(x)=m_A(x)=p(x)$$

2) V_τ *is cyclic if and only if* τ *can be represented by a companion matrix, in which case the representing basis is* τ*-cyclic.*

Proof. For part 1), let $\mathcal{E}=(e_1,\ldots,e_n)$ be the standard basis for F^n. Since $e_i=A^{i-1}e_1$ for $i\geq 2$, it follows that for any polynomial $f(x)$,

$$f(A)=0\quad\Leftrightarrow\quad f(A)e_i=0\text{ for all }i\quad\Leftrightarrow\quad f(A)e_1=0$$

If $p(x)=a_0+a_1x+\cdots+a_{n-1}x^{n-1}+x^n$, then

$$p(A)e_1=\sum_{i=0}^{n-1}a_iA^ie_1+A^ne_1=\sum_{i=0}^{n-1}a_ie_{i+1}-\sum_{i=0}^{n-1}a_ie_{i+1}=0$$

and so $p(A)e_1=0$, whence $p(A)=0$. Also, if

$$q(x)=b_0+b_1x+\cdots+b_{m-1}x^{m-1}+b_mx^m$$

is nonzero and has degree $m<n$, then

$$q(A)e_1=b_0e_1+b_1e_2+\cdots+b_{m-1}e_m+b_me_{m+1}\neq 0$$

since $\mathcal{E}$ is linearly independent. Hence, $p(x)$ has smallest degree among all polynomials satisfied by A and so $p(x) = m_A(x)$. Finally,

$$\deg(m_A(x)) = \deg(p(x)) = \deg(c_A(x))$$

For part 2), we have already proved that if V_τ is cyclic with τ-cyclic basis $\mathcal{B}$, then $[\tau]_\mathcal{B} = C[p(x)]$. For the converse, if $[\tau]_\mathcal{B} = C[p(x)]$, then part 1) implies that τ is nonderogatory. Hence, Theorem 7.11 implies that V_τ is cyclic. It is clear from the form of $C[p(x)]$ that $\mathcal{B}$ is a τ-cyclic basis for V.$\square$

The Big Picture

If $\sigma, \tau \in \mathcal{L}(V)$, then Theorem 7.2 and the fact that the elementary divisors form a complete invariant for isomorphism imply that

$$\sigma \sim \tau \quad \Leftrightarrow \quad V_\sigma \approx V_\tau \quad \Leftrightarrow \quad \mathrm{ElemDiv}(\tau) = \mathrm{ElemDiv}(\sigma)$$

Hence, the multiset of elementary divisors is a complete invariant for similarity of operators. Of course, the same is true for matrices:

$$A \sim B \quad \Leftrightarrow \quad F_A^n \approx F_B^n \quad \Leftrightarrow \quad \mathrm{ElemDiv}(A) = \mathrm{ElemDiv}(B)$$

where we write F_A^n in place of $F_{\tau_A}^n$.

The connection between the elementary divisors of an operator τ and the elementary divisors of the matrix representations of τ is described as follows. If $A = [\tau]_\mathcal{B}$, then the coordinate map $\phi_\mathcal{B}: V \approx F^n$ is also a *module* isomorphism $\phi_\mathcal{B}: V_\tau \to F_A^n$. Specifically, we have

$$\phi_\mathcal{B}(p(\tau)v) = [p(\tau)v]_\mathcal{B} = p([\tau]_\mathcal{B})[v]_\mathcal{B} = p(A)\phi_\mathcal{B}(v)$$

and so $\phi_\mathcal{B}$ preserves $F[x]$-scalar multiplication. Hence,

$$A = [\tau]_\mathcal{B} \text{ for some } \mathcal{B} \quad \Rightarrow \quad V_\tau \approx F_A^n$$

For the converse, suppose that $\sigma: V_\tau \approx F_A^n$. If we define $b_i \in V$ by $\sigma b_i = e_i$, where e_i is the ith standard basis vector, then $\mathcal{B} = (b_1, \ldots, b_n)$ is an ordered basis for V and $\sigma = \phi_\mathcal{B}$ is the coordinate map for $\mathcal{B}$. Hence, $\phi_\mathcal{B}$ is a module isomorphism and so

$$\phi_\mathcal{B}(\tau v) = \tau_A(\phi_\mathcal{B} v)$$

for all $v \in V$, that is,

$$[\tau v]_\mathcal{B} = \tau_A([v]_\mathcal{B})$$

which shows that $A = [\tau]_\mathcal{B}$.

Theorem 7.13 *Let V be a finite-dimensional vector space over F. Let $\sigma, \tau \in \mathcal{L}(V)$ and let $A, B \in \mathcal{M}_n(F)$.*

1) *The multiset of elementary divisors (or invariant factors) is a complete invariant for similarity of operators, that is,*

$$\begin{aligned}\sigma \sim \tau \Leftrightarrow V_\sigma \approx V_\tau \\ &\Leftrightarrow \mathrm{ElemDiv}(\tau) = \mathrm{ElemDiv}(\sigma) \\ &\Leftrightarrow \mathrm{InvFact}(\tau) = \mathrm{InvFact}(\sigma)\end{aligned}$$

A similar statement holds for matrices:

$$\begin{aligned}A \sim B \Leftrightarrow F_A^n \approx F_B^n \\ &\Leftrightarrow \mathrm{ElemDiv}(A) = \mathrm{ElemDiv}(B) \\ &\Leftrightarrow \mathrm{InvFact}(A) = \mathrm{InvFact}(B)\end{aligned}$$

2) *The connection between operators and their representing matrices is*

$$\begin{aligned}A = [\tau]_{\mathcal{B}} \textit{ for some } \mathcal{B} \Leftrightarrow V_\tau \approx F_A^n \\ &\Leftrightarrow \mathrm{ElemDiv}(\tau) = \mathrm{ElemDiv}(A) \\ &\Leftrightarrow \mathrm{InvFact}(\tau) = \mathrm{InvFact}(A)\end{aligned}$$ □

Theorem 7.13 can be summarized in Figure 7.1, which shows the big picture.

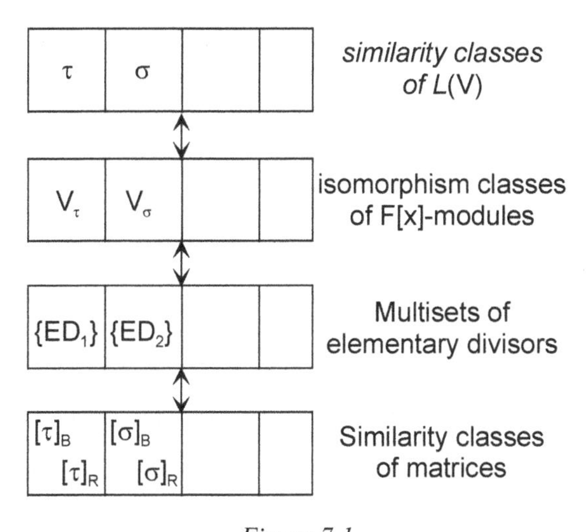

Figure 7.1

Figure 7.1 shows that the similarity classes of $\mathcal{L}(V)$ are in one-to-one correspondence with the isomorphism classes of $F[x]$-modules V_τ and that these are in one-to-one correspondence with the multisets of elementary divisors, which, in turn, are in one-to-one correspondence with the similarity classes of matrices.

We will see shortly that any multiset of prime power polynomials is the multiset of elementary divisors for some operator (or matrix) and so the third family in the figure could be replaced by the family of all multisets of prime power polynomials.

The Rational Canonical Form

We are now ready to determine a set of canonical forms for similarity. Let $\tau \in \mathcal{L}(V)$. The elementary divisor basis $\mathcal{R}$ for V_τ that gives the primary cyclic decomposition of V_τ,

$$V_\tau = (\langle\langle v_{1,1} \rangle\rangle \oplus \cdots \oplus \langle\langle v_{1,k_1} \rangle\rangle) \oplus \cdots \oplus (\langle\langle v_{n,1} \rangle\rangle \oplus \cdots \oplus \langle\langle v_{n,k_n} \rangle\rangle)$$

is the union of the bases

$$\mathcal{B}_{i,j} = (v_{i,j}, \tau v_{i,j}, \ldots, \tau^{d_{i,j}-1} v_{i,j})$$

and so the matrix of τ with respect to $\mathcal{R}$ is the block diagonal matrix

$$[\tau]_{\mathcal{R}} = \operatorname{diag}(C[p_1^{e_{1,1}}(x)], \ldots, C[p_1^{e_{1,k_1}}(x)], \ldots, C[p_n^{e_{n,1}}(x)], \ldots, C[p_n^{e_{n,k_n}}(x)])$$

with companion matrices on the block diagonal. This matrix has the following form.

Definition *A matrix A is in the* **elementary divisor form** *of* **rational canonical form** *if*

$$A = \operatorname{diag}\Big(C[r_1^{e_1}(x)], \ldots, C[r_n^{e_n}(x)]\Big)$$

where the $r_i(x)$ are monic prime polynomials. □

Thus, as shown in Figure 7.1, each similarity class $\mathcal{S}$ contains at least one matrix in the elementary divisor form of rational canonical form.

On the other hand, suppose that M is a rational canonical matrix

$$M = \operatorname{diag}(C[q_1^{f_{1,1}}(x)], \ldots, C[q_1^{f_{1,j_1}}(x)], \ldots, C[q_m^{f_{m,1}}(x)], \ldots, C[q_m^{f_{m,j_m}}(x)])$$

of size $d \times d$. Then M represents the matrix multiplication operator τ_M under the standard basis $\mathcal{E}$ on F^d. The basis $\mathcal{E}$ can be partitioned into blocks $\mathcal{E}_{i,k}$ corresponding to the position of each of the companion matrices on the block diagonal of M. Since

$$[\tau_M|_{\langle \mathcal{E}_{i,k} \rangle}]_{\mathcal{E}_{i,k}} = C[q_i^{f_{i,k}}(x)]$$

it follows from Theorem 7.12 that each subspace $\langle \mathcal{E}_{i,k} \rangle$ is τ_M-cyclic with monic order $q_i^{f_{i,k}}(x)$ and so Theorem 7.9 implies that the multiset of elementary divisors of τ_M is $\{q_i^{f_{i,k}}(x)\}$.

This shows two important things. First, any multiset of prime power polynomials is the multiset of elementary divisors for some matrix. Second, M

lies in the similarity class that is associated with the elementary divisors $\{q_i^{f_{i,j}}(x)\}$. Hence, two matrices in the elementary divisor form of rational canonical form lie in the same similarity class if and only if they have the same multiset of elementary divisors. In other words, the elementary divisor form of rational canonical form is a set of canonical forms for similarity, up to order of blocks on the block diagonal.

Theorem 7.14 (The rational canonical form: elementary divisor version) *Let V be a finite-dimensional vector space and let $\tau \in \mathcal{L}(V)$ have minimal polynomial*

$$m_\tau(x) = p_1^{e_1}(x) \cdots p_n^{e_n}(x)$$

where the $p_i(x)$'s are distinct monic prime polynomials.

1) *If $\mathcal{R}$ is an elementary divisor basis for V_τ, then $[\tau]_\mathcal{R}$ is in the elementary divisor form of rational canonical form:*

$$[\tau]_\mathcal{R} = \operatorname{diag}\Big(C[p_1^{e_{1,1}}(x)], \ldots, C[p_1^{e_{1,k_1}}(x)], \ldots, C[p_n^{e_{n,1}}(x)], \ldots, C[p_n^{e_{n,k_n}}(x)]\Big)$$

where $p_k^{e_{k,i}}(x)$ are the elementary divisors of τ. This block diagonal matrix is called an **elementary divisor version** *of a* **rational canonical form** *of τ.*

2) *Each similarity class $\mathcal{S}$ of matrices contains a matrix R in the elementary divisor form of rational canonical form. Moreover, the set of matrices in $\mathcal{S}$ that have this form is the set of matrices obtained from M by reordering the block diagonal matrices. Any such matrix is called an* **elementary divisor verison** *of a* **rational canonical form** *of A.*

3) *The dimension of V is the sum of the degrees of the elementary divisors of τ, that is,*

$$\dim(V) = \sum_{i=1}^{n} \sum_{j=1}^{k_i} \deg(p_i^{e_{i,j}}) \qquad \square$$

Example 7.1 Let τ be a linear operator on the vector space $\mathbb{R}^7$ and suppose that τ has minimal polynomial

$$m_\tau(x) = (x-1)(x^2+1)^2$$

Noting that $x-1$ and $(x^2+1)^2$ are elementary divisors and that the sum of the degrees of all elementary divisors must equal 7, we have two possibilities:

1) $x-1,\ (x^2+1)^2,\ x^2+1$
2) $x-1,\ x-1,\ x-1,\ (x^2+1)^2$

These correspond to the following rational canonical forms:

1) $\begin{bmatrix} 1 & 0 & 0 & 0 & 0 & 0 & 0 \\ 0 & 0 & 0 & 0 & -1 & 0 & 0 \\ 0 & 1 & 0 & 0 & 0 & 0 & 0 \\ 0 & 0 & 1 & 0 & -2 & 0 & 0 \\ 0 & 0 & 0 & 1 & 0 & 0 & 0 \\ 0 & 0 & 0 & 0 & 0 & 0 & -1 \\ 0 & 0 & 0 & 0 & 0 & 1 & 0 \end{bmatrix}$

2) $\begin{bmatrix} 1 & 0 & 0 & 0 & 0 & 0 & 0 \\ 0 & 1 & 0 & 0 & 0 & 0 & 0 \\ 0 & 0 & 1 & 0 & 0 & 0 & 0 \\ 0 & 0 & 0 & 0 & 0 & 0 & -1 \\ 0 & 0 & 0 & 1 & 0 & 0 & 0 \\ 0 & 0 & 0 & 0 & 1 & 0 & -2 \\ 0 & 0 & 0 & 0 & 0 & 1 & 0 \end{bmatrix}$ □

The rational canonical form may be far from the ideal of simplicity that we had in mind for a set of simple canonical forms. Indeed, the rational canonical form can be important as a theoretical tool, more so than a practical one.

The Invariant Factor Version

There is also an invariant factor version of the rational canonical form. We begin with the following simple result.

Theorem 7.15 *If $p(x), q(x) \in F[x]$ are relatively prime polynomials, then*

$$C[p(x)q(x)] \sim \begin{pmatrix} C[p(x)] & 0 \\ 0 & C[q(x)] \end{pmatrix}_{\text{block}}$$

Proof. Speaking in general terms, if an $m \times m$ matrix A has minimal polynomial

$$m_\tau(x) = p_1^{e_1}(x) \cdots p_n^{e_n}(x)$$

of degree equal to the size m of the matrix, then Theorem 7.14 implies that the elementary divisors of A are precisely

$$p_1^{e_1}(x), \ldots, p_n^{e_n}(x)$$

Since the matrices $C[p(x)q(x)]$ and $\operatorname{diag}(C[p(x)], C[q(x)])$ have the same size $m \times m$ and the same minimal polynomial $p(x)q(x)$ of degree m, it follows that they have the same multiset of elementary divisors and so are similar.□

Definition *A matrix A is in the* **invariant factor form** *of* **rational canonical form** *if*

$$A = \text{diag}\Big(C[s_1(x)], \ldots, C[s_n(x)]\Big)$$

where $s_{k+1}(x) \mid s_k(x)$ *for* $k = 1, \ldots, n-1$.□

Theorem 7.15 can be used to rearrange and combine the companion matrices in an elementary divisor version of a rational canonical form R to produce an invariant factor version of rational canonical form that is similar to R. Also, this process is reversible.

Theorem 7.16 (The rational canonical form: invariant factor version) *Let* $\dim(V) < \infty$ *and suppose that* $\tau \in \mathcal{L}(V)$ *has minimal polynomial*

$$m_\tau(x) = p_1^{e_1}(x) \cdots p_n^{e_n}(x)$$

where the monic polynomials $p_i(x)$ *are distinct prime (irreducible) polynomials*

1) *V has an* **invariant factor basis** $\mathcal{B}$, *that is, a basis for which*

$$[\tau]_{\mathcal{B}} = \text{diag}\Big(C[s_1(x)], \ldots, C[s_n(x)]\Big)$$

where the polynomials $s_k(x)$ *are the invariant factors of* τ *and* $s_{k+1}(x) \mid s_k(x)$. *This block diagonal matrix is called an* **invariant factor version** *of a* **rational canonical form** *of* τ.

2) *Each similarity class* $\mathcal{S}$ *of matrices contains a matrix* R *in the invariant factor form of rational canonical form. Moreover, the set of matrices in* $\mathcal{S}$ *that have this form is the set of matrices obtained from* M *by reordering the block diagonal matrices. Any such matrix is called an* **invariant factor verison** *of a* **rational canonical form** *of* A.

3) *The dimension of* V *is the sum of the degrees of the invariant factors of* τ, *that is,*

$$\dim(V) = \sum_{i=1}^{n} \deg(s_i)$$ □

The Determinant Form of the Characteristic Polynomial

In general, the minimal polynomial of an operator τ is hard to find. One of the virtues of the characteristic polynomial is that it is comparatively easy to find. This also provides a nice example of the theoretical value of the rational canonical form.

Let us first take the case of a companion matrix. If $A = C[p_n(x)]$ is the companion matrix of a monic polynomial

$$p_n(x; a_0, \ldots, a_{n-1}) = a_0 + a_1 x + \cdots + a_{n-1}x^{n-1} + x^n$$

then how can we recover $p(x) = c_A(x)$ from $C[p(x)]$ by arithmetic operations?

When $n = 2$, we can write $p_2(x)$ as

$$p_2(x; a_0, a_1) = a_0 + a_1 x + x^2 = x(x + a_1) + a_0$$

which looks suspiciously like a determinant:

$$\begin{aligned} p_2(x; a_0, a_1) &= \det \begin{bmatrix} x & a_0 \\ -1 & x + a_1 \end{bmatrix} \\ &= \det\left(xI - \begin{bmatrix} 0 & -a_0 \\ 1 & -a_1 \end{bmatrix} \right) \\ &= \det(xI - C[p_2(x)]) \end{aligned}$$

So, let us define

$$\begin{aligned} A(x; a_0, \ldots, a_{n-1}) &= xI - C[p_n(x)] \\ &= \begin{bmatrix} x & 0 & \cdots & 0 & a_0 \\ -1 & x & \cdots & 0 & a_1 \\ 0 & -1 & \ddots & & \vdots \\ \vdots & \vdots & \ddots & x & a_{n-2} \\ 0 & 0 & \cdots & -1 & x + a_{n-1} \end{bmatrix} \end{aligned}$$

where x is an independent variable. The determinant of this matrix is a polynomial in x whose degree equals the number of parameters $a_0, \ldots, a_{n-1}$. We have just seen that

$$\det(A(x; a_0, a_1)) = p_2(x; a_0, a_1)$$

and this is also true for $n = 1$. As a basis for induction, if

$$\det(A(x; a_0, \ldots, a_{n-1})) = p_n(x; a_0, \ldots, a_{n-1})$$

then expanding along the first row gives

$$\begin{aligned} &\det(A(x, a_0, \ldots, a_n)) \\ &\quad = x \det(A(x, a_1, \ldots, a_n)) + (-1)^n a_0 \det \begin{bmatrix} -1 & x & \cdots & 0 \\ 0 & -1 & \ddots & \\ \vdots & \vdots & \ddots & x \\ 0 & 0 & \cdots & -1 \end{bmatrix}_{n \times n} \\ &\quad = x \det(A(x, a_1, \ldots, a_n)) + a_0 \\ &\quad = x\, p_n(x; a_1, \ldots, a_n) + a_0 \\ &\quad = a_1 x + a_2 x^2 + \cdots + a_n x^n + x^{n+1} + a_0 \\ &\quad = p_{n+1}(x; a_0, \ldots, a_n) \end{aligned}$$

We have proved the following.

Lemma 7.17 *For any* $p(x) \in F[x]$,

$$\det(xI - C[p(x)]) = p(x) \qquad \square$$

Now suppose that R is a matrix in the elementary divisor form of rational canonical form. Since the determinant of a block diagonal matrix is the product of the determinants of the blocks on the diagonal, it follows that

$$\det(xI - R) = \prod_{i,j} p_i^{e_{i,j}}(x) = c_R(x)$$

Moreover, if $A \sim R$, say $A = PRP^{-1}$, then

$$\begin{aligned}\det(xI - A) &= \det(xI - PRP^{-1}) \\ &= \det\left[P(xI - R)P^{-1}\right] \\ &= \det(P)\det(xI - R)\det(P^{-1}) \\ &= \det(xI - R)\end{aligned}$$

and so

$$\det(xI - A) = \det(xI - R) = c_R(x) = c_A(x)$$

Hence, the fact that all matrices have a rational canonical form allows us to deduce the following theorem.

Theorem 7.18 *Let* $\tau \in \mathcal{L}(V)$. *If* A *is any matrix that represents* τ, *then*

$$c_\tau(x) = c_A(x) = \det(xI - A) \qquad \square$$

Changing the Base Field

A change in the base field will generally change the primeness of polynomials and therefore has an effect on the multiset of elementary divisors. It is perhaps a surprising fact that a change of base field has *no effect* on the invariant factors—hence the adjective *invariant.*

Theorem 7.19 *Let* F *and* K *be fields with* $F \subseteq K$. *Suppose that the elementary divisors of a matrix* $A \in \mathcal{M}_n(F)$ *are*

$$\mathcal{A} = \{p_1^{e_{1,1}}, \ldots, p_1^{e_{1,k_1}}, \ldots, p_n^{e_{n,1}}, \ldots, p_n^{e_{n,k_n}}\}$$

Suppose also that the polynomials p_i *can be further factored over* K, *say*

$$p_i = a_{i,1}^{d_{i,1}} \cdots a_{i,m_i}^{d_{i,m_i}}$$

where $a_{i,j}$ *is prime over* K. *Then the prime powers*

$$\mathcal{B} = \{a_{1,1}^{d_{1,1}e_{1,1}}, \ldots, a_{1,m_1}^{d_{1,m_1}e_{1,1}}, \ldots, \ldots, a_{n,1}^{d_{n,1}e_{n,k_n}}, \ldots, a_{n,m_n}^{d_{n,m_n}e_{n,k_n}}\}$$

are the elementary divisors of A *over* K.

Proof. Consider the companion matrix $C[p_i^{e_{i,j}}(x)]$ in the rational canonical form of A over F. This is a matrix over K as well and Theorem 7.15 implies that

$$C[p_i^{e_{i,j}}(x)] \sim \text{diag}(C[a_{i,1}^{d_{i,1}e_{i,j}}],\ldots,C[a_{i,m_i}^{d_{i,m_i}e_{i,j}}])$$

Hence, $\mathcal{B}$ is an elementary divisor basis for A over K.□

As mentioned, unlike the elementary divisors, the invariant factors are *field independent*. This is equivalent to saying that the invariant factors of a matrix $A \in M_n(F)$ are polynomials over the *smallest* subfield of F that contains the entries of A.

Theorem 7.20 *Let $A \in \mathcal{M}_n(F)$ and let $E \subseteq F$ be the smallest subfield of F that contains the entries of A.*

1) *The invariant factors of A are polynomials over E.*
2) *Two matrices $A, B \in \mathcal{M}_n(F)$ are similar over F if and only if they are similar over E.*

Proof. Part 1) follows immediately from Theorem 7.19, since using either $\mathcal{A}$ or $\mathcal{B}$ to compute invariant factors gives the same result. Part 2) follows from the fact that two matrices are similar over a given field if and only if they have the same multiset of *invariant factors* over that field.□

Example 7.2 Over the real field, the matrix

$$A = \begin{pmatrix} 0 & -1 \\ 1 & 0 \end{pmatrix}$$

is the companion matrix for the polynomial $x^2 + 1$, and so

$$\text{ElemDiv}_{\mathbb{R}}(A) = \{x^2 + 1\} = \text{InvFact}_{\mathbb{R}}(A)$$

However, as a complex matrix, the rational canonical form for A is

$$A = \begin{pmatrix} i & 0 \\ 0 & -i \end{pmatrix}$$

and so

$$\text{ElemDiv}_{\mathbb{C}}(A) = \{x - i, x + i\} \quad \text{and} \quad \text{InvFact}_{\mathbb{C}}(A) = \{x^2 + 1\} \qquad \square$$

Exercises

1. We have seen that any $\tau \in \mathcal{L}(V)$ can be used to make V into an $F[x]$-module. Does every module V over $F[x]$ come from some $\tau \in \mathcal{L}(V)$? Explain.
2. Let $\tau \in \mathcal{L}(V)$ have minimal polynomial

$$m_\tau(x) = p_1^{e_1}(x) \cdots p_n^{e_n}(x)$$

where $p_i(x)$ are distinct monic primes. Prove that the following are equivalent:

a) V_τ is τ-cyclic.
b) $\deg(m_\tau(x)) = \dim(V)$.
c) The elementary divisors of τ are the prime power factors $p_i^{e_i}(x)$ and so

$$V_\tau = \langle\langle v_1 \rangle\rangle \oplus \cdots \oplus \langle\langle v_k \rangle\rangle$$

is a direct sum of τ-cyclic submodules $\langle\langle v_i \rangle\rangle$ of order $p_i^{e_i}(x)$.

3. Prove that a matrix $A \in \mathcal{M}_n(F)$ is nonderogatory if and only if it is similar to a companion matrix.
4. Show that if A and B are block diagonal matrices with the same blocks, but in possibly different order, then A and B are similar.
5. Let $A \in \mathcal{M}_n(F)$. Justify the statement that the entries of any invariant factor version of a rational canonical form for A are "rational" expressions in the coefficients of A, hence the origin of the term *rational canonical form*. Is the same true for the elementary divisor version?
6. Let $\tau \in \mathcal{L}(V)$ where V is finite-dimensional. If $p(x) \in F[x]$ is irreducible and if $p(\tau)$ is not one-to-one, prove that $p(x)$ divides the minimal polynomial of τ.
7. Prove that the minimal polynomial of $\tau \in \mathcal{L}(V)$ is the least common multiple of its elementary divisors.
8. Let $\tau \in \mathcal{L}(V)$ where V is finite-dimensional. Describe conditions on the minimal polynomial of τ that are equivalent to the fact that the elementary divisor version of the rational canonical form of τ is diagonal. What can you say about the elementary divisors?
9. Verify the statement that the multiset of elementary divisors (or invariant factors) is a complete invariant for similarity of matrices.
10. Prove that given any multiset of monic prime power polynomials

$$M = \{p_1^{e_{1,1}}(x), \ldots, p_1^{e_{1,k_1}}(x), \ldots, \ldots, p_n^{e_{n,1}}(x), \ldots, p_n^{e_{n,k_n}}(x)\}$$

and given any vector space V of dimension equal to the sum of the degrees of these polynomials, there is an operator $\tau \in \mathcal{L}(V)$ whose multiset of elementary divisors is M.

11. Find all rational canonical forms (up to the order of the blocks on the diagonal) for a linear operator on $\mathbb{R}^6$ having minimal polynomial $(x-1)^2(x+1)^2$.
12. How many possible rational canonical forms (up to order of blocks) are there for linear operators on $\mathbb{R}^6$ with minimal polynomial $(x-1)(x+1)^2$?
13. a) Show that if A and B are $n \times n$ matrices, at least one of which is invertible, then AB and BA are similar.

b) What do the matrices

$$A = \begin{bmatrix} 1 & 0 \\ 0 & 0 \end{bmatrix} \quad \text{and} \quad B = \begin{bmatrix} 0 & 1 \\ 0 & 0 \end{bmatrix}$$

have to do with this issue?

c) Show that even without the assumption on invertibility the matrices AB and BA have the same characteristic polynomial. *Hint*: Write

$$A = PI_{n,r}Q$$

where P and Q are invertible and $I_{n,r}$ is an $n \times n$ matrix that has the $r \times r$ identity in the upper left-hand corner and 0's elsewhere. Write $B' = QBP$. Compute AB and BA and find their characteristic polynomials.

14. Let τ be a linear operator on F^4 with minimal polynomial $m_\tau(x) = (x^2 + 1)(x^2 - 2)$. Find the rational canonical form for τ if $F = \mathbb{Q}$, $F = \mathbb{R}$ or $F = \mathbb{C}$.
15. Suppose that the minimal polynomial of $\tau \in \mathcal{L}(V)$ is irreducible. What can you say about the dimension of V?
16. Let $\tau \in \mathcal{L}(V)$ where V is finite-dimensional. Suppose that $p(x)$ is an irreducible factor of the minimal polynomial $m(x)$ of τ. Suppose further that $u, v \in V$ have the property that $o(u) = o(v) = p(x)$. Prove that $u = f(\tau)v$ for some polyjomial $f(x)$ if and only if $v = g(\tau)u$ for some polynomial $g(x)$.

Chapter 8
Eigenvalues and Eigenvectors

Unless otherwise noted, we will assume throughout this chapter that all vector spaces are finite-dimensional.

Eigenvalues and Eigenvectors

We have seen that for any $\tau \in \mathcal{L}(V)$, the minimal and characteristic polynomials have the same set of roots (but not generally the same *multiset* of roots). These roots are of vital importance.

Let $A = [\tau]_{\mathcal{B}}$ be a matrix that represents τ. A scalar $\lambda \in F$ is a root of the characteristic polynomial $c_\tau(x) = c_A(x) = \det(xI - A)$ if and only if

$$\det(\lambda I - A) = 0 \tag{8.1}$$

that is, if and only if the matrix $\lambda I - A$ is singular. In particular, if $\dim(V) = n$, then (8.1) holds if and only if there exists a nonzero vector $x \in F^n$ for which

$$(\lambda I - A)x = 0$$

or equivalently,

$$\tau_A x = \lambda x$$

If $[v]_{\mathcal{B}} = x$, then this is equivalent to

$$[\tau]_{\mathcal{B}}[v]_{\mathcal{B}} = \lambda[v]_{\mathcal{B}}$$

or in operator language,

$$\tau v = \lambda v$$

This prompts the following definition.

Definition *Let V be a vector space over a field F and let $\tau \in \mathcal{L}(V)$.*

1) *A scalar $\lambda \in F$ is an* **eigenvalue** *(or* **characteristic value***) of τ if there exists a* nonzero *vector $v \in V$ for which*

$$\tau v = \lambda v$$

In this case, v is called an **eigenvector** (*or* **characteristic vector**) *of τ associated with λ.*

2) *A scalar $\lambda \in F$ is an* **eigenvalue** *for a matrix A if there exists a* nonzero *column vector x for which*

$$Ax = \lambda x$$

In this case, x is called an **eigenvector** (*or* **characteristic vector**) *for A associated with λ.*

3) *The set of all eigenvectors associated with a given eigenvalue λ, together with the zero vector, forms a subspace of V, called the* **eigenspace** *of λ and denoted by $\mathcal{E}_\lambda$. This applies to both linear operators and matrices.*
4) *The set of all eigenvalues of an operator or matrix is called the* **spectrum** *of the operator or matrix. We denote the spectrum of τ by* $\operatorname{Spec}(\tau)$.□

Theorem 8.1 *Let $\tau \in \mathcal{L}(V)$ have minimal polynomial $m_\tau(x)$ and characteristic polynomial $c_\tau(x)$.*

1) *The spectrum of τ is the set of all roots of $m_\tau(x)$ or of $c_\tau(x)$, not counting multiplicity.*
2) *The eigenvalues of a matrix are invariants under similarity.*
3) *The eigenspace $\mathcal{E}_\lambda$ of the matrix A is the solution space to the homogeneous system of equations*

$$(\lambda I - A)(x) = 0$$ □

One way to compute the eigenvalues of a linear operator τ is to first represent τ by a matrix A and then solve the **characteristic equation**

$$\det(xI - A) = 0$$

Unfortunately, it is quite likely that this equation cannot be solved when $\dim(V) \geq 5$. As a result, the art of approximating the eigenvalues of a matrix is a very important area of applied linear algebra.

The following theorem describes the relationship between eigenspaces and eigenvectors of distinct eigenvalues.

Theorem 8.2 *Suppose that $\lambda_1, \ldots, \lambda_k$ are distinct eigenvalues of a linear operator $\tau \in \mathcal{L}(V)$.*

1) *Eigenvectors associated with distinct eigenvalues are linearly independent; that is, if $v_i \in \mathcal{E}_{\lambda_i}$, then the set $\{v_1, \ldots, v_k\}$ is linearly independent.*
2) *The sum $\mathcal{E}_{\lambda_1} + \cdots + \mathcal{E}_{\lambda_k}$ is direct; that is, $\mathcal{E}_{\lambda_1} \oplus \cdots \oplus \mathcal{E}_{\lambda_k}$ exists.*

Proof. For part 1), if $\{v_1, \ldots, v_k\}$ is linearly dependent, then by renumbering if necessary, we may assume that among all nontrivial linear combinations of these vectors that equal 0, the equation

$$r_1v_1 + \cdots + r_jv_j = 0 \tag{8.2}$$

has the fewest number of terms. Applying τ gives

$$r_1\lambda_1v_1 + \cdots + r_j\lambda_jv_j = 0 \tag{8.3}$$

Multiplying (8.2) by λ_1 and subtracting from (8.3) gives

$$r_2(\lambda_2 - \lambda_1)v_2 + \cdots + r_j(\lambda_j - \lambda_1)v_j = 0$$

But this equation has fewer terms than (8.2) and so all of its coefficients must equal 0. Since the λ_i's are distinct, $r_i = 0$ for $i \geq 2$ and so $r_1 = 0$ as well. This contradiction implies that the v_i's are linearly independent.□

The next theorem describes the spectrum of a polynomial $p(\tau)$ in τ.

Theorem 8.3 (The spectral mapping theorem) *Let V be a vector space over an algebraically closed field F. Let $\tau \in \mathcal{L}(V)$ and let $p(x) \in F[x]$. Then*

$$\text{Spec}(p(\tau)) = p(\text{Spec}(\tau)) = \{p(\lambda) \mid \lambda \in \text{Spec}(\tau)\}$$

Proof. We leave it as an exercise to show that if λ is an eigenvalue of τ, then $p(\lambda)$ is an eigenvalue of $p(\tau)$. Hence, $p(\text{Spec}(\tau)) \subseteq \text{Spec}(p(\tau))$. For the reverse inclusion, let $\lambda \in \text{Spec}(p(\tau))$, that is,

$$(p(\tau) - \lambda)v = 0$$

for $v \neq 0$. If

$$p(x) - \lambda = (x - r_1)^{e_1} \cdots (x - r_n)^{e_n}$$

where $r_i \in F$, then writing this as a product of (not necessarily distinct) linear factors, we have

$$(\tau - r_1) \cdots (\tau - r_1) \cdots (\tau - r_n) \cdots (\tau - r_n)v = 0$$

(The operator $r_k\iota$ is written r_k for convenience.) We can remove factors from the left end of this equation one by one until we arrive at an operator σ (perhaps the identity) for which $\sigma v \neq 0$ but $(\tau - r_k)\sigma v = 0$. Then σv is an eigenvector for τ with eigenvalue r_k. But since $p(r_k) - \lambda = 0$, it follows that $\lambda = p(r_k) \in p(\text{Spec}(\tau))$. Hence, $\text{Spec}(p(\tau)) \subseteq p(\text{Spec}(\tau))$.□

The Trace and the Determinant

Let F be algebraically closed and let $A \in \mathcal{M}_n(F)$ have characteristic polynomial

$$\begin{aligned} c_A(x) &= x^n + c_{n-1}x^{n-1} + \cdots + c_1x + c_0 \\ &= (x - \lambda_1) \cdots (x - \lambda_n) \end{aligned}$$

where $\lambda_1, \ldots, \lambda_n$ are the eigenvalues of A. Then

$$c_A(x) = \det(xI - A)$$

and setting $x = 0$ gives

$$\det(A) = -c_0 = (-1)^{n-1}\lambda_1 \cdots \lambda_n$$

Hence, if F is algebraically closed then, *up to sign*, $\det(A)$ is the constant term of $c_A(x)$ and the product of the eigenvalues of A, including multiplicity.

The *sum* of the eigenvalues of a matrix over an algebraically closed field is also an interesting quantity. Like the determinant, this quantity is one of the coefficients of the characteristic polynomial (up to sign) and can also be computed directly from the entries of the matrix, without knowing the eigenvalues explicitly.

Definition *The* **trace** *of a matrix $A \in \mathcal{M}_n(F)$, denoted by* $\operatorname{tr}(A)$*, is the sum of the elements on the main diagonal of A.*□

Here are the basic propeties of the trace. Proof is left as an exercise.

Theorem 8.4 *Let $A, B \in \mathcal{M}_n(F)$.*

1) $\operatorname{tr}(rA) = r\operatorname{tr}(A)$*, for $r \in F$.*
2) $\operatorname{tr}(A + B) = \operatorname{tr}(A) + \operatorname{tr}(B)$.
3) $\operatorname{tr}(AB) = \operatorname{tr}(BA)$.
4) $\operatorname{tr}(ABC) = \operatorname{tr}(CAB) = \operatorname{tr}(BCA)$. *However,* $\operatorname{tr}(ABC)$ *may not equal* $\operatorname{tr}(ACB)$.
5) *The trace is an invariant under similarity.*
6) *If F is algebraically closed, then* $\operatorname{tr}(A)$ *is the sum of the eigenvalues of A, including multiplicity, and so*

$$\operatorname{tr}(A) = -c_{n-1}$$

where $c_A(x) = x^n + c_{n-1}x^{n-1} + \cdots + c_1x + c_0$.□

Since the trace is invariant under similarity, we can make the following definition.

Definition *The* **trace** *of a linear operator $\tau \in \mathcal{L}(V)$ is the trace of any matrix that represents τ.*□

As an aside, the reader who is familar with symmetric polynomials knows that the coefficients of any polynomial

$$\begin{aligned} p(x) &= x^n + c_{n-1}x^{n-1} + \cdots + c_1x + c_0 \\ &= (x - \lambda_1)\cdots(x - \lambda_n) \end{aligned}$$

are the **elementary symmetric functions** of the roots:

$$\begin{aligned} c_{n-1} &= (-1)^1 \sum_i \lambda_i \\ c_{n-2} &= (-1)^2 \sum_{i<j} \lambda_i \lambda_j \\ c_{n-3} &= (-1)^3 \sum_{i<j<k} \lambda_i \lambda_j \lambda_k \\ &\vdots \\ c_0 &= (-1)^n \prod_{i=1}^{n} \lambda_i \end{aligned}$$

The most important elementary symmetric functions of the eigenvalues are the first and last ones:

$$c_{n-1} = -\lambda_1 + \cdots + \lambda_n = \mathrm{tr}(A) \quad \text{and} \quad c_0 = (-1)^n \lambda_1 \cdots \lambda_n = \det(A)$$

Geometric and Algebraic Multiplicities

Eigenvalues actually have two forms of multiplicity, as described in the next definition.

Definition *Let λ be an eigenvalue of a linear operator $\tau \in \mathcal{L}(V)$.*
1) *The* **algebraic multiplicity** *of λ is the multiplicity of λ as a root of the characteristic polynomial $c_\tau(x)$.*
2) *The* **geometric multiplicity** *of λ is the dimension of the eigenspace $\mathcal{E}_\lambda$.*□

Theorem 8.5 *The geometric multiplicity of an eigenvalue λ of $\tau \in \mathcal{L}(V)$ is less than or equal to its algebraic multiplicity.*
Proof. We can extend any basis $\mathcal{B}_1 = \{v_1, \ldots, v_k\}$ of $\mathcal{E}_\lambda$ to a basis $\mathcal{B}$ for V. Since $\mathcal{E}_\lambda$ is invariant under τ, the matrix of τ with respect to $\mathcal{B}$ has the block form

$$[\tau]_{\mathcal{B}} = \begin{pmatrix} \lambda I_k & A \\ 0 & B \end{pmatrix}_{\text{block}}$$

where A and B are matrices of the appropriate sizes and so

$$\begin{aligned} c_\tau(x) &= \det(xI - [\tau]_{\mathcal{B}}) \\ &= \det(xI_k - \lambda I_k)\det(xI_{n-k} - B) \\ &= (x-\lambda)^k \det(xI_{n-k} - B) \end{aligned}$$

(Here n is the dimension of V.) Hence, the algebraic multiplicity of λ is at least equal to the the geometric multiplicity k of τ.□

The Jordan Canonical Form

One of the virtues of the rational canonical form is that every linear operator on a finite-dimensional vector space has a rational canonical form. However, as mentioned earlier, the rational canonical form may be far from the ideal of simplicity that we had in mind for a set of simple canonical forms and is really more of a theoretical tool than a practical tool.

When the minimal polynomial $m_\tau(x)$ of τ splits over F,

$$m_\tau(x) = (x - \lambda_1)^{e_1} \cdots (x - \lambda_n)^{e_n}$$

there is another set of canoncial forms that is arguably simpler than the set of rational canonical forms.

In some sense, the complexity of the rational canonical form comes from the choice of basis for the cyclic submodules $\langle\langle v_{i,j} \rangle\rangle$. Recall that the τ-cyclic bases have the form

$$\mathcal{B}_{i,j} = \left(v_{i,j}, \tau v_{i,j}, \ldots, \tau^{d_{i,j}-1} v_{i,j} \right)$$

where $d_{i,j} = \deg(p_i^{e_{i,j}})$. With this basis, all of the complexity comes at the end, so to speak, when we attempt to express

$$\tau(\tau^{d_{i,j}-1}(v_{i,j})) = \tau^{d_{i,j}}(v_{i,j})$$

as a linear combination of the basis vectors.

However, since $\mathcal{B}_{i,j}$ has the form

$$\left(v, \tau v, \tau^2 v, \ldots, \tau^{d-1} v \right)$$

any ordered set of the form

$$(p_0(\tau)v, p_1(\tau)v, \ldots, p_{d-1}(\tau)v)$$

where $\deg(p_k(x)) = k$ will also be a basis for $\langle\langle v_{i,j} \rangle\rangle$. In particular, when $m_\tau(x)$ splits over F, the elementary divisors are

$$p_i^{e_{i,j}}(x) = (x - \lambda_i)^{e_{i,j}}$$

and so the set

$$\mathcal{C}_{i,j} = \left(v_{i,j}, (\tau - \lambda_i) v_{i,j}, \ldots, (\tau - \lambda_i)^{e_{i,j}-1} v_{i,j} \right)$$

is also a basis for $\langle\langle v_{i,j} \rangle\rangle$.

If we temporarily denote the kth basis vector in $\mathcal{C}_{i,j}$ by b_k, then for $k = 0, \ldots, e_{i,j} - 2$,

$$\begin{aligned}\tau b_k &= \tau[(\tau-\lambda_i)^k(v_{i,j})] \\ &= (\tau-\lambda_i+\lambda_i)[(\tau-\lambda_i)^k(v_{i,j})] \\ &= (\tau-\lambda_i)^{k+1}(v_{i,j})+\lambda_i(\tau-\lambda_i)^k(v_{i,j}) \\ &= b_{k+1}+\lambda_i b_k\end{aligned}$$

For $k = e_{i,j} - 1$, a similar computation, using the fact that

$$(\tau-\lambda_i)^{k+1}(v_{i,j}) = (\tau-\lambda_i)^{e_{i,j}}(v_{i,j}) = 0$$

gives

$$\tau(b_{e_{i,j}-1}) = \lambda_i b_{e_{i,j}-1}$$

Thus, for this basis, the complexity is more or less spread out evenly, and the matrix of $\tau|_{\langle\langle v_{i,j}\rangle\rangle}$ with respect to $\mathcal{C}_{i,j}$ is the $e_{i,j}\times e_{i,j}$ matrix

$$\mathcal{J}(\lambda_i, e_{i,j}) = \begin{bmatrix} \lambda_i & 0 & \cdots & \cdots & 0 \\ 1 & \lambda_i & \ddots & & \vdots \\ 0 & 1 & \ddots & \ddots & \vdots \\ \vdots & \ddots & \ddots & \ddots & 0 \\ 0 & \cdots & 0 & 1 & \lambda_i \end{bmatrix}$$

which is called a **Jordan block** associated with the scalar λ_i. Note that a Jordan block has λ_i's on the main diagonal, 1's on the subdiagonal and 0's elsewhere. Let us refer to the basis

$$\mathcal{C} = \bigcup \mathcal{C}_{i,j}$$

as a **Jordan basis** for τ.

Theorem 8.6 (The Jordan canonical form) *Suppose that the minimal polynomial of $\tau \in \mathcal{L}(V)$ splits over the base field F, that is,*

$$m_\tau(x) = (x-\lambda_1)^{e_1}\cdots(x-\lambda_n)^{e_n}$$

where $\lambda_i \in F$.

1) *The matrix of τ with respect to a Jordan basis $\mathcal{C}$ is*

$$\operatorname{diag}(\mathcal{J}(\lambda_1, e_{1,1}), \ldots, \mathcal{J}(\lambda_1, e_{1,k_1}), \ldots, \mathcal{J}(\lambda_n, e_{n,1}), \ldots, \mathcal{J}(\lambda_n, e_{n,k_n}))$$

   where the polynomials $(x-\lambda_i)^{e_{i,j}}$ are the elementary divisors of τ. This block diagonal matrix is said to be in **Jordan canonical form** *and is called the* **Jordan canonical form of** *τ.*
2) *If F is algebraically closed, then up to order of the block diagonal matrices, the set of matrices in Jordan canonical form constitutes a set of canonical forms for similarity.*

Proof. For part 2), the companion matrix and corresponding Jordan block are similar:

$$C[(x-\lambda_i)^{e_{i,j}}] \sim \mathcal{J}(\lambda_i, e_{i,j})$$

since they both represent the same operator τ on the subspace $\langle\langle v_{i,j}\rangle\rangle$. It follows that the rational canonical matrix and the Jordan canonical matrix for τ are similar.□

Note that the diagonal elements of the Jordan canonical form $\mathcal{J}$ of τ are precisely the eigenvalues of τ, each appearing a number of times equal to its algebraic multiplicity. In general, the rational canonical form does not "expose" the eigenvalues of the matrix, even when these eigenvalues lie in the base field.

Triangularizability and Schur's Lemma

We have discussed two different canonical forms for similarity: the rational canonical form, which applies in all cases and the Jordan canonical form, which applies only when the base field is algebraically closed. Moreover, there is an annoying sense in which these sets of canoncial forms leave something to be desired: One is too complex and the other does not always exist.

Let us now drop the rather strict requirements of canonical forms and look at two classes of matrices that are too large to be canonical forms (the upper triangular matrices and the almost upper triangular matrices) and one class of matrices that is too small to be a canonical form (the diagonal matrices).

The upper triangular matrices (or lower triangular matrices) have some nice algebraic properties and it is of interest to know when an arbitrary matrix is similar to a triangular matrix. We confine our attention to upper triangular matrices, since there are direct analogs for lower triangular matrices as well.

Definition *A linear operator $\tau \in \mathcal{L}(V)$ is* **upper triangularizable** *if there is an ordered basis $\mathcal{B} = (v_1, \ldots, v_n)$ of V for which the matrix $[\tau]_\mathcal{B}$ is upper triangular, or equivalently, if*

$$\tau v_i \in \langle v_1, \ldots, v_i \rangle$$

for all $i = 1, \ldots, n$.□

As we will see next, when the base field is algebraically closed, all operators are upper triangularizable. However, since two distinct upper triangular matrices can be similar, the class of upper triangular matrices is not a canonical form for similarity. Simply put, there are just too many upper triangular matrices.

Theorem 8.7 (Schur's theorem) *Let V be a finite-dimensional vector space over a field F.*

1) *If the characteristic polynomial (or minimal polynomial) of $\tau \in \mathcal{L}(V)$ splits over F, then τ is upper triangularizable.*
2) *If F is algebraically closed, then all operators are upper triangularizable.*

Proof. Part 2) follows from part 1). The proof of part 1) is most easily accomplished by matrix means, namely, we prove that every square matrix $A \in M_n(F)$ whose characteristic polynomial splits over F is similar to an upper triangular matrix. If $n = 1$ there is nothing to prove, since all 1×1 matrices are upper triangular. Assume the result is true for $n-1$ and let $A \in M_n(F)$.

Let v_1 be an eigenvector associated with the eigenvalue $\lambda_1 \in F$ of A and extend $\{v_1\}$ to an ordered basis $\mathcal{B} = (v_1, \ldots, v_n)$ for $\mathbb{R}^n$. The matrix of τ_A with respect to $\mathcal{B}$ has the form

$$[\tau_A]_{\mathcal{B}} = \begin{bmatrix} \lambda_1 & * \\ 0 & A_1 \end{bmatrix}_{\text{block}}$$

for some $A_1 \in M_{n-1}(F)$. Since $[\tau_A]_{\mathcal{B}}$ and A are similar, we have

$$\det(xI - A) = \det(xI - [\tau_A]_{\mathcal{B}}) = (x - \lambda_1)\det(xI - A_1)$$

Hence, the characteristic polynomial of A_1 also splits over F and the induction hypothesis implies that there is an invertible matrix $P \in M_{n-1}(F)$ for which

$$U = PA_1P^{-1}$$

is upper triangular. Hence, if

$$Q = \begin{bmatrix} 1 & 0 \\ 0 & P \end{bmatrix}_{\text{block}}$$

then Q is invertible and

$$Q[A]_{\mathcal{B}}Q^{-1} = \begin{bmatrix} 1 & 0 \\ 0 & P \end{bmatrix} \begin{bmatrix} \lambda_1 & * \\ 0 & A_1 \end{bmatrix} \begin{bmatrix} 1 & 0 \\ 0 & P^{-1} \end{bmatrix} = \begin{bmatrix} \lambda_1 & * \\ 0 & U \end{bmatrix}$$

is upper triangular.□

The Real Case

When the base field is $F = \mathbb{R}$, an operator τ is upper triangularizable if and only if its characteristic polynomial splits over $\mathbb{R}$. (Why?) We can, however, always achieve a form that is close to triangular by permitting values on the first subdiagonal.

Before proceeding, let us recall Theorem 7.11, which says that for a module W_τ of prime order $m_\tau(x)$, the following are equivalent:

1) W_τ is cyclic
2) W_τ is indecomposable
3) $c_\tau(x)$ is irreducible
4) τ is nonderogatory, that is, $c_\tau(x) = m_\tau(x)$
5) $\dim(W_\tau) = \deg(p(x))$.

Now suppose that $F = \mathbb{R}$ and $c_\tau(x) = x^2 + sx + t$ is an irreducible quadratic. If $\mathcal{B}$ is a τ-cyclic basis for W_τ, then

$$[\tau]_\mathcal{B} = \begin{bmatrix} 0 & -t \\ 1 & -s \end{bmatrix}$$

However, there is a more appealing matrix representation of τ. To this end, let A be the matrix above. As a complex matrix, A has two distinct eigenvalues:

$$\lambda = -\frac{s}{2} \pm i\frac{\sqrt{4t - s^2}}{2}$$

Now, a matrix of the form

$$B = \begin{bmatrix} a & -b \\ b & a \end{bmatrix}$$

has characteristic polynomial $q(x) = (x - a)^2 + b^2$ and eigenvalues $a \pm ib$. So if we set

$$a = -\frac{s}{2} \quad \text{and} \quad b = -\frac{\sqrt{4t - s^2}}{2}$$

then B has the same two distinct eigenvalues as A and so A and B have the same Jordan canonical form over $\mathbb{C}$. It follows that A and B are similar over $\mathbb{C}$ and therefore also over $\mathbb{R}$, by Theorem 7.20. Thus, there is an ordered basis $\mathcal{C}$ for which $[\tau]_\mathcal{C} = B$.

Theorem 8.8 *If $F = \mathbb{R}$ and W_τ is cyclic and* $\deg(c_\tau(x)) = 2$, *then there is an ordered basis $\mathcal{C}$ for which*

$$[\tau]_\mathcal{C} = \begin{bmatrix} a & -b \\ b & a \end{bmatrix}$$ □

Now we can proceed with the real version of Schur's theorem. For the sake of the exposition, we make the following definition.

Definition *A matrix $A \in M_n(F)$ is* **almost upper triangular** *if it has the form*

$$A = \begin{bmatrix} A_1 & & * \\ & A_2 & \\ & & \ddots & \\ & 0 & & A_k \end{bmatrix}_{\text{block}}$$

where

$$A_i = [a] \quad or \quad A_i = \begin{bmatrix} a & -b \\ b & a \end{bmatrix}$$

for $a, b \in F$. A linear operator $\tau \in \mathcal{L}(V)$ is **almost upper triangularizable** *if there is an ordered basis $\mathcal{B}$ for which $[\tau]_{\mathcal{B}}$ is almost upper triangular.*□

To see that every real linear operator is almost upper triangularizable, we use Theorem 7.19, which states that if $p(x)$ is a prime factor of $c_\tau(x)$, then V_τ has a cyclic submodule W_τ of order $p(x)$. Hence, W is a τ-cyclic subspace of dimension $\deg(p(x))$ and $\tau|_W$ has characteristic polynomial $p(x)$.

Now, the minimal polynomial of a real operator $\tau \in \mathcal{L}(V)$ factors into a product of linear and irreducible quadratic factors. If $c_\tau(x)$ has a linear factor over F, then V_τ has a one-dimensional τ-invariant subspace W. If $c_\tau(x)$ has an irreducible quadratic factor $p(x)$, then V_τ has a cyclic submodule W_τ of order $p(x)$ and so a matrix representation of τ on W_τ is given by the matrix

$$A = \begin{bmatrix} a & -b \\ b & a \end{bmatrix}$$

This is the basis for an inductive proof, as in the complex case.

Theorem 8.9 (Schur's theorem: real case) *If V is a real vector space, then every linear operator on V is almost upper triangularizable.*

Proof. As with the complex case, it is simpler to proceed using matrices, by showing that any $n \times n$ real matrix A is similar to an almost upper triangular matrix. The result is clear if $n = 1$. Assume for the purposes of induction that any square matrix of size less than $n \times n$ is almost upper triangularizable.

We have just seen that F^n has a one-dimensional τ_A-invariant subspace W or a two-dimensional τ_A-cyclic subspace W, where τ_A has irreducible characteristic polynomial on W. Hence, we may choose a basis $\mathcal{B}$ for F^n for which the first one or first two vectors are a basis for W. Then

$$[\tau_A]_{\mathcal{B}} = \begin{bmatrix} A_1 & * \\ 0 & A_2 \end{bmatrix}_{\text{block}}$$

where

$$A_1 = [a] \quad \text{or} \quad A_1 = \begin{bmatrix} a & -b \\ b & a \end{bmatrix}$$

and A_2 has size $k \times k$. The induction hypothesis applied to A_2 gives an invertible matrix $P \in M_k$ for which

$$U = PA_2P^{-1}$$

is almost upper triangular. Hence, if

$$Q = \begin{bmatrix} I_{n-k} & 0 \\ 0 & P \end{bmatrix}_{\text{block}}$$

then Q is invertible and

$$Q[A]_{\mathcal{B}}Q^{-1} = \begin{bmatrix} I_{n-k} & 0 \\ 0 & P \end{bmatrix} \begin{bmatrix} A_1 & * \\ 0 & A_2 \end{bmatrix} \begin{bmatrix} I_{n-k} & 0 \\ 0 & P^{-1} \end{bmatrix} = \begin{bmatrix} A_1 & * \\ 0 & U \end{bmatrix}$$

is almost upper triangular.□

Unitary Triangularizability

Although we have not yet discussed inner product spaces and orthonormal bases, the reader may very well be familiar with these concepts. For those who are, we mention that when V is a real or complex inner product space, then if an operator τ on V can be triangularized (or almost triangularized) using an ordered basis $\mathcal{B}$, it can also be triangularized (or almost triangularized) using an *orthonormal* ordered basis $\mathcal{O}$.

To see this, suppose we apply the Gram–Schmidt orthogonalization process to a basis $\mathcal{B} = (v_1, \ldots, v_n)$ that triangularizes (or almost triangularizes) τ. The resulting ordered orthonormal basis $\mathcal{O} = (u_1, \ldots, u_n)$ has the property that

$$\langle v_1, \ldots, v_i \rangle = \langle u_1, \ldots, u_i \rangle$$

for all $i \le n$. Since $[\tau]_{\mathcal{B}}$ is (almost) upper triangular, that is,

$$\tau v_i \in \langle v_1, \ldots, v_i \rangle$$

for all $i \le n$, it follows that

$$\tau u_i \in \langle \tau v_1, \ldots, \tau v_i \rangle \subseteq \langle v_1, \ldots, v_i \rangle = \langle u_1, \ldots, u_i \rangle$$

and so the matrix $[\tau]_{\mathcal{O}}$ is also (almost) upper triangular.

A linear operator τ is **unitarily upper triangularizable** if there is an ordered *orthonormal* basis with respect to which τ is upper triangular. Accordingly, when V is an inner product space, we can replace the term "upper triangularizable" with "unitarily upper triangularizable" in Schur's theorem. (A similar statement holds for almost upper triangular matrices.)

Diagonalizable Operators

Definition *A linear operator $\tau \in \mathcal{L}(V)$ is* **diagonalizable** *if there is an ordered basis $\mathcal{B} = (v_1, \ldots, v_n)$ of V for which the matrix $[\tau]_{\mathcal{B}}$ is diagonal, or equivalently, if*

$$\tau v_i = \lambda_i v_i$$

for all $i = 1, \ldots, n$.□

The previous definition leads immediately to the following simple characterization of diagonalizable operators.

Theorem 8.10 *Let* $\tau \in \mathcal{L}(V)$. *The following are equivalent:*
1) τ *is diagonalizable.*
2) V *has a basis consisting entirely of eigenvectors of* τ.
3) V *has the form*

$$V = \mathcal{E}_{\lambda_1} \oplus \cdots \oplus \mathcal{E}_{\lambda_k}$$

where $\lambda_1, \ldots, \lambda_k$ *are the distinct eigenvalues of* τ.□

Diagonalizable operators can also be characterized in a simple way via their minimal polynomials.

Theorem 8.11 *A linear operator* $\tau \in \mathcal{L}(V)$ *on a finite-dimensional vector space is diagonalizable if and only if its minimal polynomial is the product of* distinct *linear factors.*
Proof. If τ is diagonalizable, then

$$V = \mathcal{E}_{\lambda_1} \oplus \cdots \oplus \mathcal{E}_{\lambda_k}$$

and Theorem 7.7 implies that $m_\tau(x)$ is the least common multiple of the minimal polynomials $x - \lambda_i$ of τ restricted to $\mathcal{E}_i$. Hence, $m_\tau(x)$ is a product of distinct linear factors. Conversely, if $m_\tau(x)$ is a product of distinct linear factors, then the primary decomposition of V has the form

$$V = V_1 \oplus \cdots \oplus V_k$$

where

$$V_i = \{v \in V \mid (\tau - \lambda_i)v = 0\} = \mathcal{E}_{\lambda_i}$$

and so τ is diagonalizable.□

Spectral Resolutions

We have seen (Theorem 2.25) that resolutions of the identity on a vector space V correspond to direct sum decompositions of V. We can do something similar for any *diagonalizable* linear operator τ on V (not just the identity operator). Suppose that τ has the form

$$\tau = \lambda_1 \rho_1 + \cdots + \lambda_k \rho_k$$

where $\rho_1 + \cdots + \rho_k = \iota$ is a resolution of the identity and the $\lambda_i \in F$ are distinct. This is referred to as a **spectral resolution** of τ.

We claim that the λ_i's are the eigenvalues of τ and $\operatorname{im}(\rho_i) = \mathcal{E}_{\lambda_i}$. Theorem 2.25 implies that

$$V = \operatorname{im}(\rho_1) \oplus \cdots \oplus \operatorname{im}(\rho_k)$$

If $\rho_i v \in \operatorname{im}(\rho_i)$, then

$$\tau(\rho_i v) = (\lambda_1 \rho_1 + \cdots + \lambda_k \rho_k)\rho_i v = \lambda_i(\rho_i v)$$

and so $\rho_i v \in \mathcal{E}_{\lambda_i}$. Hence, $\operatorname{im}(\rho_j) \subseteq \mathcal{E}_{\lambda_j}$ and so

$$V = \operatorname{im}(\rho_1) \oplus \cdots \oplus \operatorname{im}(\rho_k) \subseteq \mathcal{E}_{\lambda_1} \oplus \cdots \oplus \mathcal{E}_{\lambda_k} \subseteq V$$

which implies that $\operatorname{im}(\rho_i) = \mathcal{E}_{\lambda_i}$ and

$$V = \mathcal{E}_{\lambda_1} \oplus \cdots \oplus \mathcal{E}_{\lambda_k}$$

The converse also holds, for if $V = \mathcal{E}_{\lambda_1} \oplus \cdots \oplus \mathcal{E}_{\lambda_k}$ and if ρ_i is projection onto $\mathcal{E}_{\lambda_i}$ along the direct sum of the other eigenspaces, then

$$\rho_1 + \cdots + \rho_k = \iota$$

and since $\tau\rho_i = \lambda_i \rho_i$, it follows that

$$\tau = \tau(\rho_1 + \cdots + \rho_k) = \lambda_1 \rho_1 + \cdots + \lambda_k \rho_k$$

Theorem 8.12 *A linear operator $\tau \in \mathcal{L}(V)$ is diagonalizable if and only if it has a spectral resolution*

$$\tau = \lambda_1 \rho_1 + \cdots + \lambda_k \rho_k$$

In this case, $\{\lambda_1, \ldots, \lambda_k\}$ is the spectrum of τ and

$$\operatorname{im}(\rho_i) = \mathcal{E}_{\lambda_i} \quad \textit{and} \quad \ker(\rho_i) = \bigoplus_{j \neq i} \mathcal{E}_{\lambda_j}$$ □

Exercises

1. Let J be the $n \times n$ matrix all of whose entries are equal to 1. Find the minimal polynomial and characteristic polynomial of J and the eigenvalues.
2. Prove that the eigenvalues of a matrix do not form a complete set of invariants under similarity.
3. Show that $\tau \in \mathcal{L}(V)$ is invertible if and only if 0 is not an eigenvalue of τ.
4. Let A be an $n \times n$ matrix over a field F that contains all roots of the characteristic polynomial of A. Prove that $\det(A)$ is the product of the eigenvalues of A, counting multiplicity.
5. Show that if λ is an eigenvalue of τ, then $p(\lambda)$ is an eigenvalue of $p(\tau)$, for any polynomial $p(x)$. Also, if $\lambda \neq 0$, then λ^{-1} is an eigenvalue for τ^{-1}.
6. An operator $\tau \in \mathcal{L}(V)$ is **nilpotent** if $\tau^n = 0$ for some positive $n \in \mathbb{N}$.

   a) Show that if τ is nilpotent, then the spectrum of τ is $\{0\}$.
   b) Find a nonnilpotent operator τ with spectrum $\{0\}$.
7. Show that if $\sigma, \tau \in \mathcal{L}(V)$ and one of σ and τ is invertible, then $\sigma\tau \sim \tau\sigma$ and so $\sigma\tau$ and $\tau\sigma$ have the same eigenvalues, counting multiplicty.
8. (Halmos)
   a) Find a linear operator τ that is not idempotent but for which $\tau^2(\iota - \tau) = 0$.
   b) Find a linear operator τ that is not idempotent but for which $\tau(\iota - \tau)^2 = 0$.
   c) Prove that if $\tau^2(\iota - \tau) = \tau(\iota - \tau)^2 = 0$, then τ is idempotent.
9. An **involution** is a linear operator θ for which $\theta^2 = \iota$. If τ is idempotent what can you say about $2\tau - \iota$? Construct a one-to-one correspondence between the set of idempotents on V and the set of involutions.
10. Let $A, B \in M_2(\mathbb{C})$ and suppose that $A^2 = B^3 = I, ABA = B^{-1}$ but $A \neq I$ and $B \neq I$. Show that if $C \in M_2(\mathbb{C})$ commutes with both A and B, then $C = rI$ for some scalar $r \in \mathbb{C}$.
11. Let $\tau \in \mathcal{L}(V)$ and let

$$S = \langle v, \tau v, \ldots, \tau^{de-1} v \rangle$$

be a τ-cyclic submodule of V_τ with minimal polynomial $p(x)^e$ where $p(x)$ is prime of degree d. Let $\sigma = p(\tau)$ restricted to $\langle v \rangle$. Show that S is the direct sum of d σ-cyclic submodules each of dimension e, that is,

$$S = T_1 \oplus \cdots \oplus T_d$$

Hint: For each $0 \leq i < d$, consider the set

$$\mathcal{B}_i = \{\tau^i v, p(\tau)\tau^i v, \ldots, p(\tau)^{e-1}\tau^i v\rangle$$

12. Fix $\epsilon > 0$. Show that any complex matrix is similar to a matrix that looks just like a Jordan matrix except that the entries that are equal to 1 are replaced by entries with value ϵ, where ϵ is any complex number. Thus, any complex matrix is similar to a matrix that is "almost" diagonal. *Hint*: consider the fact that

$$\begin{bmatrix} 1 & 0 & 0 \\ 0 & \epsilon & 0 \\ 0 & 0 & \epsilon^2 \end{bmatrix} \begin{bmatrix} \lambda & 0 & 0 \\ 1 & \lambda & 0 \\ 0 & 1 & \lambda \end{bmatrix} \begin{bmatrix} 1 & 0 & 0 \\ 0 & \epsilon^{-1} & 0 \\ 0 & 0 & \epsilon^{-2} \end{bmatrix} = \begin{bmatrix} \lambda & 0 & 0 \\ \epsilon & \lambda & 0 \\ 0 & \epsilon & \lambda \end{bmatrix}$$

13. Show that the Jordan canonical form is not very robust in the sense that a small change in the entries of a matrix A may result in a large jump in the entries of the Jordan form J. *Hint*: consider the matrix

$$A_\epsilon = \begin{bmatrix} \epsilon & 0 \\ 1 & 0 \end{bmatrix}$$

What happens to the Jordan form of A_ϵ as $\epsilon \to 0$?

14. Give an example of a complex nonreal matrix all of whose eigenvalues are real. Show that any such matrix is similar to a real matrix. What about the type of the invertible matrices that are used to bring the matrix to Jordan form?
15. Let $J = [\tau]_{\mathcal{B}}$ be the Jordan form of a linear operator $\tau \in \mathcal{L}(V)$. For a given Jordan block of $J(\lambda, e)$ let U be the subspace of V spanned by the basis vectors of $\mathcal{B}$ associated with that block.
    a) Show that $\tau|_U$ has a single eigenvalue λ with geometric multiplicity 1. In other words, there is essentially only one eigenvector (up to scalar multiple) associated with each Jordan block. Hence, the geometric multiplicity of λ for τ is the number of Jordan blocks for λ. Show that the algebraic multiplicity is the sum of the dimensions of the Jordan blocks associated with λ.
    b) Show that the number of Jordan blocks in J is the maximum number of linearly independent eigenvectors of τ.
    c) What can you say about the Jordan blocks if the algebraic multiplicity of every eigenvalue is equal to its geometric multiplicity?
16. Assume that the base field F is algebraically closed. Then assuming that the eigenvalues of a matrix A are known, it is possible to determine the Jordan form J of A by looking at the rank of various matrix powers. A matrix B is **nilpotent** if $B^n = 0$ for some $n > 0$. The smallest such exponent is called the **index of nilpotence**.
    a) Let $J = J(\lambda, n)$ be a single Jordan block of size $n \times n$. Show that $J - \lambda I$ is nilpotent of index n. Thus, n is the smallest integer for which $\text{rk}(J - \lambda I)^n = 0$.

    Now let J be a matrix in Jordan form but possessing only one eigenvalue λ.
    b) Show that $J - \lambda I$ is nilpotent. Let m be its index of nilpotence. Show that m is the maximum size of the Jordan blocks of J and that $\text{rk}(J - \lambda I)^{m-1}$ is the number of Jordan blocks in J of maximum size.
    c) Show that $\text{rk}(J - \lambda I)^{m-2}$ is equal to 2 times the number of Jordan blocks of maximum size plus the number of Jordan blocks of size one less than the maximum.
    d) Show that the sequence $\text{rk}(J - \lambda I)^k$ for $k = 1, \ldots, m$ uniquely determines the number and size of all of the Jordan blocks in J, that is, it uniquely determines J up to the order of the blocks.
    e) Now let J be an arbitrary Jordan matrix. If λ is an eigenvalue for J show that the sequence $\text{rk}(J - \lambda I)^k$ for $k = 1, \ldots, m$ where m is the first integer for which $\text{rk}(J - \lambda I)^m = \text{rk}(J - \lambda I)^{m+1}$ uniquely determines J up to the order of the blocks.
    f) Prove that for any matrix A with spectrum $\{\lambda_1, \ldots, \lambda_s\}$ the sequence $\text{rk}(A - \lambda_i I)^k$ for $i = 1, \ldots, s$ and $k = 1, \ldots, m$ where m is the first integer for which $\text{rk}(A - \lambda_i I)^m = \text{rk}(A - \lambda_i I)^{m+1}$ uniquely determines the Jordan matrix J for A up to the order of the blocks.
17. Let $A \in \mathcal{M}_n(F)$.

a) If all the roots of the characteristic polynomial of A lie in F prove that A is similar to its transpose A^t. Hint: Let B be the matrix

$$B = \begin{bmatrix} 0 & \cdots & 0 & 1 \\ \vdots & \iddots & 1 & 0 \\ 0 & \iddots & \iddots & \vdots \\ 1 & 0 & \cdots & 0 \end{bmatrix}$$

with 1's on the diagonal that moves up from left to right and 0's elsewhere. Let J be a Jordan block of the same size as B. Show that $BJB^{-1} = J^t$.

b) Let $A, B \in \mathcal{M}_n(F)$. Let K be a field containing F. Show that if A and B are similar over K, that is, if $B = PAP^{-1}$ where $P \in \mathcal{M}_n(K)$, then A and B are also similar over F, that is, there exists $Q \in \mathcal{M}_n(F)$ for which $B = QAQ^{-1}$.

c) Show that any matrix is similar to its transpose.

The Trace of a Matrix

18. Let $A \in \mathcal{M}_n(F)$. Verify the following statements.
    a) $\operatorname{tr}(rA) = r\operatorname{tr}(A)$, for $r \in F$.
    b) $\operatorname{tr}(A + B) = \operatorname{tr}(A) + \operatorname{tr}(B)$.
    c) $\operatorname{tr}(AB) = \operatorname{tr}(BA)$.
    d) $\operatorname{tr}(ABC) = \operatorname{tr}(CAB) = \operatorname{tr}(BCA)$. Find an example to show that $\operatorname{tr}(ABC)$ may not equal $\operatorname{tr}(ACB)$.
    e) The trace is an invariant under similarity.
    f) If F is algebraically closed, then the trace of A is the sum of the eigenvalues of A.
19. Use the concept of the trace of a matrix, as defined in the previous exercise, to prove that there are no matrices A, $B \in \mathcal{M}_n(\mathbb{C})$ for which

$$AB - BA = I$$

20. Let $T: \mathcal{M}_n(F) \to F$ be a function with the following properties. For all matrices A, $B \in \mathcal{M}_n(F)$ and $r \in F$,
    1) $T(rA) = rT(A)$
    2) $T(A + B) = T(A) + T(B)$
    3) $T(AB) = T(BA)$

    Show that there exists $s \in F$ for which $T(A) = s\operatorname{tr}(A)$, for all $A \in \mathcal{M}_n(F)$.

Commuting Operators

Let

$$\mathcal{F} = \{\tau_i \in \mathcal{L}(V) \mid i \in \mathcal{I}\}$$

be a family of operators on a vector space V. Then $\mathcal{F}$ is a **commuting family** if every pair of operators commutes, that is, $\sigma\tau = \tau\sigma$ for all $\sigma, \tau \in \mathcal{F}$. A subspace

U of V is $\mathcal{F}$**-invariant** if it is τ-invariant for every $\tau \in \mathcal{F}$. It is often of interest to know whether a family $\mathcal{F}$ of linear operators on V has a **common eigenvector**, that is, a single vector $v \in V$ that is an eigenvector for every $\sigma \in \mathcal{F}$ (the corresponding eigenvalues may be different for each operator, however).

21. A pair of linear operators $\sigma, \tau \in \mathcal{L}(V)$ is **simultaneously diagonalizable** if there is an ordered basis $\mathcal{B}$ for V for which $[\tau]_{\mathcal{B}}$ and $[\sigma]_{\mathcal{B}}$ are both diagonal, that is, $\mathcal{B}$ is an ordered basis of eigenvectors for both τ and σ. Prove that two diagonalizable operators σ and τ are simultaneously diagonalizable if and only if they commute, that is, $\sigma\tau = \tau\sigma$. *Hint*: If $\sigma\tau = \tau\sigma$, then the eigenspaces of τ are invariant under σ.
22. Let $\sigma, \tau \in \mathcal{L}(V)$. Prove that if σ and τ commute, then every eigenspace of σ is τ-invariant. Thus, if $\mathcal{F}$ is a commuting family, then every eigenspace of any member of $\mathcal{F}$ is $\mathcal{F}$-invariant.
23. Let $\mathcal{F}$ be a family of operators in $\mathcal{L}(V)$ with the property that each operator in $\mathcal{F}$ has a full set of eigenvalues in the base field F, that is, the characteristic polynomial splits over F. Prove that if $\mathcal{F}$ is a commuting family, then $\mathcal{F}$ has a common eigenvector $v \in V$.
24. What do the real matrices
$$A = \begin{bmatrix} 1 & 1 \\ -1 & 1 \end{bmatrix} \text{ and } B = \begin{bmatrix} 1 & 2 \\ -2 & 1 \end{bmatrix}$$
have to do with the issue of common eigenvectors?

Geršgorin Disks

It is generally impossible to determine precisely the eigenvalues of a given complex operator or matrix $A \in \mathcal{M}_n(\mathbb{C})$, for if $n \geq 5$, then the characteristic equation has degree 5 and cannot in general be solved. As a result, the approximation of eigenvalues is big business. Here we consider one aspect of this approximation problem, which also has some interesting theoretical consequences.

Let $A \in \mathcal{M}_n(\mathbb{C})$ and suppose that $Av = \lambda v$ where $v = (b_1, \ldots, b_n)^t$. Comparing kth rows gives
$$\sum_{i=1}^{n} A_{ki} b_i = \lambda b_k$$
which can also be written in the form
$$b_k(\lambda - A_{kk}) = \sum_{\substack{i=1 \\ i \neq k}}^{n} A_{ki} b_i$$
If k has the property that $|b_k| \geq |b_i|$ for all i, we have

$$|b_k||\lambda - A_{kk}| \leq \sum_{\substack{i=1 \\ i \neq k}}^{n} |A_{ki}||b_i| \leq |b_k| \sum_{\substack{i=1 \\ i \neq k}}^{n} |A_{ki}|$$

and thus

$$|\lambda - A_{kk}| \leq \sum_{\substack{i=1 \\ i \neq k}}^{n} |A_{ki}| \tag{8.7}$$

The right-hand side is the sum of the absolute values of all entries in the kth row of A except the diagonal entry A_{kk}. This sum $R_k(A)$ is the kth **deleted absolute row sum** of A. The inequality (8.7) says that, in the complex plane, the eigenvalue λ lies in the disk centered at the diagonal entry A_{kk} with radius equal to $R_k(A)$. This disk

$$\text{GR}_k(A) = \{z \in \mathbb{C} \mid |z - A_{kk}| \leq R_k(A)\}$$

is called the **Geršgorin row disk** for the kth row of A. The union of all of the Geršgorin row disks is called the **Geršgorin row region** for A.

Since there is no way to know in general which is the index k for which $|b_k| \geq |b_i|$, the best we can say in general is that the eigenvalues of A lie in the union of all Geršgorin row disks, that is, in the Geršgorin row region of A.

Similar definitions can be made for columns and since a matrix has the same eigenvalues as its transpose, we can say that the eigenvalues of A lie in the Geršgorin column region of A. The **Geršgorin region** $G(A)$ of a matrix $A \in M_n(F)$ is the intersection of the Geršgorin row region and the Geršgorin column region and we can say that all eigenvalues of A lie in the Geršgorin region of A. In symbols, $\sigma A \subseteq GA$.

25. Find and sketch the Geršgorin region and the eigenvalues for the matrix

$$A = \begin{bmatrix} 1 & 2 & 3 \\ 4 & 5 & 6 \\ 7 & 8 & 9 \end{bmatrix}$$

26. A matrix $A \in M_n(\mathbb{C})$ is **diagonally dominant** if for each $k = 1, \ldots, n$,

$$|A_{kk}| \geq R_k(A)$$

and it is **strictly diagonally dominant** if strict inequality holds. Prove that if A is strictly diagonally dominant, then it is invertible.

27. Find a matrix $A \in M_n(\mathbb{C})$ that is diagonally dominant but not invertible.
28. Find a matrix $A \in M_n(\mathbb{C})$ that is invertible but not strictly diagonally dominant.

Chapter 9
Real and Complex Inner Product Spaces

We now turn to a discussion of real and complex vector spaces that have an additional function defined on them, called an *inner product*, as described in the following definition. *In this chapter, F will denote either the real or complex field.* Also, the complex conjugate of $r \in \mathbb{C}$ is denoted by $\overline{r}$.

Definition *Let V be a vector space over $F = \mathbb{R}$ or $F = \mathbb{C}$. An* **inner product** *on V is a function $\langle , \rangle : V \times V \to F$ with the following properties:*

1) **(Positive definiteness)** *For all $v \in V$,*

$$\langle v, v\rangle \geq 0 \quad \textit{and} \quad \langle v, v\rangle = 0 \Leftrightarrow v = 0$$

2) *For $F = \mathbb{C}$:* **(Conjugate symmetry)**

$$\langle u, v\rangle = \overline{\langle v, u\rangle}$$

For $F = \mathbb{R}$: **(Symmetry)**

$$\langle u, v\rangle = \langle v, u\rangle$$

3) **(Linearity in the first coordinate)** *For all $u, v \in V$ and $r, s \in F$*

$$\langle ru + sv, w\rangle = r\langle u, w\rangle + s\langle v, w\rangle$$

A real (or complex) vector space V, together with an inner product, is called a **real** (*or* **complex**) **inner product space.** □

If $X, Y \subseteq V$, then we let

$$\langle X, Y\rangle = \{\langle x, y\rangle \mid x \in X, y \in Y\}$$

and

$$\langle v, X\rangle = \{\langle v, x\rangle \mid x \in X\}$$

Note that a vector subspace S of an inner product space V is also an inner product space under the restriction of the inner product of V to S.

We will study bilinear forms (also called *inner products*) on vector spaces over fields other than $\mathbb{R}$ or $\mathbb{C}$ in Chapter 11. Note that property 1) implies that $\langle v, v\rangle$ is always real, even if V is a complex vector space.

If $F = \mathbb{R}$, then properties 2) and 3) imply that the inner product is linear in both coordinates, that is, the inner product is **bilinear**. However, if $F = \mathbb{C}$, then

$$\langle w, ru + sv\rangle = \overline{\langle ru + sv, w\rangle} = \overline{r}\overline{\langle u, w\rangle} + \overline{s}\overline{\langle v, w\rangle} = \overline{r}\langle w, u\rangle + \overline{s}\langle w, v\rangle$$

This is referred to as **conjugate linearity** in the second coordinate. Specifically, a function $f: V \to W$ between complex vector spaces is **conjugate linear** if

$$f(u + v) = f(u) + f(v)$$

and

$$f(ru) = \overline{r}f(u)$$

for all $u, v \in V$ and $r \in \mathbb{C}$. Thus, a complex inner product is linear in its first coordinate and conjugate linear in its second coordinate. This is often described by saying that a complex inner product is **sesquilinear**. (Sesqui means "one and a half times.")

Example 9.1

1) The vector space $\mathbb{R}^n$ is an inner product space under the **standard inner product**, or **dot product**, defined by

$$\langle (r_1, \ldots, r_n), (s_1, \ldots, s_n)\rangle = r_1 s_1 + \cdots + r_n s_n$$

The inner product space $\mathbb{R}^n$ is often called ***n*-dimensional Euclidean space**.

2) The vector space $\mathbb{C}^n$ is an inner product space under the **standard inner product** defined by

$$\langle (r_1, \ldots, r_n), (s_1, \ldots, s_n)\rangle = r_1 \overline{s}_1 + \cdots + r_n \overline{s}_n$$

This inner product space is often called ***n*-dimensional unitary space**.

3) The vector space $C[a, b]$ of all continuous complex-valued functions on the closed interval $[a, b]$ is a complex inner product space under the inner product

$$\langle f, g\rangle = \int_a^b f(x)\overline{g(x)}\, dx$$

□

Example 9.2 One of the most important inner product spaces is the vector space ℓ^2 of all real (or complex) sequences (s_n) with the property that

$$\sum |s_n|^2 < \infty$$

under the inner product

$$\langle (s_n), (t_n) \rangle = \sum_{n=0}^{\infty} s_n \overline{t_n}$$

Such sequences are called **square summable**. Of course, for this inner product to make sense, the sum on the right must converge. To see this, note that if $(s_n), (t_n) \in \ell^2$, then

$$0 \leq (|s_n| - |t_n|)^2 = |s_n|^2 - 2|s_n||t_n| + |t_n|^2$$

and so

$$2|s_n t_n| \leq |s_n|^2 + |t_n|^2$$

which implies that $(s_n t_n) \in \ell^2$. We leave it to the reader to verify that ℓ^2 is an inner product space.□

The following simple result is quite useful.

Lemma 9.1 *If V is an inner product space and $\langle u, x \rangle = \langle v, x \rangle$ for all $x \in V$, then $u = v$.*□

The next result points out one of the main differences between real and complex inner product spaces and will play a key role in later work.

Theorem 9.2 *Let V be an inner product space and let $\tau \in \mathcal{L}(V)$.*
1)

$$\langle \tau v, w \rangle = 0 \text{ for all } v, w \in V \quad \Rightarrow \quad \tau = 0$$

2) If V is a complex inner product space, then

$$\langle \tau v, v \rangle = 0 \text{ for all } v \in V \quad \Rightarrow \quad \tau = 0$$

but this does not hold in general for real inner product spaces.

Proof. Part 1) follows directly from Lemma 9.1. As for part 2), let $v = rx + y$, for $x, y \in V$ and $r \in F$. Then

$$\begin{aligned} 0 &= \langle \tau(rx + y), rx + y \rangle \\ &= |r|^2 \langle \tau x, x \rangle + \langle \tau y, y \rangle + r\langle \tau x, y \rangle + \overline{r} \langle \tau y, x \rangle \\ &= r\langle \tau x, y \rangle + \overline{r} \langle \tau y, x \rangle \end{aligned}$$

Setting $r = 1$ gives

$$\langle \tau x, y \rangle + \langle \tau y, x \rangle = 0$$

and setting $r = i$ gives

$$\langle \tau x, y\rangle - \langle \tau y, x\rangle = 0$$

These two equations imply that $\langle \tau x, y\rangle = 0$ for all $x, y \in V$ and so part 1) implies that $\tau = 0$. For the last statement, rotation by 90 degrees in the real plane $\mathbb{R}^2$ has the property that $\langle \tau v, v\rangle = 0$ for all v.□

Norm and Distance

If V is an inner product space, the **norm**, or **length** of $v \in V$ is defined by

$$\|v\| = \sqrt{\langle v, v\rangle} \tag{9.1}$$

A vector v is a **unit vector** if $\|v\| = 1$. Here are the basic properties of the norm.

Theorem 9.3

1) $\|v\| \geq 0$ *and* $\|v\| = 0$ *if and only if* $v = 0$.

2) *For all* $r \in F$ *and* $v \in V$,

$$\|rv\| = |r|\|v\|$$

3) **(The Cauchy–Schwarz inequality)** *For all* $u, v \in V$,

$$|\langle u, v\rangle| \leq \|u\|\|v\|$$

with equality if and only if one of u *and* v *is a scalar multiple of the other.*

4) **(The triangle inequality)** *For all* $u, v \in V$,

$$\|u + v\| \leq \|u\| + \|v\|$$

with equality if and only if one of u *and* v *is a scalar multiple of the other.*

5) *For all* $u, v, x \in V$,

$$\|u - v\| \leq \|u - x\| + \|x - v\|$$

6) *For all* $u, v \in V$,

$$|\|u\| - \|v\|| \leq \|u - v\|$$

7) **(The parallelogram law)** *For all* $u, v \in V$,

$$\|u + v\|^2 + \|u - v\|^2 = 2\|u\|^2 + 2\|v\|^2$$

Proof. We prove only Cauchy–Schwarz and the triangle inequality. For Cauchy–Schwarz, if either u or v is zero the result follows, so assume that $u, v \neq 0$. Then, for any scalar $r \in F$,

$$\begin{aligned} 0 &\leq \|u - rv\|^2 \\ &= \langle u - rv, u - rv\rangle \\ &= \langle u, u\rangle - \overline{r}\langle u, v\rangle - r[\langle v, u\rangle - \overline{r}\langle v, v\rangle] \end{aligned}$$

Choosing $\overline{r} = \langle v, u\rangle / \langle v, v\rangle$ makes the value in the square brackets equal to 0

and so

$$0 \le \langle u, u\rangle - \frac{\langle v, u\rangle\langle u, v\rangle}{\langle v, v\rangle} = \|u\|^2 - \frac{|\langle u, v\rangle|^2}{\|v\|^2}$$

which is equivalent to the Cauchy–Schwarz inequality. Furthermore, equality holds if and only if $\|u - rv\|^2 = 0$, that is, if and only if $u - rv = 0$, which is equivalent to u and v being scalar multiples of one another.

To prove the triangle inequality, the Cauchy–Schwarz inequality gives

$$\begin{aligned}\|u+v\|^2 &= \langle u+v, u+v\rangle \\ &= \langle u, u\rangle + \langle u, v\rangle + \langle v, u\rangle + \langle v, v\rangle \\ &\le \|u\|^2 + 2\|u\|\|v\| + \|v\|^2 \\ &= (\|u\| + \|v\|)^2\end{aligned}$$

from which the triangle inequality follows. The proof of the statement concerning equality is left to the reader.□

Any vector space V, together with a function $\|\cdot\|: V \to \mathbb{R}$ that satisfies properties 1), 2) and 4) of Theorem 9.3, is called a **normed linear space** and the function $\|\cdot\|$ is called a **norm**. Thus, any inner product space is a normed linear space, under the norm given by (9.1).

It is interesting to observe that the inner product on V can be recovered from the norm. Thus, knowing the length of all vectors in V is equivalent to knowing all inner products of vectors in V.

Theorem 9.4 (The polarization identities)

1) If V is a real inner product space, then

$$\langle u, v\rangle = \frac{1}{4}(\|u+v\|^2 - \|u-v\|^2)$$

2) If V is a complex inner product space, then

$$\langle u, v\rangle = \frac{1}{4}(\|u+v\|^2 - \|u-v\|^2) + \frac{1}{4}i(\|u+iv\|^2 - \|u-iv\|^2)$$

The norm can be used to define the distance between any two vectors in an inner product space.

Definition *Let V be an inner product space. The* **distance** $d(u, v)$ *between any two vectors u and v in V is*

$$d(u, v) = \|u - v\| \qquad (9.2)$$ □

Here are the basic properties of distance.

Theorem 9.5

1) $d(u,v) \geq 0$ *and* $d(u,v) = 0$ *if and only if* $u = v$

2) **(Symmetry)**

$$d(u,v) = d(v,u)$$

3) **(The triangle inequality)**

$$d(u,v) \leq d(u,w) + d(w,v)$$

□

Any nonempty set V, together with a function $d: V \times V \to \mathbb{R}$ that satisfies the properties of Theorem 9.5, is called a **metric space** and the function d is called a **metric** on V. Thus, any inner product space is a metric space under the metric (9.2).

Before continuing, we should make a few remarks about our goals in this and the next chapter. The presence of an inner product, and hence a metric, permits the definition of a topology on V, and in particular, convergence of infinite sequences. A sequence (v_n) of vectors in V **converges** to $v \in V$ if

$$\lim_{n \to \infty} \|v_n - v\| = 0$$

Some of the more important concepts related to convergence are closedness and closures, completeness and the continuity of linear operators and linear functionals.

In the finite-dimensional case, the situation is very straightforward: All subspaces are closed, all inner product spaces are complete and all linear operators and functionals are continuous. However, in the infinite-dimensional case, things are not as simple.

Our goals in this chapter and the next are to describe some of the basic properties of inner product spaces—both finite and infinite-dimensional—and then discuss certain special types of operators (normal, unitary and self-adjoint) in the finite-dimensional case only. To achieve the latter goal as rapidly as possible, we will postpone a discussion of convergence-related properties until Chapter 12. This means that we must state some results only for the finite-dimensional case in this chapter.

Isometries

An isomorphism of vector spaces preserves the vector space operations. The corresponding concept for inner product spaces is the *isometry*.

Definition *Let* V *and* W *be inner product spaces and let* $\tau \in \mathcal{L}(V,W)$.

1) τ *is an* **isometry** *if it preserves the inner product, that is, if*

$$\langle \tau u, \tau v \rangle = \langle u, v \rangle$$

for all $u, v \in V$.

2) A bijective isometry is called an **isometric isomorphism**. *When* $\tau: V \to W$ *is an isometric isomorphism, we say that* V *and* W *are* **isometrically isomorphic**.□

It is clear that an isometry is injective and so it is an isometric isomorphism provided it is surjective. Moreover, if

$$\dim(V) = \dim(W) < \infty$$

injectivity implies surjectivity and τ is an isometry if and only if τ is an isometric isomorphism. On the other hand, the following simple example shows that this is not the case for infinite-dimensional inner product spaces.

Example 9.3 The map $\tau: \ell^2 \to \ell^2$ defined by

$$\tau(x_1, x_2, x_3, \ldots) = (0, x_1, x_2, \ldots)$$

is an isometry, but it is clearly not surjective.□

Since the norm determines the inner product, the following should not come as a surprise.

Theorem 9.6 *A linear transformation* $\tau \in \mathcal{L}(V, W)$ *is an isometry if and only if it preserves the norm, that is, if and only if*

$$\|\tau v\| = \|v\|$$

for all $v \in V$.

Proof. Clearly, an isometry preserves the norm. The converse follows from the polarization identities. In the real case, we have

$$\begin{aligned}\langle \tau u, \tau v \rangle &= \frac{1}{4}(\|\tau u + \tau v\|^2 - \|\tau u - \tau v\|^2) \\ &= \frac{1}{4}(\|\tau(u + v)\|^2 - \|\tau(u - v)\|^2) \\ &= \frac{1}{4}(\|u + v\|^2 - \|u - v\|^2) \\ &= \langle u, v \rangle\end{aligned}$$

and so τ is an isometry. The complex case is similar.□

Orthogonality

The presence of an inner product allows us to define the concept of orthogonality.

Definition *Let V be an inner product space.*

1) *Two vectors $u,v\in V$ are* **orthogonal**, *written $u\perp v$, if*
$$\langle u,v\rangle = 0$$
2) *Two subsets $X,Y\subseteq V$ are* **orthogonal**, *written $X\perp Y$, if $\langle X,Y\rangle=\{0\}$, that is, if $x\perp y$ for all $x\in X$ and $y\in Y$. We write $v\perp X$ in place of $\{v\}\perp X$.*
3) *The* **orthogonal complement** *of a subset $X\subseteq V$ is the set*
$$X^{\perp}=\{v\in V\mid v\perp X\}$$ □

The following result is easily proved.

Theorem 9.7 *Let V be an inner product space.*

1) *The orthogonal complement $X^{\perp}$ of any subset $X\subseteq V$ is a subspace of V.*
2) *For any subspace S of V,*
$$S\cap S^{\perp}=\{0\}$$ □

Definition *An inner product space V is the* **orthogonal direct sum** *of subspaces S and T if*
$$V=S\oplus T,\quad S\perp T$$
In this case, we write
$$S\odot T$$
More generally, V is the **orthogonal direct sum** *of the subspaces $S_1,\dots,S_n$, written*
$$S=S_1\odot\cdots\odot S_n$$
if
$$V=S_1\oplus\cdots\oplus S_n \quad \text{and} \quad S_i\perp S_j \text{ for } i\neq j$$ □

Theorem 9.8 *Let V be an inner product space. The following are equivalent.*

1) *$V=S\odot T$*
2) *$V=S\oplus T$ and $T=S^{\perp}$*

Proof. If $V=S\odot T$, then by definition, $T\subseteq S^{\perp}$. However, if $v\in S^{\perp}$, then $v=s+t$ where $s\in S$ and $t\in T$. Then s is orthogonal to both t and v and so s is orthogonal to itself, which implies that $s=0$ and so $v\in T$. Hence, $T=S^{\perp}$. The converse is clear.□

Orthogonal and Orthonormal Sets

Definition *A nonempty set $\mathcal{O}=\{u_i\mid i\in K\}$ of vectors in an inner product space is said to be an* **orthogonal set** *if $u_i\perp u_j$ for all $i\neq j\in K$. If, in addition, each vector u_i is a unit vector, then $\mathcal{O}$ is an* **orthonormal set**. *Thus, a* *set is orthonormal if*

$$\langle u_i, u_j \rangle = \delta_{i,j}$$

for all $i, j \in K$, where $\delta_{i,j}$ is the Kronecker delta function.□

Of course, given any nonzero vector $v \in V$, we may obtain a unit vector u by multiplying v by the reciprocal of its norm:

$$u = \frac{1}{\|v\|} v$$

This process is referred to as **normalizing** the vector v. Thus, it is a simple matter to construct an orthonormal set from an orthogonal set of *nonzero* vectors.

Note that if $u \perp v$, then

$$\|u + v\|^2 = \|u\|^2 + \|v\|^2$$

and the converse holds if $F = \mathbb{R}$.

Orthogonality is stronger than linear independence.

Theorem 9.9 *Any orthogonal set of nonzero vectors in V is linearly independent.*
Proof. If $\mathcal{O} = \{u_i \mid i \in K\}$ is an orthogonal set of nonzero vectors and

$$r_1 u_1 + \cdots + r_n u_n = 0$$

then

$$0 = \langle r_1 u_1 + \cdots + r_n u_n, u_k \rangle = r_k \langle u_k, u_k \rangle$$

and so $r_k = 0$, for all k. Hence, $\mathcal{O}$ is linearly independent.□

Gram–Schmidt Orthogonalization

The Gram–Schmidt process can be used to transform a sequence of vectors into an orthogonal sequence. We begin with the following.

Theorem 9.10 (Gram–Schmidt augmentation) *Let V be an inner product space and let $\mathcal{O} = \{u_1, \ldots, u_n\}$ be an orthogonal set of vectors in V. If $v \notin \langle u_1, \ldots, u_n \rangle$, then there is a nonzero $u \in V$ for which $\{u_1, \ldots, u_n, u\}$ is orthogonal and*

$$\langle u_1, \ldots, u_n, u \rangle = \langle u_1, \ldots, u_n, v \rangle$$

In particular,

$$u = v - \sum_{i=1}^{n} r_i u_i$$

where

$$r_i = \begin{cases} 0 & \text{if } u_i = 0 \\ \frac{\langle v, u_i \rangle}{\langle u_i, u_i \rangle} & \text{if } u_i \neq 0 \end{cases}$$

Proof. We simply set

$$u = v - r_1 u_1 - \cdots - r_n u_n$$

and force $u \perp u_i$ for all i, that is,

$$0 = \langle u, u_i \rangle = \langle v - r_1 u_1 - \cdots - r_n u_n, u_i \rangle = \langle v, u_i \rangle - r_i \langle u_i, u_i \rangle$$

Thus, if $u_i = 0$, take $r_i = 0$ and if $u_i \neq 0$, take

$$r_i = \frac{\langle v, u_i \rangle}{\langle u_i, u_i \rangle}$$

□

The Gram–Schmidt augmentation is traditionally applied to a sequence of linearly independent vectors, but it also applies to any sequence of vectors.

Theorem 9.11 (The Gram–Schmidt orthogonalization process) *Let $\mathcal{B} = (v_1, v_2, \ldots)$ be a sequence of vectors in an inner product space V. Define a sequence $\mathcal{O} = (u_1, u_2, \ldots)$ by repeated Gram–Schmidt augmentation, that is,*

$$u_k = v_k - \sum_{i=1}^{k-1} r_{k,i} u_i$$

where $u_1 = v_1$ and

$$r_{k,i} = \begin{cases} 0 & \text{if } u_i = 0 \\ \frac{\langle v_k, u_i \rangle}{\langle u_i, u_i \rangle} & \text{if } u_i \neq 0 \end{cases}$$

Then $\mathcal{O}$ is an orthogonal sequence in V with the property that

$$\langle u_1, \ldots, u_k \rangle = \langle v_1, \ldots, v_k \rangle$$

for all $k > 0$. Also, $u_k = 0$ if and only if $v_k \in \langle v_1, \ldots, v_{k-1} \rangle$.

Proof. The result holds for $k = 1$. Assume it holds for $k - 1$. If $v_k \in \langle v_1, \ldots, v_{k-1} \rangle$, then

$$v_k \in \langle v_1, \ldots, v_{k-1} \rangle = \langle u_1, \ldots, u_{k-1} \rangle$$

Writing

$$v_k = \sum_{i=1}^{k-1} a_i u_i$$

we have

$$\langle v_k, u_i \rangle = \begin{cases} 0 & \text{if } u_i = 0 \\ a_i \langle u_i, u_i \rangle & \text{if } u_i \neq 0 \end{cases}$$

Therefore, $a_i = r_{k,i}$ when $u_i \neq 0$ and so $u_k = 0$. Hence,

$$\langle u_1, \ldots, u_k \rangle = \langle u_1, \ldots, u_{k-1}, 0 \rangle = \langle v_1, \ldots, v_{k-1} \rangle = \langle v_1, \ldots, v_k \rangle$$

If $v_k \notin \langle v_1, \ldots, v_{k-1} \rangle$ then

$$\langle u_1, \ldots, u_k \rangle = \langle v_1, \ldots, v_{k-1}, u_k \rangle = \langle v_1, \ldots, v_{k-1}, v_k \rangle$$ □

Example 9.4 Consider the inner product space $\mathbb{R}[x]$ of real polynomials, with inner product defined by

$$\langle p(x), q(x) \rangle = \int_{-1}^{1} p(x)q(x)dx$$

Applying the Gram–Schmidt process to the sequence $\mathcal{B} = (1, x, x^2, x^3, \ldots)$ gives

$$\begin{aligned}
u_1(x) &= 1 \\
u_2(x) &= x - \frac{\int_{-1}^{1} x\,dx}{\int_{-1}^{1} dx} \cdot 1 = x \\
u_3(x) &= x^2 - \frac{\int_{-1}^{1} x^2\,dx}{\int_{-1}^{1} dx} \cdot 1 - \frac{\int_{-1}^{1} x^3\,dx}{\int_{-1}^{1} x\,dx} \cdot x = x^2 - \frac{1}{3} \\
u_4(x) &= x^3 - \frac{\int_{-1}^{1} x^3\,dx}{\int_{-1}^{1} dx} \cdot 1 - \frac{\int_{-1}^{1} x^4\,dx}{\int_{-1}^{1} x\,dx} \cdot x - \frac{\int_{-1}^{1} x^3(x^2 - \frac{1}{3})dx}{\int_{-1}^{1} (x^2 - \frac{1}{3})^2 dx} \cdot \left[x^2 - \frac{1}{3}\right] \\
&= x^3 - \frac{3}{5}x
\end{aligned}$$

and so on. The polynomials in this sequence are (at least up to multiplicative constants) the **Legendre polynomials.**□

The QR Factorization

The Gram–Schmidt process can be used to factor any real or complex matrix into a product of a matrix with orthogonal columns and an upper triangular matrix. Suppose that $A = (v_1 \mid v_2 \mid \cdots \mid v_n)$ is an $m \times n$ matrix with columns v_i, where $n \leq m$. The Gram–Schmidt process applied to these columns gives orthogonal vectors $O = (u_1 \mid u_2 \mid \cdots \mid u_n)$ for which

$$\langle u_1, \dots, u_k \rangle = \langle v_1, \dots, v_k \rangle$$

for all $k \le n$. In particular,

$$v_k = u_k + \sum_{i=1}^{k-1} r_{k,i} u_i$$

where

$$r_{k,i} = \begin{cases} 0 & \text{if } u_i = 0 \\ \frac{\langle v_k, u_i \rangle}{\langle u_i, u_i \rangle} & \text{if } u_i \neq 0 \end{cases}$$

In matrix terms,

$$(v_1 \mid v_2 \mid \cdots \mid v_n) = (u_1 \mid u_2 \mid \cdots \mid u_n) \begin{bmatrix} 1 & r_{2,1} & \cdots & r_{n,1} \\ & 1 & \cdots & r_{n,2} \\ & & \ddots & \\ & & & 1 \end{bmatrix}$$

that is, $A = OB$ where O has orthogonal columns and B is upper triangular. We may normalize the nonzero columns u_i of O and move the positive constants to B. In particular, if $a_i = \|u_i\|$ for $u_i \neq 0$ and $a_i = 1$ for $u_i = 0$, then

$$(v_1 \mid v_2 \mid \cdots \mid v_n) = \left(\frac{u_1}{a_1} \middle| \frac{u_2}{a_2} \middle| \cdots \middle| \frac{u_n}{a_n} \right) \begin{bmatrix} a_1 & a_1 r_{2,1} & \cdots & a_1 r_{n,1} \\ & a_2 & \cdots & a_2 r_{n,2} \\ & & \ddots & \\ & & & a_n \end{bmatrix}$$

and so

$$A = QR$$

where the columns of Q are orthogonal and each column is either a unit vector or the zero vector and R is upper triangular with positive entries on the main diagonal. Moreover, if the vectors $v_1, \dots, v_n$ are linearly independent, then the columns of Q are nonzero. Also, if $m = n$ and A is nonsingular, then Q is unitary/orthogonal.

If the columns of A are not linearly independent, we can make one final adjustment to this matrix factorization. If a column u_i / a_i is zero, then we may replace this column by any vector as long as we replace the (i, i)th entry a_i in R by 0. Therefore, we can take nonzero columns of Q, extend to an orthonormal basis for the span of the columns of Q and replace the zero columns of Q by the additional members of this orthonormal basis. In this way, Q is replaced by a unitary/orthogonal matrix Q' and R is replaced by an upper triangular matrix R' that has nonnegative entries on the main diagonal.

Theorem 9.12 *Let $A \in \mathcal{M}_{m,n}(F)$, where $F = \mathbb{C}$ or $F = \mathbb{R}$. There exists a matrix $Q \in \mathcal{M}_{m,n}(F)$ with orthonormal columns and an upper triangular matrix $R \in \mathcal{M}_n(F)$ with nonnegative real entries on the main diagonal for which*

$$A = QR$$

Moreover, if $m = n$, then Q is unitary/orthogonal. If A is nonsingular, then R can be chosen to have positive entries on the main diagonal, in which case the factors Q and R are unique. The factorization $A = QR$ is called the ***QR* factorization** *of the matrix A. If A is real, then Q and R may be taken to be real.*

Proof. As to uniqueness, if A is nonsingular and $QR = Q_1R_1$ then

$$Q_1^{-1}Q = R_1R^{-1}$$

and the right side is upper triangular with nonzero entries on the main diagonal and the left side is unitary. But an upper triangular matrix with positive entries on the main diagonal is unitary if and only if it is the identity and so $Q_1 = Q$ and $R_1 = R$. Finally, if A is real, then all computations take place in the real field and so Q and R are real.□

The QR decomposition has important applications. For example, a system of linear equations $Ax = u$ can be written in the form

$$QRx = u$$

and since $Q^{-1} = Q^*$, we have

$$Rx = Q^*u$$

This is an upper triangular system, which is easily solved by back substitution; that is, starting from the bottom and working up.

We mention also that the QR factorization is associated with an algorithm for approximating the eigenvalues of a matrix, called the ***QR* algorithm**. Specifically, if $A = A_0$ is an $n \times n$ matrix, define a sequence of matrices as follows:

1) Let $A_0 = Q_0R_0$ be the QR factorization of A_0 and let $A_1 = R_0Q_0$.
2) Once A_k has been defined, let $A_k = Q_kR_k$ be the QR factorization of A_k and let $A_{k+1} = R_kQ_k$.

Then A_k is unitarily/orthogonally similar to A, since

$$Q_{k-1}A_kQ_{k-1}^* = Q_{k-1}(R_{k-1}Q_{k-1})Q_{k-1}^* = Q_{k-1}R_{k-1} = A_{k-1}$$

For complex matrices, it can be shown that under certain circumstances, such as when the eigenvalues of A have distinct norms, the sequence A_k converges (entrywise) to an upper triangular matrix U, which therefore has the eigenvalues of A on its main diagonal. Results can be obtained in the real case as well. For more details, we refer the reader to [48], page 115.

Hilbert and Hamel Bases

Definition *A* **maximal orthonormal set** *in an inner product space* V *is called a* **Hilbert basis** *for* V.□

Zorn's lemma can be used to show that any nontrivial inner product space has a Hilbert basis. We leave the details to the reader.

Some care must be taken not to confuse the concepts of a basis for a vector space and a Hilbert basis for an inner product space. To avoid confusion, a vector space basis, that is, a maximal linearly independent set of vectors, is referred to as a **Hamel basis**. We will refer to an orthonormal Hamel basis as an **orthonormal basis**.

To be perfectly clear, there are maximal linearly independent sets called (Hamel) bases and maximal orthonormal sets (called Hilbert bases). If a maximal linearly independent set (basis) is orthonormal, it is called an orthonormal basis.

Moreover, since every orthonormal set is linearly independent, it follows that an orthonormal basis is a Hilbert basis, since it cannot be properly contained in an orthonormal set. For *finite-dimensional* inner product spaces, the two types of bases are the same.

Theorem 9.13 *Let* V *be an inner product space. A finite subset* $\mathcal{O} = \{u_1, \ldots, u_k\}$ *of* V *is an orthonormal (Hamel) basis for* V *if and only if it is a Hilbert basis for* V.
Proof. We have seen that any orthonormal basis is a Hilbert basis. Conversely, if $\mathcal{O}$ is a finite maximal orthonormal set and $\mathcal{O} \subset \mathcal{P}$, where $\mathcal{P}$ is linearly independent, then we may apply part 1) to extend $\mathcal{O}$ to a strictly larger orthonormal set, in contradiction to the maximality of $\mathcal{O}$. Hence, $\mathcal{O}$ is maximal linearly independent.□

The following example shows that the previous theorem fails for infinite-dimensional inner product spaces.

Example 9.5 Let $V = \ell^2$ and let M be the set of all vectors of the form

$$e_i = (0, \ldots, 0, 1, 0, \ldots)$$

where e_i has a 1 in the ith coordinate and 0's elsewhere. Clearly, M is an orthonormal set. Moreover, it is maximal. For if $v = (x_n) \in \ell^2$ has the property that $v \perp M$, then

$$x_i = \langle v, e_i \rangle = 0$$

for all i and so $v = 0$. Hence, no nonzero vector $v \notin M$ is orthogonal to M. This shows that M is a Hilbert basis for the inner product space ℓ^2.

On the other hand, the vector space span of M is the subspace S of all sequences in ℓ^2 that have finite support, that is, have only a finite number of nonzero terms and since $\operatorname{span}(M) = S \neq \ell^2$, we see that M is not a Hamel basis for the vector space ℓ^2.$\square$

The Projection Theorem and Best Approximations

Orthonormal bases have a great practical advantage over arbitrary bases. From a computational point of view, if $\mathcal{B} = \{v_1, \ldots, v_n\}$ is a basis for V, then each $v \in V$ has the form

$$v = r_1 v_1 + \cdots + r_n v_n$$

In general, determining the coordinates r_i requires solving a system of linear equations of size $n \times n$.

On the other hand, if $\mathcal{O} = \{u_1, \ldots, u_n\}$ is an orthonormal basis for V and

$$v = r_1 u_1 + \cdots + r_n u_n$$

then the coefficients r_i are quite easily computed:

$$\langle v, u_i \rangle = \langle r_1 u_1 + \cdots + r_n u_n, u_i \rangle = r_i \langle u_i, u_i \rangle = r_i$$

Even if $\mathcal{O} = \{u_1, \ldots, u_n\}$ is not a basis (but just an orthonormal set), we can still consider the expansion

$$\widehat{v} = \langle v, u_1 \rangle u_1 + \cdots + \langle v, u_n \rangle u_n$$

Theorem 9.14 *Let $\mathcal{O} = \{u_1, \ldots, u_k\}$ be an orthonormal subset of an inner product space V and let $S = \langle \mathcal{O} \rangle$. The* **Fourier expansion** *with respect to $\mathcal{O}$ of a vector $v \in V$ is*

$$\widehat{v} = \langle v, u_1 \rangle u_1 + \cdots + \langle v, u_k \rangle u_k$$

Each coefficient $\langle v, u_i \rangle$ is called a **Fourier coefficient** *of v with respect to $\mathcal{O}$. The vector $\widehat{v}$ can be characterized as follows:*

1) *$\widehat{v}$ is the unique vector $s \in S$ for which $(v - s) \perp S$.*
2) *$\widehat{v}$ is the* **best approximation** *to v from within S, that is, $\widehat{v}$ is the unique vector $s \in S$ that is closest to v, in the sense that*

$$\|v - \widehat{v}\| < \|v - s\|$$

for all $s \in S \setminus \{\widehat{v}\}$.

3) **Bessel's inequality** *holds for all* $v \in V$, *that is*

$$\|\widehat{v}\| \le \|v\|$$

Proof. For part 1), since

$$\langle v - \widehat{v}, u_i \rangle = \langle v, u_i \rangle - \langle \widehat{v}, u_i \rangle = 0$$

it follows that $v - \widehat{v} \in S^\perp$. Also, if $v - s \in S^\perp$ for $s \in S$, then $s - \widehat{v} \in S$ and

$$s - \widehat{v} = (v - \widehat{v}) - (v - s) \in S^\perp$$

and so $s = \widehat{v}$. For part 2), if $s \in S$, then $v - \widehat{v} \in S^\perp$ implies that $(v - \widehat{v}) \perp (\widehat{v} - s)$ and so

$$\|v - s\|^2 = \|v - \widehat{v} + \widehat{v} - s\|^2 = \|v - \widehat{v}\|^2 + \|\widehat{v} - s\|^2$$

Hence, $\|v - s\|$ is smallest if and only if $s = \widehat{v}$ and the smallest value is $\|v - \widehat{v}\|$. We leave proof of Bessel's inequality as an exercise.□

Theorem 9.15 (The projection theorem) *If* S *is a finite-dimensional subspace of an inner product space* V, *then*

$$S = S \odot S^\perp$$

In particular, if $v \in V$, *then*

$$v = \widehat{v} + (v - \widehat{v}) \in S \odot S^\perp$$

It follows that

$$\dim(V) = \dim(S) + \dim(S^\perp)$$

Proof. We have seen that $v - \widehat{v} \in S^\perp$ and so $V = S + S^\perp$. But $S \cap S^\perp = \{0\}$ and so $V = S \odot S^\perp$.□

The following example shows that the projection theorem may fail if S is not finite-dimensional. Indeed, in the infinite-dimensional case, S must be a *complete* subspace, but we postpone a discussion of this case until Chapter 13.

Example 9.6 As in Example 9.5, let $V = \ell^2$ and let S be the subspace of all sequences with finite support, that is, S is spanned by the vectors

$$e_i = (0, \ldots, 0, 1, 0, \ldots)$$

where e_i has a 1 in the ith coordinate and 0's elsewhere. If $x = (x_n) \in S^\perp$, then $x_i = \langle x, e_i \rangle = 0$ for all i and so $x = 0$. Therefore, $S^\perp = \{0\}$. However,

$$S \odot S^\perp = S \ne \ell^2$$

□

The projection theorem has a variety of uses.

Theorem 9.16 *Let V be an inner product space and let S be a finite-dimensional subspace of V.*
1) $S^{\perp\perp} = S$
2) *If $X \subseteq V$ and* $\dim(\langle X \rangle) < \infty$*, then*

$$X^{\perp\perp} = \langle X \rangle$$

Proof. For part 1), it is clear that $S \subseteq S^{\perp\perp}$. On the other hand, if $v \in S^{\perp\perp}$, then the projection theorem implies that $v = s + s'$ where $s \in S$ and $s' \in S^{\perp}$. Then s' is orthogonal to both s and v and so s' is orthogonal to itself. Hence, $s' = 0$ and $v = s \in S$ and so $S = S^{\perp\perp}$. We leave the proof of part 2) as an exercise.□

Characterizing Orthonormal Bases

We can characterize orthonormal bases using Fourier expansions.

Theorem 9.17 *Let $\mathcal{O} = \{u_1, \ldots, u_k\}$ be an orthonormal subset of an inner product space V and let $S = \langle \mathcal{O} \rangle$. The following are equivalent:*
1) *$\mathcal{O}$ is an orthonormal basis for V.*
2) $\langle \mathcal{O} \rangle^{\perp} = \{0\}$
3) *Every vector is equal to its Fourier expansion, that is, for all $v \in V$,*

$$\widehat{v} = v$$

4) **Bessel's identity** *holds for all $v \in V$, that is,*

$$\|\widehat{v}\| = \|v\|$$

5) **Parseval's identity** *holds for all $v, w \in V$, that is,*

$$\langle v, w \rangle = [\widehat{v}]_{\mathcal{O}} \cdot [\widehat{w}]_{\mathcal{O}}$$

where

$$[\widehat{v}]_{\mathcal{O}} \cdot [\widehat{w}]_{\mathcal{O}} = \langle v, u_1 \rangle \overline{\langle w, u_1 \rangle} + \cdots + \langle v, u_k \rangle \overline{\langle w, u_k \rangle}$$

is the standard dot product in F^k.

Proof. To see that 1) implies 2), if $v \in \langle \mathcal{O} \rangle^{\perp}$ is nonzero, then $\mathcal{O} \cup \{v/\|v\|\}$ is orthonormal and so $\mathcal{O}$ is not maximal. Conversely, if $\mathcal{O}$ is not maximal, there is an orthonormal set $\mathcal{P}$ for which $\mathcal{O} \subset \mathcal{P}$. Then any nonzero $v \in \mathcal{P} \setminus \mathcal{O}$ is in $\langle \mathcal{O} \rangle^{\perp}$. Hence, 2) implies 1). We leave the rest of the proof as an exercise.□

The Riesz Representation Theorem

We have been dealing with linear maps for some time. We now have a need for conjugate linear maps.

Definition *A function $\sigma: V \to W$ on complex vector spaces is* **conjugate linear** *if it is additive,*

$$\sigma(v_1 + v_2) = \sigma v_1 + \sigma v_2$$

and

$$\sigma(rv) = \overline{r}\sigma v$$

for all $r \in \mathbb{C}$. *A* **conjugate isomorphism** *is a bijective conjugate linear map.*□

If $x \in V$, then the inner product function $\langle \cdot , x \rangle: V \to F$ defined by

$$\langle \cdot , x \rangle v = \langle v, x \rangle$$

is a linear functional on V. Thus, the linear map $\tau: V \to V^*$ defined by

$$\tau x = \langle \cdot , x \rangle$$

is conjugate linear. Moreover, since $\langle \cdot , x \rangle = \langle \cdot , y \rangle$ implies $x = y$, it follows that τ is injective and therefore a conjugate isomorphism (since V is finite-dimensional).

Theorem 9.18 (The Riesz representation theorem) *Let* V *be a finite-dimensional inner product space.*

1) *The map* $\tau: V \to V^*$ *defined by*

$$\tau x = \langle \cdot , x \rangle$$

is a conjugate isomorphism. In particular, for each $f \in V^*$, *there exists a unique vector* $x \in V$ *for which* $f = \langle \cdot , x \rangle$, *that is,*

$$fv = \langle v, x \rangle$$

for all $v \in V$. *We call* x *the* **Riesz vector** *for* f *and denote it by* R_f.

2) *The map* $R: V^* \to V$ *defined by*

$$Rf = R_f$$

is also a conjugate isomorphism, being the inverse of τ. *We will call this map the* **Riesz map**.

Proof. Here is the usual proof that τ is surjective. If $f = 0$, then $R_f = 0$, so let us assume that $f \neq 0$. Then $K = \ker(f)$ has codimension 1 and so

$$V = \langle w \rangle \odot K$$

for $w \in K^\perp$. Letting $x = \alpha w$ for $\alpha \in F$, we require that

$$f(v) = \langle v, \alpha w \rangle$$

and since this clearly holds for any $v \in K$, it is sufficient to show that it holds for $v = w$, that is,

$$f(w) = \langle w, \alpha w \rangle = \overline{\alpha}\langle w, w \rangle$$

Thus, $\alpha = \overline{f(w)}/\|w\|^2$ and

$$R_f = \frac{\overline{f(w)}}{\|w\|^2} w$$

For part 2), we have

$$\begin{aligned}\langle v, R_{rf+sg}\rangle &= (rf + sg)(v) \\ &= rf(v) + sg(v) \\ &= \langle v, \overline{r}R_f\rangle + \langle v, \overline{s}R_g\rangle \\ &= \langle v, \overline{r}R_f + \overline{s}R_g\rangle\end{aligned}$$

for all $v \in V$ and so

$$R_{rf+sg} = \overline{r}R_f + \overline{s}R_g$$

□

Note that if $V = \mathbb{R}^n$, then $R_f = (f(e_1), \ldots, f(e_n))$, where $(e_1, \ldots, e_n)$ is the standard basis for $\mathbb{R}^n$.

Exercises

1. Prove that if a matrix M is unitary, upper triangular and has positive entries on the main diagonal, must be the identity matrix.
2. Use the QR factorization to show that any triangularizable matrix is unitarily (orthogonally) triangularizable.
3. Verify the statement concerning equality in the triangle inequality.
4. Prove the parallelogram law.
5. Prove the **Apollonius identity**

$$\|w - u\|^2 + \|w - v\|^2 = \frac{1}{2}\|u - v\|^2 + 2\left\|w - \frac{1}{2}(u + v)\right\|^2$$

6. Let V be an inner product space with basis $\mathcal{B}$. Show that the inner product is uniquely defined by the values $\langle u, v\rangle$, for all $u, v \in \mathcal{B}$.
7. Prove that two vectors u and v in a real inner product space V are orthogonal if and only if

$$\|u + v\|^2 = \|u\|^2 + \|v\|^2$$

8. Show that an isometry is injective.
9. Use Zorn's lemma to show that any nontrivial inner product space has a Hilbert basis.
10. Prove Bessel's inequality.
11. Prove that an orthonormal set $\mathcal{O}$ is a Hilbert basis for a finite-dimensional vector space V if and only if $\widehat{v} = v$, for all $v \in V$.
12. Prove that an orthonormal set $\mathcal{O}$ is a Hilbert basis for a finite-dimensional vector space V if and only if Bessel's identity holds for all $v \in V$, that is, if and only if

$$\|\widehat{v}\| = \|v\|$$

for all $v \in V$.

13. Prove that an orthonormal set $\mathcal{O}$ is a Hilbert basis for a finite-dimensional vector space V if and only if Parseval's identity holds for all $v, w \in V$, that is, if and only if

$$\langle v, w \rangle = [\widehat{v}]_{\mathcal{O}} \cdot [\widehat{w}]_{\mathcal{O}}$$

for all $v, w \in V$.

14. Let $u = (r_1, \ldots, r_n)$ and $v = (s_1, \ldots, s_n)$ be in $\mathbb{R}^n$. The Cauchy–Schwarz inequality states that

$$|r_1 s_1 + \cdots + r_n s_n|^2 \le (r_1^2 + \cdots + r_n^2)(s_1^2 + \cdots + s_n^2)$$

Prove that we can do better:

$$(|r_1 s_1| + \cdots + |r_n s_n|)^2 \le (r_1^2 + \cdots + r_n^2)(s_1^2 + \cdots + s_n^2)$$

15. Let V be a finite-dimensional inner product space. Prove that for any subset X of V, we have $X^{\perp\perp} = \operatorname{span}(X)$.
16. Let $\mathcal{P}_3$ be the inner product space of all polynomials of degree at most 3, under the inner product

$$\langle p(x), q(x) \rangle = \int_{-\infty}^{\infty} p(x)q(x)e^{-x^2}\,dx$$

Apply the Gram–Schmidt process to the basis $\{1, x, x^2, x^3\}$, thereby computing the first four **Hermite polynomials** (at least up to a multiplicative constant).

17. Verify uniqueness in the Riesz representation theorem.
18. Let V be a complex inner product space and let S be a subspace of V. Suppose that $v \in V$ is a vector for which $\langle v, s \rangle + \langle s, v \rangle \le \langle s, s \rangle$ for all $s \in S$. Prove that $v \in S^{\perp}$.
19. If V and W are inner product spaces, consider the function on $V \boxplus W$ defined by

$$\langle (v_1, w_1), (v_2, w_2) \rangle = \langle v_1, v_2 \rangle + \langle w_1, w_2 \rangle$$

Is this an inner product on $V \boxplus W$?

20. A **normed vector space** over $\mathbb{R}$ or $\mathbb{C}$ is a vector space (over $\mathbb{R}$ or $\mathbb{C}$) together with a function $\|\,\|\colon V \to \mathbb{R}$ for which for all $u, v \in V$ and scalars r we have
    a) $\|rv\| = |r|\|v\|$
    b) $\|u + v\| \le \|u\| + \|v\|$
    c) $\|v\| = 0$ if and only if $v = 0$

    If V is a real normed space (over $\mathbb{R}$) and if the norm satisfies the parallelogram law

$$\|u+v\|^2 + \|u-v\|^2 = 2\|u\|^2 + 2\|v\|^2$$

prove that the polarization identity

$$\langle u, v \rangle = \frac{1}{4}(\|u+v\|^2 - \|u-v\|^2)$$

defines an inner product on V. *Hint*: Evaluate $8\langle u, x\rangle + 8\langle v, x\rangle$ to show that $\langle u, 2x\rangle = 2\langle u, x\rangle$ and $\langle u, x\rangle + \langle v, x\rangle = \langle u+v, x\rangle$. Then complete the proof that $\langle u, rx\rangle = r\langle u, x\rangle$.

21. Let S be a subspace of a finite-dimensional inner product space V. Prove that each coset in V/S contains *exactly one* vector that is orthogonal to S.

Extensions of Linear Functionals

22. Let f be a linear functional on a subspace S of a finite-dimensional inner product space V. Let $f(v) = \langle v, R_f\rangle$. Suppose that $g \in V^*$ is an extension of f, that is, $g|_S = f$. What is the relationship between the Riesz vectors R_f and R_g?
23. Let f be a nonzero linear functional on a subspace S of a finite-dimensional inner product space V and let $K = \ker(f)$. Show that if $g \in V^*$ is an extension of f, then $R_g \in K^\perp \setminus S^\perp$. Moreover, for each vector $u \in K^\perp \setminus S^\perp$ there is exactly one scalar λ for which the linear functional $g(X) = \langle X, \lambda u\rangle$ is an extension of f.

Positive Linear Functionals on $\mathbb{R}^n$

A vector $v = (a_1, \ldots, a_n)$ in $\mathbb{R}^n$ is **nonnegative** (also called **positive**), written $v \geq 0$, if $a_i \geq 0$ for all i. The vector v is **strictly positive**, written $v > 0$, if v is nonnegative but not 0. The set $\mathbb{R}^n_+$ of all strictly positive vectors in $\mathbb{R}^n$ is called the **nonnegative orthant** in $\mathbb{R}^n$. The vector v is **strongly positive**, written $v \gg 0$, if $a_i > 0$ for all i. The set $\mathbb{R}^n_{++}$, of all strongly positive vectors in $\mathbb{R}^n$ is the **strongly positive orthant** in $\mathbb{R}^n$.

Let $f: S \to \mathbb{R}$ be a linear functional on a subspace S of $\mathbb{R}^n$. Then f is **nonnegative** (also called **positive**), written $f \geq 0$, if

$$v > 0 \Rightarrow f(v) \geq 0$$

for all $v \in S$ and f is **strictly positive**, written $f > 0$, if

$$v > 0 \Rightarrow f(v) > 0$$

for all $v \in S$.

24. Prove that a linear functional f on $\mathbb{R}^n$ is positive if and only if $R_f > 0$ and strictly positive if and only if $R_f \gg 0$. If S is a subspace of $\mathbb{R}^n$ is it true that a linear functional f on S is nonnegative if and only if $R_f > 0$?

25. Let $f: S \to \mathbb{R}$ be a strictly positive linear functional on a subspace S of $\mathbb{R}^n$. Prove that f has a strictly positive extension to $\mathbb{R}^n$. Use the fact that if $U \cap \mathbb{R}^m_+ = \{0\}$, where

$$\mathbb{R}^n_+ = \{(a_1, \ldots, a_n) \mid a_i \geq 0 \text{ all } i\}$$

and U is a subspace of $\mathbb{R}^n$, then $U^\perp$ contains a strongly positive vector.

26. If V is a real inner product space, then we can define an inner product on its complexification $V^\mathbb{C}$ as follows (this is the same formula as for the ordinary inner product on a complex vector space):

$$\langle u + vi, x + yi \rangle = \langle u, x \rangle + \langle v, y \rangle + (\langle v, x \rangle - \langle u, y \rangle)i$$

Show that

$$\|(u + vi)\|^2 = \|u\|^2 + \|v\|^2$$

where the norm on the left is induced by the inner product on $V^\mathbb{C}$ and the norm on the right is induced by the inner product on V.

Chapter 10
Structure Theory for Normal Operators

Throughout this chapter, all vector spaces are assumed to be finite-dimensional unless otherwise noted. Also, the field F is either $\mathbb{R}$ or $\mathbb{C}$.

The Adjoint of a Linear Operator

The purpose of this chapter is to study the structure of certain special types of linear operators on finite-dimensional real and complex inner product spaces. In order to define these operators, we introduce another type of adjoint (different from the operator adjoint of Chapter 3).

Theorem 10.1 *Let V and W be finite-dimensional inner product spaces over F and let $\tau \in \mathcal{L}(V,W)$. Then there is a unique function $\tau^*: W \to V$, defined by the condition*

$$\langle \tau v, w \rangle = \langle v, \tau^* w \rangle$$

for all $v \in V$ and $w \in W$. This function is in $\mathcal{L}(W,V)$ and is called the **adjoint** *of τ.*
Proof. If τ^* exists, then it is unique, for if

$$\langle \tau v, w \rangle = \langle v, \sigma w \rangle$$

then $\langle v, \sigma w \rangle = \langle v, \tau^* w \rangle$ for all v and w and so $\sigma = \tau^*$.

We seek a linear map $\tau^*: W \to V$ for which

$$\langle v, \tau^* w \rangle = \langle \tau v, w \rangle$$

By way of motivation, the vector $\tau^* w$, if it exists, looks very much like a linear map sending v to $\langle \tau v, w \rangle$. The only problem is that $\tau^* v$ is supposed to be a vector, not a linear map. But the Riesz representation theorem tells us that linear maps can be represented by vectors.

Specifically, for each $w \in W$, the linear functional $f_w \in V^*$ defined by

$$f_w v = \langle \tau v, w \rangle$$

has the form

$$f_w v = \langle v, R_{f_w} \rangle$$

where $R_{f_w} \in V$ is the Riesz vector for f_w. If $\tau^*: W \to V$ is defined by

$$\tau^* w = R_{f_w} = R(f_w)$$

where R is the Riesz map, then

$$\langle v, \tau^* w \rangle = \langle v, R_{f_w} \rangle = f_w v = \langle \tau v, w \rangle$$

Finally, since $\tau^* = R \circ f$ is the composition of the Riesz map R and the map $f: w \mapsto f_w$ and since both of these maps are conjugate linear, their composition is linear.□

Here are some of the basic properties of the adjoint.

Theorem 10.2 *Let V and W be finite-dimensional inner product spaces. For every $\sigma, \tau \in \mathcal{L}(V, W)$ and $r \in F$,*
1) $(\sigma + \tau)^* = \sigma^* + \tau^*$
2) $(r\tau)^* = \overline{r}\tau^*$
3) $\tau^{**} = \tau$ *and so*

$$\langle \tau^* v, w \rangle = \langle v, \tau w \rangle$$

4) *If $V = W$, then $(\sigma\tau)^* = \tau^*\sigma^*$*
5) *If τ is invertible, then $(\tau^{-1})^* = (\tau^*)^{-1}$*
6) *If $V = W$ and $p(x) \in \mathbb{R}[x]$, then $p(\tau)^* = p(\tau^*)$.*

Moreover, if $\tau \in \mathcal{L}(V)$ and S is a subspace of V, then
7) *S is τ-invariant if and only if $S^\perp$ is τ^*-invariant.*
8) *$(S, S^\perp)$ reduces τ if and only if S is both τ-invariant and τ^*-invariant, in which case*

$$(\tau|_S)^* = (\tau^*)|_S$$

Proof. For part 7), let $s \in S$ and $z \in S^\perp$ and write

$$\langle \tau^* z, s \rangle = \langle z, \tau s \rangle$$

Now, if S is τ-invariant, then $\langle \tau^* z, s \rangle = 0$ for all $s \in S$ and so $\tau^* z \in S^\perp$ and $S^\perp$ is τ^*-invariant. Conversely, if $S^\perp$ is τ^*-invariant, then $\langle z, \tau s \rangle = 0$ for all $z \in S^\perp$ and so $\tau s \in S^{\perp\perp} = S$, whence S is τ-invariant.

The first statement in part 8) follows from part 7) applied to both S and $S^\perp$. For the second statement, since S is both τ-invariant and τ^*-invariant, if $s, t \in S$,

then

$$\langle s, (\tau^*)|_S(t)\rangle = \langle s, \tau^* t\rangle = \langle \tau s, t\rangle = \langle \tau|_S(s), t\rangle$$

Hence, by definition of adjoint, $(\tau^*)|_S = (\tau|_S)^*$.□

Now let us relate the kernel and image of a linear transformation to those of its adjoint.

Theorem 10.3 *Let $\tau \in \mathcal{L}(V, W)$, where V and W are finite-dimensional inner product spaces.*
1)

$$\ker(\tau^*) = \operatorname{im}(\tau)^\perp \quad \textit{and} \quad \operatorname{im}(\tau^*) = \ker(\tau)^\perp$$

and so

$$\begin{aligned} \tau \textit{ surjective} \quad &\Leftrightarrow \quad \tau^* \textit{ injective} \\ \tau \textit{ injective} \quad &\Leftrightarrow \quad \tau^* \textit{ surjective} \end{aligned}$$

2)

$$\ker(\tau^*\tau) = \ker(\tau) \quad \textit{and} \quad \ker(\tau\tau^*) = \ker(\tau^*)$$

3)

$$\operatorname{im}(\tau^*\tau) = \operatorname{im}(\tau^*) \quad \textit{and} \quad \operatorname{im}(\tau\tau^*) = \operatorname{im}(\tau)$$

4)

$$(\rho_{S,T})^* = \rho_{T^\perp, S^\perp}$$

Proof. For part 1),

$$\begin{aligned} u \in \ker(\tau^*) &\Leftrightarrow \tau^* u = 0 \\ &\Leftrightarrow \langle \tau^* u, V\rangle = \{0\} \\ &\Leftrightarrow \langle u, \tau V\rangle = \{0\} \\ &\Leftrightarrow u \in \operatorname{im}(\tau)^\perp \end{aligned}$$

and so $\ker(\tau^*) = \operatorname{im}(\tau)^\perp$. The second equation in part 1) follows by replacing τ by τ^* and taking complements.

For part 2), it is clear that $\ker(\tau) \subseteq \ker(\tau^*\tau)$. For the reverse inclusion, we have

$$\tau^*\tau u = 0 \quad \Rightarrow \quad \langle \tau^*\tau u, u\rangle = 0 \quad \Rightarrow \quad \langle \tau u, \tau u\rangle = 0 \quad \Rightarrow \quad \tau u = 0$$

and so $\ker(\tau^*\tau) \subseteq \ker(\tau)$. The second equation follows from the first by replacing τ with τ^*. We leave the rest of the proof for the reader.□

The Operator Adjoint and the Hilbert Space Adjoint

We should make some remarks about the relationship between the operator adjoint $\tau^\times$ of τ, as defined in Chapter 3 and the adjoint τ^* that we have just defined, which is sometimes called the **Hilbert space adjoint**. In the first place, if $\tau: V \to W$, then $\tau^\times$ and τ^* have different domains and ranges:

$$\tau^\times: W^* \to V^* \quad \text{and} \quad \tau^*: W \to V$$

The two maps are shown in Figure 10.1, along with the conjugate Riesz isomorphisms $R^V: V^* \to V$ and $R^W: W^* \to W$.

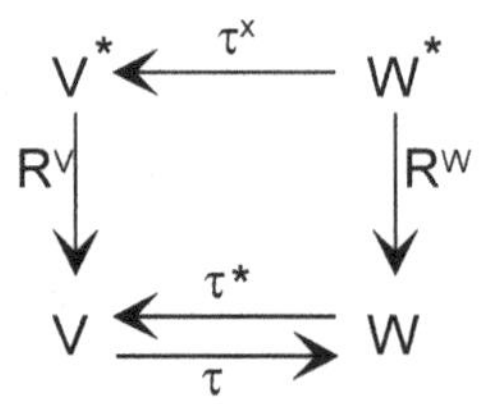

Figure 10.1

The composite map $\sigma: W^* \to V^*$ defined by

$$\sigma = (R^V)^{-1} \circ \tau^* \circ R^W$$

is linear. Moreover, for all $f \in W^*$ and $v \in V$,

$$\begin{aligned}(\tau^\times(f))v &= f(\tau v)\\ &= \langle \tau v, R^W(f)\rangle\\ &= \langle v, \tau^* R^W(f)\rangle\\ &= [(R^V)^{-1}(\tau^* R^W(f))](v)\\ &= (\sigma f)v\end{aligned}$$

and so $\sigma = \tau^\times$. Hence, the relationship between $\tau^\times$ and τ^* is

$$\tau^\times = (R^V)^{-1} \circ \tau^* \circ R^W$$

Loosely speaking, the Riesz functions are like "change of variables" functions from linear functionals to vectors, and we can say that τ^* does to Riesz vectors what $\tau^\times$ does to the corresponding linear functionals. Put another way (and just as loosely), τ and τ^* are the same, up to conjugate Riesz isomorphism.

In Chapter 3, we showed that the matrix of the operator adjoint $\tau^\times$ is the transpose of the matrix of the map τ. For Hilbert space adjoints, the situation is slightly different (due to the conjugate linearity of the inner product). Suppose that $\mathcal{B} = (b_1, \ldots, b_n)$ and $\mathcal{C} = (c_1, \ldots, c_m)$ are ordered orthonormal bases for V and W, respectively. Then

$$([\tau^*]_{\mathcal{C},\mathcal{B}})_{i,j} = \langle \tau^* c_j, b_i \rangle = \langle c_j, \tau b_i \rangle = \overline{\langle \tau b_i, c_j \rangle} = \overline{([\tau]_{\mathcal{B},\mathcal{C}})_{j,i}}$$

and so $[\tau^*]_{\mathcal{C},\mathcal{B}}$ and $[\tau]_{\mathcal{B},\mathcal{C}}$ are conjugate transposes. The conjugate transpose of a matrix $A = (a_{i,j})$ is

$$A^* = (\overline{a}_{i,j})^t$$

and is called the **adjoint** of A.

Theorem 10.4 *Let $\tau \in \mathcal{L}(V, W)$, where V and W are finite-dimensional inner product spaces.*

1) *The operator adjoint $\tau^\times$ and the Hilbert space adjoint τ^* are related by*

$$\tau^\times = (R^V)^{-1} \circ \tau^* \circ R^W$$

where R^V and R^W are the conjugate Riesz isomorphisms on V and W, respectively.

2) *If $\mathcal{B}$ and $\mathcal{C}$ are ordered* orthonormal bases *for V and W, respectively, then*

$$[\tau^*]_{\mathcal{C},\mathcal{B}} = ([\tau]_{\mathcal{B},\mathcal{C}})^*$$

In words, the matrix of the adjoint τ^* is the adjoint (conjugate transpose) of the matrix of τ.□

Orthogonal Projections

In an inner product space, we can single out some special projection operators.

Definition *A projection of the form $\rho_{S,S^\perp}$ is said to be* **orthogonal**. *Equivalently, a projection ρ is orthogonal if* $\ker(\rho) \perp \operatorname{im}(\rho)$.□

Some care must be taken to avoid confusion between orthogonal projections and two projections that are orthogonal to each other, that is, for which $\rho\sigma = \sigma\rho = 0$.

We have seen that an operator ρ is a projection operator if and only if it is idempotent. Here is the analogous characterization of orthogonal projections.

Theorem 10.5 *Let V be a finite-dimensional inner product space. The following are equivalent for an operator ρ on V:*

1) *ρ is an orthogonal projection*
2) *ρ is idempotent and self-adjoint*
3) *ρ is idempotent and does not expand lengths, that is*

$$\|\rho v\| \le \|v\|$$

for all $v \in V$.

Proof. Since

$$(\rho_{S,T})^* = \rho_{T^\perp,S^\perp}$$

it follows that $\rho = \rho^*$ if and only if $S^\perp = T$, that is, if and only if ρ is orthogonal. Hence, 1) and 2) are equivalent.

To prove that 1) implies 3), let $\rho = \rho_{S,S^\perp}$. Then if $v = s + t$ for $s \in S$ and $t \in S^\perp$, it follows that

$$\|v\|^2 = \|s\|^2 + \|t\|^2 \geq \|s\|^2 = \|\rho v\|^2$$

Now suppose that 3) holds. Then

$$\operatorname{im}(\rho) \oplus \ker(\rho) = V = \ker(\rho)^\perp \odot \ker(\rho)$$

and we wish to show that the first sum is orthogonal. If $w \in \operatorname{im}(\rho)$, then $w = x + y$, where $x \in \ker(\rho)$ and $y \in \ker(\rho)^\perp$. Hence,

$$w = \rho w = \rho x + \rho y = \rho y$$

and so the orthogonality of x and y implies that

$$\|x\|^2 + \|y\|^2 = \|w\|^2 = \|\rho y\|^2 \leq \|y\|^2$$

Hence, $x = 0$ and so $\operatorname{im}(\rho) \subseteq \ker(\rho)^\perp$, which implies that $\operatorname{im}(\rho) = \ker(\rho)^\perp$.□

Orthogonal Resolutions of the Identity

We have seen (Theorem 2.25) that resolutions of the identity

$$\rho_1 + \cdots + \rho_k = \iota$$

on V correspond to direct sum decompositions of V. If, in addition, the projections are orthogonal, then the direct sum is an orthogonal sum.

Definition *An* **orthogonal resolution of the identity** *is a resolution of the identity $\rho_1 + \cdots + \rho_k = \iota$ in which each projection ρ_i is orthogonal.*□

The following theorem displays a correspondence between orthogonal direct sum decompositions of V and orthogonal resolutions of the identity.

Theorem 10.6 *Let V be an inner product space. Orthogonal resolutions of the identity on V correspond to orthogonal direct sum decompositions of V as follows:*

1) If $\rho_1 + \cdots + \rho_k = \iota$ is an orthogonal resolution of the identity, then

$$V = \operatorname{im}(\rho_1) \odot \cdots \odot \operatorname{im}(\rho_k)$$

and ρ_i is orthogonal projection onto $\operatorname{im}(\rho_i)$.

2) *Conversely, if*

$$V = S_1 \odot \cdots \odot S_k$$

and if ρ_i *is orthogonal projection onto* S_i*, then* $\rho_1 + \cdots + \rho_k = \iota$ *is an orthogonal resolution of the identity.*

Proof. To prove 1), if $\rho_1 + \cdots + \rho_k = \iota$ is an orthogonal resolution of the identity, Theorem 2.25 implies that

$$V = \operatorname{im}(\rho_1) \oplus \cdots \oplus \operatorname{im}(\rho_k)$$

However, since the ρ_i's are pairwise orthogonal and self-adjoint, it follows that

$$\langle \rho_i v, \rho_j w \rangle = \langle v, \rho_i \rho_j w \rangle = \langle v, 0 \rangle = 0$$

and so

$$V = \operatorname{im}(\rho_1) \odot \cdots \odot \operatorname{im}(\rho_k)$$

For the converse, Theorem 2.25 implies that $\rho_1 + \cdots + \rho_k = \iota$ is a resolution of the identity where ρ_i is projection onto $\operatorname{im}(\rho_i)$ along

$$\ker(\rho_i) = \bigodot_{j \neq i} \operatorname{im}(\rho_j) = \operatorname{im}(\rho_i)^{\perp}$$

Hence, ρ_i is orthogonal.□

Unitary Diagonalizability

We have seen (Theorem 8.10) that a linear operator $\tau \in \mathcal{L}(V)$ on a finite-dimensional vector space V is diagonalizable if and only if

$$V = \mathcal{E}_{\lambda_1} \oplus \cdots \oplus \mathcal{E}_{\lambda_k}$$

Of course, each eigenspace $\mathcal{E}_{\lambda_i}$ has an orthonormal basis $\mathcal{O}_i$, but the union of these bases need not be an *orthonormal* basis for V.

Definition *A linear operator* $\tau \in \mathcal{L}(V)$ *is* **unitarily diagonalizable** (*when* V *is complex*) *and* **orthogonally diagonalizable** (*when* V *is real*) *if there is an ordered orthonormal basis* $\mathcal{O} = (u_1, \ldots, u_n)$ *of* V *for which the matrix* $[\tau]_{\mathcal{O}}$ *is diagonal, or equivalently, if*

$$\tau u_i = \lambda_i u_i$$

for all $i = 1, \ldots, n$.□

Here is the counterpart of Theorem 8.10 for inner product spaces.

Theorem 10.7 *Let* V *be a finite-dimensional inner product space and let* $\tau \in \mathcal{L}(V)$*. The following are equivalent:*

1) τ *is unitarily (orthogonally) diagonalizable.*
2) V *has an orthonormal basis that consists entirely of eigenvectors of* τ.

3) *V has the form*

$$V = \mathcal{E}_{\lambda_1} \odot \cdots \odot \mathcal{E}_{\lambda_k}$$

where $\lambda_1, \ldots, \lambda_k$ *are the distinct eigenvalues of* τ.□

For simplicity in exposition, we will tend to use the term unitarily diagonalizable for both cases. Since unitarily diagonalizable operators are so well behaved, it is natural to seek a characterization of such operators. Remarkably, there is a simple one, as we will see next.

Normal Operators

Operators that commute with their own adjonts are very special.

Definition

1) *A linear operator* τ *on an inner product space* V *is* **normal** *if it commutes with its adjoint:*

$$\tau\tau^* = \tau^*\tau$$

2) *A matrix* $A \in \mathcal{M}_n(F)$ *is* **normal** *if* A *commutes with its adjoint* A^*.□

If τ is normal and $\mathcal{O}$ is an ordered orthonormal basis of V, then

$$[\tau]_{\mathcal{O}}[\tau]_{\mathcal{O}}^* = [\tau]_{\mathcal{O}}[\tau^*]_{\mathcal{O}} = [\tau\tau^*]_{\mathcal{O}}$$

and

$$[\tau]_{\mathcal{O}}^*[\tau]_{\mathcal{O}} = [\tau^*]_{\mathcal{O}}[\tau]_{\mathcal{O}} = [\tau^*\tau]_{\mathcal{O}}$$

and so τ is normal if and only if $[\tau]_{\mathcal{O}}$ is normal for some, and hence all, *orthonormal* bases for V. Note that this does not hold for bases that are not orthonormal.

Normal operators have some very special properties.

Theorem 10.8 *Let* $\tau \in \mathcal{L}(V)$ *be normal.*

1) *The following are also normal:*
   a) $\tau|_S$, *if* τ *reduces* $(S, S^\perp)$
   b) τ^*
   c) τ^{-1}, *if* τ *is invertible*
   d) $p(\tau)$, *for any polynomial* $p(x) \in F[x]$
2) *For any* $v, w \in V$,

$$\langle \tau v, \tau w \rangle = \langle \tau^* v, \tau^* w \rangle$$

and, in particular,

$$\|\tau v\| = \|\tau^* v\|$$

and so

$$\ker(\tau^*) = \ker(\tau)$$

3) *For any integer* $k \geq 1$,

$$\ker(\tau^k) = \ker(\tau)$$

4) *The minimal polynomial* $m_\tau(x)$ *is a product of distinct prime monic polynomials.*
5)

$$\tau v = \lambda v \quad \Leftrightarrow \quad \tau^* v = \overline{\lambda} v$$

6) *If* S *and* T *are submodules of* V_τ *with relatively prime orders, then* $S \perp T$.
7) *If* λ *and* μ *are distinct eigenvalues of* τ, *then* $\mathcal{E}_\lambda \perp \mathcal{E}_\mu$.

Proof. We leave part 1) for the reader. For part 2), normality implies that

$$\langle \tau v, \tau w \rangle = \langle \tau^* \tau v, v \rangle = \langle \tau \tau^* v, v \rangle = \langle \tau^* v, \tau^* v \rangle$$

We prove part 3) first for the operator $\sigma = \tau^* \tau$, which is *self-adjoint*, that is,

$$\sigma^* = (\tau^* \tau)^* = \tau^* \tau = \sigma$$

If $\sigma^k v = 0$ for $k > 1$, then

$$0 = \langle \sigma^k v, \sigma^{k-2} v \rangle = \langle \sigma^{k-1} v, \sigma^{k-1} v \rangle$$

and so $\sigma^{k-1} v = 0$. Continuing in this way gives $\sigma v = 0$. Now, if $\tau^k v = 0$ for $k > 1$, then

$$\sigma^k v = (\tau^* \tau)^k v = (\tau^*)^k \tau^k v = 0$$

and so $\sigma v = 0$. Hence,

$$0 = \langle \sigma v, v \rangle = \langle \tau^* \tau v, v \rangle = \langle \tau v, \tau v \rangle$$

and so $\tau v = 0$.

For part 4), suppose that

$$m_\tau(x) = p^e(x) q(x)$$

where $p(x)$ is monic and prime. Then for any $v \in V$,

$$p^e(\tau)[q(\tau) v] = 0$$

and since $p(\tau)$ is also normal, part 3) implies that

$$p(\tau)[q(\tau) v] = 0$$

for all $v \in V$. Hence, $p(\tau) q(\tau) = 0$, which implies that $e = 1$. Thus, the prime factors of $m_\tau(x)$ appear only to the first power.

Part 5) follows from part 2):

$$\ker(\tau - \lambda) = \ker[(\tau - \lambda)^*] = \ker(\tau^* - \overline{\lambda})$$

For part 6), if $o(S) = p(x)$ and $o(T) = q(x)$, then there are polynomials $a(x)$ and $b(x)$ for which $a(x)p(x) + b(x)q(x) = 1$ and so

$$a(\tau)p(\tau) + b(\tau)q(\tau) = \iota$$

Now, $\alpha = a(\tau)p(\tau)$ annihilates S and $\beta = b(\tau)q(\tau)$ annihilates T. Therefore β^* also annihilates T and so

$$\langle S, T\rangle = \langle(\alpha + \beta)S, T\rangle = \langle \beta S, T\rangle = \langle S, \beta^* T\rangle = \{0\}$$

Part 7) follows from part 6), since $o(\mathcal{E}_\lambda) = x - \lambda$ and $o(\mathcal{E}_\mu) = x - \mu$ are relatively prime when $\lambda \neq \mu$. Alternatively, for $v \in \mathcal{E}_\lambda$ and $w \in \mathcal{E}_\mu$, we have

$$\lambda\langle v, w\rangle = \langle \tau v, w\rangle = \langle v, \tau^* w\rangle = \langle v, \overline{\mu} w\rangle = \mu\langle v, w\rangle$$

and so $\lambda \neq \mu$ implies that $\langle v, w\rangle = 0$.□

The Spectral Theorem for Normal Operators

Theorem 10.8 implies that when $F = \mathbb{C}$, the minimal polynomial $m_\tau(x)$ splits into distinct linear factors and so Theorem 8.11 implies that τ is diagonalizable, that is,

$$V_\tau = \mathcal{E}_{\lambda_1} \oplus \cdots \oplus \mathcal{E}_{\lambda_k}$$

Moreover, since distinct eigenspaces of a normal operator are orthogonal, we have

$$V_\tau = \mathcal{E}_{\lambda_1} \odot \cdots \odot \mathcal{E}_{\lambda_k}$$

and so τ is unitarily diagonalizable.

The converse of this is also true. If V has an orthonormal basis $\mathcal{O} = \{v_1, \ldots, v_n\}$ of eigenvectors for τ, then since $[\tau]_\mathcal{O}$ and $[\tau^*]_\mathcal{O} = [\tau]_\mathcal{O}^*$ are diagonal, these matrices commute and therefore so do τ^* and τ.

Theorem 10.9 (The spectral theorem for normal operators: complex case) *Let V be a finite-dimensional complex inner product space and let $\tau \in \mathcal{L}(V)$. The following are equivalent:*

1) *τ is normal.*
2) *τ is unitarily diagonalizable, that is,*

$$V_\tau = \mathcal{E}_{\lambda_1} \odot \cdots \odot \mathcal{E}_{\lambda_k}$$

3) τ *has an* **orthogonal spectral resolution**

$$\tau = \lambda_1\rho_1 + \cdots + \lambda_k\rho_k \tag{10.1}$$

where $\rho_1 + \cdots + \rho_n = \iota$ *and* ρ_i *is orthogonal for all* i*, in which case,* $\{\lambda_1, \ldots, \lambda_k\}$ *is the spectrum of* τ *and*

$$\operatorname{im}(\rho_i) = \mathcal{E}_{\lambda_i} \quad \textit{and} \quad \ker(\rho_i) = \bigodot_{j \neq i} \mathcal{E}_{\lambda_j}$$

Proof. We have seen that 1) and 2) are equivalent. To see that 2) and 3) are equivalent, Theorem 8.12 says that

$$V_\tau = \mathcal{E}_{\lambda_1} \oplus \cdots \oplus \mathcal{E}_{\lambda_k}$$

if and only if

$$\tau = \lambda_1\rho_1 + \cdots + \lambda_k\rho_k$$

and in this case,

$$\operatorname{im}(\rho_i) = \mathcal{E}_{\lambda_i} \quad \text{and} \quad \ker(\rho_i) = \bigoplus_{j \neq i} \mathcal{E}_{\lambda_j}$$

But $\mathcal{E}_{\lambda_i} \perp \mathcal{E}_{\lambda_j}$ for $i \neq j$ if and only if

$$\operatorname{im}(\rho_i) \perp \ker(\rho_i)$$

that is, if and only if each ρ_i is orthogonal. Hence, the direct sum $V_\tau = \mathcal{E}_{\lambda_1} \oplus \cdots \oplus \mathcal{E}_{\lambda_k}$ is an orthogonal sum if and only if each projection is orthogonal.□

The Real Case

If $F = \mathbb{R}$, then $m_\tau(x)$ has the form

$$m_\tau(x) = (x - \lambda_1)\cdots(x - \lambda_k)p_1(x)\cdots p_m(x)$$

where each $p_i(x)$ is an irreducible monic quadratic. Hence, the primary cyclic decomposition of V_τ gives

$$V_\tau = \mathcal{E}_{\lambda_1} \odot \cdots \odot \mathcal{E}_{\lambda_k} \odot W_1 \odot \cdots \odot W_m$$

where W_i is cyclic with prime quadratic order $p_i(x)$. Therefore, Theorem 8.8 implies that there is an ordered basis $\mathcal{B}_i$ for which

$$[\tau|_{W_i}]_{\mathcal{B}_i} = \begin{bmatrix} a_i & -b_i \\ b_i & a_i \end{bmatrix}$$

Theorem 10.10 (The spectral theorem for normal operators: real case) *A linear operator* τ *on a finite-dimensional real inner product space is normal if and only if*

$$V = \mathcal{E}_{\lambda_1} \odot \cdots \odot \mathcal{E}_{\lambda_k} \odot W_1 \odot \cdots \odot W_m$$

where $\{\lambda_1, \ldots, \lambda_k\}$ *is the spectrum of* τ *and each* W_i *is an indecomposable two-dimensional* τ*-invariant subspace with an ordered basis* $\mathcal{B}_i$ *for which*

$$[\tau]_{\mathcal{B}_i} = \begin{bmatrix} a_i & -b_i \\ b_i & a_i \end{bmatrix}$$

Proof. We need only show that if V has such a decomposition, then τ is normal. But

$$[\tau]_{\mathcal{B}_i}[\tau]^t_{\mathcal{B}_i} = (a_i^2 + b_i^2)I_2 = [\tau]^t_{\mathcal{B}_i}[\tau]_{\mathcal{B}_i}$$

and so $[\tau]_{\mathcal{B}_i}$ is normal. It follows easily that τ is normal.□

Special Types of Normal Operators

We now want to introduce some special types of normal operators.

Definition *Let* V *be an inner product space.*

1) $\tau \in \mathcal{L}(V)$ *is* **self-adjoint** (*also called* **Hermitian** *in the complex case and* **symmetric** *in the real case*) *if*

$$\tau^* = \tau$$

2) $\tau \in \mathcal{L}(V)$ *is* **skew self-adjoint** (*also called* **skew-Hermitian** *in the complex case and* **skew-symmetric** *in the real case*) *if*

$$\tau^* = -\tau$$

3) $\tau \in \mathcal{L}(V)$ *is* **unitary** *in the complex case and* **orthogonal** *in the real case if* τ *is invertible and*

$$\tau^* = \tau^{-1}$$ □

There are also matrix versions of these definitions, obtained simply by replacing the operator τ by a matrix A. Moreover, the operator τ is self-adjoint if and only if any matrix that represents τ with respect to an ordered *orthonormal* basis $\mathcal{O}$ is self-adjoint. Similar statements hold for the other types of operators in the previous definition.

In some sense, square complex matrices are a generalization of complex numbers and the adjoint (conjugate transpose) is a generalization of the complex conjugate. In looking for a better analogy, we could consider just the diagonal matrices, but this is a bit too restrictive. The next logical choice is the set $\mathcal{N}$ of normal matrices.

Indeed, among the complex numbers, there are some special subsets: the real numbers, the positive numbers and the numbers on the unit circle. We will soon see that a complex matrix A is self-adjoint if and only if its complex eigenvalues are real. This would suggest that the analog of the set of real numbers is the set of self-adjoint matrices. Also, we will see that a complex matrix is unitary if and only if its eigenvalues have norm 1, so numbers on the unit circle seem to correspond to the set of unitary matrices. This leaves open the question of which normal matrices correspond to the positive real numbers. These are the *positive definite* matrices, which we will discuss later in the chapter.

Self-Adjoint Operators

Let us consider the basic properties of self-adjoint operators. The **quadratic form** associated with the linear operator τ is the function $Q_\tau: V \to F$ defined by

$$Q_\tau(v) = \langle \tau v, v \rangle$$

We have seen (Theorem 9.2) that in a *complex* inner product space, $\tau = 0$ if and only if $Q_\tau = 0$ but this does not hold, in general, for real inner product spaces. However, it does hold for symmetric operators on a real inner product space.

Theorem 10.11 *Let V be a finite-dimensional inner product space and let $\sigma, \tau \in \mathcal{L}(V)$.*

1) *If τ and σ are self-adjoint, then so are the following:*
   a) *$\sigma + \tau$*
   b) *τ^{-1}, if τ is invertible*
   c) *$p(\tau)$, for any real polynomial $p(x) \in \mathbb{R}[x]$*
2) *A complex operator τ is Hermitian if and only if $Q_\tau(v)$ is real for all $v \in V$.*
3) *If τ is a complex operator or a real symmetric operator, then*

$$\tau = 0 \quad \Leftrightarrow \quad Q_\tau = 0$$

4) *The characteristic polynomial $c_\tau(x)$ of a self-adjoint operator τ splits over $\mathbb{R}$, that is, all complex roots of $c_\tau(x)$ are real. Hence, the minimal polynomial $m_\tau(x)$ of τ is the product of distinct monic linear factors over $\mathbb{R}$.*

Proof. For part 2), if τ is Hermitian, then

$$\langle \tau v, v \rangle = \langle v, \tau v \rangle = \overline{\langle \tau v, v \rangle}$$

and so $Q_\tau(v) = \langle \tau v, v \rangle$ is real. Conversely, if $\langle \tau v, v \rangle \in \mathbb{R}$, then

$$\langle v, \tau v \rangle = \langle \tau v, v \rangle = \langle v, \tau^* v \rangle$$

and so $\tau = \tau^*$.

For part 3), we need only prove that $Q_\tau = 0$ implies $\tau = 0$ when $F = \mathbb{R}$. But if $Q_\tau = 0$, then

$$\begin{aligned} 0 &= \langle \tau(x+y), x+y\rangle \\ &= \langle \tau x, x\rangle + \langle \tau y, y\rangle + \langle \tau x, y\rangle + \langle \tau y, x\rangle \\ &= \langle \tau x, y\rangle + \langle \tau y, x\rangle \\ &= \langle \tau x, y\rangle + \langle x, \tau y\rangle \\ &= \langle \tau x, y\rangle + \langle \tau x, y\rangle \\ &= 2\langle \tau x, y\rangle \end{aligned}$$

and so $\tau = 0$.

For part 4), if τ is Hermitian ($F = \mathbb{C}$) and $\tau v = \lambda v$, then

$$\lambda v = \tau v = \tau^* v = \overline{\lambda} v$$

and so $\lambda = \overline{\lambda}$ is real. If τ is symmetric ($F = \mathbb{R}$), we must be a bit careful, since a nonreal root of $c_\tau(x)$ is *not* an eigenvalue of τ. However, matrix techniques can come to the rescue here. If $A = [\tau]_{\mathcal{O}}$ for any ordered orthonormal basis $\mathcal{O}$ for V, then $c_\tau(x) = c_A(x)$. Now, A is a real symmetric matrix, but can be thought of as a complex Hermitian matrix with real entries. As such, it represents a Hermitian linear operator on the complex space $\mathbb{C}^n$ and so, by what we have just shown, all (complex) roots of its characteristic polynomial are real. But the characteristic polynomial of A is the same, whether we think of A as a real or a complex matrix and so the result follows.□

Unitary Operators and Isometries

We now turn to the basic properties of unitary operators. These are the workhorse operators, in that a unitary operator is precisely a normal operator that maps orthonormal bases to orthonormal bases.

Note that τ is unitary if and only if

$$\langle \tau v, w\rangle = \langle v, \tau^{-1} w\rangle$$

for all $v, w \in V$.

Theorem 10.12 *Let V be a finite-dimensional inner product space and let $\sigma, \tau \in \mathcal{L}(V)$.*

1) *If τ and σ are unitary/orthogonal, then so are the following:*
   a) *$r\tau$, for $r \in \mathbb{C}$, $|r| = 1$*
   b) *$\sigma\tau$*
   c) *τ^{-1}, if τ is invertible.*
2) *τ is unitary/orthogonal if and only it is an isometric isomorphism.*
3) *τ is unitary/orthogonal if and only if it takes some orthonormal basis to an orthonormal basis, in which case it takes all orthonormal bases to orthonormal bases.*
4) *If τ is unitary/orthogonal, then the eigenvalues of τ have absolute value* 1.

Proof. We leave the proof of part 1) to the reader. For part 2), a unitary/orthogonal map is injective and since V is finite-dimensional, it is bijective. Moreover, for a bijective linear map τ, we have

$$\begin{aligned}\tau \text{ is an isometry} &\Leftrightarrow \langle \tau v, \tau w\rangle = \langle v, w\rangle \text{ for all } v, w \in V\\ &\Leftrightarrow \langle v, \tau^*\tau w\rangle = \langle v, w\rangle \text{ for all } v, w \in V\\ &\Leftrightarrow \tau^*\tau = \iota\\ &\Leftrightarrow \tau^* = \tau^{-1}\\ &\Leftrightarrow \tau \text{ is unitary/orthogonal}\end{aligned}$$

For part 3), suppose that τ is unitary/orthogonal and that $\mathcal{O} = \{u_1, \ldots, u_n\}$ is an orthonormal basis for V. Then

$$\langle \tau u_i, \tau u_j\rangle = \langle u_i, u_j\rangle = \delta_{i,j}$$

and so $\tau\mathcal{O}$ is an orthonormal basis for V. Conversely, suppose that $\mathcal{O}$ and $\tau\mathcal{O}$ are orthonormal bases for V. Then

$$\langle \tau u_i, \tau u_j\rangle = \delta_{i,j} = \langle u_i, u_j\rangle$$

which implies that $\langle \tau v, \tau w\rangle = \langle v, w\rangle$ for all $v, w \in V$ and so τ is unitary/orthogonal.

For part 4), if τ is unitary and $\tau v = \lambda v$, then

$$\lambda\overline{\lambda}\langle v, v\rangle = \langle \lambda v, \lambda v\rangle = \langle \tau v, \tau v\rangle = \langle v, v\rangle$$

and so $|\lambda|^2 = \lambda\overline{\lambda} = 1$, which implies that $|\lambda| = 1$.□

We also have the following theorem concerning unitary (and orthogonal) matrices.

Theorem 10.13 *Let A be an $n \times n$ matrix over $F = \mathbb{C}$ or $F = \mathbb{R}$.*

1) *The following are equivalent:*
   a) *A is unitary/orthogonal.*
   b) *The columns of A form an orthonormal set in F^n.*
   c) *The rows of A form an orthonormal set in F^n.*
2) *If A is unitary, then* $|\det(A)| = 1$. *If A is orthogonal, then* $\det(A) = \pm 1$.

Proof. The matrix A is unitary if and only if $AA^* = I$, which is equivalent to the rows of A being orthonormal. Similarly, A is unitary if and only if $A^*A = I$, which is equivalent to the columns of A being orthonormal. As for part 2),

$$AA^* = I \quad\Rightarrow\quad \det(A)\det(A^*) = 1 \quad\Rightarrow\quad \det(A)\overline{\det(A)} = 1$$

from which the result follows.□

Unitary/orthogonal matrices play the role of change of basis matrices when we restrict attention to orthonormal bases. Let us first note that if $\mathcal{B} = (u_1, \ldots, u_n)$ is an ordered orthonormal basis and

$$\begin{aligned} v &= a_1 u_1 + \cdots + a_n u_n \\ w &= b_1 u_1 + \cdots + b_n u_n \end{aligned}$$

then

$$\langle v, w \rangle = a_1 b_1 + \cdots + a_n b_n = [v]_{\mathcal{B}} \cdot [w]_{\mathcal{B}}$$

where the right hand side is the standard inner product in F^n and so $v \perp w$ if and only if $[v]_{\mathcal{B}} \perp [w]_{\mathcal{B}}$. We can now state the analog of Theorem 2.9.

Theorem 10.14 *If we are given any two of the following:*
1) A unitary/orthogonal $n \times n$ matrix A,
2) An ordered orthonormal basis $\mathcal{B}$ for F^n,
3) An ordered orthonormal basis $\mathcal{C}$ for F^n,
then the third is uniquely determined by the equation

$$A = M_{\mathcal{B},\mathcal{C}}$$

Proof. Let $\mathcal{B} = \{b_i\}$ be a basis for V. If $\mathcal{C}$ is an orthonormal basis for V, then

$$\langle b_i, b_j \rangle = [b_i]_{\mathcal{C}} \cdot [b_j]_{\mathcal{C}}$$

where $[b_i]_{\mathcal{C}}$ is the ith column of $A = M_{\mathcal{B},\mathcal{C}}$. Hence, A is unitary if and only if $\mathcal{B}$ is orthonormal. We leave the rest of the proof to the reader.□

Unitary Similarity

We have seen that the change of basis formula for operators is given by

$$[\tau]_{\mathcal{B}'} = P[\tau]_{\mathcal{B}} P^{-1}$$

where P is an invertible matrix. What happens when the bases are orthonormal?

Definition

1) *Two complex matrices A and B are* **unitarily similar** (*also called* **unitarily equivalent**) *if there exists a unitary matrix U for which*

$$B = UAU^{-1} = UAU^*$$

The equivalence classes associated with unitary similarity are called **unitary similarity classes**.

2) *Similarly, two real matrices A and B are* **orthogonally similar** (*also called* **orthogonally equivalent**) *if there exists an orthogonal matrix O for which*

$$B = OAO^{-1} = OAO^t$$

The equivalence classes associated with orthogonal similarity are called **orthogonal similarity classes**.□

The analog of Theorem 2.19 is the following.

Theorem 10.15 *Let V be an inner product space of dimension n. Then two $n \times n$ matrices A and B are unitarily/orthogonally similar if and only if they represent the same linear operator $\tau \in \mathcal{L}(V)$ with respect to (possibly different) ordered orthonormal bases. In this case, A and B represent exactly the same set of linear operators in $\mathcal{L}(V)$ with respect to ordered* orthonormal *bases.*

Proof. If A and B represent $\tau \in \mathcal{L}(V)$, that is, if

$$A = [\tau]_{\mathcal{B}} \quad \text{and} \quad B = [\tau]_{\mathcal{C}}$$

for ordered orthonormal bases $\mathcal{B}$ and $\mathcal{C}$, then

$$B = M_{\mathcal{B},\mathcal{C}} A M_{\mathcal{C},\mathcal{B}}$$

and according to Theorem 10.14, $M_{\mathcal{B},\mathcal{C}}$ is unitary/orthogonal. Hence, A and B are unitarily/orthogonally similar.

Now suppose that A and B are unitarily/orthogonally similar, say

$$B = UAU^{-1}$$

where U is unitary/orthogonal. Suppose also that A represents a linear operator $\tau \in \mathcal{L}(V)$ for some ordered orthonormal basis $\mathcal{B}$, that is,

$$A = [\tau]_{\mathcal{B}}$$

Theorem 10.14 implies that there is a unique ordered orthonormal basis $\mathcal{C}$ for V for which $U = M_{\mathcal{B},\mathcal{C}}$. Hence

$$B = M_{\mathcal{B},\mathcal{C}}[\tau]_{\mathcal{B}} M_{\mathcal{B},\mathcal{C}}^{-1} = [\tau]_{\mathcal{C}}$$

and so B also represents τ. By symmetry, we see that A and B represent the same set of linear operators, under all possible ordered orthonormal bases.□

We have shown (see the discussion of Schur's theorem) that any complex matrix A is unitarily similar to an upper triangular matrix, that is, that A is unitarily upper triangularizable. However, upper triangular matrices do not form a set of canonical forms under unitary similarity. Indeed, the subject of canonical forms for unitary similarity is rather complicated and we will not discuss it in this book, but instead refer the reader to the survey article [28].

Reflections

The following defines a very special type of unitary operator.

Definition *For a nonzero $v \in V$, the unique operator H_v for which*

$$H_v v = -v, \; H_v w = w \text{ for all } w \in \langle v \rangle^{\perp}$$

is called a **reflection** *or a* **Householder transformation**.$\square$

It is easy to verify that

$$H_v x = x - \frac{2\langle x, v \rangle}{\langle v, v \rangle} v$$

Moreover, $H_v x = -x$ for $x \neq 0$ if and only if $x = \alpha v$ for some $\alpha \in F$ and so we can uniquely identify v by the behavior of the reflection on V.

If H_v is a reflection and if we extend v to an ordered orthonormal basis $\mathcal{B}$ for V, then $[H_v]_{\mathcal{B}}$ is the matrix obtained from the identity matrix by replacing the upper left entry by -1,

$$[H_v]_{\mathcal{B}} = \begin{bmatrix} -1 & & & \\ & 1 & & \\ & & \ddots & \\ & & & 1 \end{bmatrix}$$

Thus, a reflection is both unitary and Hermitian, that is,

$$H_v^* = H_v^{-1} = H_v$$

Given two nonzero vectors of equal length, there is precisely one reflection that interchanges these vectors.

Theorem 10.16 *Let $v, w \in V$ be distinct nonzero vectors of equal length. Then H_{v-w} is the unique reflection sending v to w and w to v.*
Proof. If $\|v\| = \|w\|$, then $(v - w) \perp (v + w)$ and so

$$H_{v-w}(v - w) = w - v$$
$$H_{v-w}(v + w) = v + w$$

from which it follows that $H_{v-w}(v) = w$ and $H_{v-w}(w) = v$. As to uniqueness, suppose H_x is a reflection for which $H_x(v) = w$. Since $H_x^{-1} = H_x$, we have $H_x(w) = v$ and so

$$H_x(v - w) = -(v - w)$$

which implies that $H_x = H_{v-w}$.$\square$

Reflections can be used to characterize unitary operators.

Theorem 10.17 *Let V be a finite-dimensional inner product space. The following are equivalent for an operator $\tau \in \mathcal{L}(V)$:*

1) τ *is unitary/orthogonal*
2) τ *is a product of reflections.*

Proof. Since reflections are unitary/orthogonal and the product of unitary/orthogonal operators is unitary, it follows that 2) implies 1). For the converse, let τ be unitary. Let $\mathcal{B} = (u_1, \ldots, u_n)$ be an orthonormal basis for V. Then

$$H_{\tau_1 u_1 - u_1}(\tau u_1) = u_1$$

and so if $x_1 = \tau u_1 - u_1$ then

$$(H_{x_1}\tau)u_1 = u_1$$

that is, $\tau_1 := H_{x_1}\tau$ is the identity on $\langle u_1 \rangle$. Suppose that we have found reflections $H_{x_{k-1}}, \ldots, H_{x_1}$ for which $\tau_{k-1} := H_{x_{k-1}} \cdots H_{x_1}\tau$ is the identity on $\langle u_1, \ldots, u_{k-1} \rangle$. Then

$$H_{\tau_{k-1}u_k - u_k}(\tau_{k-1}u_k) = u_k$$

Moreover, we claim that $(\tau_{k-1}u_k - u_k) \perp u_i$ for $i < k$, since

$$\begin{aligned} \langle \tau_{k-1}u_k - u_k, u_i \rangle &= \langle (H_{x_{k-1}} \cdots H_{x_1}\tau)u_k, u_i \rangle \\ &= \langle \tau u_k, H_{x_1} \cdots H_{x_{k-1}} u_i \rangle \\ &= \langle \tau u_k, \tau u_i \rangle \\ &= \langle u_k, u_i \rangle \\ &= 0 \end{aligned}$$

Hence, if $x_k = \tau_{k-1}u_k - u_k$, then

$$(H_{x_k} \cdots H_{x_1}\tau)u_i = H_{x_k}u_i = u_i$$

and so $\tau_k := H_{x_k} \cdots H_{x_1}\tau$ is the identity on $\langle u_1, \ldots, u_k \rangle$. Thus, for $k = n$ we have $H_{x_n} \cdots H_{x_1}\tau = \iota$ and so $\tau = H_{x_1} \cdots H_{x_n}$, as desired.□

The Structure of Normal Operators

The following theorem includes the spectral theorems stated above for real and complex normal operators, along with some further refinements related to self-adjoint and unitary/orthogonal operators.

Theorem 10.18 (The structure theorem for normal operators)

1) **(Complex case)** *Let* V *be a finite-dimensional complex inner product space.*
  a) *The following are equivalent for* $\tau \in \mathcal{L}(V)$*:*
    i) τ *is normal*
    ii) τ *is unitarily diagonalizable*
    iii) τ *has an orthogonal spectral resolution*

$$\tau = \lambda_1 \rho_1 + \cdots + \lambda_k \rho_k$$

b) *Among the normal operators, the Hermitian operators are precisely those for which all complex eigenvalues are real.*
c) *Among the normal operators, the unitary operators are precisely those for which all complex eigenvalues have norm* 1.

2) **(Real case)** *Let V be a finite-dimensional real inner product space.*
a) $\tau \in \mathcal{L}(V)$ *is normal if and only if*

$$V = \mathcal{E}_{\lambda_1} \odot \cdots \odot \mathcal{E}_{\lambda_k} \odot W_1 \odot \cdots \odot W_m$$

where $\{\lambda_1, \ldots, \lambda_k\}$ *is the spectrum of* τ *and each* W_j *is a two-dimensional indecomposable* τ*-invariant subspace with an ordered basis* $\mathcal{B}_j$ *for which*

$$[\tau]_{\mathcal{B}_j} = \begin{bmatrix} a_j & -b_j \\ b_j & a_j \end{bmatrix}$$

b) *Among the real normal operators, the symmetric operators are those for which there are no subspaces* W_i *in the decomposition of part 2a). Hence, the following are equivalent for* $\tau \in \mathcal{L}(V)$:
i) τ *is symmetric.*
ii) τ *is orthogonally diagonalizable.*
iii) τ *has the orthogonal spectral resolution*

$$\tau = \lambda_1\rho_1 + \cdots + \lambda_k\rho_k$$

c) *Among the real normal operators, the orthogonal operators are precisely those for which the eigenvalues are equal to* ± 1 *and the matrices* $[\tau]_{\mathcal{B}_i}$ *described in part 2a) have rows (and columns) of norm* 1, *that is,*

$$[\tau]_{\mathcal{B}_i} = \begin{bmatrix} \sin\theta & -\cos\theta \\ \cos\theta & \sin\theta \end{bmatrix}$$

for some $\theta \in \mathbb{R}$.

Proof. We have proved part 1a). As to part 1b), it is only necessary to look at a diagonal matrix A representing τ. This matrix has the eigenvalues of τ on its main diagonal and so it is Hermitian if and only if the eigenvalues of τ are real. Similarly, A is unitary if and only if the eigenvalues of τ have absolute value equal to 1.

We have proved part 2a). Parts 2b) and 2c) follow by looking at the matrix $A = [\tau]_{\mathcal{B}}$ where $\mathcal{B} = \bigcup \mathcal{B}_j$. This matrix is symmetric if and only if A is diagonal, and A is orthogonal if and only if $\lambda_i = \pm 1$ and the matrices $[\tau]_{\mathcal{B}_j}$ have orthonormal rows.□

Matrix Versions

We can formulate matrix versions of the structure theorem for normal operators.

Theorem 10.19 (The structure theorem for normal matrices)

1) **(Complex case)**

a) A complex matrix A is normal if and only if it is unitarily diagonalizable, that is, if and only if there is a unitary matrix U for which

$$UAU^* = \operatorname{diag}(\lambda_1, \ldots, \lambda_k)$$

b) A complex matrix A is Hermitian if and only if 1a) holds, where all eigenvalues λ_i are real.

c) A complex matrix A is unitary if and only if 1a) holds, where all eigenvalues λ_i have norm 1.

2) **(Real case)**

a) A real matrix A is normal if and only if there is an orthogonal matrix O for which

$$OAO^t = \operatorname{diag}\left(\lambda_1, \ldots, \lambda_k, \begin{bmatrix} a_1 & -b_1 \\ b_1 & a_1 \end{bmatrix}, \ldots, \begin{bmatrix} a_m & -b_m \\ b_m & a_m \end{bmatrix}\right)$$

b) A real matrix A is symmetric if and only if it is orthogonally diagonalizable, that is, if and only if there is an orthogonal matrix O for which

$$OAO^t = \operatorname{diag}(\lambda_1, \ldots, \lambda_k)$$

c) A real matrix A is orthogonal if and only if there is an orthogonal matrix O for which

$$OAO^t = \operatorname{diag}\left(\lambda_1, \ldots, \lambda_k, \begin{bmatrix} \sin\theta_1 & -\cos\theta_1 \\ \cos\theta_1 & \sin\theta_1 \end{bmatrix}, \ldots, \begin{bmatrix} \sin\theta_m & -\cos\theta_m \\ \cos\theta_m & \sin\theta_m \end{bmatrix}\right)$$

for some $\theta_1, \ldots, \theta_m \in \mathbb{R}$.$\square$

Functional Calculus

Let τ be a normal operator on a finite-dimensional inner product space V and let τ have spectral resolution

$$\tau = \lambda_1\rho_1 + \cdots + \lambda_k\rho_k$$

Since each ρ_i is idempotent, we have $\rho_i^m = \rho_i$ for all $m \geq 1$. The pairwise orthogonality of the projections implies that

$$\tau^n = (\lambda_1\rho_1 + \cdots + \lambda_k\rho_k)^n = \lambda_1^n\rho_1 + \cdots + \lambda_k^n\rho_k$$

More generally, for any polynomial $p(x)$ over F,

$$p(\tau) = p(\lambda_1)\rho_1 + \cdots + p(\lambda_k)\rho_k$$

Note that a polynomial of degree $k - 1$ is uniquely determined by specifying an arbitrary set of k of its values at the distinct points $\alpha_1, \ldots, \alpha_k$. This follows from the **Lagrange interpolation formula**

$$p(x) = \sum_{i=0}^{k-1} p(\alpha_i) \left[\prod_{j \neq i} \frac{x - \alpha_j}{\alpha_i - \alpha_j} \right]$$

Therefore, we can define a unique polynomial $p(x)$ by specifying the values $p(\lambda_i)$, for $i = 1, \ldots, k$.

For example, for a given $1 \leq j \leq k$, if $p_j(x)$ is a polynomial for which

$$p_j(\lambda_i) = \delta_{i,j}$$

for $i = 1, \ldots, k$, then

$$p_j(\tau) = \rho_j$$

and so each projection ρ_j is a polynomial function of τ. As another example, if τ is invertible and $p(\lambda_i) = \lambda_i^{-1}$, then

$$p(\tau) = \lambda_1^{-1}\rho_1 + \cdots + \lambda_k^{-1}\rho_k = \tau^{-1}$$

as can easily be verified by direct calculation. Finally, if $p(\lambda_i) = \overline{\lambda}_i$, then since each ρ_i is self-adjoint, we have

$$p(\tau) = \overline{\lambda}_1\rho_1 + \cdots + \overline{\lambda}_k\rho_k = \tau^*$$

and so τ^* is a polynomial in τ.

We can extend this idea further by *defining*, for *any* function

$$f\colon \{\lambda_1, \ldots, \lambda_k\} \to F$$

the linear operator $f(\tau)$ by

$$f(\tau) = f(\lambda_1)\rho_1 + \cdots + f(\lambda_k)\rho_k$$

For example, we may define $\sqrt{\tau}$, τ^{-1}, e^τ and so on. Notice, however, that since the spectral resolution of τ is a finite sum, we gain nothing (but convenience) by using functions other than polynomials, for we can always find a polynomial $p(x)$ for which $p(\lambda_i) = f(\lambda_i)$ for $i = 1, \ldots, k$ and so $f(\tau) = p(\tau)$. The study of the properties of functions of an operator τ is referred to as the **functional calculus** of τ.

According to the spectral theorem, if V is complex and τ is normal, then $f(\tau)$ is a normal operator whose eigenvalues are $f(\lambda_i)$. Similarly, if V is real and τ is symmetric, then $f(\tau)$ is symmetric, with eigenvalues $f(\lambda_i)$.

Commutativity

The functional calculus can be applied to the study of the commutativity properties of operators. Here are two simple examples.

Theorem 10.20 *Let V be a finite-dimensional complex inner product space. For $\tau,\sigma \in \mathcal{L}(V)$, we write $\tau \leftrightarrow \sigma$ to denote the fact that τ and σ commute. Let τ and σ have spectral resolutions*

$$\begin{aligned}\tau &= \lambda_1\rho_1 + \cdots + \lambda_k\rho_k \\ \sigma &= \mu_1\nu_1 + \cdots + \mu_m\nu_m\end{aligned}$$

Then

1) *For any $\mu \in \mathcal{L}(V)$,*

$$\mu \leftrightarrow \tau \quad \Leftrightarrow \quad \mu \leftrightarrow \rho_i \textit{ for all } i$$

2)

$$\tau \leftrightarrow \sigma \quad \Leftrightarrow \quad \rho_i \leftrightarrow \nu_j, \textit{ for all } i,j$$

3) *If $f\colon \{\lambda_1,\ldots,\lambda_k\} \to F$ and $g\colon \{\mu_1,\ldots,\mu_m\} \to F$ are injective functions, then*

$$f(\tau) \leftrightarrow g(\sigma) \quad \Leftrightarrow \quad \tau \leftrightarrow \sigma$$

Proof. For 1), if $\mu \leftrightarrow \rho_i$ for all i, then $\mu \leftrightarrow \tau$ and the converse follows from the fact that ρ_i is a polynomial in τ. Part 2) is similar. For part 3), $\tau \leftrightarrow \sigma$ clearly implies $f(\tau) \leftrightarrow g(\sigma)$. For the converse, let $\Lambda = \{\lambda_1,\ldots,\lambda_k\}$. Since f is injective, the inverse function $f^{-1}\colon f(\Lambda) \to \Lambda$ is well-defined and $f^{-1}(f(\tau)) = \tau$. Thus, τ is a function of $f(\tau)$. Similarly, σ is a function of $g(\sigma)$. It follows that $f(\tau) \leftrightarrow g(\sigma)$ implies $\tau \leftrightarrow \sigma$.□

Theorem 10.21 *Let τ and σ be normal operators on a finite-dimensional complex inner product space V. Then τ and σ commute if and only if they have the form*

$$\begin{aligned}\tau &= p(r(\tau,\sigma)) \\ \sigma &= q(r(\tau,\sigma))\end{aligned}$$

where $p(x)$, $q(x)$ and $r(x,y)$ are polynomials.

Proof. If τ and σ are polynomials in $\theta = r(\tau,\sigma)$, then they clearly commute. For the converse, suppose that $\tau\sigma = \sigma\tau$ and let

$$\tau = \lambda_1\rho_1 + \cdots + \lambda_k\rho_k$$

and

$$\sigma = \mu_1\nu_1 + \cdots + \mu_m\nu_m$$

be the orthogonal spectral resolutions of τ and σ.

Then Theorem 10.20 implies that $\rho_i\nu_j = \nu_j\rho_i$. Hence,

$$\begin{aligned}\tau^r\sigma^s &= (\lambda_1\rho_1 + \cdots + \lambda_k\rho_k)^r(\mu_1\nu_1 + \cdots + \mu_m\nu_m)^s \\ &= (\lambda_1^r\rho_1 + \cdots + \lambda_k^r\rho_k)(\mu_1^s\nu_1 + \cdots + \mu_m^s\nu_m) \\ &= \sum_{i,j} \lambda_i^r\mu_j^s\rho_i\nu_j\end{aligned}$$

It follows that for any polynomial $r(x,y)$ in two variables,

$$r(\tau,\sigma) = \sum_{i,j} r(\lambda_i,\mu_j)\rho_i\nu_j$$

So if we choose $r(x,y)$ with the property that $\alpha_{i,j} = r(\lambda_i,\mu_j)$ are distinct, then

$$r(\tau,\sigma) = \sum_{i,j} \alpha_{i,j}\rho_i\nu_j$$

and we can also choose $p(x)$ and $q(x)$ so that $p(\alpha_{i,j}) = \lambda_i$ for all j and $q(\alpha_{i,j}) = \mu_j$ for all i. Then

$$\begin{aligned}p(r(\tau,\sigma)) &= \sum_{i,j} p(\alpha_{i,j})\rho_i\nu_j = \sum_{i,j} \lambda_i\rho_i\nu_j \\ &= \left(\sum_i \lambda_i\rho_i\right)\left(\sum_j \nu_j\right) = \sum_i \lambda_i\rho_i = \tau\end{aligned}$$

and similarly, $q(r(\tau,\sigma)) = \sigma$.□

Positive Operators

One of the most important cases of the functional calculus is $f(x) = \sqrt{x}$. Recall that the quadratic form associated with a linear operator τ is

$$Q_\tau(v) = \langle \tau v, v\rangle$$

Definition *A self-adjoint linear operator $\tau \in \mathcal{L}(V)$ is*
1) **positive** *if $Q_\tau(v) \geq 0$ for all $v \in V$*
2) **positive definite** *if $Q_\tau(v) > 0$ for all $v \neq 0$.*□

Theorem 10.22 *A self-adjoint operator τ on a finite-dimensional inner product space is*
1) *positive if and only if all of its eigenvalues are nonnegative*
2) *positive definite if and only if all of its eigenvalues are positive.*

Proof. If $Q_\tau(v) \geq 0$ and $\tau v = \lambda v$, then

$$0 \leq \langle \tau v, v\rangle = \lambda\langle v, v\rangle$$

and so $\lambda \geq 0$. Conversely, if all eigenvalues of τ are nonnegative, then

$$\tau = \lambda_1 \rho_1 + \cdots + \lambda_k \rho_k, \; \lambda_i \geq 0$$

and since $\iota = \rho_1 + \cdots + \rho_k$,

$$\langle \tau v, v \rangle = \sum_{i,j} \lambda_i \langle \rho_i v, \rho_j v \rangle = \sum_i \lambda_i \|\rho_i v\|^2 \geq 0$$

and so τ is positive. Part 2) is proved similarly.□

If τ is a positive operator, with spectral resolution

$$\tau = \lambda_1 \rho_1 + \cdots + \lambda_k \rho_k, \; \lambda_i \geq 0$$

then we may take the **positive square root** of τ,

$$\sqrt{\tau} = \sqrt{\lambda_1} \rho_1 + \cdots + \sqrt{\lambda_k} \rho_k$$

where $\sqrt{\lambda_i}$ is the nonnegative square root of λ_i. It is clear that

$$(\sqrt{\tau})^2 = \tau$$

and it is not hard to see that $\sqrt{\tau}$ is the only positive operator whose square is τ. In other words, every positive operator has a unique positive square root. Conversely, if τ has a positive square root, that is, if $\tau = \sigma^2$, for some positive operator σ, then τ is positive. Hence, an operator τ is positive if and only if it has a positive square root.

If τ is positive, then $\sqrt{\tau}$ is self-adjoint and so

$$(\sqrt{\tau})^* \sqrt{\tau} = \tau$$

Conversely, if $\tau = \sigma^* \sigma$ for some operator σ, then τ is positive, since it is clearly self-adjoint and

$$\langle \tau v, v \rangle = \langle \sigma^* \sigma v, v \rangle = \langle \sigma v, \sigma v \rangle \geq 0$$

Thus, τ is positive if and only if it has the form $\tau = \sigma^* \sigma$ for some operator σ. (A complex number z is nonnegative if and only if has the form $z = \overline{w} w$ for some complex number w.)

Theorem 10.23 *Let* $\tau \in \mathcal{L}(V)$.

1) τ *is positive if and only if it has a positive square root.*
2) τ *is positive if and only if it has the form* $\tau = \sigma^* \sigma$ *for some operator* σ.□

Here is an application of square roots.

Theorem 10.24 *If* τ *and* σ *are positive operators and* $\tau\sigma = \sigma\tau$, *then* $\tau\sigma$ *is positive.*

Proof. Since τ is a positive operator, it has a positive square root $\sqrt{\tau}$, which is a polynomial in τ. A similar statement holds for σ. Therefore, since τ and σ commute, so do $\sqrt{\tau}$ and $\sqrt{\sigma}$. Hence,

$$(\sqrt{\tau}\sqrt{\sigma})^2 = (\sqrt{\tau})^2(\sqrt{\sigma})^2 = \tau\sigma$$

Since $\sqrt{\tau}$ and $\sqrt{\sigma}$ are self-adjoint and commute, their product is self-adjoint and so $\tau\sigma$ is positive.□

The Polar Decomposition of an Operator

It is well known that any nonzero complex number z can be written in the *polar form* $z = re^{i\theta}$, where r is a positive number and θ is real. We can do the same for any nonzero linear operator τ on a finite-dimensional complex inner product space.

Theorem 10.25 *Let τ be a nonzero linear operator on a finite-dimensional complex inner product space V.*

1) *There exist a positive operator ρ and a unitary operator ν for which $\tau = \nu\rho$. Moreover, ρ is unique and if τ is invertible, then ν is also unique.*
2) *Similarly, there exist a positive operator σ and a unitary operator μ for which $\tau = \sigma\mu$. Moreover, σ is unique and if τ is invertible, then μ is also unique.*

Proof. Let us suppose for a moment that $\tau = \nu\rho$. Then

$$\tau^* = (\nu\rho)^* = \rho^*\nu^* = \rho\nu^{-1}$$

and so

$$\tau^*\tau = \rho\nu^{-1}\nu\rho = \rho^2$$

Also, if $v \in V$, then

$$\tau v = \nu(\rho v)$$

These equations give us a clue as to how to define ρ and ν.

Let us define ρ to be the unique positive square root of the positive operator $\tau^*\tau$. Then

$$\|\rho v\|^2 = \langle \rho v, \rho v\rangle = \langle \rho^2 v, v\rangle = \langle \tau^*\tau v, v\rangle = \|\tau v\|^2 \tag{10.2}$$

Define ν on $\mathrm{im}(\rho)$ by

$$\nu(\rho v) = \tau v$$

for all $v \in V$. Equation (10.2) shows that $\rho x = \rho y$ implies that $\tau x = \tau y$ and so this definition of ν on $\mathrm{im}(\rho)$ is well-defined.

Moreover, ν is an isometry on $\mathrm{im}(\rho)$, since (10.2) gives

$$\|\nu(\rho v)\| = \|\tau v\| = \|\rho v\|$$

Thus, if $\mathcal{B} = \{b_1, \ldots, b_k\}$ is an orthonormal basis for $\mathrm{im}(\rho)$, then $\nu\mathcal{B} = \{\nu b_1, \ldots, \nu b_k\}$ is an orthonormal basis for $\nu(\mathrm{im}(\rho)) = \mathrm{im}(\tau)$. Finally, we may extend both orthonormal bases to orthonormal bases for V and then extend the definition of ν to an isometry on V for which $\tau = \nu\rho$.

As for the uniqueness, we have seen that ρ must satisfy $\rho^2 = \tau^*\tau$ and since ρ^2 has a unique positive square root, we deduce that ρ is uniquely defined. Finally, if τ is invertible, then so is ρ since $\ker(\rho) \subseteq \ker(\tau)$. Hence, $\nu = \tau\rho^{-1}$ is uniquely determined by τ.

Part 2) can be proved by applying the previous theorem to the map τ^*, to get

$$\tau = (\tau^*)^* = (\nu\rho)^* = \rho\nu^{-1} = \rho\mu$$

where μ is unitary.□

We leave it as an exercise to show that any unitary operator μ has the form $\mu = e^{i\sigma}$, where σ is a self-adjoint operator. This gives the following corollary.

Corollary 10.26 (Polar decomposition) *Let τ be a nonzero linear operator on a finite-dimensional complex inner product space. Then there is a positive operator ρ and a self-adjoint operator σ for which τ has the* **polar decomposition**

$$\tau = \rho e^{i\sigma}$$

Moreover, ρ is unique and if τ is invertible, then σ is also unique.□

Normal operators can be characterized using the polar decomposition.

Theorem 10.27 *Let $\tau = \rho e^{i\sigma}$ be a polar decomposition of a nonzero linear operator τ. Then τ is normal if and only if $\rho\sigma = \sigma\rho$.*
Proof. Since

$$\tau\tau^* = \rho e^{i\sigma} e^{-i\sigma} \rho = \rho^2$$

and

$$\tau^*\tau = e^{-i\sigma} \rho\rho e^{i\sigma} = e^{-i\sigma} \rho^2 e^{i\sigma}$$

we see that τ is normal if and only if

$$e^{-i\sigma} \rho^2 e^{i\sigma} = \rho^2$$

or equivalently,

$$\rho^2 e^{i\sigma} = e^{i\sigma}\rho^2$$

Now, ρ is a polynomial in ρ^2 and σ is a polynomial in $e^{i\sigma}$ and so this holds if and only if $\rho\sigma = \sigma\rho$.$\square$

Exercises

1. Let $\tau \in \mathcal{L}(U,V)$. If τ is surjective, find a formula for the right inverse of τ in terms of τ^*. If τ is injective, find a formula for a left inverse of τ in terms of τ^*. *Hint*: Consider $\tau\tau^*$ and $\tau^*\tau$.
2. Let $\tau \in \mathcal{L}(V)$ where V is a complex vector space and let
$$\tau_1 = \frac{1}{2}(\tau + \tau^*) \text{ and } \tau_2 = \frac{1}{2i}(\tau - \tau^*)$$
Show that τ_1 and τ_2 are self-adjoint and that
$$\tau = \tau_1 + i\tau_2 \text{ and } \tau^* = \tau_1 - i\tau_2$$
What can you say about the uniqueness of these representations of τ and τ^*?
3. Prove that all of the roots of the characteristic polynomial of a skew-Hermitian matrix are pure imaginary.
4. Give an example of a normal operator that is neither self-adjoint nor unitary.
5. Prove that if $\|\tau v\| = \|\tau^*(v)\|$ for all $v \in V$, where V is complex, then τ is normal.
6. Let τ be a normal operator on a complex finite-dimensional inner product space V or a self-adjoint operator on a real finite-dimensional inner product space.
   a) Show that $\tau^* = p(\tau)$, for some polynomial $p(x) \in \mathbb{C}[x]$.
   b) Show that for any $\sigma \in \mathcal{L}(V)$, $\sigma\tau = \tau\sigma$ implies $\sigma\tau^* = \tau^*\sigma$. In other words, τ^* commutes with all operators that commute with τ.
7. Show that a linear operator τ on a finite-dimensional complex inner product space V is normal if and only if whenever S is an invariant subspace under τ, so is $S^\perp$.
8. Let V be a finite-dimensional inner product space and let τ be a normal operator on V.
   a) Prove that if τ is idempotent, then it is also self-adjoint.
   b) Prove that if τ is nilpotent, then $\tau = 0$.
   c) Prove that if $\tau^2 = \tau^3$, then τ is idempotent.
9. Show that if τ is a normal operator on a finite-dimensional complex inner product space, then the algebraic multiplicity is equal to the geometric multiplicity for all eigenvalues of τ.
10. Show that two orthogonal projections σ and ρ are orthogonal to each other if and only if $\operatorname{im}(\sigma) \perp \operatorname{im}(\rho)$.

11. Let τ be a normal operator and let σ be any operator on V. If the eigenspaces of τ are σ-invariant, show that τ and σ commute.
12. Prove that if τ and σ are normal operators on a finite-dimensional complex inner product space and if $\tau\theta = \theta\sigma$ for some operator θ then $\tau^*\theta = \theta\sigma^*$.
13. Prove that if two normal $n \times n$ complex matrices are similar, then they are *unitarily similar*, that is, similar via a unitary matrix.
14. If ν is a unitary operator on a complex inner product space, show that there exists a self-adjoint operator σ for which $\nu = e^{i\sigma}$.
15. Show that a positive operator has a unique positive square root.
16. Prove that if τ has a square root, that is, if $\tau = \sigma^2$, for some positive operator σ, then τ is positive.
17. Prove that if $\sigma \le \tau$ (that is, $\tau - \sigma$ is positive) and if θ is a positive operator that commutes with both σ and τ then $\sigma\theta \le \tau\theta$.
18. Using the QR factorization, prove the following result, known as the **Cholsky decomposition**. An invertible linear operator $\tau \in \mathcal{L}(V)$ is positive if and only if it has the form $\tau = \rho^*\rho$ where ρ is upper triangularizable. Moreover, ρ can be chosen with positive eigenvalues, in which case the factorization is unique.
19. Does every self-adjoint operator on a finite-dimensional real inner product space have a square root?
20. Let τ be a linear operator on $\mathbb{C}^n$ and let $\lambda_1, \ldots, \lambda_n$ be the eigenvalues of τ, each one written a number of times equal to its algebraic multiplicity. Show that

$$\sum_i |\lambda_i|^2 \le \mathrm{tr}(\tau^*\tau)$$

where tr is the trace. Show also that equality holds if and only if τ is normal.
21. If $\tau \in \mathcal{L}(V)$ where V is a real inner product space, show that the Hilbert space adjoint satisfies $(\tau^*)^{\mathbb{C}} = (\tau^{\mathbb{C}})^*$.

Part II—Topics

Chapter 11
Metric Vector Spaces: The Theory of Bilinear Forms

In this chapter, we study vector spaces over arbitrary fields that have a bilinear form defined on them.

Unless otherwise mentioned, all vector spaces are assumed to be finite-dimensional. The symbol F denotes an arbitrary field and F_q denotes a finite field of size q.

Symmetric, Skew-Symmetric and Alternate Forms

We begin with the basic definition.

Definition *Let V be a vector space over F. A mapping $\langle , \rangle : V \times V \to F$ is called a* **bilinear form** *if it is linear in each coordinate, that is, if*

$$\langle \alpha x + \beta y, z \rangle = \alpha \langle x, z \rangle + \beta \langle y, z \rangle$$

and

$$\langle z, \alpha x + \beta y \rangle = \alpha \langle z, x \rangle + \beta \langle z, y \rangle$$

A bilinear form is

1) **symmetric** *if*

$$\langle x, y \rangle = \langle y, x \rangle$$

for all $x,\ y \in V$.

2) **skew-symmetric** (*or* **antisymmetric**) *if*

$$\langle x, y \rangle = -\langle y, x \rangle$$

for all $x, y \in V$.

3) **alternate** (*or* **alternating**) *if*

$$\langle x, x\rangle = 0$$

for all $x \in V$.

A bilinear form that is either symmetric, skew-symmetric, or alternate is referred to as an **inner product** *and a pair* $(V, \langle,\rangle)$*, where* V *is a vector space and* $\langle,\rangle$ *is an inner product on* V*, is called a* **metric vector space** *or* **inner product space**. *As usual, we will refer to* V *as a metric vector space when the form is understood.*

4) *A metric vector space* V *with a symmetric form is called an* **orthogonal geometry** *over* F.
5) *A metric vector space* V *with an alternate form is called a* **symplectic geometry** *over* F.□

The term *symplectic*, from the Greek for "intertwined," was introduced in 1939 by the famous mathematician Hermann Weyl in his book *The Classical Groups*, as a substitute for the term *complex*. According to the dictionary, symplectic means "relating to or being an intergrowth of two different minerals." An example is *ophicalcite*, which is marble spotted with green serpentine.

Example 11.1 Minkowski space M_4 is the four-dimensional real orthogonal geometry $\mathbb{R}^4$ with inner product defined by

$$\begin{aligned}\langle e_1, e_1\rangle &= \langle e_2, e_2\rangle = \langle e_3, e_3\rangle = 1\\ \langle e_4, e_4\rangle &= -1\\ \langle e_i, e_j\rangle &= 0 \text{ for } i \neq j\end{aligned}$$

where $e_1, \ldots, e_4$ is the standard basis for $\mathbb{R}^4$.□

As is traditional, when the inner product is understood, we will use the phrase "let V be a metric vector space."

The real inner products discussed in Chapter 9 are inner products in the present sense and have the additional property of being *positive definite*—a notion that does not even make sense if the base field is not ordered. Thus, a real inner product space is an orthogonal geometry. On the other hand, the complex inner products of Chapter 9, being sesquilinear, are not inner products in the present sense. For this reason, we use the term *metric vector space* in this chapter, rather than *inner product space*.

If S is a vector subspace of a metric vector space V, then S inherits the metric structure from V. With this structure, we refer to S as a **subspace** of V.

The concepts of being symmetric, skew-symmetric and alternate are not independent. However, their relationship depends on the characteristic of the base field F, as do many other properties of metric vector spaces. In fact, the next theorem tells us that we do not need to consider skew-symmetric forms per se, since skew-symmetry is always equivalent to either symmetry or alternateness.

Theorem 11.1 *Let V be a vector space over a field F.*
1) *If* $\operatorname{char}(F) = 2$, *then*

$$\textit{alternate} \Rightarrow \textit{symmetric} \Leftrightarrow \textit{skew-symmetric}$$

2) *If* $\operatorname{char}(F) \neq 2$, *then*

$$\textit{alternate} \Leftrightarrow \textit{skew-symmetric}$$

Also, the only form that is both alternate and symmetric is the zero form: $\langle x, y \rangle = 0$ for all $x, y \in V$.

Proof. First note that for an alternating form over any base field,

$$0 = \langle x + y, x + y \rangle = \langle x, y \rangle + \langle y, x \rangle$$

and so

$$\langle x, y \rangle = -\langle y, x \rangle$$

which shows that the form is skew-symmetric. Thus, alternate always implies skew-symmetric.

If $\operatorname{char}(F) = 2$, then $-1 = 1$ and so the definitions of symmetric and skew-symmetric are equivalent, which proves 1). If $\operatorname{char}(F) \neq 2$ and the form is skew-symmetric, then for any $x \in V$, we have $\langle x, x \rangle = -\langle x, x \rangle$ or $2\langle x, x \rangle = 0$, which implies that $\langle x, x \rangle = 0$. Hence, the form is alternate. Finally, if the form is alternate and symmetric, then it is also skew-symmetric and so $\langle u, v \rangle = -\langle u, v \rangle$ for all $u, v \in V$, that is, $\langle u, v \rangle = 0$ for all $u, v \in V$.□

Example 11.2 The standard inner product on $V(n, q)$, defined by

$$(x_1, \ldots, x_n) \cdot (y_1, \ldots, y_n) = x_1 y_1 + \cdots + x_n y_n$$

is symmetric, but not alternate, since

$$(1, 0, \ldots, 0) \cdot (1, 0, \ldots, 0) = 1 \neq 0$$

□

The Matrix of a Bilinear Form

If $\mathcal{B} = (b_1, \ldots, b_n)$ is an ordered basis for a metric vector space V, then a bilinear form is completely determined by the $n \times n$ matrix of values

$$M_{\mathcal{B}} = (a_{i,j}) = (\langle b_i, b_j \rangle)$$

This is referred to as the **matrix of the form** (or the matrix of V) with respect to the ordered basis $\mathcal{B}$. Moreover, any $n \times n$ matrix over F is the matrix of some bilinear form on V.

Note that if $x = \Sigma r_i b_i$ then

$$M_{\mathcal{B}}[x]_{\mathcal{B}} = \begin{bmatrix} \langle b_1, x\rangle \\ \vdots \\ \langle b_n, x\rangle \end{bmatrix}$$

and

$$[x]_{\mathcal{B}}^t M_{\mathcal{B}} = (\,\langle x, b_1\rangle \quad \cdots \quad \langle x, b_n\rangle\,)$$

It follows that if $y = \sum s_i b_i$, then

$$[x]_{\mathcal{B}}^t M_{\mathcal{B}}[y]_{\mathcal{B}} = (\,\langle x, b_1\rangle \quad \cdots \quad \langle x, b_n\rangle\,)\begin{bmatrix} s_1 \\ \vdots \\ s_n \end{bmatrix} = \langle x, y\rangle$$

and this uniquely defines the matrix $M_{\mathcal{B}}$, that is, if $[x]_{\mathcal{B}}^t A[y]_{\mathcal{B}} = \langle x, y\rangle$ for all $x, y \in V$, then $A = M_{\mathcal{B}}$.

A matrix is **alternate** if it is skew-symmetric and has 0's on the main diagonal. Thus, we can say that a form is symmetric (skew-symmetric, alternate) if and only if the matrix $M_{\mathcal{B}}$ is symmetric (skew-symmetric, alternate).

Now let us see how the matrix of a form behaves with respect to a change of basis. Let $\mathcal{C} = (c_1, \ldots, c_n)$ be an ordered basis for V. Recall from Chapter 2 that the change of basis matrix $M_{\mathcal{C},\mathcal{B}}$, whose ith column is $[c_i]_{\mathcal{B}}$, satisfies

$$[v]_{\mathcal{B}} = M_{\mathcal{C},\mathcal{B}}[v]_{\mathcal{C}}$$

Hence,

$$\begin{aligned} \langle x, y\rangle &= [x]_{\mathcal{B}}^t\, M_{\mathcal{B}}[y]_{\mathcal{B}} \\ &= ([x]_{\mathcal{C}}^t M_{\mathcal{C},\mathcal{B}}^t\,)M_{\mathcal{B}}(M_{\mathcal{C},\mathcal{B}}[y]_{\mathcal{C}}\,) \\ &= [x]_{\mathcal{C}}^t(M_{\mathcal{C},\mathcal{B}}^t\, M_{\mathcal{B}} M_{\mathcal{C},\mathcal{B}})[y]_{\mathcal{C}} \end{aligned}$$

and so

$$M_{\mathcal{C}} = M_{\mathcal{C},\mathcal{B}}^t\, M_{\mathcal{B}} M_{\mathcal{C},\mathcal{B}}$$

This prompts the following definition.

Definition *Two matrices $A, B \in \mathcal{M}_n(F)$ are* **congruent** *if there exists an invertible matrix P for which*

$$A = P^t BP$$

The equivalence classes under congruence are called **congruence classes.** □

Thus, if two matrices represent the same bilinear form on V, they must be congruent. Conversely, if $B = M_{\mathcal{B}}$ represents a bilinear form on V and

$$A = P^t BP$$

where P is invertible, then there is an ordered basis $\mathcal{C}$ for V for which

$$P = M_{\mathcal{C},\mathcal{B}}$$

and so

$$A = M_{\mathcal{C},\mathcal{B}}^t M_{\mathcal{B}} M_{\mathcal{C},\mathcal{B}}$$

Thus, $A = M_{\mathcal{C}}$ represents the same form with respect to $\mathcal{C}$.

Theorem 11.2 *Let $\mathcal{B} = (b_1, \ldots, b_n)$ be an ordered basis for an inner product space V, with matrix*

$$M_{\mathcal{B}} = (\langle b_i, b_j \rangle)$$

1) The form can be recovered from the matrix by the formula

$$\langle x, y \rangle = [x]_{\mathcal{B}}^t M_{\mathcal{B}} [y]_{\mathcal{B}}$$

2) If $\mathcal{C} = (c_1, \ldots, c_n)$ is also an ordered basis for V, then

$$M_{\mathcal{C}} = M_{\mathcal{C},\mathcal{B}}^t M_{\mathcal{B}} M_{\mathcal{C},\mathcal{B}}$$

where $M_{\mathcal{C},\mathcal{B}}$ is the change of basis matrix from $\mathcal{C}$ to $\mathcal{B}$.

3) Two matrices A and B represent the same bilinear form on a vector space V if and only if they are congruent, in which case they represent the same set of bilinear forms on V. □

In view of the fact that congruent matrices have the same rank, we may define the rank of a bilinear form (or of V) to be the rank of any matrix that represents that form.

The Discriminant of a Form

If A and B are congruent matrices, then

$$\det(A) = \det(P^t BP) = \det(P)^2 \det(B)$$

and so $\det(A)$ and $\det(B)$ differ by a square factor. The **discriminant** Δ of a bilinear form is the set of determinants of all of the matrices that represent the form. Thus, if $\mathcal{B}$ is an ordered basis for V, then

$$\Delta = F^2 \det(M_{\mathcal{B}}) = \{r^2 \det(M_{\mathcal{B}}) \mid 0 \neq r \in F\}$$

Quadratic Forms

There is a close link between symmetric bilinear forms on V and quadratic forms on V.

Definition *A* **quadratic form** *on a vector space V is a map $Q\colon V \to F$ with the following properties:*
1) For all $r \in F$, $v \in V$,

$$Q(rv) = r^2 Q(v)$$

2) The map

$$\langle u, v \rangle_Q = Q(u+v) - Q(u) - Q(v)$$

is a (symmetric) bilinear form.□

Thus, every quadratic form Q on V defines a symmetric bilinear form $\langle u, v \rangle_Q$ on V. Conversely, if $\operatorname{char}(F) \neq 2$ and if $\langle , \rangle$ is a symmetric bilinear form on V, then the function

$$Q(x) = \frac{1}{2}\langle x, x \rangle$$

is a quadratic form Q. Moreover, the bilinear form associated with Q is the original bilinear form:

$$\begin{aligned}\langle u, v \rangle_Q &= Q(u+v) - Q(u) - Q(v) \\ &= \frac{1}{2}\langle u+v, u+v \rangle - \frac{1}{2}\langle u, u \rangle - \frac{1}{2}\langle v, v \rangle \\ &= \frac{1}{2}\langle u, v \rangle + \frac{1}{2}\langle v, u \rangle = \langle u, v \rangle\end{aligned}$$

Thus, the maps $\langle , \rangle \to Q$ and $Q \to \langle , \rangle_Q$ are inverses and so there is a one-to-one correspondence between symmetric bilinear forms on V and quadratic forms on V. Put another way, knowing the quadratic form is equivalent to knowing the corresponding bilinear form.

Again assuming that $\operatorname{char}(F) \neq 2$, if $\mathcal{B} = (v_1, \ldots, v_n)$ is an ordered basis for an orthogonal geometry V and if the matrix of the symmetric form on V is $M_{\mathcal{B}} = (a_{i,j})$, then for $x = \Sigma x_i v_i$,

$$Q(x) = \frac{1}{2}\langle x, x \rangle = \frac{1}{2}[x]_{\mathcal{B}}^t M_{\mathcal{B}} [x]_{\mathcal{B}} = \sum_{i,j} \frac{1}{2} a_{i,j} x_i x_j$$

and so $Q(x)$ is a homogeneous polynomial of degree 2 in the coordinates x_i. (The term "form" means *homogeneous polynomial*—hence the term quadratic *form*.)

Orthogonality

As we will see, not all metric vector spaces behave as nicely as real inner product spaces and this necessitates the introduction of a new set of terminology to cover various types of behavior. (The base field F is the culprit, of course.) The most striking differences stem from the possibility that $\langle x,x\rangle=0$ for a nonzero vector $x\in V$.

The following terminology should be familiar.

Definition *Let V be a metric vector space. A vector x is* **orthogonal** *to a vector y, written $x\perp y$, if $\langle x,y\rangle=0$. A vector $x\in V$ is* **orthogonal** *to a subset S of V, written $x\perp S$, if $\langle x,s\rangle=0$ for all $s\in S$. A subset S of V is* **orthogonal** *to a subset T of V, written $S\perp T$, if $\langle s,t\rangle=0$ for all $s\in S$ and $t\in T$. The* **orthogonal complement** *$X^{\perp}$ of a subset X of V is the subspace*

$$X^{\perp}=\{v\in V\mid v\perp X\}$$ □

Note that regardless of whether the form is symmetric or alternate (and hence skew-symmetric), orthogonality is a symmetric relation, that is, $x\perp y$ implies $y\perp x$. Indeed, this is precisely why we restrict attention to these two types of bilinear forms.

There are two types of degenerate behaviors that a vector may possess: It may be orthogonal to itself or, worse yet, it may be orthogonal to *every* vector in V. With respect to the former, we have the following terminology.

Definition *Let V be a metric vector space.*

1) *A nonzero $x\in V$ is* **isotropic** *(or* **null***) if $\langle x,x\rangle=0$; otherwise it is* **nonisotropic**.
2) *V is* **isotropic** *if it contains at least one isotropic vector. Otherwise, V is* **nonisotropic** *(or* **anisotropic***).*
3) *V is* **totally isotropic** *(that is, symplectic) if all vectors in V are isotropic.*□

Note that if v is an isotropic vector, then so is av for all $a\in F$. This can be expressed by saying that the set I of isotropic vectors in V is a *cone* in V. (A **cone** in V is a nonempty subset that is closed under scalar multiplication.)

With respect to the more severe forms of degeneracy, we have the following terminology.

Definition *Let V be a metric vector space.*

1) *A vector* $v \in V$ *is* **degenerate** *if* $v \perp V$. *The set* $V^\perp$ *of all degenerate vectors is called the* **radical** *of* V *and denoted by* $\text{rad}(V)$. *Thus,*

$$\text{rad}(V) = V^\perp$$

2) V *is* **nonsingular**, *or* **nondegenerate**, *if* $\text{rad}(V) = \{0\}$.
3) V *is* **singular**, *or* **degenerate**, *if* $\text{rad}(V) \neq \{0\}$.
4) V *is* **totally singular**, *or* **totally degenerate**, *if* $\text{rad}(V) = V$.□

Some of the above terminology is not entirely standard, so care should be exercised in reading the literature.

Theorem 11.3 *A metric vector space* V *is nonsingular if and only if all representing matrices* $M_\mathcal{B}$ *are nonsingular.*□

A note of caution is in order. If S is a subspace of a metric vector space V, then $\text{rad}(S)$ denotes the set of vectors in S that are degenerate in S, that is, $\text{rad}(S)$ is the radical of S, as a metric vector space in its own right. However, $S^\perp$ denotes the set of all vectors in V that are orthogonal to S. Thus,

$$\text{rad}(S) = S \cap S^\perp$$

Note also that

$$\text{rad}(S) = S \cap S^\perp \subseteq S^{\perp\perp} \cap S^\perp = \text{rad}(S^\perp)$$

and so if S is singular, then so is $S^\perp$.

Example 11.3 Recall that $V(n,q)$ is the set of all ordered n-tuples whose components come from the finite field F_q. (See Example 11.2.) It is easy to see that the subspace

$$S = \{0000, 1100, 0011, 1111\}$$

of $V(4,2)$ has the property that $S = S^\perp$. Note also that $V(4,2)$ is nonsingular and yet the subspace S is *totally* singular.□

The following result explains why we restrict attention to symmetric or alternate forms (which includes skew-symmetric forms).

Theorem 11.4 *Let* V *be a vector space with a bilinear form. The following are equivalent:*
1) *Orthogonality is a symmetric relation, that is,*

$$x \perp y \Rightarrow y \perp x$$

2) *The form on* V *is symmetric or alternate, that is,* V *is a metric vector space.*

Proof. It is clear that orthogonality is symmetric if the form is symmetric or alternate, since in the latter case, the form is also skew-symmetric.

For the converse, assume that orthogonality is symmetric. For convenience, let $x \bowtie y$ mean that $\langle x,y\rangle = \langle y,x\rangle$ and let $x \bowtie V$ mean that $\langle x,v\rangle = \langle v,x\rangle$ for all $v \in V$. If $x \bowtie V$ for all $x \in V$, then V is orthogonal and we are done. So let us examine vectors x with the property that $x \not\bowtie V$.

We wish to show that

$$x \not\bowtie V \quad \Rightarrow \quad x \text{ is isotropic} \quad \text{and} \quad (x \bowtie y \Rightarrow x \perp y) \tag{11.1}$$

Note that if the second conclusion holds, then since $x \bowtie x$, it follows that x is isotropic. So suppose that $x \bowtie y$. Since $x \not\bowtie V$, there is a $z \in V$ for which $\langle x,z\rangle \neq \langle z,x\rangle$ and so $x \perp y$ if and only if

$$\langle x,y\rangle(\langle x,z\rangle - \langle z,x\rangle) = 0$$

Now,

$$\begin{aligned} \langle x,y\rangle(\langle x,z\rangle - \langle z,x\rangle) &= \langle x,y\rangle\langle x,z\rangle - \langle x,y\rangle\langle z,x\rangle \\ &= \langle y,x\rangle\langle x,z\rangle - \langle x,y\rangle\langle z,x\rangle \\ &= \langle x, \langle y,x\rangle z - y\langle z,x\rangle\rangle \end{aligned}$$

But reversing the coordinates in the last expression gives

$$\langle \langle y,x\rangle z - y\langle z,x\rangle, x\rangle = \langle y,x\rangle\langle z,x\rangle - \langle y,x\rangle\langle z,x\rangle = 0$$

and so the symmetry of orthogonality implies that the last expression is 0 and so we have proven (11.1).

Let us assume that V is not orthogonal and show that all vectors in V are isotropic, whence V is symplectic. Since V is not orthogonal, there exist $u,v \in V$ for which $u \not\bowtie v$ and so $u \not\bowtie V$ and $v \not\bowtie V$. Hence, the vectors u and v are isotropic and for all $y \in V$,

$$\begin{aligned} y \bowtie u \quad &\Rightarrow \quad y \perp u \\ y \bowtie v \quad &\Rightarrow \quad y \perp v \end{aligned}$$

Since all vectors w for which $w \not\bowtie V$ are isotropic, let $w \bowtie V$. Then $w \bowtie u$ and $w \bowtie v$ and so $w \perp u$ and $w \perp v$. Now write

$$w = (w-u) + u$$

where $w - u \perp u$, since u is isotropic. Since the sum of two orthogonal isotropic vectors is isotropic, it follows that w is isotropic if $w - u$ is isotropic. But

$$\langle w+u, v\rangle = \langle u,v\rangle \neq \langle v,u\rangle = \langle v, w+u\rangle$$

and so $(w+u) \not\perp V$, which implies that $w+u$ is isotropic. Thus, w is also isotropic and so all vectors in V are isotropic.$\square$

Orthogonal and Symplectic Geometries

If a metric vector space is both orthogonal and symplectic, then the form is both symmetric and skew-symmetric and so

$$\langle u, v\rangle = \langle v, u\rangle = -\langle u, v\rangle$$

Therefore, when $\text{char}(F) \neq 2$, V is orthogonal and symplectic if and only if V is totally degenerate.

However, if $\text{char}(F) = 2$, then there are orthogonal symplectic geometries that are not totally degenerate. For example, let $V = \text{span}(u, v)$ be a two-dimensional vector space and define a form on V whose matrix is

$$M = \begin{bmatrix} 0 & 1 \\ 1 & 0 \end{bmatrix}$$

Since M is both symmetric and alternate, so is the form.

Linear Functionals

The Riesz representation theorem says that every linear functional f on a finite-dimensional real or complex inner product space V is represented by a Riesz vector $R_f \in V$, in the sense that

$$f(v) = \langle v, R_f\rangle$$

for all $v \in V$. A similar result holds for *nonsingular* metric vector spaces.

Let V be a metric vector space over F. Let $x \in V$ and define the inner product map $\langle \cdot\,, x\rangle: V \to F$ by

$$\langle \cdot\,, x\rangle v = \langle v, x\rangle$$

This is easily seen to be a linear functional and so we can define a linear map $\tau: V \to V^*$ by

$$\tau x = \langle \cdot\,, x\rangle$$

The bilinearity of the form ensures that τ is linear and the kernel of τ is

$$\ker(\tau) = \{x \in V \mid \langle V, x\rangle = \{0\}\} = V^{\perp} = \text{rad}(V)$$

Hence, τ is injective (and therefore an isomorphism) if and only if V is nonsingular.

Theorem 11.5 (The Riesz representation theorem) *Let V be a finite-dimensional nonsingular metric vector space. The map $\tau: V \to V^*$ defined by*

$$\tau x = \langle \cdot , x \rangle$$

is an isomorphism from V to V^*. It follows that for each $f \in V^*$ there exists a unique vector $x \in V$ for which

$$fv = \langle v, x \rangle$$

for all $v \in V$.□

The requirement that V be nonsingular is necessary. As a simple example, if V is totally singular, then no nonzero linear functional could possibly be represented by an inner product.

The Riesz representation theorem applies to nonsingular metric vector spaces. However, we can also achieve something useful for *singular* subspaces S of a *nonsingular* metric vector space. The reason is that any linear functional $f \in S^*$ can be extended to a linear functional $\overline{f}$ on V, where it has a Riesz vector, that is,

$$\overline{f}v = \langle v, R_{\overline{f}} \rangle = \langle \cdot , R_{\overline{f}} \rangle v$$

Hence, f also has this form, where its "Riesz vector" is an element of V, but is not necessarily in S.

Theorem 11.6 (The Riesz representation theorem for subspaces) *Let S be a subspace of a metric vector space V. If either V or S is nonsingular, the linear map $\tau: V \to S^*$ defined by*

$$\tau x = \langle \cdot , x \rangle|_S$$

is surjective and has kernel $S^\perp$. Hence, for any linear functional $f \in S^*$, there is a (not necessarily unique) vector $x \in V$ for which $fs = \langle s, x \rangle$ for all $s \in S$. Moreover, if S is nonsingular, then x can be taken from S, in which case it is unique.□

Orthogonal Complements and Orthogonal Direct Sums

Definition *A metric vector space V is the* **orthogonal direct sum** *of the subspaces S and T, written*

$$V = S \odot T$$

if $V = S \oplus T$ and $S \perp T$.□

If S is a subspace of a real inner product space, the projection theorem says that the orthogonal complement $S^\perp$ of S is a true vector space complement of S, that is,

$$V = S \odot S^\perp$$

However, in general metric vector spaces, an orthogonal complement may not be a vector space complement. In fact, Example 11.3 shows that in some cases $S^\perp = S$. In other cases, for example, if v is degenerate, then $\langle v \rangle^\perp = V$. However, as we will see, the orthogonal complement of S is a vector space complement if and only if either the sum is correct, $V = S + S^\perp$, or the intersection is correct, $S \cap S^\perp = \{0\}$. Note that the latter is equivalent to the nonsingularity of S.

Many nice properties of orthogonality in real inner product spaces do carry over to *nonsingular* metric vector spaces. Moreover, the next result shows that the restriction to nonsingular spaces is not that severe.

Theorem 11.7 *Let V be a metric vector space. Then*

$$V = \operatorname{rad}(V) \odot S$$

where S is nonsingular and $\operatorname{rad}(V)$ *is totally singular.*

Proof. If S is any vector space complement of $\operatorname{rad}(V)$, then $\operatorname{rad}(V) \perp S$ and so

$$V = \operatorname{rad}(V) \odot S$$

Also, S is nonsingular since $\operatorname{rad}(S) \subseteq \operatorname{rad}(V)$.$\square$

Here are some properties of orthogonality in nonsingular metric vector spaces. In particular, if either V or S is nonsingular, then the orthogonal complement of S always has the expected dimension,

$$\dim(S^\perp) = \dim(V) - \dim(S)$$

even if $S^\perp$ is not well behaved with respect to its intersection with S.

Theorem 11.8 *Let S be a subspace of a finite-dimensional metric vector space V.*

1) *If either V or S is nonsingular, then*

$$\dim(S) + \dim(S^\perp) = \dim(V)$$

   Hence, the following are equivalent:
   a) $V = S + S^\perp$
   b) *S is nonsingular, that is, $S \cap S^\perp = \{0\}$*
   c) $V = S \odot S^\perp$.
2) *If V is nonsingular, then*
   a) $S^{\perp\perp} = S$
   b) $\operatorname{rad}(S) = \operatorname{rad}(S^\perp)$
   c) *S is nonsingular if and only if $S^\perp$ is nonsingular.*

Proof. For part 1), the map $\tau\colon V \to S^*$ of Theorem 11.6 is surjective and has kernel $S^\perp$. Thus, the rank-plus-nullity theorem implies that

$$\dim(S^*) + \dim(S^\perp) = \dim(V)$$

However, $\dim(S^*) = \dim(S)$ and so part 1) follows. For part 2), since

$$\text{rad}(S) = S \cap S^\perp \subseteq S^{\perp\perp} \cap S^\perp = \text{rad}(S^\perp)$$

the nonsingularity of $S^\perp$ implies the nonsingularity of S. Then part 1) implies that

$$\dim(S) + \dim(S^\perp) = \dim(V)$$

and

$$\dim(S^\perp) + \dim(S^{\perp\perp}) = \dim(V)$$

Hence, $S^{\perp\perp} = S$ and $\text{rad}(S) = \text{rad}(S^\perp)$.□

The previous theorem cannot in general be strengthened. Consider the two-dimensional metric vector space $V = \text{span}(u, v)$ where

$$\langle u, u\rangle = 1, \langle u, v\rangle = 0, \langle v, v\rangle = 0$$

If $S = \text{span}(u)$, then $S^\perp = \text{span}(v)$. Now, S is nonsingular but $S^\perp$ is singular and so 2c) does not hold. Also, $\text{rad}(S) = \{0\}$ and $\text{rad}(S^\perp) = S^\perp$ and so 2b) fails. Finally, $S^{\perp\perp} = V \neq S$ and so 2a) fails.

Isometries

We now turn to a discussion of structure-preserving maps on metric vector spaces.

Definition *Let V and W be metric vector spaces. We use the same notation $\langle , \rangle$ for the bilinear form on each space. A bijective linear map $\tau: V \to W$ is called an* **isometry** *if*

$$\langle \tau u, \tau v\rangle = \langle u, v\rangle$$

for all vectors u and v in V. If an isometry exists from V to W, we say that V and W are **isometric** *and write $V \approx W$. It is evident that the set of all isometries from V to V forms a group under composition.*

If V is a nonsingular orthogonal geometry, an isometry of V is called an **orthogonal transformation**. *The set $\mathcal{O}(V)$ of all orthogonal transformations on V is a group under composition, known as the* **orthogonal group** *of V.*

If V is a nonsingular symplectic geometry, an isometry of V is called a **symplectic transformation**. *The set $\text{Sp}(V)$ of all symplectic transformations on V is a group under composition, known as the* **symplectic group** *of V.*□

Note that, in contrast to the case of real inner product spaces, we must include the requirement that τ be bijective since this does not follow automatically if V is singular. Here are a few of the basic properties of isometries.

Theorem 11.9 *Let $\tau \in \mathcal{L}(V,W)$ be a linear transformation between finite-dimensional metric vector spaces V and W.*

1) *Let $\mathcal{B} = \{v_1, \ldots, v_n\}$ be a basis for V. Then τ is an isometry if and only if τ is bijective and*

$$\langle \tau v_i, \tau v_j \rangle = \langle v_i, v_j \rangle$$

for all i, j.

2) *If V is orthogonal and* $\operatorname{char}(F) \neq 2$, *then τ is an isometry if and only if it is bijective and*

$$\langle \tau v, \tau v \rangle = \langle v, v \rangle$$

for all $v \in V$.

3) *Suppose that $\tau: V \approx W$ is an isometry and*

$$V = S \odot S^{\perp} \quad \textit{and} \quad W = T \odot T^{\perp}$$

If $\tau S = T$, then $\tau(S^{\perp}) = T^{\perp}$.

Proof. We prove part 3) only. To see that $\tau(S^{\perp}) = T^{\perp}$, if $z \in S^{\perp}$ and $t \in T$, then since $T = \tau S$, we can write $t = \tau s$ for some $s \in S$ and so

$$\langle \tau z, t \rangle = \langle \tau z, \tau s \rangle = \langle z, s \rangle = 0$$

whence $\tau(S^{\perp}) \subseteq T^{\perp}$. But since the dimensions are equal, it follows that $\tau(S^{\perp}) = T^{\perp}$.$\square$

Hyperbolic Spaces

A special type of two-dimensional metric vector space plays an important role in the structure theory of metric vector spaces.

Definition *Let V be a metric vector space. A* **hyperbolic pair** *is a pair of vectors $u, v \in V$ for which*

$$\langle u, u \rangle = \langle v, v \rangle = 0, \ \langle u, v \rangle = 1$$

Note that $\langle v, u \rangle = 1$ if V is orthogonal and $\langle v, u \rangle = -1$ if V is symplectic. In either case, the subspace $H =$ $\operatorname{span}(u, v)$ *is called a* **hyperbolic plane** *and any space of the form*

$$\mathcal{H} = H_1 \odot \cdots \odot H_k$$

where each H_i is a hyperbolic plane, is called a **hyperbolic space**. *If (u_i, v_i) is a hyperbolic pair for H_i, then we refer to the basis*

$$(u_1, v_1, \ldots, u_k, v_k)$$

for $\mathcal{H}$ as a **hyperbolic basis**. (*In the symplectic case, the usual term is* **symplectic basis**.)□

Note that any hyperbolic space $\mathcal{H}$ is nonsingular.

In the orthogonal case, hyperbolic planes can be characterized by their degree of isotropy, so to speak. (In the symplectic case, all spaces are totally isotropic by definition.) Indeed, we leave it as an exercise to prove that a two-dimensional nonsingular orthogonal geometry V is a hyperbolic plane if and only if V contains exactly two one-dimensional totally isotropic (equivalently, totally degenerate) subspaces. Put another way, the cone of isotropic vectors is the union of two one-dimensional subspaces of V.

Nonsingular Completions of a Subspace

Let U be a subspace of a nonsingular metric vector space V. If U is singular, it is of interest to find a *minimal* nonsingular subspace of V containing U.

Definition *Let V be a nonsingular metric vector space and let U be a subspace of V. A subspace S of V for which $U \leq S$ is called an* **extension** *of U. A* **nonsingular completion** of U is an extension of U that is minimal in the family of all nonsingular extensions of U.□

Theorem 11.10 *Let V be a nonsingular finite-dimensional metric vector space over F. We assume that* $\operatorname{char}(F) \neq 2$ *when V is orthogonal.*

1) Let S be a subspace of V. If v is isotropic and the orthogonal direct sum

$$\operatorname{span}(v) \odot S$$

exists, then there is a hyperbolic plane $H = \operatorname{span}(v, z)$ for which

$$H \odot S$$

exists. In particular, if v is isotropic, then there is a hyperbolic plane containing v.

2) Let U be a subspace of V and let

$$U = \operatorname{span}(v_1, \ldots, v_k) \odot W$$

where W is nonsingular and $\{v_1, \ldots, v_k\}$ are linearly independent in $\operatorname{rad}(U)$. *Then there is a hyperbolic space $\mathcal{H}_k = H_1 \odot \cdots \odot H_k$ with hyperbolic basis $(v_1, z_1, \ldots, v_k, z_k)$ for which*

$$\overline{U} = \mathcal{H}_k \odot W$$

is a nonsingular proper extension of U. If $\{v_1, \ldots, v_k\}$ is a basis for $\operatorname{rad}(U)$, *then*

$$\dim(\overline{U}) = \dim(U) + \dim(\operatorname{rad}(U))$$

and we refer to $\overline{U}$ as a **hyperbolic extension** *of U. If U is nonsingular, we say that U is a hyperbolic extension of itself.*

Proof. For part 1), the nonsingularity of V implies that $S^{\perp\perp} = S$. Hence, $v \notin S = S^{\perp\perp}$ and so there is an $x \in S^{\perp}$ for which $\langle v, x \rangle \neq 0$. If V is symplectic, then all vectors are isotropic and so we can take $z = (1/\langle v, x \rangle)x$. If V is orthogonal, let $z = rv + sx$. The conditions defining (v, z) as a hyperbolic pair are (since v is isotropic)

$$1 = \langle v, z \rangle = \langle v, rv + sx \rangle = s\langle v, x \rangle$$

and

$$0 = \langle z, z \rangle = \langle rv + sx, rv + sx \rangle = 2rs\langle v, x \rangle + s^2\langle x, x \rangle = 2r + s^2\langle x, x \rangle$$

Since $\langle v, x \rangle \neq 0$, the first of these equations can be solved for s and since $\operatorname{char}(F) \neq 2$, the second equation can then be solved for r. Thus, in either case, there is a vector $z \in S^{\perp}$ for which $H = \operatorname{span}(v, z) \subseteq S^{\perp}$ is hyperbolic. Hence, $S \subseteq S^{\perp\perp} \subseteq H^{\perp}$ and since H is nonsingular, that is, $H \cap H^{\perp} = \{0\}$, we have $H \cap S = \{0\}$ and so $H \odot S$ exists.

Part 2) is proved by induction on k. Note first that all of the vectors v_i are isotropic. If $k = 1$, then $\operatorname{span}(v_1) \odot W$ exists and so part 1) implies that there is a hyperbolic plane $H = \operatorname{span}(v_1, z)$ for which $H \odot W$ exists.

Assume that the result is true for independent sets of size less than $k \geq 2$. Since

$$\operatorname{span}(v_1) \odot (\operatorname{span}(v_2, \ldots, v_k) \odot W)$$

exists, part 1) implies that there exists a hyperbolic plane $H_1 = \operatorname{span}(v_1, z_1)$ for which

$$H_1 \odot (\operatorname{span}(v_2, \ldots, v_k) \odot W)$$

exists. Since $v_2, \ldots, v_k$ are in the radical of $\operatorname{span}(v_2, \ldots, v_k) \odot W$, the inductive hypothesis implies that there is a hyperbolic space $H_2 \odot \cdots \odot H_k$ with hyperbolic basis $(v_2, z_2, \ldots, v_k, z_k)$ for which the orthogonal direct sum

$$H_2 \odot \cdots \odot H_k \odot W$$

exists. Hence, $H_1 \odot \cdots \odot H_k \odot W$ also exists.□

We can now prove that the hyperbolic extensions of U are precisely the minimal nonsingular extensions of U.

Theorem 11.11 (Nonsingular extension theorem) *Let U be a subspace of a nonsingular finite-dimensional metric vector space V. The following are equivalent:*

1) $T = \mathcal{H} \odot W$ is a hyperbolic extension of U

2) T is a minimal nonsingular extension of U

3) *T is a nonsingular extension of U and*

$$\dim(T) = \dim(U) + \dim(\mathrm{rad}(U))$$

Thus, any two nonsingular completions of U are isometric.

Proof. If $U \leq X \leq V$ where X is nonsingular, then we may apply Theorem 11.10 to U as a subspace of X, to obtain a hyperbolic extension $\mathcal{K} \odot W$ of U for which

$$U \subseteq \mathcal{K} \odot W \subseteq X$$

Thus, every nonsingular extension of U contains a hyperbolic extension of U. Moreover, all hyperbolic extensions of U have the same dimension:

$$\dim(\mathcal{H} \odot W) = \dim(U) + \dim(\mathrm{rad}(U))$$

and so no hyperbolic extension of U is properly contained in another hyperbolic extension of U. This proves that 1)–3) are equivalent. The final statement follows from the fact that hyperbolic spaces of the same dimension are isometric.□

Extending Isometries to Nonsingular Completions

Let V and V' be isometric nonsingular metric vector spaces and let $U = \mathrm{rad}(U) \odot W$ be a subspace of V, with nonsingular completion $\overline{U} = \mathcal{H} \odot W$.

If $\tau: U \to \tau U$ is an isometry, then it is a simple matter to extend τ to an isometry $\overline{\tau}$ from $\overline{U}$ onto a nonsingular completion of τU. To see this, let $(u_1, z_1, \ldots, u_k, z_k)$ be a hyperbolic basis for $\mathcal{H}$. Since $(u_1, \ldots, u_k)$ is a basis for $\mathrm{rad}(U)$, it follows that $(\tau u_1, \ldots, \tau u_k)$ is a basis for $\mathrm{rad}(\tau U)$.

Hence, we can hyperbolically extend $\tau U = \mathrm{rad}(\tau W) \odot \tau W$ to get

$$\overline{\tau U} = \mathcal{H}' \odot \tau W$$

where $\mathcal{H}'$ has hyperbolic basis $(\tau u_1, x_1, \ldots, \tau u_k, x_k)$. To extend τ, simply set $\overline{\tau} z_i = x_i$ for all $i = 1, \ldots, k$.

Theorem 11.12 *Let V and V' be isometric nonsingular metric vector spaces and let U be a subspace of V, with nonsingular completion $\overline{U}$. Any isometry $\tau: U \to \tau U$ can be extended to an isometry from $\overline{U}$ onto a nonsingular completion of τU.*□

The Witt Theorems: A Preview

There are two important theorems that are quite easy to prove in the case of real inner product spaces, but require more work in the case of metric vector spaces in general. Let V and V' be isometric nonsingular metric vector spaces over a field F. We assume that $\mathrm{char}(F) \neq 2$ if V is orthogonal.

The *Witt extension theorem* says that if S is a subspace of V, then any isometry

$$\tau: S \to \tau S \subseteq V'$$

can be extended to an isometry from V to V'. The *Witt cancellation theorem* says that if

$$V = S \odot S^{\perp} \quad \text{and} \quad V' = T \odot T^{\perp}$$

then

$$S \approx T \Rightarrow S^{\perp} \approx T^{\perp}$$

We will prove these theorems in both the orthogonal and symplectic cases a bit later in the chapter. For now, we simply want to show that it is easy to prove one Witt theorem using the other.

Suppose that the Witt extension theorem holds and assume that

$$V = S \odot S^{\perp} \quad \text{and} \quad V' = T \odot T^{\perp}$$

and $S \approx T$. Then any isometry $\tau: S \to T$ can be extended to an isometry $\overline{\tau}$ from V to V'. According to Theorem 11.9, we have $\overline{\tau}(S^{\perp}) = T^{\perp}$ and so $S^{\perp} \approx T^{\perp}$. Hence, the Witt cancellation theorem holds.

Conversely, suppose that the Witt cancellation theorem holds and let $\tau: S \to \tau S \subseteq V'$ be an isometry. Since τ can be extended to a nonsingular completion of S, we may assume that S is nonsingular. Then

$$V = S \odot S^{\perp}$$

Since τ is an isometry, τS is also nonsingular and we can write

$$V' = \tau S \odot (\tau S)^{\perp}$$

Since $S \approx \tau S$, Witt's cancellation theorem implies that $S^{\perp} \approx (\tau S)^{\perp}$. If $\mu: S^{\perp} \to (\tau S)^{\perp}$ is an isometry, then the map $\sigma: V \to V'$ defined by

$$\sigma(u + v) = \tau u + \mu v$$

for $u \in S$ and $v \in S^{\perp}$ is an isometry that extends τ. Hence Witt's extension theorem holds.

The Classification Problem for Metric Vector Spaces

The **classification problem** for a class of metric vector spaces (such as the orthogonal or symplectic spaces) is the problem of determining when two metric vector spaces in the class are isometric. The classification problem is considered "solved," at least in a theoretical sense, by finding a set of canonical forms or a complete set of invariants for matrices under congruence.

To see why, suppose that $\tau: V \to W$ is an isometry and $\mathcal{B} = (v_1, \ldots, v_n)$ is an ordered basis for V. Then $\mathcal{C} = (\tau v_1, \ldots, \tau v_n)$ is an ordered basis for W and

$$M_{\mathcal{B}}(V) = (\langle v_i, v_j \rangle) = (\langle \tau v_i, \tau v_j \rangle) = M_{\mathcal{C}}(W)$$

Thus, the congruence class of matrices representing V is identical to the congruence class of matrices representing W.

Conversely, suppose that V and W are metric vector spaces with the same congruence class of representing matrices. Then if $\mathcal{B} = (v_1, \ldots, v_n)$ is an ordered basis for V, there is an ordered basis $\mathcal{C} = (w_1, \ldots, w_n)$ for W for which

$$(\langle v_i, v_j \rangle) = M_{\mathcal{B}}(V) = M_{\mathcal{C}}(W) = (\langle w_i, w_j \rangle)$$

Hence, the map $\tau: V \to W$ defined by $\tau v_i = w_i$ is an isometry from V to W.

We have shown that two metric vector spaces are isometric if and only if they have the same congruence class of representing matrices. Thus, we can determine whether any two metric vector spaces are isometric by representing each space with a matrix and determining whether these matrices are congruent, using a set of canonical forms or a set of complete invariants.

Symplectic Geometry

We now turn to a study of the structure of orthogonal and symplectic geometries and their isometries. Since the study of the structure (and the structure itself) of symplectic geometries is simpler than that of orthogonal geometries, we begin with the symplectic case. The reader who is interested only in the orthogonal case may omit this section.

Throughout this section, let V be a nonsingular symplectic geometry.

The Classification of Symplectic Geometries

Among the simplest types of metric vector spaces are those that possess an orthogonal basis. However, it is easy to see that a symplectic geometry V has an orthogonal basis if and only if it is totally degenerate and so no "interesting" symplectic geometries have orthogonal bases.

Thus, in searching for an orthogonal decomposition of V, we turn to two-dimensional subspaces and this puts us in mind of hyperbolic spaces. Let $\mathcal{F}$ be the family of all hyperbolic subspaces of V, which is nonempty since the zero subspace $\{0\}$ is singular and so has a nonzero hyperbolic extension. Since V is finite-dimensional, $\mathcal{F}$ has a maximal member $\mathcal{H}$. Since $\mathcal{H}$ is nonsingular, if $\mathcal{H} \neq V$, then

$$V = \mathcal{H} \odot \mathcal{H}^{\perp}$$

where $\mathcal{H}^{\perp} \neq \{0\}$. But then if $v \in \mathcal{H}^{\perp}$ is nonzero, there is a hyperbolic extension

$H \odot \mathcal{H}$ of $\mathcal{H}$ containing v, which contradicts the maximality of $\mathcal{H}$. Hence, $V = \mathcal{H}$.

This proves the following structure theorem for symplectic geometries.

Theorem 11.13

1) *A symplectic geometry has an orthogonal basis if and only if it is totally degenerate.*
2) *Any nonsingular symplectic geometry V is a hyperbolic space, that is,*

$$V = H_1 \odot H_2 \odot \cdots \odot H_k$$

where each H_i is a hyperbolic plane. Thus, there is a hyperbolic basis for V, that is, a basis $\mathcal{B}$ for which the matrix of the form is

$$Y_{2k} = \begin{bmatrix} 0 & 1 & & & & & \\ -1 & 0 & & & & & \\ & & 0 & 1 & & & \\ & & -1 & 0 & & & \\ & & & & \ddots & & \\ & & & & & 0 & 1 \\ & & & & & -1 & 0 \end{bmatrix}$$

In particular, the dimension of V is even.

3) *Any symplectic geometry V has the form*

$$V = \operatorname{rad}(V) \odot \mathcal{H}$$

where $\mathcal{H}$ is a hyperbolic space and $\operatorname{rad}(V)$ is a totally degenerate space. The rank of the form is $\dim(\mathcal{H})$ and V is uniquely determined up to isometry by its rank and its dimension. Put another way, up to isometry, there is precisely one symplectic geometry of each rank and dimension. □

Symplectic forms are represented by alternate matrices, that is, skew-symmetric matrices with zero diagonal. Moreover, according to Theorem 11.13, each $n \times n$ alternate matrix is congruent to a matrix of the form

$$X_{2k,n-2k} = \begin{bmatrix} Y_{2k} & 0 \\ 0 & 0_{n-2k} \end{bmatrix}_{\text{block}}$$

Since the rank of $X_{2k,n-2k}$ is $2k$, no two such matrices are congruent.

Theorem 11.14 *The set of $n \times n$ matrices of the form $X_{2k,n-2k}$ is a set of canonical forms for alternate matrices under congruence.* □

The previous theorems solve the classification problem for symplectic geometries by stating that the rank and dimension of V form a complete set of invariants under congruence and that the set of all matrices of the form $X_{2k,n-2k}$ is a set of canonical forms.

Witt's Extension and Cancellation Theorems

We now prove the Witt theorems for symplectic geometries.

Theorem 11.15 (Witt's extension theorem) *Let V and V' be isometric nonsingular symplectic geometries over a field F. Then any isometry*

$$\tau: S \to \tau S \subseteq V'$$

on a subspace S of V can be extended to an isometry from V to V'.

Proof. According to Theorem 11.12, we can extend τ to a nonsingular completion of S, so we may simply assume that S and τS are nonsingular. Hence,

$$V = S \odot S^{\perp}$$

and

$$V' = \tau S \odot (\tau S)^{\perp}$$

To complete the extension of τ to V, we need only choose a hyperbolic basis

$$(e_1, f_1, \ldots, e_p, f_p)$$

for $S^{\perp}$ and a hyperbolic basis

$$(e_1', f_1', \ldots, e_p', f_p')$$

for $(\tau S)^{\perp}$ and define the extension by setting $\tau e_i = e_i'$ and $\tau f_i = f_i'$.□

As a corollary to Witt's extension theorem, we have Witt's cancellation theorem.

Theorem 11.16 (Witt's cancellation theorem) *Let V and V' be isometric nonsingular symplectic geometries over a field F. If*

$$V = S \odot S^{\perp} \quad \textit{and} \quad V' = T \odot T^{\perp}$$

then

$$S \approx T \Rightarrow S^{\perp} \approx T^{\perp}$$ □

The Structure of the Symplectic Group: Symplectic Transvections

Let us examine the nature of symplectic transformations (isometries) on a nonsingular symplectic geometry V. Recall that for a real vector space, an isometric isomorphism, which corresponds to an isometry in the present context, is the same as an orthogonal map and orthogonal maps are products of reflections (Theorem 10.17). Recall also that a reflection H_v is defined as an operator for which

$$H_v v = -v,\ H_v w = w \textit{ for all } w \in \langle v \rangle^{\perp}$$

and that

$$H_v x = x - \frac{2\langle x, v\rangle}{\langle v, v\rangle} v$$

In the present context, we do not dare divide by $\langle v, v\rangle$, since all vectors are isotropic. So here is the next-best thing.

Definition *Let V be a nonsingular symplectic geometry over F. Let $v \in V$ be nonzero and let $a \in F$. The map $\tau_{v,a}: V \to V$ defined by*

$$\tau_{v,a}(x) = x + a\langle x, v\rangle v$$

is called the **symplectic transvection** *determined by v and a.*□

Note that if $a = 0$, then $\tau_{v,a} = \iota$ and if $a \neq 0$, then $\tau_{v,a}$ is the identity precisely on the subspace $\mathrm{span}(v)^{\perp}$ of codimension 1. In the case of a reflection, H_v is the identity precisely on $\mathrm{span}(v)^{\perp}$ and

$$V = \mathrm{span}(v)^{\perp} \odot \mathrm{span}(v)$$

However, for a symplectic transvection, $\tau_{v,a}$ is the identity precisely on $\mathrm{span}(v)^{\perp}$ (for $a \neq 0$) but $\mathrm{span}(v) \subseteq \mathrm{span}(v)^{\perp}$. Here are the basic properties of symplectic transvections.

Theorem 11.17 *Let $\tau_{v,a}$ be a symplectic transvection on V. Then*

1) *$\tau_{v,a}$ is a symplectic transformation (isometry).*
2) *$\tau_{v,a} = \iota$ if and only if $a = 0$.*
3) *If $x \perp v$, then $\tau_{v,a}(x) = x$. For $a \neq 0$, $x \perp v$ if and only if $\tau_{v,a}(x) = x$.*
4) *$\tau_{v,a}\tau_{v,b} = \tau_{v,a+b}$.*
5) *$\tau_{v,a}^{-1} = \tau_{v,-a}$.*
6) *For any symplectic transformation σ,*

$$\sigma\tau_{v,a}\sigma^{-1} = \tau_{\sigma v,a}$$

7) *For $b \in F^*$,*

$$\tau_{bv,a} = \tau_{v,ab^2}$$ □

Note that if U is a subspace of V and if $\tau_{u,a}$ is a symplectic transvection on U, then, by definition, $u \in U$. However, the formula

$$\tau_{u,a}(x) = x + a\langle x, u\rangle u$$

also defines a symplectic transvection on V, where x ranges over V. Moreover, for any $z \in U^{\perp}$, we have $\tau_{u,a} z = z$ and so $\tau_{u,a}$ is the identity on $U^{\perp}$.

We now wish to prove that any symplectic transformation on a nonsingular symplectic geometry V is the product of symplectic transvections. The proof is not difficult, but it is a bit lengthy, so we break it up into parts. Our first goal is to show that we can get from any hyperbolic pair to any other hyperbolic pair using a product of symplectic transvections.

Let us say that two *hyperbolic pairs* (x,y) and (w,z) are **connected** if there is a product μ of symplectic transvections that carries x to w and y to z and write

$$\mu\colon (x,y) \mapsto (w,z)$$

or $(x,y) \leftrightarrow (w,z)$. It is clear that connectedness is an equivalence relation on the set of hyperbolic pairs.

Theorem 11.18 *In a nonsingular symplectic geometry V, every pair of hyperbolic pairs are connected.*
Proof. Note first that if $\langle s,t\rangle \neq 0$, then $s \neq t$ and so

$$\tau_{s-t,a}s = s + a\langle s, s-t\rangle(s-t) = s - a\langle s,t\rangle(s-t)$$

Taking $a = 1/\langle s,t\rangle$ gives $\tau_{s-t,a}s = t$. Therefore, if (s,u) is hyperbolic, then we can always find a vector x for which

$$(s,u) \leftrightarrow (t,x)$$

namely, $x = \tau_{s-t,a}u$. Also, if both (s,u) and (t,u) are hyperbolic, then

$$(s,u) \leftrightarrow (t,u)$$

since $\langle s-t,u\rangle = 0$ and so $\tau_{s-t,a}u = u$.

Actually, these statements are still true if $\langle s,t\rangle = 0$. For in this case, there is a nonzero vector y for which $\langle s,y\rangle \neq 0$ and $\langle t,y\rangle \neq 0$. This follows from the fact that there is an $f \in V^*$ for which $fs \neq 0$ and $ft \neq 0$ and so the Riesz vector R_f is such a vector. Therefore, if (s,u) is hyperbolic, then

$$(s,u) \leftrightarrow (y, \tau_{s-y,a}u) \leftrightarrow (t, \tau_{y-t,a'}\tau_{s-y,a}u)$$

and if both (s,u) and (t,u) are hyperbolic, then

$$(s,u) \leftrightarrow (y,u) \leftrightarrow (t,u)$$

Hence, transitivity gives the same result as in the case $\langle s,t\rangle \neq 0$.

Finally, if (u_1,u_2) and (v_1,v_2) are hyperbolic, then there is a y for which

$$(u_1,u_2) \leftrightarrow (v_1,y) \leftrightarrow (v_1,v_2)$$

and so transitivity shows that $(u_1,u_2) \leftrightarrow (v_1,v_2)$.$\square$

We can now show that the symplectic transvections generate the symplectic group.

Theorem 11.19 *Every symplectic transformation on a nonsingular symplectic geometry V is the product of symplectic transvections.*

Proof. Let μ be a symplectic transformation on V. We proceed by induction on $d = \dim(V)$. If $d = 2$, then $V = H = \operatorname{span}(u, z)$ is a hyperbolic plane and Theorem 11.18 implies that there is a product τ of symplectic transvections on V for which

$$\tau\colon (u, z) \mapsto (\mu u, \mu z)$$

This proves the result if $d = 2$. Assume that the result holds for all dimensions less than d and let $\dim(V) = d$.

Now,

$$V = H \odot \mathcal{K}$$

where $H = \operatorname{span}(u, z)$ and $\mathcal{K}$ is a symplectic geometry of dimension less than that of V. As before, there is a product τ of symplectic transvections on V for which

$$\tau\colon (u, z) \mapsto (\mu u, \mu z)$$

and so

$$\tau|_H = \mu|_H$$

Note that $\tau^{-1}\mu H = H$ and so Theorem 11.9 implies that $\tau^{-1}\mu(H^{\perp}) = H^{\perp}$. Since $\dim(H^{\perp}) < \dim(H)$, the inductive hypothesis applied to the symplectic transformation $\tau^{-1}\mu$ on $H^{\perp}$ implies that there is a product π of symplectic transvections on $H^{\perp}$ for which $\pi = \tau^{-1}\mu$. As remarked earlier, π is also a product of symplectic transvections on V that is the identity on H and so

$$\pi|_H = \iota_H \quad \text{and} \quad \mu = \tau\pi \text{ on } H^{\perp}$$

Thus, $\mu = \tau\pi$ on both H and on $H^{\perp}$ and so $\mu = \tau\pi$ is a product of symplectic transvections on V.$\square$

The Structure of Orthogonal Geometries: Orthogonal Bases

We have seen that no interesting (that is, not totally degenerate) symplectic geometries have orthogonal bases. By contrast, almost all interesting orthogonal geometries V have orthogonal bases.

To understand why, it is convenient to group the orthogonal geometries into two classes: those that are also symplectic and those that are not. The reason is that all orthogonal *nonsymplectic* geometries have orthogonal bases, as we will see. However, an orthogonal *symplectic* geometry has an orthogonal basis if and only if it is totally degenerate. Furthermore, we have seen that if $\operatorname{char}(F) \neq 2$, then all orthogonal symplectic geometries are totally degenerate and so all such geometries have orthogonal bases. But if $\operatorname{char}(F) = 2$, then there are orthogonal symplectic geometries that are not totally degenerate and therefore do not have orthogonal bases.

Thus, if we exclude orthogonal symplectic geometries when $\operatorname{char}(F) = 2$, we can say that every orthogonal geometry has an orthogonal basis.

If a metric vector space V has an orthogonal basis, the natural next step is to look for an orthonormal basis. However, if V is singular, then there is a nonzero vector $v \in V^{\perp}$ and such a vector can never be a linear combination of vectors from an orthonormal basis $\{u_1, \ldots, u_n\}$, since the coefficients in such a linear combination are $\langle v, u_i \rangle = 0$.

However, even if V is nonsingular, orthonormal bases do not always exist and the question of how close we can come to such an orthonormal basis depends on the nature of the base field. We will examine this issue in three cases: algebraically closed fields, the field of real numbers and finite fields.

We should also mention that even when V has an orthogonal basis, the Gram–Schmidt orthogonalization process may not apply to produce such a basis, because even nonsingular orthogonal geometries may have isotropic vectors, and so division by $\langle u, u \rangle$ is problematic.

For example, consider an orthogonal hyperbolic plane $H = \operatorname{span}(u, v)$ and assume that $\operatorname{char}(F) \neq 2$. Thus, u and v are isotropic and $\langle u, v \rangle = 1$. The vector u cannot be extended to an orthogonal basis using the Gram–Schmidt process, since $\{u, au + bv\}$ is orthogonal if and only if $b = 0$. However, H does have an orthogonal basis, namely, $\{u + v, u - v\}$.

Orthogonal Bases

Let V be an orthogonal geometry. As we have discussed, if V is also symplectic, then V has an orthogonal basis if and only if it is totally degenerate. Moreover, when $\operatorname{char}(F) \neq 2$, all orthogonal symplectic geometries are totally degenerate and so all orthogonal symplectic geometries have an orthogonal basis.

If V is orthogonal but not symplectic, then V contains a nonisotropic vector u_1, the subspace $\operatorname{span}(u_1)$ is nonsingular and

$$V = \operatorname{span}(u_1) \odot V_1$$

where $V_1 = \operatorname{span}(u_1)^{\perp}$. If V_1 is not symplectic, then we may decompose it to get

$$V = \operatorname{span}(u_1) \odot \operatorname{span}(u_2) \odot V_2$$

This process may be continued until we reach a decomposition

$$V = \text{span}(u_1) \odot \cdots \odot \text{span}(u_k) \odot U$$

where U is symplectic as well as orthogonal. (This includes the case $U = \{0\}$.) Let $\mathcal{B} = (u_1, \ldots, u_k)$.

If $\text{char}(F) \neq 2$, then U is totally degenerate. Thus, if $\mathcal{C}$ is a basis for U, then the union $\mathcal{B} \cup \mathcal{C}$ is an orthogonal basis for V. If $\text{char}(F) = 2$, then $U = \mathcal{H} \odot \text{rad}(U)$, where $\mathcal{H}$ is hyperbolic and so

$$V = \text{span}(u_1) \odot \cdots \odot \text{span}(u_k) \odot \mathcal{H} \odot \text{rad}(U)$$

where $\text{rad}(U)$ is totally degenerate and the u_i are nonisotropic. If $\mathcal{C} = (x_1, y_1, \ldots, x_m, y_m)$ is a hyperbolic basis for $\mathcal{H}$ and $\mathcal{D} = (z_1, \ldots, z_m)$ is an ordered basis for $\text{rad}(U)$, then the union

$$\mathcal{E} = \mathcal{B} \cup \mathcal{C} \cup \mathcal{D} = (u_1, \ldots, u_k, x_1, y_1, \ldots, x_m, y_m, z_1, \ldots, z_m)$$

is an ordered orthogonal basis for V. However, we can do better (in some sense).

The following lemma says that when $\text{char}(F) = 2$, a pair of isotropic basis vectors, such as x_i, y_i, can be replaced by a pair of nonisotropic basis vectors, when coupled with a nonisotropic basis vector, such as u_k.

Lemma 11.20 *Suppose that* $\text{char}(F) = 2$. *Let* W *be a three-dimensional orthogonal geometry of the form*

$$W = \text{span}(v) \odot \text{span}(x, y)$$

where v *is nonisotropic and* $H_1 = \text{span}(x, y)$ *is a hyperbolic plane. Then*

$$W = \text{span}(v_1) \odot \text{span}(v_2) \odot \text{span}(v_3)$$

where each v_i *is nonisotropic.*

Proof. It is straightforward to check that if $\langle v, v \rangle = a$, then the vectors

$$\begin{aligned} v_1 &= u + x + y \\ v_2 &= u + ax \\ v_3 &= u + (1 - a)x + y \end{aligned}$$

are linearly independent and mutually orthogonal. Details are left to the reader.□

Using the previous lemma, we can replace the vectors $\{u_k, x_1, y_1\}$ with the nonisotropic vectors $\{v_k, v_{k+1}, v_{k+2}\}$, while retaining orthogonality. Moreover, the replacement process can continue until the isotropic vectors are absorbed, leaving an orthogonal basis of nonisotropic vectors.

Let us summarize.

Theorem 11.21 *Let V be an orthogonal geometry.*

1) *If V is also symplectic, then V has an orthogonal basis if and only if it is totally degenerate. When $\operatorname{char}(F) \neq 2$, all orthogonal symplectic geometries have an orthogonal basis, but this is not the case when $\operatorname{char}(F) = 2$.*
2) *If V is not symplectic, then V has an ordered orthogonal basis $\mathcal{B} = (u_1, \ldots, u_k, z_1, \ldots, z_m)$ for which $\langle u_i, u_i \rangle = a_i \neq 0$ and $\langle z_i, z_i \rangle = 0$. Hence, $M_{\mathcal{B}}$ has the diagonal form*

$$M_{\mathcal{B}} = \begin{bmatrix} a_1 & & & & & \\ & \ddots & & & & \\ & & a_k & & & \\ & & & 0 & & \\ & & & & \ddots & \\ & & & & & 0 \end{bmatrix}$$

with $k = \operatorname{rk}(M_{\mathcal{B}})$ nonzero entries on the diagonal.□

As a corollary, we get a nice theorem about symmetric matrices.

Corollary 11.22 *Let M be a symmetric matrix and assume that M is not alternate if $\operatorname{char}(F) = 2$. Then M is congruent to a diagonal matrix.*□

The Classification of Orthogonal Geometries: Canonical Forms

We now want to consider the question of improving upon Theorem 11.21. The diagonal matrices of this theorem do not form a set of canonical forms for congruence. In fact, if $r_1, \ldots, r_k$ are nonzero scalars, then the matrix of V with respect to the basis $\mathcal{C} = (r_1 u_1, \ldots, r_k u_n, z_1, \ldots, z_m)$ is

$$M_{\mathcal{C}} = \begin{bmatrix} r_1^2 a_1 & & & & & \\ & \ddots & & & & \\ & & r_k^2 a_k & & & \\ & & & 0 & & \\ & & & & \ddots & \\ & & & & & 0 \end{bmatrix} \tag{11.2}$$

Hence, $M_{\mathcal{B}}$ and $M_{\mathcal{C}}$ are congruent diagonal matrices. Thus, by a simple change of basis, we can multiply any diagonal entry by a nonzero square in F.

The determination of a set of canonical forms for symmetric (nonalternate when $\operatorname{char}(F) = 2$) matrices under congruence depends on the properties of the base field. Our plan is to consider three types of base fields: algebraically closed fields, the real field $\mathbb{R}$ and finite fields. Here is a preview of the forthcoming results.

1) When the base field F is algebraically closed, there is an ordered basis $\mathcal{B}$ for which

$$M_{\mathcal{B}} = Z_{k,m} = \begin{bmatrix} 1 & & & & & \\ & \ddots & & & & \\ & & 1 & & & \\ & & & 0 & & \\ & & & & \ddots & \\ & & & & & 0 \end{bmatrix}$$

If V is nonsingular, then $M_{\mathcal{B}}$ is an identity matrix and V has an orthonormal basis.

2) Over the real base field, there is an ordered basis $\mathcal{B}$ for which

$$M_{\mathcal{B}} = \mathrm{Z}_{p,m,k} = \begin{bmatrix} 1 & & & & & & & & \\ & \ddots & & & & & & & \\ & & 1 & & & & & & \\ & & & -1 & & & & & \\ & & & & \ddots & & & & \\ & & & & & -1 & & & \\ & & & & & & 0 & & \\ & & & & & & & \ddots & \\ & & & & & & & & 0 \end{bmatrix}$$

3) If F is a finite field, there is an ordered basis $\mathcal{B}$ for which

$$M_{\mathcal{B}} = \mathrm{Z}_{k,m}(d) = \begin{bmatrix} 1 & & & & & & \\ & \ddots & & & & & \\ & & 1 & & & & \\ & & & d & & & \\ & & & & 0 & & \\ & & & & & \ddots & \\ & & & & & & 0 \end{bmatrix}$$

where d is unique up to multiplication by a square and if $\mathrm{char}(F) = 2$, then we can take $d = 1$.

Now let us turn to the details.

Algebraically Closed Fields

If F is algebraically closed, then for every $r \in F$, the polynomial $x^2 - r$ has a root in F, that is, every element of F has a square root in F. Therefore, we may choose $r_i = 1/\sqrt{a_i}$ in (11.2), which leads to the following result.

Theorem 11.23 *Let V be an orthogonal geometry over an algebraically closed field F. Provided that V is not symplectic as well when* $\mathrm{char}(F)=2$, *then V has an ordered orthogonal basis $\mathcal{B}=(u_1,\dots,u_k,z_1,\dots,z_m)$ for which $\langle u_i,u_i\rangle=1$ and $\langle z_i,z_i\rangle=0$. Hence, $M_{\mathcal{B}}$ has the diagonal form*

$$M_{\mathcal{B}}=Z_{k,m}=\begin{bmatrix}1 & & & & & \\ & \ddots & & & & \\ & & 1 & & & \\ & & & 0 & & \\ & & & & \ddots & \\ & & & & & 0\end{bmatrix}$$

with k ones and m zeros on the diagonal. In particular, if V is nonsingular, then V has an orthonormal basis.□

The matrix version of Theorem 11.23 follows.

Theorem 11.24 *Let $\mathcal{S}_n$ be the set of all $n\times n$ symmetric matrices over an algebraically closed field F. If* $\mathrm{char}(F)=2$, *we restrict $\mathcal{S}_n$ to the set of all symmetric matrices with at least one nonzero entry on the main diagonal.*

1) *Any matrix M in $\mathcal{S}_n$ is congruent to a unique matrix of the form $Z_{k,m}$, in fact, $k=\mathrm{rk}(M)$ and $m=n-\mathrm{rk}(M)$.*
2) *The set of all matrices of the form $Z_{k,m}$ for $k+m=n$ is a set of canonical forms for congruence on $\mathcal{S}_n$.*
3) *The rank of a matrix is a complete invariant for congruence on $\mathcal{S}_n$.*□

The Real Field $\mathbb{R}$

If $F=\mathbb{R}$, we can choose $r_i=1/\sqrt{|a_i|}$, so that all nonzero diagonal elements in (11.2) will be either 0, 1 or -1.

Theorem 11.25 (Sylvester's law of inertia) *Any real orthogonal geometry V has an ordered orthogonal basis*

$$\mathcal{B}=(u_1,\dots,u_p,v_1,\dots,v_m,z_1,\dots,z_k)$$

for which $\langle u_i,u_i\rangle=1$, $\langle v_i,v_i\rangle=-1$ and $\langle z_i,z_i\rangle=0$. Hence, the matrix $M_{\mathcal{B}}$ has the diagonal form

$$M_{\mathcal{B}} = Z_{p,m,k} = \begin{bmatrix} 1 & & & & & & & \\ & \ddots & & & & & & \\ & & 1 & & & & & \\ & & & -1 & & & & \\ & & & & \ddots & & & \\ & & & & & -1 & & \\ & & & & & & 0 & & \\ & & & & & & & \ddots & \\ & & & & & & & & 0 \end{bmatrix}$$

with p ones, m negative ones and k zeros on the diagonal. $\square$

Here is the matrix version of Theorem 11.25.

Theorem 11.26 *Let $\mathcal{S}_n$ be the set of all $n \times n$ symmetric matrices over the real field $\mathbb{R}$.*

1) *Any matrix in $\mathcal{S}_n$ is congruent to a unique matrix of the form $Z_{p,m,k}$, for some p, m and $k = n - p - m$.*
2) *The set of all matrices of the form $Z_{p,m,k}$ for $p + m + k = n$ is a set of canonical forms for congruence on $\mathcal{S}_n$.*
3) *Let $M \in \mathcal{S}_n$ and let M be congruent to $Z_{p,m,k}$. Then $p + m$ is the rank of M and $p - m$ is the* **signature** *of M and the triple (p, m, k) is the* **inertia** *of M. The pair (p, m), or equivalently the pair $(p + m, p - m)$, is a complete invariant under congruence on $\mathcal{S}_n$.*

Proof. We need only prove the uniqueness statement in part 1). Let

$$\mathcal{B} = (u_1, \ldots, u_p, v_1, \ldots, v_m, z_1, \ldots, z_k)$$

and

$$\mathcal{C} = (u_1', \ldots, u_{p'}', v_1', \ldots, v_{m'}' z_1', \ldots, z_{k'}')$$

be ordered bases for which the matrices $M_{\mathcal{B}}$ and $M_{\mathcal{C}}$ have the form shown in Theorem 11.25. Since the rank of these matrices must be equal, we have $p + m = p' + m'$ and so $k = k'$.

If $x \in \text{span}(u_1, \ldots, u_p)$ and $x \neq 0$, then

$$\langle x, x \rangle = \left\langle \sum r_i u_i, \sum r_j u_j \right\rangle = \sum_{i,j} r_i r_j \langle u_i, u_j \rangle = \sum_{i,j} r_i r_j \delta_{i,j} = \sum r_i^2 > 0$$

On the other hand, if $y \in \text{span}(v_1', \ldots, v_{m'}')$ and $y \neq 0$, then

$$\langle y, y \rangle = \left\langle \sum s_i v_i', \sum s_j v_j' \right\rangle = \sum_{i,j} s_i s_j \langle v_i', v_j' \rangle = -\sum_{i,j} s_i s_j \delta_{i,j} = -\sum s_i^2 < 0$$

Hence, if $y \in \text{span}(v_1', \ldots, v_{m'}', z_1', \ldots, z_{k'}')$ then $\langle y, y \rangle \leq 0$. It follows that

$$\operatorname{span}(u_1,\ldots,u_p)\cap\operatorname{span}(v'_1,\ldots,v'_{m'},z'_1,\ldots,z'_{k'})=\{0\}$$

and so

$$p+(n-p')\leq n$$

that is, $p\leq p'$. By symmetry, $p'\leq p$ and so $p=p'$. Finally, since $k=k'$, it follows that $m=m'$.□

Finite Fields

To deal with the case of finite fields, we must know something about the distribution of squares in finite fields, as well as the possible values of the scalars $\langle v,v\rangle$.

Theorem 11.27 *Let F_q be a finite field with q elements.*
1) *If* $\operatorname{char}(F_q)=2$*, then every element of F_q is a square.*
2) *If* $\operatorname{char}(F_q)\neq 2$*, then exactly half of the nonzero elements of F_q are squares, that is, there are $(q-1)/2$ nonzero squares in F_q. Moreover, if x is any nonsquare in F_q, then all nonsquares have the form r^2x, for some $r\in F_q$.*

Proof. Write $F=F_q$, let F^* be the subgroup of all nonzero elements in F and let

$$(F^*)^2=\{a^2\mid a\in F^*\}$$

be the subgroup of all nonzero squares in F. The *Frobenius map* $\phi\colon F^*\to(F^*)^2$ defined by $\phi(a)=a^2$ is a surjective group homomorphism, with kernel

$$\ker(\phi)=\{a\in F\mid a^2=1\}=\{-1,1\}$$

If $\operatorname{char}(F)=2$, then $\ker(\phi)=\{1\}$ and so ϕ is bijective and $|F^*|=|(F^*)^2|$, which proves part 1). If $\operatorname{char}(F)\neq 2$, then $|\ker(\phi)|=2$ and so $|F^*|=2|(F^*)^2|$, which proves the first part of part 2). We leave proof of the last statement to the reader.□

Definition *A bilinear form on V is* **universal** *if for any nonzero $c\in F$ there exists a vector $v\in V$ for which $\langle v,v\rangle=c$.*□

Theorem 11.28 *Let V be an orthogonal geometry over a finite field F with* $\operatorname{char}(F)\neq 2$ *and assume that V has a nonsingular subspace of dimension at least* 2*. Then the bilinear form of V is universal.*

Proof. Theorem 11.21 implies that V contains two linearly independent vectors u and v for which

$$\langle u,u\rangle=a\neq 0,\ \langle v,v\rangle=b\neq 0,\ \langle u,v\rangle=0$$

Given any $c \in F$, we want to find α and β for which

$$c = \langle \alpha u + \beta v, \alpha u + \beta v \rangle = a\alpha^2 + b\beta^2$$

or

$$a\alpha^2 = c - b\beta^2$$

If $A = \{a\alpha^2 \mid \alpha \in F\}$, then $|A| = (q+1)/2$, since there are $(q-1)/2$ nonzero squares α^2, along with $\alpha = 0$. If $B = \{c - b\beta^2 \mid \beta \in F\}$, then for the same reasons $|B| = (q+1)/2$. It follows that $A \cap B$ cannot be the empty set and so there exist α and β for which $a\alpha^2 = c - b\beta^2$.□

Now we can proceed with the business at hand.

Theorem 11.29 *Let V be an orthogonal geometry over a finite field F and assume that V is not symplectic if* $\operatorname{char}(F) = 2$. *If* $\operatorname{char}(F) \neq 2$, *then let d be a fixed nonsquare in F. For any nonzero $a \in F$, write*

$$X_k(a) = \begin{bmatrix} 1 & & & & & & \\ & \ddots & & & & & \\ & & 1 & & & & \\ & & & a & & & \\ & & & & 0 & & \\ & & & & & \ddots & \\ & & & & & & 0 \end{bmatrix}$$

where $\operatorname{rk}(X_k(a)) = k$.

1) *If* $\operatorname{char}(F) = 2$, *then there is an ordered basis $\mathcal{B}$ for which $M_{\mathcal{B}} = X_k(1)$.*
2) *If* $\operatorname{char}(F) \neq 2$, *then there is an ordered basis $\mathcal{B}$ for which $M_{\mathcal{B}}$ equals $X_k(1)$ or $X_k(d)$.*

Proof. We can dispose of the case $\operatorname{char}(F) = 2$ quite easily: Referring to (11.2), since every element of F has a square root, we may take $r_i = (\sqrt{a_i})^{-1}$.

If $\operatorname{char}(F) \neq 2$, then Theorem 11.21 implies that there is an ordered orthogonal basis

$$\mathcal{B} = (u_1, \ldots, u_k, z_1, \ldots, z_m)$$

for which $\langle u_i, u_i \rangle = a_i \neq 0$ and $\langle z_i, z_i \rangle = 0$. Hence, $M_{\mathcal{B}}$ has the diagonal form

$$M_{\mathcal{B}} = \begin{bmatrix} a_1 & & & & & \\ & \ddots & & & & \\ & & a_k & & & \\ & & & 0 & & \\ & & & & \ddots & \\ & & & & & 0 \end{bmatrix}$$

Now consider the nonsingular orthogonal geometry $V_1 = \text{span}(u_1, u_2)$. According to Theorem 11.28, the form is universal when restricted to V_1. Hence, there exists a $v_1 \in V_1$ for which $\langle v_1, v_1 \rangle = 1$.

Now, $v_1 = ru_1 + su_2$ for $r, s \in F$ not both 0, and we may swap u_1 and u_2 if necessary to ensure that $r \neq 0$. Hence,

$$\mathcal{B}_1 = (v_1, u_2, \ldots, u_k, z_1, \ldots, z_m)$$

is an ordered basis for V for which the matrix $M_{\mathcal{B}_1}$ is diagonal and has a 1 in the upper left entry. We can repeat the process with the subspace $V_2 = \text{span}(v_2, v_3)$. Continuing in this way, we can find an ordered basis

$$\mathcal{C} = (v_1, v_2, \ldots, v_k, z_1, \ldots, z_m)$$

for which $M_{\mathcal{C}} = X_k(a)$ for some nonzero $a \in F$. Now, if a is a square in F, then we can replace v_k by $(1/\sqrt{a})v_k$ to get a basis $\mathcal{D}$ for which $M_{\mathcal{D}} = X_k(1)$. If a is not a square in F, then $a = r^2 d$ for some $r \in F$ and so replacing v_k by $(1/r)v_k$ gives a basis $\mathcal{D}$ for which $M_{\mathcal{D}} = X_k(d)$.□

Theorem 11.30 *Let $\mathcal{S}_n$ be the set of all $n \times n$ symmetric matrices over a finite field F. If $\text{char}(F) = 2$, we restrict $\mathcal{S}_n$ to the set of all symmetric matrices with at least one nonzero entry on the main diagonal.*

1) *If $\text{char}(F) = 2$, then any matrix in $\mathcal{S}_n$ is congruent to a unique matrix of the form $X_k(1)$ and the matrices $\{X_k(1) \mid k = 0, \ldots, n\}$ form a set of canonical forms for $\mathcal{S}_n$ under congruence. Also, the rank is a complete invariant.*
2) *If $\text{char}(F) \neq 2$, let d be a fixed nonsquare in F. Then any matrix $\mathcal{S}_n$ is congruent to a unique matrix of the form $X_k(1)$ or $X_k(d)$. The set $\{X_k(1), X_k(d) \mid k = 0, \ldots, n\}$ is a set of canonical forms for congruence on $\mathcal{S}_n$. (Thus, there are exactly two congruence classes for each rank k.)*□

The Orthogonal Group

Having "settled" the classification question for orthogonal geometries over certain types of fields, let us turn to a discussion of the structure-preserving maps, that is, the isometries.

Rotations and Reflections

We begin by examining the matrix of an orthogonal transformation. If $\mathcal{B}$ is an ordered basis for V, then for any $x, y \in V$,

$$\langle x, y \rangle = [x]_{\mathcal{B}}^t M_{\mathcal{B}} [y]_{\mathcal{B}}$$

and so if $\tau \in \mathcal{L}(V)$, then

$$\langle \tau x, \tau y \rangle = [\tau x]_{\mathcal{B}}^t M_{\mathcal{B}} [\tau y]_{\mathcal{B}} = [x]_{\mathcal{B}}^t ([\tau]_{\mathcal{B}}^t M_{\mathcal{B}} [\tau]_{\mathcal{B}}) [y]_{\mathcal{B}}$$

Hence, τ is an orthogonal transformation if and only if

$$[\tau]_{\mathcal{B}}^t M_{\mathcal{B}} [\tau]_{\mathcal{B}} = M_{\mathcal{B}}$$

Taking determinants gives

$$\det(M_{\mathcal{B}}) = \det([\tau]_{\mathcal{B}})^2 \det(M_{\mathcal{B}})$$

Therefore, if V is nonsingular, then

$$\det([\tau]_{\mathcal{B}}) = \pm 1$$

Since the determinant is an invariant under similarity, we have the following theorem.

Theorem 11.31 *Let τ be an orthogonal transformation on a nonsingular orthogonal geometry V.*

1) $\det([\tau]_{\mathcal{B}})$ *is the same for all ordered bases $\mathcal{B}$ for V and*

$$\det([\tau]_{\mathcal{B}}) = \pm 1$$

This determinant is called the **determinant** *of τ and denoted by* $\det(\tau)$.

2) *If* $\det(\tau) = 1$, *then τ is called a* **rotation** *and if* $\det(\tau) = -1$, *then τ is called a* **reflection**.
3) *The set $\mathcal{O}^+(V)$ of rotations is a subgroup of the orthogonal group $\mathcal{O}(V)$ and the determinant map* $\det\colon \mathcal{O}(V) \to \{-1, 1\}$ *is an epimorphism with kernel $\mathcal{O}^+(V)$. Hence, if* $\operatorname{char}(F) \neq 2$, *then $\mathcal{O}^+(V)$ is a normal subgroup of $\mathcal{O}(V)$ of index* 2.□

Symmetries

Recall again that for a real inner product space, a reflection H_u is defined as an operator for which

$$H_u u = -u,\ H_u w = w \text{ for all } w \in \langle u \rangle^{\perp}$$

and that

$$H_u x = x - \frac{2\langle x, u \rangle}{\langle u, u \rangle} u$$

In particular, if $\operatorname{char}(F) \neq 2$ and $u \in V$ is nonisotropic, then $\operatorname{span}(u)$ is nonsingular and so

$$V = \operatorname{span}(u) \odot \operatorname{span}(u)^{\perp}$$

Then the reflection H_u is well-defined and, in the context of general orthogonal geometries, is called the **symmetry** determined by u and we will denote it by σ_u. We can also write $\sigma_u = -\iota \odot \iota$, that is,

$$\sigma_u(x + y) = -x + y$$

for all $x \in \operatorname{span}(u)$ and $y \in \operatorname{span}(u)^{\perp}$.

For real inner product spaces, Theorem 10.16 says that if $\|v\| = \|w\| \neq 0$, then H_{v-w} is the unique reflection sending v to w, that is, $H_{v-w}(v) = w$. In the present context, we must be careful, since symmetries are defined for nonisotropic vectors only. Here is what we can say.

Theorem 11.32 *Let V be a nonsingular orthogonal geometry over a field F, with $\mathrm{char}(F) \neq 2$. If $u, v \in V$ are nonisotropic vectors with the same (nonzero) "length," that is, if*

$$\langle u, u \rangle = \langle v, v \rangle \neq 0$$

then there exists a symmetry σ for which

$$\sigma u = v \quad \text{or} \quad \sigma u = -v$$

Proof. Since u and v are nonisotropic, one of $u - v$ or $u + v$ must also be nonisotropic, for otherwise, since $u - v$ and $u + v$ are orthogonal, their sum $2u$ would also be isotropic. If $u + v$ is nonisotropic, then

$$\sigma_{u+v}(u + v) = -(u + v)$$

and

$$\sigma_{u+v}(u - v) = u - v$$

and so $\sigma_{u+v}u = -v$. On the other hand, if $u - v$ is nonisotropic, then

$$\sigma_{u-v}(u - v) = -(u - v)$$

and

$$\sigma_{u-v}(u + v) = u + v$$

and so $\sigma_{u-v}u = v$.□

Recall that an operator on a real inner product space is unitary if and only if it is a product of reflections. Here is the generalization to nonsingular orthogonal geometries.

Theorem 11.33 *Let V be a nonsingular orthogonal geometry over a field F with $\mathrm{char}(F) \neq 2$. A linear transformation τ on V is an orthogonal transformation if and only if τ is the product of symmetries on V.*

Proof. The proof is by induction on $d = \dim(V)$. If $d = 1$, then $V = \mathrm{span}(v)$ where $\langle v, v \rangle \neq 0$. Let $\tau v = \alpha v$ for $\alpha \in F$. Since τ is unitary

$$\alpha^2 \langle v, v \rangle = \langle \alpha v, \alpha v \rangle = \langle \tau v, \tau v \rangle = \langle v, v \rangle$$

and so $\alpha = \pm 1$. If $\alpha = 1$, then τ is the identity, which is equal to σ_v^2. On the other hand, if $\alpha = -1$ then $\tau = \sigma_v$. In either case, τ is a product of symmetries.

Assume now that the theorem is true for dimensions less than d and let $\dim(V) = d$. Let $v \in V$ be nonisotropic. Since $\langle \tau v, \tau v \rangle = \langle v, v \rangle \neq 0$, Theorem 11.32 implies the existence of a symmetry σ on V for which

$$\sigma(\tau v) = \epsilon v$$

where $\epsilon = \pm 1$. Thus, $\sigma\tau = \pm\iota$ on $\text{span}(v)$. Since Theorem 11.9 implies that $\text{span}(v)^{\perp}$ is $\sigma\tau$-invariant, we may apply the induction hypothesis to $\sigma\tau$ on $\text{span}(v)^{\perp}$ to get

$$\sigma\tau|_{\text{span}(v)^{\perp}} = \sigma_{w_1} \cdots \sigma_{w_k} = \rho$$

where $w_i \in \text{span}(v)^{\perp}$ and each σ_{w_i} is a symmetry on $\text{span}(v)^{\perp}$. But each σ_{w_i} can be extended to a symmetry on V by setting $\sigma_{w_i} v = v$. Assume that $\overline{\rho}$ is the extension of ρ to V, where $\rho = \iota$ on $\text{span}(v)$. Hence, $\sigma\tau = \overline{\rho}$ on $\text{span}(v)^{\perp}$ and $\sigma\tau = \epsilon\overline{\rho}$ on $\text{span}(v)$.

If $\epsilon = 1$, then $\sigma\tau = \overline{\rho}$ on V and so $\tau = \sigma\overline{\rho}$, which completes the proof. If $\epsilon = -1$, then $\sigma\tau = \sigma_v\overline{\rho}$ on $\text{span}(v)^{\perp}$ since σ_v is the identity on $\text{span}(v)^{\perp}$ and $\sigma\tau = \sigma_v\overline{\rho}$ on $\text{span}(v)$. Hence, $\sigma\tau = \sigma_v\overline{\rho}$ on V and so $\tau = \sigma\sigma_v\overline{\rho}$ on V.□

The Witt Theorems for Orthogonal Geometries

We are now ready to consider the Witt theorems for orthogonal geometries.

Theorem 11.34 (Witt's cancellation theorem) *Let V and W be isometric nonsingular orthogonal geometries over a field F with* $\text{char}(F) \neq 2$. *Suppose that*

$$V = S \odot S^{\perp} \quad \textit{and} \quad W = T \odot T^{\perp}$$

Then

$$S \approx T \Rightarrow S^{\perp} \approx T^{\perp}$$

Proof. First, we prove that it is sufficient to consider the case $V = W$. Suppose that the result holds when $V = W$ and that $\mu: V \to W$ is an isometry. Then

$$\mu(S) \odot \mu(S^{\perp}) = \mu(S \odot S^{\perp}) = \mu V = W = T \odot T^{\perp}$$

Furthermore, $\mu S \approx S \approx T$. We can therefore apply the theorem to W to get

$$S^{\perp} \approx \mu(S^{\perp}) \approx T^{\perp}$$

as desired. To prove the theorem when $V = W$, assume that

$$V = S \odot S^{\perp} = T \odot T^{\perp}$$

where S and T are nonsingular and $S \approx T$. Let $\tau: S \to T$ be an isometry. We proceed by induction on $\dim(S)$.

Suppose first that $\dim(S) = 1$ and that $S = \text{span}(s)$. Since

$$\langle \tau s, \tau s \rangle = \langle s, s \rangle \neq 0$$

Theorem 11.32 implies that there is a symmetry σ for which $\sigma s = \epsilon \tau s$ where $\epsilon = \pm 1$. Hence, σ is an isometry of V for which $T = \sigma S$ and Theorem 11.9 implies that $T^\perp = \sigma(S^\perp)$. Thus, $\sigma|_{S^\perp}$ is the desired isometry.

Now suppose the theorem is true for $\dim(S) < k$ and let $\dim(S) = k$. Let $\tau: S \to T$ be an isometry. Since S is nonsingular, we can choose a nonisotropic vector $s \in S$ and write $S = \text{span}(s) \odot U$, where U is nonsingular. It follows that

$$V = S \odot S^\perp = \text{span}(s) \odot U \odot S^\perp$$

and

$$V = T \odot T^\perp = \tau(\text{span}(s)) \odot \tau U \odot T^\perp$$

Now we may apply the one-dimensional case to deduce that

$$U \odot S^\perp \approx \tau U \odot T^\perp$$

If $\sigma: U \odot S^\perp \to \tau U \odot T^\perp$ is an isometry, then

$$\sigma U \odot \sigma(S^\perp) = \sigma(U \odot S^\perp) = \tau U \odot T^\perp$$

But $\sigma U \approx \tau U$ and since $\dim(\sigma U) = \dim(U) < k$, the induction hypothesis implies that $S^\perp \approx \sigma(S^\perp) \approx T^\perp$.□

As we have seen, Witt's extension theorem is a corollary of Witt's cancellation theorem.

Theorem 11.35 (Witt's extension theorem) *Let V and V' be isometric nonsingular orthogonal geometries over a field F, with* $\text{char}(F) \neq 2$. *Suppose that U is a subspace of V and*

$$\tau: U \to \tau U \subseteq V'$$

is an isometry. Then τ can be extended to an isometry from V to V'.□

Maximal Hyperbolic Subspaces of an Orthogonal Geometry

We have seen that any orthogonal geometry V can be written in the form

$$V = U \odot \text{rad}(V)$$

where U is nonsingular. Nonsingular spaces are better behaved than singular ones, but they can still possess isotropic vectors.

We can improve upon the preceding decomposition by noticing that if $u \in U$ is isotropic, then Theorem 11.10 implies that $\text{span}(u)$ can be "captured" in a hyperbolic plane $H = \text{span}(u, x)$. Then we can write

$$V = H \odot H^{\perp_U} \odot \text{rad}(V)$$

where $H^{\perp_U}$ is the orthogonal complement of H in U and has "one fewer" isotropic vector. In order to generalize this process, we first discuss maximal totally degenerate subspaces.

Maximal Totally Degenerate Subspaces

Let V be a nonsingular orthogonal geometry over a field F, with $\text{char}(F) \neq 2$. Suppose that U and U' are maximal totally degenerate subspaces of V. We claim that $\dim(U) = \dim(U')$. For if $\dim(U) \leq \dim(U')$, then there is a vector space isomorphism $\tau: U \to \tau U \subseteq U'$, which is also an isometry, since U and U' are totally degenerate. Thus, Witt's extension theorem implies the existence of an isometry $\overline{\tau}: V \to V$ that extends τ. In particular, $\overline{\tau}^{-1}(U')$ is a totally degenerate space that contains U and so $\overline{\tau}^{-1}(U') = U$, which shows that $\dim(U) = \dim(U')$.

Theorem 11.36 *Let V be a nonsingular orthogonal geometry over a field F, with* $\text{char}(F) \neq 2$.
1) *All maximal totally degenerate subspaces of V have the same dimension, which is called the* **Witt index** *of V and is denoted by $w(V)$.*
2) *Any totally degenerate subspace of V of dimension $w(V)$ is maximal.*□

Maximal Hyperbolic Subspaces

We can prove by a similar argument that all maximal hyperbolic subspaces of V have the same dimension. Let

$$\mathcal{H}_{2k} = H_1 \odot \cdots \odot H_k$$

and

$$\mathcal{K}_{2m} = K_1 \odot \cdots \odot K_m$$

be maximal hyperbolic subspaces of V and suppose that $H_i = \text{span}(u_i, v_i)$ and $K_i = \text{span}(x_i, y_i)$. We may assume that $\dim(\mathcal{H}) \leq \dim(\mathcal{K})$.

The linear map $\tau: \mathcal{H} \to \mathcal{K}$ defined by

$$\tau u_i = x_i, \; \tau v_i = y_i$$

is clearly an isometry from $\mathcal{H}$ to $\tau\mathcal{H}$. Thus, Witt's extension theorem implies the existence of an isometry $\overline{\tau}: V \to V$ that extends τ. In particular, $\overline{\tau}^{-1}(\mathcal{K})$ is a hyperbolic space that contains $\mathcal{H}$ and so $\overline{\tau}^{-1}(\mathcal{K}) = \mathcal{H}$. It follows that $\dim(\mathcal{K}) = \dim(\mathcal{H})$.

It is not hard to see that the maximum dimension $h(V)$ of a hyperbolic subspace of V is $2w(V)$, where $w(V)$ is the Witt index of V. First, the nonsingular extension of a maximal totally degenerate subspace U_w of V is a hyperbolic space of dimension $2w(V)$ and so $h(V) \geq 2w(V)$. On the other hand, there is a totally degenerate subspace U_k contained in any hyperbolic space $\mathcal{H}_{2k}$ and so $k \leq w(V)$, that is, $\dim(\mathcal{H}_{2k}) \leq 2w(V)$. Hence $h(V) \leq 2w(V)$ and so $h(V) = 2w(V)$.

Theorem 11.37 *Let V be a nonsingular orthogonal geometry over a field F, with* $\operatorname{char}(F) \neq 2$.

1) *All maximal hyperbolic subspaces of V have dimension $2w(V)$.*
2) *Any hyperbolic subspace of dimension $2w(V)$ must be maximal.*
3) *The Witt index of a hyperbolic space $\mathcal{H}_{2k}$ is k.*□

The Anisotropic Decomposition of an Orthogonal Geometry

If $\mathcal{H}$ is a maximal hyperbolic subspace of V, then

$$V = \mathcal{H} \odot \mathcal{H}^{\perp}$$

Since $\mathcal{H}$ is maximal, $\mathcal{H}^{\perp}$ is anisotropic, for if $u \in \mathcal{H}^{\perp}$ were isotropic, then the nonsingular extension of $\mathcal{H} \odot \operatorname{span}(u)$ would be a hyperbolic space strictly larger than $\mathcal{H}$.

Thus, we arrive at the following decomposition theorem for orthogonal geometries.

Theorem 11.38 (The anisotropic decomposition of an orthogonal geometry) *Let $V = U \odot \operatorname{rad}(V)$ be an orthogonal geometry over F, with* $\operatorname{char}(F) \neq 2$. *Let $\mathcal{H}$ be a maximal hyperbolic subspace of U, where $\mathcal{H} = \{0\}$ if U has no isotropic vectors. Then*

$$V = S \odot \mathcal{H} \odot \operatorname{rad}(V)$$

where S is anisotropic, $\mathcal{H}$ is hyperbolic of dimension $2w(V)$ and $\operatorname{rad}(V)$ *is totally degenerate.*□

Exercises

1. Let U, W be subspaces of a metric vector space V. Show that
   a) $U \subseteq W \Rightarrow W^{\perp} \subseteq U^{\perp}$
   b) $U \subseteq U^{\perp\perp}$
   c) $U^{\perp} = U^{\perp\perp\perp}$
2. Let U, W be subspaces of a metric vector space V. Show that
   a) $(U + W)^{\perp} = U^{\perp} \cap W^{\perp}$
   b) $(U \cap W)^{\perp} = U^{\perp} + W^{\perp}$
3. Prove that the following are equivalent:
   a) V is nonsingular
   b) $\langle u, x \rangle = \langle v, x \rangle$ for all $x \in V$ implies $u = v$

4. Show that a metric vector space V is nonsingular if and only if the matrix $M_{\mathcal{B}}$ of the form is nonsingular, for every ordered basis $\mathcal{B}$.
5. Let V be a finite-dimensional vector space with a bilinear form $\langle , \rangle$. We do *not* assume that the form is symmetric or alternate. Show that the following are equivalent:
   a) $\{v \in V \mid \langle v, w \rangle = 0 \text{ for all } w \in V\} = 0$
   b) $\{v \in V \mid \langle w, v \rangle = 0 \text{ for all } w \in V\} = 0$

   Hint: Consider the singularity of the matrix of the form.
6. Find a diagonal matrix congruent to

$$\begin{bmatrix} 1 & 2 & 3 \\ 2 & 0 & 1 \\ 3 & 1 & -1 \end{bmatrix}$$

7. Prove that the matrices

$$I_2 = \begin{bmatrix} 1 & 0 \\ 0 & 1 \end{bmatrix} \text{ and } M = \begin{bmatrix} 5 & 0 \\ 0 & 5 \end{bmatrix}$$

   are congruent over the base field $F = \mathbb{Q}$ of rational numbers. Find an invertible matrix P such that $P^t I_2 P = M$.
8. Let V be an orthogonal geometry over a field F with $\text{char}(F) \neq 2$. We wish to construct an orthogonal basis $\mathcal{O} = (u_1, \ldots, u_n)$ for V, starting with any generating set $\mathcal{G} = (v_1, \ldots, v_n)$. Justify the following steps, essentially due to Lagrange. We may assume that V is not totally degenerate.
   a) If $\langle v_i, v_i \rangle \neq 0$ for some i, then let $u_1 = v_i$. Otherwise, there are indices $i \neq j$ for which $\langle v_i, v_j \rangle \neq 0$. Let $u_1 = v_i + v_j$.
   b) Assume we have found an ordered set of vectors $\mathcal{O}_k = (u_1, \ldots, u_k)$ that form an orthogonal basis for a subspace V_k of V and that none of the u_i's are isotropic. Then $V = V_k \odot V_k^{\perp}$.
   c) For each $v_i \in \mathcal{G}$, let

$$w_i = v_i - \sum_{j=1}^{k} \frac{\langle v_i, u_j \rangle}{\langle u_j, u_j \rangle} u_j$$

   Then the vectors w_i span $V_k^{\perp}$. If $V_k^{\perp}$ is totally degenerate, take any basis for $V_k^{\perp}$ and append it to $\mathcal{O}_k$. Otherwise, repeat step a) on $V_k^{\perp}$ to get another vector u_{k+1} and let $\mathcal{O}_{k+1} = (u_1, \ldots, u_{k+1})$. Eventually, we arrive at an orthogonal basis $\mathcal{O}_n$ for V.
9. Prove that orthogonal hyperbolic planes may be characterized as two-dimensional nonsingular orthogonal geometries that have exactly two one-dimensional totally isotropic (equivalently: totally degenerate) subspaces.
10. Prove that a two-dimensional nonsingular orthogonal geometry is a hyperbolic plane if and only if its discriminant is $F^2(-1)$.
11. Does Minkowski space contain any isotropic vectors? If so, find them.
12. Is Minkowski space isometric to Euclidean space $\mathbb{R}^4$?

13. If $\langle , \rangle$ is a symmetric bilinear form on V and $\operatorname{char}(F) \neq 2$, show that $Q(x) = \langle x, x \rangle / 2$ is a quadratic form.
14. Let V be a vector space over a field F, with ordered basis $\mathcal{B} = (v_1, \ldots, v_n)$. Let $p(x_1, \ldots, x_n)$ be a *homogeneous* polynomial of degree d over F, that is, a polynomial each of whose terms has degree d. The ***d*-form** defined by p is the function from V to F defined as follows. If $v = \Sigma a_i v_i$, then

$$p(v) = p(a_1, \ldots, a_n)$$

(We use the same notation for the form and the polynomial.) Prove that 2-forms are the same as quadratic forms.
15. Show that τ is an isometry on V if and only if $Q(\tau v) = Q(v)$ where Q is the quadratic form associated with the bilinear form on V. (Assume that $\operatorname{char}(F) \neq 2$.)
16. Show that a quadratic form Q on V satisfies the parallelogram law:

$$Q(x+y) + Q(x-y) = 2[Q(x) + Q(y)]$$

17. Show that if V is a nonsingular orthogonal geometry over a field F, with $\operatorname{char}(F) \neq 2$, then any totally isotropic subspace of V is also a totally degenerate space.
18. Is it true that $V = \operatorname{rad}(V) \odot \operatorname{rad}(V)^{\perp}$?
19. Let V be a nonsingular symplectic geometry and let $\tau_{v,a}$ be a symplectic transvection. Prove that
    a) $\tau_{v,a}\tau_{v,b} = \tau_{v,a+b}$
    b) For any symplectic transformation σ,

$$\sigma \tau_{v,a} \sigma^{-1} = \tau_{\sigma v,a}$$

    c) For $b \in F^*$,

$$\tau_{bv,a} = \tau_{v,ab^2}$$

    d) For a fixed $v \neq 0$, the map $a \mapsto \tau_{v,a}$ is an isomorphism from the additive group of F onto the group $\{\tau_{v,a} \mid a \in F\} \subseteq \operatorname{Sp}(V)$.
20. Prove that if x is any nonsquare in a finite field F_q, then all nonsquares have the form $r^2 x$, for some $r \in F$. Hence, the product of any two nonsquares in F_q is a square.
21. Formulate Sylvester's law of inertia in terms of quadratic forms on V.
22. Show that a two-dimensional space is a hyperbolic plane if and only if it is nonsingular and contains an isotropic vector. Assume that $\operatorname{char}(F) \neq 2$.
23. Prove directly that a hyperbolic plane in an orthogonal geometry cannot have an orthogonal basis when $\operatorname{char}(F) = 2$.
24. a) Let U be a subspace of V. Show that the inner product $\langle x+U, y+U \rangle = \langle x, y \rangle$ on the quotient space V/U is well-defined if and only if $U \subseteq \operatorname{rad}(V)$.
    b) If $U \subseteq \operatorname{rad}(V)$, when is V/U nonsingular?
25. Let $V = N \odot S$, where N is a totally degenerate space.

a) Prove that $N = \text{rad}(V)$ if and only if S is nonsingular.
b) If S is nonsingular, prove that $S \approx V/\text{rad}(V)$.

26. Let $\dim(V) = \dim(W)$. Prove that $V/\text{rad}(V) \approx W/\text{rad}(W)$ implies $V \approx W$.
27. Let $V = S \odot T$. Prove that
    a) $\text{rad}(V) = \text{rad}(S) \odot \text{rad}(T)$
    b) $V/\text{rad}(V) \approx S/\text{rad}(S) \odot T/\text{rad}(T)$
    c) $\dim(\text{rad}(V)) = \dim(\text{rad}(S)) + \dim(\text{rad}(T))$
    d) V is nonsingular if and only if S and T are both nonsingular.
28. Let V be a nonsingular metric vector space. Because the Riesz representation theorem is valid in V, we can define the adjoint τ^* of a linear map $\tau \in \mathcal{L}(V)$ exactly as in the case of real inner product spaces. Prove that τ is an isometry if and only if it is bijective and unitary (that is, $\tau\tau^* = \iota$).
29. If $\text{char}(F) \neq 2$, prove that $\tau \in \mathcal{L}(V, W)$ is an isometry if and only if it is bijective and $\langle \tau v, \tau v \rangle = \langle v, v \rangle$ for all $v \in V$.
30. Let $\mathcal{B} = \{v_1, \ldots, v_n\}$ be a basis for V. Prove that $\tau \in \mathcal{L}(V, W)$ is an isometry if and only if it is bijective and $\langle \tau v_i, \tau v_j \rangle = \langle v_i, v_j \rangle$ for all i, j.
31. Let τ be a linear operator on a metric vector space V. Let $\mathcal{B} = (v_1, \ldots, v_n)$ be an ordered basis for V and let $M_{\mathcal{B}}$ be the matrix of the form relative to $\mathcal{B}$. Prove that τ is an isometry if and only if
$$[\tau]_{\mathcal{B}}^t M_{\mathcal{B}} [\tau]_{\mathcal{B}} = M_{\mathcal{B}}$$
32. Let V be a nonsingular orthogonal geometry and let $\tau \in \mathcal{L}(V)$ be an isometry.
    a) Show that $\dim(\ker(\iota - \tau)) = \dim(\text{im}(\iota - \tau)^{\perp})$.
    b) Show that $\ker(\iota - \tau) = \text{im}(\iota - \tau)^{\perp}$. How would you describe $\ker(\iota - \tau)$ in words?
    c) If τ is a symmetry, what is $\dim(\ker(\iota - \tau))$?
    d) Can you characterize symmetries by means of $\dim(\ker(\iota - \tau))$?
33. A linear transformation $\tau \in \mathcal{L}(V)$ is called **unipotent** if $\tau - \iota$ is nilpotent. Suppose that V is a nonisotropic metric vector space and that τ is unipotent and isometric. Show that $\tau = \iota$.
34. Let V be a hyperbolic space of dimension $2m$ and let U be a hyperbolic subspace of V of dimension $2k$. Show that for each $k \le j \le m$, there is a hyperbolic subspace $\mathcal{H}_{2j}$ of V for which $U \subseteq \mathcal{H}_{2j} \subseteq V$.
35. Let $\text{char}(F) \neq 2$. Prove that if X is a totally degenerate subspace of an orthogonal geometry V, then $\dim(X) \le \dim(V)/2$.
36. Prove that an orthogonal geometry V of dimension n is a hyperbolic space if and only if V is nonsingular, n is even and V contains a totally degenerate subspace of dimension $n/2$.
37. Prove that a symplectic transformation has determinant equal to 1.

Chapter 12
Metric Spaces

The Definition

In Chapter 9, we studied the basic properties of real and complex inner product spaces. Much of what we did does not depend on whether the space in question is finite-dimensional or infinite-dimensional. However, as we discussed in Chapter 9, the presence of an inner product and hence a metric, on a vector space, raises a host of new issues related to convergence. In this chapter, we discuss briefly the concept of a metric space. This will enable us to study the convergence properties of real and complex inner product spaces.

A metric space is not an algebraic structure. Rather it is designed to model the abstract properties of distance.

Definition *A* **metric space** *is a pair* (M, d)*, where* M *is a nonempty set and* $d: M \times M \to \mathbb{R}$ *is a real-valued function, called a* **metric** *on* M*, with the following properties. The expression* $d(x, y)$ *is read "the distance from* x *to* y*."*

1) **(Positive definiteness)** *For all* $x, y \in M$,

$$d(x, y) \geq 0$$

and $d(x, y) = 0$ *if and only if* $x = y$.

2) **(Symmetry)** *For all* $x, y \in M$,

$$d(x, y) = d(y, x)$$

3) **(Triangle inequality)** *For all* $x, y, z \in M$,

$$d(x, y) \leq d(x, z) + d(z, y)$$

□

As is customary, when there is no cause for confusion, we simply say "let M be a metric space."

Example 12.1 Any nonempty set M is a metric space under the **discrete metric**, defined by

$$d(x,y)=\begin{cases}0 & \text{if } x=y\\ 1 & \text{if } x\neq y\end{cases}$$ □

Example 12.2

1) The set $\mathbb{R}^n$ is a metric space, under the metric defined for $x=(x_1,\ldots,x_n)$ and $y=(y_1,\ldots,y_n)$ by

$$d(x,y)=\sqrt{(x_1-y_1)^2+\cdots+(x_n-y_n)^2}$$

This is called the **Euclidean metric** on $\mathbb{R}^n$. We note that $\mathbb{R}^n$ is also a metric space under the metric

$$d_1(x,y)=|x_1-y_1|+\cdots+|x_n-y_n|$$

Of course, $(\mathbb{R}^n,d)$ and $(\mathbb{R}^n,d_1)$ are different metric spaces.

2) The set $\mathbb{C}^n$ is a metric space under the **unitary metric**

$$d(x,y)=\sqrt{|x_1-y_1|^2+\cdots+|x_n-y_n|^2}$$

where $x=(x_1,\ldots,x_n)$ and $y=(y_1,\ldots,y_n)$ are in $\mathbb{C}^n$. □

Example 12.3

1) The set $C[a,b]$ of all real-valued (or complex-valued) continuous functions on $[a,b]$ is a metric space, under the metric

$$d(f,g)=\sup_{x\in[a,b]}|f(x)-g(x)|$$

We refer to this metric as the **sup metric**.

2) The set $C[a,b]$ of all real-valued (or complex-valued) continuous functions on $[a,b]$ is a metric space, under the metric

$$d_1(f(x),g(x))=\int_a^b|f(x)-g(x)|\,dx$$ □

Example 12.4 Many important sequence spaces are metric spaces. We will often use boldface italic letters to denote sequences, as in $\boldsymbol{x}=(x_n)$ and $\boldsymbol{y}=(y_n)$.

1) The set $\ell^\infty_{\mathbb{R}}$ of all bounded sequences of real numbers is a metric space under the metric defined by

$$d(\boldsymbol{x},\boldsymbol{y})=\sup_n|x_n-y_n|$$

The set $\ell^\infty_{\mathbb{C}}$ of all bounded complex sequences, with the same metric, is also a metric space. As is customary, we will usually denote both of these spaces by ℓ^∞.

2) For $p \geq 1$, let ℓ^p be the set of all sequences $\boldsymbol{x} = (x_n)$ of real (or complex) numbers for which

$$\sum_{n=1}^{\infty} |x_n|^p < \infty$$

We define the p-**norm** of $\boldsymbol{x}$ by

$$\|\boldsymbol{x}\|_p = \left(\sum_{n=1}^{\infty} |x_n|^p \right)^{1/p}$$

Then ℓ^p is a metric space, under the metric

$$d(\boldsymbol{x}, \boldsymbol{y}) = \|\boldsymbol{x} - \boldsymbol{y}\|_p = \left(\sum_{n=1}^{\infty} |x_n - y_n|^p \right)^{1/p}$$

The fact that ℓ^p is a metric follows from some rather famous results about sequences of real or complex numbers, whose proofs we leave as (well-hinted) exercises.

Hölder's inequality Let $p, q \geq 1$ and $p + q = pq$. If $\boldsymbol{x} \in \ell^p$ and $\boldsymbol{y} \in \ell^q$, then the product sequence $\boldsymbol{xy} = (x_n y_n)$ is in ℓ^1 and

$$\|\boldsymbol{xy}\|_1 \leq \|\boldsymbol{x}\|_p \|\boldsymbol{y}\|_q$$

that is,

$$\sum_{n=1}^{\infty} |x_n y_n| \leq \left(\sum_{n=1}^{\infty} |x_n|^p \right)^{1/p} \left(\sum_{n=1}^{\infty} |y_n|^q \right)^{1/q}$$

A special case of this (with $p = q = 2$) is the **Cauchy–Schwarz inequality**

$$\sum_{n=1}^{\infty} |x_n y_n| \leq \sqrt{\sum_{n=1}^{\infty} |x_n|^2} \sqrt{\sum_{n=1}^{\infty} |y_n|^2}$$

Minkowski's inequality For $p \geq 1$, if $\boldsymbol{x}, \boldsymbol{y} \in \ell^p$ then the sum $\boldsymbol{x} + \boldsymbol{y} = (x_n + y_n)$ is in ℓ^p and

$$\|\boldsymbol{x} + \boldsymbol{y}\|_p \leq \|\boldsymbol{x}\|_p + \|\boldsymbol{y}\|_p$$

that is,

$$\left(\sum_{n=1}^{\infty} |x_n + y_n|^p \right)^{1/p} \leq \left(\sum_{n=1}^{\infty} |x_n|^p \right)^{1/p} + \left(\sum_{n=1}^{\infty} |y_n|^p \right)^{1/p} \qquad \square$$

If M is a metric space under a metric d, then any nonempty subset S of M is also a metric under the restriction of d to $S \times S$. The metric space S thus obtained is called a **subspace** of M.

Open and Closed Sets

Definition *Let M be a metric space. Let $x_0 \in M$ and let r be a positive real number.*

1) The **open ball** *centered at x_0, with radius r, is*

$$B(x_0, r) = \{x \in M \mid d(x, x_0) < r\}$$

2) The **closed ball** *centered at x_0, with radius r, is*

$$\overline{B}(x_0, r) = \{x \in M \mid d(x, x_0) \le r\}$$

3) The **sphere** *centered at x_0, with radius r, is*

$$S(x_0, r) = \{x \in M \mid d(x, x_0) = r\}$$

□

Definition *A subset S of a metric space M is said to be* **open** *if each point of S is the center of an open ball that is contained completely in S. More specifically, S is open if for all $x \in S$, there exists an $r > 0$ such that $B(x, r) \subseteq S$. Note that the empty set is open. A set $T \subseteq M$ is* **closed** *if its complement T^c in M is open.*□

It is easy to show that an open ball is an open set and a closed ball is a closed set. If $x \in M$, we refer to any open set S containing x as an **open neighborhood** of x. It is also easy to see that a set is open if and only if it contains an open neighborhood of each of its points.

The next example shows that it is possible for a set to be both open and closed, or neither open nor closed.

Example 12.5 In the metric space $\mathbb{R}$ with the usual Euclidean metric, the open balls are just the open intervals

$$B(x_0, r) = (x_0 - r, x_0 + r)$$

and the closed balls are the closed intervals

$$\overline{B}(x_0, r) = [x_0 - r, x_0 + r]$$

Consider the half-open interval $S = (a, b]$, for $a < b$. This set is not open, since it contains no open ball centered at $b \in S$ and it is not closed, since its complement $S^c = (-\infty, a] \cup (b, \infty)$ is not open, since it contains no open ball about a.

Observe also that the empty set is both open and closed, as is the entire space $\mathbb{R}$. (Although we will not do so, it is possible to show that these are the only two sets that are both open and closed in $\mathbb{R}$.)□

It is not our intention to enter into a detailed discussion of open and closed sets, the subject of which belongs to the branch of mathematics known as *topology*. In order to put these concepts in perspective, however, we have the following result, whose proof is left to the reader.

Theorem 12.1 *The collection $\mathcal{O}$ of all open subsets of a metric space M has the following properties:*
1) $\emptyset \in \mathcal{O}$, $M \in \mathcal{O}$
2) *If* $S, T \in \mathcal{O}$ *then* $S \cap T \in \mathcal{O}$
3) *If* $\{S_i \mid i \in K\}$ *is any collection of open sets, then* $\bigcup_{i \in K} S_i \in \mathcal{O}$.□

These three properties form the basis for an axiom system that is designed to generalize notions such as convergence and continuity and leads to the following definition.

Definition *Let X be a nonempty set. A collection $\mathcal{O}$ of subsets of X is called a* **topology** *for X if it has the following properties:*
1) $\emptyset \in \mathcal{O}$, $X \in \mathcal{O}$
2) *If* $S, T \in \mathcal{O}$ *then* $S \cap T \in \mathcal{O}$
3) *If* $\{S_i \mid i \in K\}$ *is any collection of sets in* $\mathcal{O}$, *then* $\bigcup\limits_{i \in K} S_i \in \mathcal{O}$.

We refer to subsets in $\mathcal{O}$ as **open sets** *and the pair $(X, \mathcal{O})$ as a* **topological space**.□

According to Theorem 12.1, the open sets (as we defined them earlier) in a metric space M form a topology for M, called the topology **induced** by the metric.

Topological spaces are the most general setting in which we can define concepts such as convergence and continuity, which is why these concepts are called topological concepts. However, since the topologies with which we will be dealing are induced by a metric, we will generally phrase the definitions of the topological properties that we will need directly in terms of the metric.

Convergence in a Metric Space

Convergence of sequences in a metric space is defined as follows.

Definition *A sequence (x_n) in a metric space M* **converges** *to $x \in M$, written $(x_n) \to x$, if*

$$\lim_{n \to \infty} d(x_n, x) = 0$$

Equivalently, $(x_n) \to x$ if for any $\epsilon > 0$, there exists an $N > 0$ such that

$$n > N \Rightarrow d(x_n, x) < \epsilon$$

or equivalently,

$$n > N \Rightarrow x_n \in B(x, \epsilon)$$

In this case, x is called the **limit** *of the sequence (x_n).* □

If M is a metric space and S is a subset of M, by a *sequence in* S, we mean a sequence whose terms all lie in S. We next characterize closed sets and therefore also open sets, using convergence.

Theorem 12.2 *Let M be a metric space. A subset $S \subseteq M$ is closed if and only if whenever (x_n) is a sequence in S and $(x_n) \to x$, then $x \in S$. In loose terms, a subset S is closed if it is closed under the taking of sequential limits.*

Proof. Suppose that S is closed and let $(x_n) \to x$, where $x_n \in S$ for all n. Suppose that $x \notin S$. Then since $x \in S^c$ and S^c is open, there exists an $\epsilon > 0$ for which $x \in B(x, \epsilon) \subseteq S^c$. But this implies that

$$B(x, \epsilon) \cap \{x_n\} = \emptyset$$

which contradicts the fact that $(x_n) \to x$. Hence, $x \in S$.

Conversely, suppose that S is closed under the taking of limits. We show that S^c is open. Let $x \in S^c$ and suppose to the contrary that no open ball about x is contained in S^c. Consider the open balls $B(x, 1/n)$, for all $n \geq 1$. Since none of these balls is contained in S^c, for each n, there is an $x_n \in S \cap B(x, 1/n)$. It is clear that $(x_n) \to x$ and so $x \in S$. But x cannot be in both S and S^c. This contradiction implies that S^c is open. Thus, S is closed.□

The Closure of a Set

Definition *Let S be any subset of a metric space M. The* **closure** *of S, denoted by* $\mathrm{cl}(S)$, *is the smallest closed set containing S.*□

We should hasten to add that, since the entire space M is closed and since the intersection of any collection of closed sets is closed (exercise), the closure of any set S does exist and is the intersection of all closed sets containing S. The following definition will allow us to characterize the closure in another way.

Definition Let S be a nonempty subset of a metric space M. An element $x \in M$ is said to be a **limit point**, or **accumulation point**, of S if every open ball centered at x meets S at a point other than x itself. Let us denote the set of all limit points of S by $\ell(S)$.□

Here are some key facts concerning limit points and closures.

Theorem 12.3 *Let S be a nonempty subset of a metric space M.*

1) *$x \in \ell(S)$ if and only if there is a sequence (x_n) in S for which $x_n \neq x$ for all n and $(x_n) \to x$.*
2) *S is closed if and only if $\ell(S) \subseteq S$. In words, S is closed if and only if it contains all of its limit points.*
3) $\mathrm{cl}(S) = S \cup \ell(S)$.
4) *$x \in \mathrm{cl}(S)$ if and only if there is a sequence (x_n) in S for which $(x_n) \to x$.*

Proof. For part 1), assume first that $x \in \ell(S)$. For each n, there exists a point $x_n \neq x$ such that $x_n \in B(x, 1/n) \cap S$. Thus, we have

$$d(x_n, x) < 1/n$$

and so $(x_n) \to x$. For the converse, suppose that $(x_n) \to x$, where $x \neq x_n \in S$. If $B(x, r)$ is any ball centered at x, then there is some N such that $n > N$ implies $x_n \in B(x, r)$. Hence, for any ball $B(x, r)$ centered at x, there is a point $x_n \neq x$ such that $x_n \in S \cap B(x, r)$. Thus, x is a limit point of S.

As for part 2), if S is closed, then by part 1), any $x \in \ell(S)$ is the limit of a sequence (x_n) in S and so must be in S. Hence, $\ell(S) \subseteq S$. Conversely, if $\ell(S) \subseteq S$, then S is closed. For if (x_n) is any sequence in S and $(x_n) \to x$, then there are two possibilities. First, we might have $x_n = x$ for some n, in which case $x = x_n \in S$. Second, we might have $x_n \neq x$ for all n, in which case $(x_n) \to x$ implies that $x \in \ell(S) \subseteq S$. In either case, $x \in S$ and so S is closed under the taking of limits, which implies that S is closed.

For part 3), let $T = S \cup \ell(S)$. Clearly, $S \subseteq T$. To show that T is closed, we show that it contains all of its limit points. So let $x \in \ell(T)$. Hence, there is a sequence $(x_n) \in T$ for which $x_n \neq x$ and $(x_n) \to x$. Of course, each x_n is either in S, or is a limit point of S. We must show that $x \in T$, that is, that x is either in S or is a limit point of S.

Suppose for the purposes of contradiction that $x \notin S$ and $x \notin \ell(S)$. Then there is a ball $B(x, r)$ for which $B(x, r) \cap S \neq \emptyset$. However, since $(x_n) \to x$, there must be an $x_n \in B(x, r)$. Since x_n cannot be in S, it must be a limit point of S. Referring to Figure 12.1, if $d(x_n, x) = d < r$, then consider the ball $B(x_n, (r-d)/2)$. This ball is completely contained in $B(x, r)$ and must contain an element y of S, since its center x_n is a limit point of S. But then $y \in S \cap B(x, r)$, a contradiction. Hence, $x \in S$ or $x \in \ell(S)$. In either case, $x \in T = S \cup \ell(S)$ and so T is closed.

Thus, T is closed and contains S and so $\mathrm{cl}(S) \subseteq T$. On the other hand, $T = S \cup \ell(S) \subseteq \mathrm{cl}(S)$ and so $\mathrm{cl}(S) = T$.

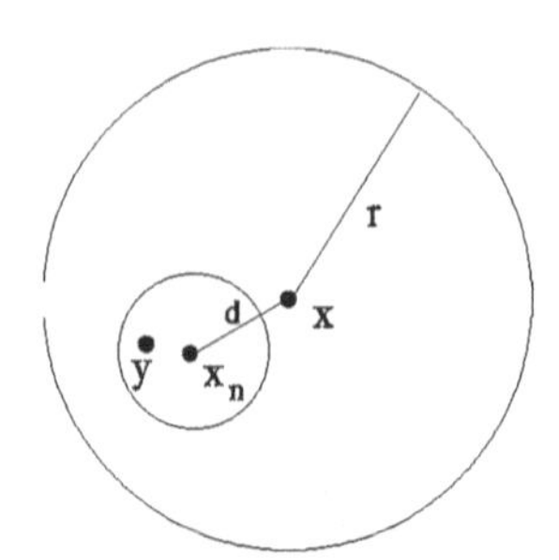

Figure 12.1

For part 4), if $x \in \mathrm{cl}(S)$, then there are two possibilities. If $x \in S$, then the constant sequence (x_n), with $x_n = x$ for all x, is a sequence in S that converges to x. If $x \notin S$, then $x \in \ell(S)$ and so there is a sequence (x_n) in S for which $x_n \neq x$ and $(x_n) \to x$. In either case, there is a sequence in S converging to x. Conversely, if there is a sequence (x_n) in S for which $(x_n) \to x$, then either $x_n = x$ for some n, in which case $x \in S \subseteq \mathrm{cl}(S)$, or else $x_n \neq x$ for all n, in which case $x \in \ell(S) \subseteq \mathrm{cl}(S)$.□

Dense Subsets

The following concept is meant to convey the idea of a subset $S \subseteq M$ being "arbitrarily close" to every point in M.

Definition *A subset S of a metric space M is* **dense** *in M if* $\mathrm{cl}(S) = M$. *A metric space is said to be* **separable** *if it contains a* countable *dense subset.*□

Thus, a subset S of M is dense if every open ball about any point $x \in M$ contains at least one point of S.

Certainly, any metric space contains a dense subset, namely, the space itself. However, as the next examples show, not every metric space contains a countable dense subset.

Example 12.6

1) The real line $\mathbb{R}$ is separable, since the rational numbers $\mathbb{Q}$ form a countable dense subset. Similarly, $\mathbb{R}^n$ is separable, since the set $\mathbb{Q}^n$ is countable and dense.
2) The complex plane $\mathbb{C}$ is separable, as is $\mathbb{C}^n$ for all n.
3) A discrete metric space is separable if and only if it is countable. We leave proof of this as an exercise.□

Example 12.7 The space ℓ^∞ is not separable. Recall that ℓ^∞ is the set of all bounded sequences of real numbers (or complex numbers) with metric

$$d(\boldsymbol{x}, \boldsymbol{y}) = \sup_n |x_n - y_n|$$

To see that this space is not separable, consider the set S of all binary sequences

$$S = \{(x_n) \mid x_i = 0 \text{ or } 1 \text{ for all } i\}$$

This set is in one-to-one correspondence with the set of all subsets of $\mathbb{N}$ and so is uncountable. (It has cardinality $2^{\aleph_0} > \aleph_0$.) Now, each sequence in S is certainly bounded and so lies in ℓ^∞. Moreover, if $\boldsymbol{x} \neq \boldsymbol{y} \in \ell^\infty$, then the two sequences must differ in at least one position and so $d(x, y) = 1$.

In other words, we have a subset S of ℓ^∞ that is uncountable and for which the distance between any two distinct elements is 1. This implies that the balls in the uncountable collection $\{B(s, 1/3) \mid s \in S\}$ are mutually disjoint. Hence, no countable set can meet every ball, which implies that no countable set can be dense in ℓ^∞.□

Example 12.8 The metric spaces ℓ^p are separable, for $p \geq 1$. The set S of all sequences of the form

$$s = (q_1, \dots, q_n, 0, \dots)$$

for all $n > 0$, where the q_i's are rational, is a countable set. Let us show that it is dense in ℓ^p. Any $x \in \ell^p$ satisfies

$$\sum_{n=1}^{\infty} |x_n|^p < \infty$$

Hence, for any $\epsilon > 0$, there exists an N such that

$$\sum_{n=N+1}^{\infty} |x_n|^p < \frac{\epsilon}{2}$$

Since the rational numbers are dense in $\mathbb{R}$, we can find rational numbers q_i for which

$$|x_i - q_i|^p < \frac{\epsilon}{2N}$$

for all $i = 1, \dots, N$. Hence, if $s = (q_1, \dots, q_N, 0, \dots)$, then

$$d(x, s)^p = \sum_{n=1}^{N} |x_n - q_n|^p + \sum_{n=N+1}^{\infty} |x_n|^p < \frac{\epsilon}{2} + \frac{\epsilon}{2} = \epsilon$$

which shows that there is an element of S arbitrarily close to any element of ℓ^p. Thus, S is dense in ℓ^p and so ℓ^p is separable.□

Continuity

Continuity plays a central role in the study of linear operators on infinite-dimensional inner product spaces.

Definition *Let $f: M \to M'$ be a function from the metric space (M, d) to the metric space (M', d'). We say that f is* **continuous at** $x_0 \in M$ *if for any $\epsilon > 0$, there exists a $\delta > 0$ such that*

$$d(x, x_0) < \delta \Rightarrow d'(f(x), f(x_0)) < \epsilon$$

or, equivalently,

$$f\left(B(x_0, \delta)\right) \subseteq B(f(x_0), \epsilon)$$

(See Figure 12.2.) A function is **continuous** *if it is continuous at every $x_0 \in M$.*□

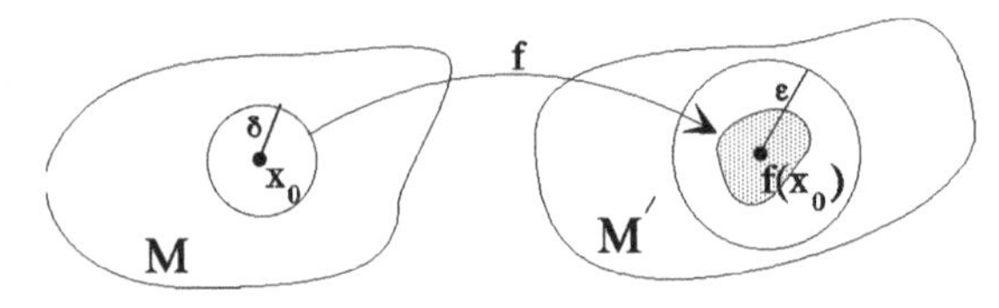

Figure 12.2

We can use the notion of convergence to characterize continuity for functions between metric spaces.

Theorem 12.4 *A function $f: M \to M'$ is continuous if and only if whenever (x_n) is a sequence in M that converges to $x_0 \in M$, then the sequence $(f(x_n))$ converges to $f(x_0)$, in short,*

$$(x_n) \to x_0 \Rightarrow (f(x_n)) \to f(x_0)$$

Proof. Suppose first that f is continuous at x_0 and let $(x_n) \to x_0$. Then, given $\epsilon > 0$, the continuity of f implies the existence of a $\delta > 0$ such that

$$f\left(B(x_0, \delta)\right) \subseteq B(f(x_0), \epsilon)$$

Since $(x_n) \to x$, there exists an $N > 0$ such that $x_n \in B(x_0, \delta)$ for $n > N$ and so

$$n > N \Rightarrow f(x_n) \in B(f(x_0), \epsilon)$$

Thus, $f(x_n) \to f(x_0)$.

Conversely, suppose that $(x_n) \to x_0$ implies $(f(x_n)) \to f(x_0)$. Suppose, for the purposes of contradiction, that f is not continuous at x_0. Then there exists an

$\epsilon > 0$ such that for all $\delta > 0$,

$$f\left(B(x_0, \delta)\right) \not\subseteq B(f(x_0), \epsilon)$$

Thus, for all $n > 0$,

$$f\left(B\left(x_0, \frac{1}{n}\right)\right) \not\subseteq B(f(x_0), \epsilon)$$

and so we may construct a sequence (x_n) by choosing each term x_n with the property that

$$x_n \in B\left(x_0, \frac{1}{n}\right), \text{ but } f(x_n) \notin B(f(x_0), \epsilon)$$

Hence, $(x_n) \to x_0$, but $f(x_n)$ does not converge to $f(x_0)$. This contradiction implies that f must be continuous at x_0.□

The next theorem says that the distance function is a continuous function in both variables.

Theorem 12.5 *Let (M, d) be a metric space. If $(x_n) \to x$ and $(y_n) \to y$, then $d(x_n, y_n) \to d(x, y)$.*
Proof. We leave it as an exercise to show that

$$|d(x_n, y_n) - d(x, y)| \le d(x_n, x) + d(y_n, y)$$

But the right side tends to 0 as $n \to \infty$ and so $d(x_n, y_n) \to d(x, y)$.□

Completeness

The reader who has studied analysis will recognize the following definitions.

Definition *A sequence (x_n) in a metric space M is a* **Cauchy sequence** *if for any $\epsilon > 0$, there exists an $N > 0$ for which*

$$n, m > N \Rightarrow d(x_n, x_m) < \epsilon$$ □

We leave it to the reader to show that any convergent sequence is a Cauchy sequence. When the converse holds, the space is said to be *complete*.

Definition *Let M be a metric space.*
1) *M is said to be* **complete** *if every Cauchy sequence in M converges in M.*
2) *A subspace S of M is* **complete** *if it is complete as a metric space. Thus, S is complete if every Cauchy sequence (s_n) in S converges to an element in S.*□

Before considering examples, we prove a very useful result about completeness of subspaces.

Theorem 12.6 *Let M be a metric space.*

1) *Any complete subspace of M is closed.*
2) *If M is complete, then a subspace S of M is complete if and only if it is closed.*

Proof. To prove 1), assume that S is a complete subspace of M. Let (x_n) be a sequence in S for which $(x_n) \to x \in M$. Then (x_n) is a Cauchy sequence in S and since S is complete, (x_n) must converge to an element of S. Since limits of sequences are unique, we have $x \in S$. Hence, S is closed.

To prove part 2), first assume that S is complete. Then part 1) shows that S is closed. Conversely, suppose that S is closed and let (x_n) be a Cauchy sequence in S. Since (x_n) is also a Cauchy sequence in the complete space M, it must converge to some $x \in M$. But since S is closed, we have $(x_n) \to x \in S$. Hence, S is complete.□

Now let us consider some examples of complete (and incomplete) metric spaces.

Example 12.9 It is well known that the metric space $\mathbb{R}$ is complete. (However, a proof of this fact would lead us outside the scope of this book.) Similarly, the complex numbers $\mathbb{C}$ are complete.□

Example 12.10 The Euclidean space $\mathbb{R}^n$ and the unitary space $\mathbb{C}^n$ are complete. Let us prove this for $\mathbb{R}^n$. Suppose that (x_k) is a Cauchy sequence in $\mathbb{R}^n$, where

$$x_k = (x_{k,1}, \ldots, x_{k,n})$$

Thus,

$$d(x_k, x_m)^2 = \sum_{i=1}^{n} (x_{k,i} - x_{m,i})^2 \to 0 \text{ as } k, m \to \infty$$

and so, for each coordinate position i,

$$(x_{k,i} - x_{m,i})^2 \le d(x_k, x_m)^2 \to 0$$

which shows that the sequence $(x_{k,i})_{k=1,2,\ldots}$ of ith coordinates is a Cauchy sequence in $\mathbb{R}$. Since $\mathbb{R}$ is complete, we must have

$$(x_{k,i}) \to y_i \text{ as } k \to \infty$$

If $y = (y_1, \ldots, y_n)$, then

$$d(x_k, y)^2 = \sum_{i=1}^{n} (x_{k,i} - y_i)^2 \to 0 \text{ as } k \to \infty$$

and so $(x_n) \to y \in \mathbb{R}^n$. This proves that $\mathbb{R}^n$ is complete.□

Example 12.11 The metric space $(C[a,b],d)$ of all real-valued (or complex-valued) continuous functions on $[a,b]$, with metric

$$d(f,g) = \sup_{x\in[a,b]} |f(x) - g(x)|$$

is complete. To see this, we first observe that the limit with respect to d is the uniform limit on $[a,b]$, that is $d(f_n,f) \to 0$ if and only if for any $\epsilon > 0$, there is an $N > 0$ for which

$$n > N \Rightarrow |f_n(x) - f(x)| \le \epsilon \text{ for all } x \in [a,b]$$

Now let (f_n) be a Cauchy sequence in $(C[a,b],d)$. Thus, for any $\epsilon > 0$, there is an N for which

$$m,n > N \Rightarrow |f_n(x) - f_m(x)| \le \epsilon \text{ for all } x \in [a,b] \tag{12.1}$$

This implies that, for each $x \in [a,b]$, the sequence $(f_n(x))$ is a Cauchy sequence of real (or complex) numbers and so it converges. We can therefore define a function f on $[a,b]$ by

$$f(x) = \lim_{n\to\infty} f_n(x)$$

Letting $m \to \infty$ in (12.1), we get

$$n > N \Rightarrow |f_n(x) - f(x)| \le \epsilon \text{ for all } x \in [a,b]$$

Thus, $f_n(x)$ converges to $f(x)$ uniformly. It is well known that the uniform limit of continuous functions is continuous and so $f(x) \in C[a,b]$. Thus, $(f_n(x)) \to f(x) \in C[a,b]$ and so $(C[a,b],d)$ is complete.□

Example 12.12 The metric space $(C[a,b],d_1)$ of all real-valued (or complex-valued) continuous functions on $[a,b]$, with metric

$$d_1(f(x),g(x)) = \int_a^b |f(x) - g(x)|dx$$

is not complete. For convenience, we take $[a,b] = [0,1]$ and leave the general case for the reader. Consider the sequence of functions $f_n(x)$ whose graphs are shown in Figure 12.3. (The definition of $f_n(x)$ should be clear from the graph.)

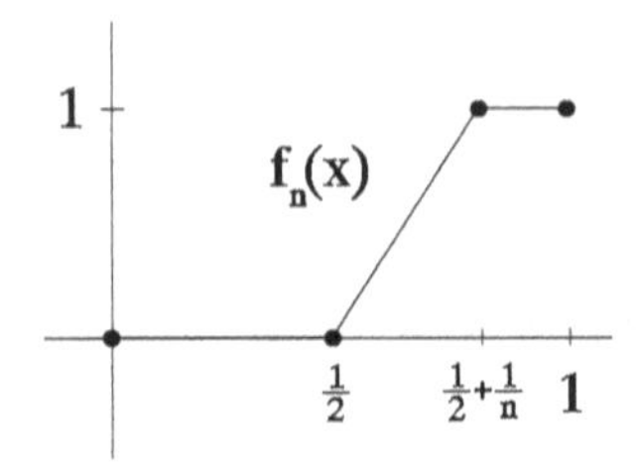

Figure 12.3

We leave it to the reader to show that the sequence $(f_n(x))$ is Cauchy, but does not converge in $(C[0,1],d_1)$. (The sequence converges to a function that is not continuous.)□

Example 12.13 The metric space ℓ^∞ is complete. To see this, suppose that (x_n) is a Cauchy sequence in ℓ^∞, where

$$x_n = (x_{n,1}, x_{n,2}, \ldots)$$

Then, for each coordinate position i, we have

$$|x_{n,i} - x_{m,i}| \le \sup_j |x_{n,j} - x_{m,j}| \to 0 \text{ as } n,m \to \infty \tag{12.2}$$

Hence, for each i, the sequence $(x_{n,i})$ of ith coordinates is a Cauchy sequence in $\mathbb{R}$ (or $\mathbb{C}$). Since $\mathbb{R}$ (or $\mathbb{C}$) is complete, we have

$$(x_{n,i}) \to y_i \text{ as } n \to \infty$$

for each coordinate position i. We want to show that $y = (y_i) \in \ell^\infty$ and that $(x_n) \to y$.

Letting $m \to \infty$ in (12.2) gives

$$\sup_j |x_{n,j} - y_j| \to 0 \text{ as } n \to \infty \tag{12.3}$$

and so, for some n,

$$|x_{n,j} - y_j| < 1 \text{ for all } j$$

and so

$$|y_j| < 1 + |x_{n,j}| \text{ for all } j$$

But since $x_n \in \ell^\infty$, it is a bounded sequence and therefore so is (y_j). That is, $y = (y_j) \in \ell^\infty$. Since (12.3) implies that $(x_n) \to y$, we see that ℓ^∞ is complete.□

Example 12.14 The metric space ℓ^p is complete. To prove this, let (x_n) be a Cauchy sequence in ℓ^p, where

$$x_n = (x_{n,1}, x_{n,2}, \dots)$$

Then, for each coordinate position i,

$$|x_{n,i} - x_{m,i}|^p \le \sum_{j=1}^{\infty} |x_{n,j} - x_{m,j}|^p = d(x_n, x_m)^p \to 0$$

which shows that the sequence $(x_{n,i})$ of ith coordinates is a Cauchy sequence in $\mathbb{R}$ (or $\mathbb{C}$). Since $\mathbb{R}$ (or $\mathbb{C}$) is complete, we have

$$(x_{n,i}) \to y_i \text{ as } n \to \infty$$

We want to show that $y = (y_i) \in \ell^p$ and that $(x_n) \to y$.

To this end, observe that for any $\epsilon > 0$, there is an N for which

$$n, m > N \Rightarrow \sum_{i=1}^{r} |x_{n,i} - x_{m,i}|^p \le \epsilon$$

for all $r > 0$. Now we let $m \to \infty$, to get

$$n > N \Rightarrow \sum_{i=1}^{r} |x_{n,i} - y_i|^p \le \epsilon$$

for all $r > 0$. Letting $r \to \infty$, we get, for any $n > N$,

$$\sum_{i=1}^{\infty} |x_{n,i} - y_i|^p < \epsilon$$

which implies that $(x_n) - y \in \ell^p$ and so $y = y - (x_n) + (x_n) \in \ell^p$ and in addition, $(x_n) \to y$.$\square$

As we will see in the next chapter, the property of completeness plays a major role in the theory of inner product spaces. Inner product spaces for which the induced metric space is complete are called **Hilbert spaces**.

Isometries

A function between two metric spaces that preserves distance is called an isometry. Here is the formal definition.

Definition Let (M, d) and (M', d') be metric spaces. A function $f: M \to M'$ is called an **isometry** if

$$d'(f(x), f(y)) = d(x, y)$$

for all $x, y \in M$. If $f: M \to M'$ is a bijective isometry from M to M', we say that M and M' are **isometric** and write $M \approx M'$.□

Theorem 12.7 *Let $f: (M, d) \to (M', d')$ be an isometry. Then*
1) f is injective
2) f is continuous
3) $f^{-1}: f(M) \to M$ is also an isometry and hence also continuous.
Proof. To prove 1), we observe that

$$f(x) = f(y) \Leftrightarrow d'(f(x), f(y)) = 0 \Leftrightarrow d(x, y) = 0 \Leftrightarrow x = y$$

To prove 2), let $(x_n) \to x$ in M. Then

$$d'(f(x_n), f(x)) = d(x_n, x) \to 0 \text{ as } n \to \infty$$

and so $(f(x_n)) \to f(x)$, which proves that f is continuous. Finally, we have

$$d(f^{-1}(f(x)), f^{-1}(f(y)) = d(x, y) = d'(f(x), f(y))$$

and so $f^{-1}: f(M) \to M$ is an isometry.□

The Completion of a Metric Space

While not all metric spaces are complete, any metric space can be embedded in a complete metric space. To be more specific, we have the following important theorem.

Theorem 12.8 *Let (M, d) be any metric space. Then there is a complete metric space (M', d') and an isometry $\tau: M \to \tau M \subseteq M'$ for which τM is dense in M'. The metric space (M', d') is called a* **completion** *of (M, d). Moreover, (M', d') is unique, up to bijective isometry.*
Proof. The proof is a bit lengthy, so we divide it into various parts. We can simplify the notation considerably by thinking of sequences (x_n) in M as functions $f: \mathbb{N} \to M$, where $f(n) = x_n$.

Cauchy Sequences in M

The basic idea is to let the elements of M' be equivalence classes of Cauchy sequences in M. So let $\text{CS}(M)$ denote the set of all Cauchy sequences in M. If $f, g \in \text{CS}(M)$, then, intuitively speaking, the terms $f(n)$ get closer together as $n \to \infty$ and so do the terms $g(n)$. Therefore, it seems reasonable that $d(f(n), g(n))$ should approach a finite limit as $n \to \infty$. Indeed, since

$$|d(f(n), g(n)) - d(f(m), g(m))| \leq d(f(n), f(m)) + d(g(n), g(m)) \to 0$$

as $n, m \to \infty$ it follows that $d(f(n), g(n))$ is a Cauchy sequence of real numbers, which implies that

$$\lim_{n\to\infty} d(f(n), g(n)) < \infty \tag{12.4}$$

(That is, the limit exists and is finite.)

Equivalence Classes of Cauchy Sequences in M

We would like to define a metric d' on the set $\mathrm{CS}(M)$ by

$$d'(f,g) = \lim_{n\to\infty} d(f(n), g(n))$$

However, it is possible that

$$\lim_{n\to\infty} d(f(n), g(n)) = 0$$

for distinct sequences f and g, so this does not define a metric. Thus, we are led to define an equivalence relation on $\mathrm{CS}(M)$ by

$$f \sim g \Leftrightarrow \lim_{n\to\infty} d(f(n), g(n)) = 0$$

Let $\overline{\mathrm{CS}(M)}$ be the set of all equivalence classes of Cauchy sequences and define, for $\overline{f}$, $\overline{g} \in \overline{\mathrm{CS}(M)}$,

$$d'(\overline{f},\overline{g}) = \lim_{n\to\infty} d(f(n), g(n)) \tag{12.5}$$

where $f \in \overline{f}$ and $g \in \overline{g}$.

To see that d' is well-defined, suppose that $f' \in \overline{f}$ and $g' \in \overline{g}$. Then since $f' \sim f$ and $g' \sim g$, we have

$$|d(f'(n), g'(n)) - d(f(n), g(n))| \le d(f'(n), f(n)) + d(g'(n), g(n)) \to 0$$

as $n \to \infty$. Thus,

$$\begin{aligned} f' \sim f \text{ and } g' \sim g &\Rightarrow \lim_{n\to\infty} d(f'(n), g'(n)) = \lim_{n\to\infty} d(f(n), g(n)) \\ &\Rightarrow d'(f', g') = d'(f, g) \end{aligned}$$

which shows that d' is well-defined. To see that d' is a metric, we verify the triangle inequality, leaving the rest to the reader. If f, g and h are Cauchy sequences, then

$$d(f(n), g(n)) \le d(f(n), h(n)) + d(h(n), g(n))$$

Taking limits gives

$$\lim_{n\to\infty} d(f(n), g(n)) \le \lim_{n\to\infty} d(f(n), h(n)) + \lim_{n\to\infty} d(h(n), g(n))$$

and so

$$d'(\overline{f},\overline{g}) \le d'(\overline{f},\overline{h}) + d'(\overline{h},\overline{g})$$

Embedding (M,d) in (M',d')

For each $x \in M$, consider the constant Cauchy sequence $[x]$, where $[x](n) = x$ for all n. The map $\tau\colon M \to M'$ defined by

$$\tau x = \overline{[x]}$$

is an isometry, since

$$d'(\tau x, \tau y) = d'(\overline{[x]}, \overline{[y]}) = \lim_{n\to\infty} d([x](n), [y](n)) = d(x,y)$$

Moreover, τM is dense in M'. This follows from the fact that we can approximate any Cauchy sequence in M by a constant sequence. In particular, let $\overline{f} \in M'$. Since $f \in \overline{f}$ is a Cauchy sequence, for any $\epsilon > 0$, there exists an N such that

$$n, m \ge N \Rightarrow d(f(n), f(m)) < \epsilon$$

Now, for the constant sequence $[f(N)]$ we have

$$d'\left(\overline{[f(N)]}, \overline{f}\right) = \lim_{n\to\infty} d(f(N), f(n)) \le \epsilon$$

and so τM is dense in M'.

(M',d') Is Complete

Suppose that

$$\overline{f_1}, \overline{f_2}, \overline{f_3}, \ldots$$

is a Cauchy sequence in M'. We wish to find a Cauchy sequence g in M for which

$$d'(\overline{f_k}, \overline{g}) = \lim_{n\to\infty} d(f_k(n), g(n)) \to 0 \text{ as } k \to \infty$$

Since $\overline{f_k} \in M'$ and since τM is dense in M', there is a constant sequence

$$[c_k] = (c_k, c_k, \ldots)$$

for which

$$d'(\overline{f_k}, \overline{[c_k]}) < \frac{1}{k}$$

We can think of c_k as a constant approximation to f_k, with error at most $1/k$. Let g be the sequence of these constant approximations:

$$g(k) = c_k$$

This is a Cauchy sequence in M. Intuitively speaking, since the f_k's get closer to each other as $k \to \infty$, so do the constant approximations. In particular, we have

$$\begin{aligned} d(c_k, c_j) &= d'(\overline{[c_k]}, \overline{[c_j]}) \\ &\le d'(\overline{[c_k]}, \overline{f_k}) + d'(\overline{f_k}, \overline{f_j}) + d'(\overline{f_j}, \overline{[c_j]}) \\ &\le \frac{1}{k} + d'(\overline{f_k}, \overline{f_j}) + \frac{1}{j} \to 0 \end{aligned}$$

as $k, j \to \infty$. To see that $\overline{f_k}$ converges to $\overline{g}$, observe that

$$\begin{aligned} d'(\overline{f_k}, \overline{g}) \le d'(\overline{f_k}, \overline{[c_k]}) + d'(\overline{[c_k]}, \overline{g}) &< \frac{1}{k} + \lim_{n \to \infty} d(c_k, g(n)) \\ &= \frac{1}{k} + \lim_{n \to \infty} d(c_k, c_n) \end{aligned}$$

Now, since g is a Cauchy sequence, for any $\epsilon > 0$, there is an N such that

$$k, n \ge N \Rightarrow d(c_k, c_n) < \epsilon$$

In particular,

$$k \ge N \Rightarrow \lim_{n \to \infty} d(c_k, c_n) \le \epsilon$$

and so

$$k \ge N \Rightarrow d'(\overline{f_k}, \overline{g}) \le \frac{1}{k} + \epsilon$$

which implies that $\overline{f_k} \to g$, as desired.

Uniqueness

Finally, we must show that if (M', d') and (M'', d'') are both completions of (M, d), then $M' \approx M''$. Note that we have bijective isometries

$$\tau: M \to \tau M \subseteq M' \text{ and } \sigma: M \to \sigma M \subseteq M''$$

Hence, the map

$$\rho = \sigma\tau^{-1}: \tau M \to \sigma M$$

is a bijective isometry from τM onto σM, where τM is dense in M'. (See Figure 12.4.)

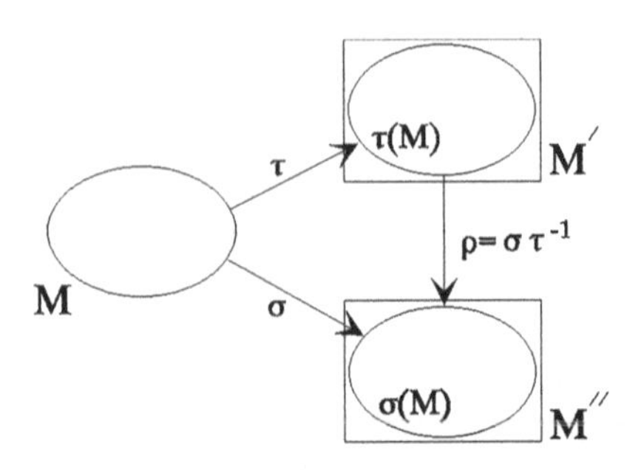

Figure 12.4

Our goal is to show that ρ can be extended to a bijective isometry $\overline{\rho}$ from M' to M''.

Let $x \in M'$. Then there is a sequence (a_n) in τM for which $(a_n) \to x$. Since (a_n) is a Cauchy sequence in τM, $(\rho(a_n))$ is a Cauchy sequence in $\sigma M \subseteq M''$ and since M'' is complete, we have $(\rho(a_n)) \to y$ for some $y \in M''$. Let us define $\overline{\rho}(x) = y$.

To see that $\overline{\rho}$ is well-defined, suppose that $(a_n) \to x$ and $(b_n) \to x$, where both sequences lie in τM. Then

$$d''(\rho(a_n), \rho(b_n)) = d'(a_n, b_n) \to 0 \text{ as } n \to \infty$$

and so $(\rho(a_n))$ and $(\rho(b_n))$ converge to the same element of M'', which implies that $\overline{\rho}(x)$ does not depend on the choice of sequence in τM converging to x. Thus, $\overline{\rho}$ is well-defined. Moreover, if $a \in \tau M$, then the constant sequence $[a]$ converges to a and so $\overline{\rho}(a) = \lim \rho(a) = \rho(a)$, which shows that $\overline{\rho}$ is an extension of ρ.

To see that $\overline{\rho}$ is an isometry, suppose that $(a_n) \to x$ and $(b_n) \to y$. Then $(\rho(a_n)) \to \overline{\rho}(x)$ and $(\rho(b_n)) \to \overline{\rho}(y)$ and since d'' is continuous, we have

$$d''(\overline{\rho}(x), \overline{\rho}(y)) = \lim_{n\to\infty} d''(\rho(a_n), \rho(b_n)) = \lim_{n\to\infty} d'(a_n, b_n) = d'(x, y)$$

Thus, we need only show that $\overline{\rho}$ is surjective. Note first that $\sigma M = \operatorname{im}(\rho) \subseteq \operatorname{im}(\overline{\rho})$. Thus, if $\operatorname{im}(\overline{\rho})$ is closed, we can deduce from the fact that σM is dense in M'' that $\operatorname{im}(\overline{\rho}) = M''$. So, suppose that $(\overline{\rho}(x_n))$ is a sequence in $\operatorname{im}(\overline{\rho})$ and $(\overline{\rho}(x_n)) \to z$. Then $(\overline{\rho}(x_n))$ is a Cauchy sequence and therefore so is (x_n). Thus, $(x_n) \to x \in M'$. But $\overline{\rho}$ is continuous and so $(\overline{\rho}(x_n)) \to \overline{\rho}(x)$, which implies that $\overline{\rho}(x) = z$ and so $z \in \operatorname{im}(\overline{\rho})$. Hence, $\overline{\rho}$ is surjective and $M' \approx M''$.□

Exercises

1. Prove the generalized triangle inequality

$$d(x_1, x_n) \le d(x_1, x_2) + d(x_2, x_3) + \cdots + d(x_{n-1}, x_n)$$

2. a) Use the triangle inequality to prove that

$$|d(x,y) - d(a,b)| \le d(x,a) + d(y,b)$$

   b) Prove that

$$|d(x,z) - d(y,z)| \le d(x,y)$$

3. Let $S \subseteq \ell^\infty$ be the subspace of all binary sequences (sequences of 0's and 1's). Describe the metric on S.
4. Let $M = \{0,1\}^n$ be the set of all binary n-tuples. Define a function $h: S \times S \to \mathbb{R}$ by letting $h(x,y)$ be the number of positions in which x and y differ. For example, $h[(11010),(01001)] = 3$. Prove that h is a metric. (It is called the **Hamming distance function** and plays an important role in the theory of error-correcting codes.)
5. Let $1 \le p < \infty$.
   a) If $\boldsymbol{x} = (x_n) \in \ell^p$ show that $x_n \to 0$
   b) Find a sequence that converges to 0 but is not an element of any ℓ^p for $1 \le p < \infty$.
6. a) Show that if $\boldsymbol{x} = (x_n) \in \ell^p$, then $\boldsymbol{x} \in \ell^q$ for all $q > p$.
   b) Find a sequence $\boldsymbol{x} = (x_n)$ that is in ℓ^p for $p > 1$, but is not in ℓ^1.
7. Show that a subset S of a metric space M is open if and only if S contains an open neighborhood of each of its points.
8. Show that the intersection of any collection of closed sets in a metric space is closed.
9. Let (M,d) be a metric space. The **diameter** of a nonempty subset $S \subseteq M$ is

$$\delta(S) = \sup_{x,y \in S} d(x,y)$$

   A set S is **bounded** if $\delta(S) < \infty$.
   a) Prove that S is bounded if and only if there is some $x \in M$ and $r \in \mathbb{R}$ for which $S \subseteq B(x,r)$.
   b) Prove that $\delta(S) = 0$ if and only if S consists of a single point.
   c) Prove that $S \subseteq T$ implies $\delta(S) \le \delta(T)$.
   d) If S and T are bounded, show that $S \cup T$ is also bounded.
10. Let (M,d) be a metric space. Let d' be the function defined by

$$d'(x,y) = \frac{d(x,y)}{1 + d(x,y)}$$

a) Show that (M, d') is a metric space and that M is bounded under this metric, even if it is not bounded under the metric d.
b) Show that the metric spaces (M, d) and (M, d') have the same open sets.

11. If S and T are subsets of a metric space (M, d), we define the **distance** between S and T by
$$\rho(S, T) = \inf_{x \in S, t \in T} d(x, y)$$
a) Is it true that $\rho(S, T) = 0$ if and only if $S = T$? Is ρ a metric?
b) Show that $x \in \text{cl}(S)$ if and only if $\rho(\{x\}, S) = 0$.
12. Prove that $x \in M$ is a limit point of $S \subseteq M$ if and only if every neighborhood of x meets S in a point other than x itself.
13. Prove that $x \in M$ is a limit point of $S \subseteq M$ if and only if every open ball $B(x, r)$ contains infinitely many points of S.
14. Prove that limits are unique, that is, $(x_n) \to x$, $(x_n) \to y$ implies that $x = y$.
15. Let S be a subset of a metric space M. Prove that $x \in \text{cl}(S)$ if and only if there exists a sequence (x_n) in S that converges to x.
16. Prove that the closure has the following properties:
a) $S \subseteq \text{cl}(S)$
b) $\text{cl}(\text{cl}(S)) = S$
c) $\text{cl}(S \cup T) = \text{cl}(S) \cup \text{cl}(T)$
d) $\text{cl}(S \cap T) \subseteq \text{cl}(S) \cap \text{cl}(T)$
Can the last part be strengthened to equality?
17. a) Prove that the closed ball $\overline{B}(x, r)$ is always a closed subset.
b) Find an example of a metric space in which the closure of an open ball $B(x, r)$ is not equal to the closed ball $\overline{B}(x, r)$.
18. Provide the details to show that $\mathbb{R}^n$ is separable.
19. Prove that $\mathbb{C}^n$ is separable.
20. Prove that a discrete metric space is separable if and only if it is countable.
21. Prove that the metric space $\mathcal{B}[a, b]$ of all bounded functions on $[a, b]$, with metric
$$d(f, g) = \sup_{x \in [a, b]} |f(x) - g(x)|$$
is not separable.
22. Show that a function $f\colon (M, d) \to (M', d')$ is continuous if and only if the inverse image of any open set is open, that is, if and only if $f^{-1}(U) = \{x \in M \mid f(x) \in U\}$ is open in M whenever U is an open set in M'.
23. Repeat the previous exercise, replacing the word open by the word closed.
24. Give an example to show that if $f\colon (M, d) \to (M', d')$ is a continuous function and U is an open set in M, it need not be the case that $f(U)$ is open in M'.

25. Show that any convergent sequence is a Cauchy sequence.
26. If $(x_n) \to x$ in a metric space M, show that any subsequence (x_{n_k}) of (x_n) also converges to x.
27. Suppose that (x_n) is a Cauchy sequence in a metric space M and that some subsequence (x_{n_k}) of (x_n) converges. Prove that (x_n) converges to the same limit as the subsequence.
28. Prove that if (x_n) is a Cauchy sequence, then the set $\{x_n\}$ is bounded. What about the converse? Is a bounded sequence necessarily a Cauchy sequence?
29. Let (x_n) and (y_n) be Cauchy sequences in a metric space M. Prove that the sequence $d_n = d(x_n, y_n)$ converges.
30. Show that the space of all convergent sequences of real numbers (or complex numbers) is complete as a subspace of ℓ^∞.
31. Let $\mathcal{P}$ denote the metric space of all polynomials over $\mathbb{C}$, with metric

$$d(p,q) = \sup_{x \in [a,b]} |p(x) - q(x)|$$

Is $\mathcal{P}$ complete?
32. Let $S \subseteq \ell^\infty$ be the subspace of all sequences with finite support (that is, with a finite number of nonzero terms). Is S complete?
33. Prove that the metric space $\mathbb{Z}$ of all integers, with metric $d(n,m) = |n - m|$, is complete.
34. Show that the subspace S of the metric space $C[a,b]$ (under the sup metric) consisting of all functions $f \in C[a,b]$ for which $f(a) = f(b)$ is complete.
35. If $M \approx M'$ and M is complete, show that M' is also complete.
36. Show that the metric spaces $C[a,b]$ and $C[c,d]$, under the sup metric, are isometric.
37. Prove Hölder's inequality

$$\sum_{n=1}^{\infty} |x_n y_n| \le \left(\sum_{n=1}^{\infty} |x_n|^p \right)^{1/p} \left(\sum_{n=1}^{\infty} |y_n|^q \right)^{1/q}$$

as follows:

a) Show that $s = t^{p-1} \Rightarrow t = s^{q-1}$

b) Let u and v be positive real numbers and consider the rectangle R in $\mathbb{R}^2$ with corners $(0,0)$, $(u,0)$, $(0,v)$ and (u,v), with area uv. Argue geometrically (that is, draw a picture) to show that

$$uv \le \int_0^u t^{p-1} dt + \int_0^v s^{q-1} ds$$

and so

$$uv \le \frac{u^p}{p} + \frac{v^q}{q}$$

c) Now let $X = \Sigma |x_n|^p < \infty$ and $Y = \Sigma |y_n|^q < \infty$. Apply the results of part b) to

$$u = \frac{|x_n|}{X^{1/p}}, \quad v = \frac{|y_n|}{Y^{1/q}}$$

and then sum on n to deduce Hölder's inequality.

38. Prove Minkowski's inequality

$$\left(\sum_{n=1}^{\infty} |x_n + y_n|^p\right)^{1/p} \le \left(\sum_{n=1}^{\infty} |x_n|^p\right)^{1/p} + \left(\sum_{n=1}^{\infty} |y_n|^p\right)^{1/p}$$

as follows:

a) Prove it for $p = 1$ first.

b) Assume $p > 1$. Show that

$$|x_n + y_n|^p \le |x_n||x_n + y_n|^{p-1} + |y_n||x_n + y_n|^{p-1}$$

c) Sum this from $n = 1$ to k and apply Hölder's inequality to each sum on the right, to get

$$\sum_{n=1}^{k} |x_n + y_n|^p$$
$$\le \left\{\left(\sum_{n=1}^{k} |x_n|^p\right)^{1/p} + \left(\sum_{n=1}^{k} |y_n|^p\right)^{1/p}\right\}\left(\sum_{n=1}^{k} |x_n + y_n|^p\right)^{1/q}$$

Divide both sides of this by the last factor on the right and let $n \to \infty$ to deduce Minkowski's inequality.

39. Prove that ℓ^p is a metric space.

Chapter 13
Hilbert Spaces

Now that we have the necessary background on the topological properties of metric spaces, we can resume our study of inner product spaces without qualification as to dimension. As in Chapter 9, we restrict attention to real and complex inner product spaces. Hence F will denote either $\mathbb{R}$ or $\mathbb{C}$.

A Brief Review

Let us begin by reviewing some of the results from Chapter 9. Recall that an inner product space V over F is a vector space V, together with an inner product $\langle , \rangle : V \times V \to F$. If $F = \mathbb{R}$, then the inner product is bilinear and if $F = \mathbb{C}$, the inner product is sesquilinear.

An inner product induces a norm on V, defined by

$$\|v\| = \sqrt{\langle v, v \rangle}$$

We recall in particular the following properties of the norm.

Theorem 13.1

1) **(The Cauchy-Schwarz inequality)** *For all* $u, v \in V$,

$$|\langle u, v \rangle| \le \|u\| \, \|v\|$$

with equality if and only if $u = rv$ *for some* $r \in F$.

2) **(The triangle inequality)** *For all* $u, v \in V$,

$$\|u + v\| \le \|u\| + \|v\|$$

with equality if and only if $u = rv$ *for some* $r \in F$.

3) **(The parallelogram law)**

$$\|u + v\|^2 + \|u - v\|^2 = 2\|u\|^2 + 2\|v\|^2$$

□

We have seen that the inner product can be recovered from the norm, as follows.

Theorem 13.2
1) If V is a real inner product space, then

$$\langle u, v\rangle = \frac{1}{4}(\|u+v\|^2 - \|u-v\|^2)$$

2) If V is a complex inner product space, then

$$\langle u, v\rangle = \frac{1}{4}(\|u+v\|^2 - \|u-v\|^2) + \frac{1}{4}i(\|u+iv\|^2 - \|u-iv\|^2) \quad \square$$

The inner product also induces a metric on V defined by

$$d(u, v) = \|u - v\|$$

Thus, any inner product space is a metric space.

Definition *Let V and W be inner product spaces and let $\tau \in \mathcal{L}(V, W)$.*
1) τ is an **isometry** *if it preserves the inner product, that is, if*

$$\langle \tau u, \tau v\rangle = \langle u, v\rangle$$

for all $u, v \in V$.
2) A bijective isometry is called an **isometric isomorphism**. *When $\tau: V \to W$ is an isometric isomorphism, we say that V and W are* **isometrically isomorphic**.$\square$

It is easy to see that an isometry is always injective but need not be surjective, even if $V = W$.

Theorem 13.3 *A linear transformation $\tau \in \mathcal{L}(V, W)$ is an isometry if and only if it preserves the norm, that is, if and only if*

$$\|\tau v\| = \|v\|$$

for all $v \in V$.$\square$

The following result points out one of the main differences between real and complex inner product spaces.

Theorem 13.4 *Let V be an inner product space and let $\tau \in \mathcal{L}(V)$.*
1) If $\langle \tau v, w\rangle = 0$ for all $v, w \in V$, then $\tau = 0$.
2) If V is a complex inner product space and $Q_\tau(v) = \langle \tau v, v\rangle = 0$ for all $v \in V$, then $\tau = 0$.
3) Part 2) does not hold in general for real inner product spaces.$\square$

Hilbert Spaces

Since an inner product space is a metric space, all that we learned about metric spaces applies to inner product spaces. In particular, if (x_n) is a sequence of vectors in an inner product space V, then

$$(x_n) \to x \text{ if and only if } \|x_n - x\| \to 0 \text{ as } n \to \infty$$

The fact that the inner product is continuous as a function of either of its coordinates is extremely useful.

Theorem 13.5 *Let V be an inner product space. Then*
1) $(x_n) \to x,\ (y_n) \to y \Rightarrow \langle x_n, y_n \rangle \to \langle x, y \rangle$
2) $(x_n) \to x \Rightarrow \|x_n\| \to \|x\|$ □

Complete inner product spaces play an especially important role in both theory and practice.

Definition *An inner product space that is complete under the metric induced by the inner product is said to be a* **Hilbert space**. □

Example 13.1 One of the most important examples of a Hilbert space is the space ℓ^2. Recall that the inner product is defined by

$$\langle \boldsymbol{x}, \boldsymbol{y} \rangle = \sum_{n=1}^{\infty} x_n \overline{y}_n$$

(In the real case, the conjugate is unnecessary.) The metric induced by this inner product is

$$d(\boldsymbol{x}, \boldsymbol{y}) = \|\boldsymbol{x} - \boldsymbol{y}\|_2 = \left(\sum_{n=1}^{\infty} |x_n - y_n|^2 \right)^{1/2}$$

which agrees with the definition of the metric space ℓ^2 given in Chapter 12. In other words, the metric in Chapter 12 is induced by this inner product. As we saw in Chapter 12, this inner product space is complete and so it is a Hilbert space. (In fact, it is the prototype of all Hilbert spaces, introduced by David Hilbert in 1912, even before the axiomatic definition of Hilbert space was given by John von Neumann in 1927.)□

The previous example raises the question whether the other metric spaces ℓ^p $(p \neq 2)$, with distance given by

$$d(\boldsymbol{x}, \boldsymbol{y}) = \|\boldsymbol{x} - \boldsymbol{y}\|_p = \left(\sum_{n=1}^{\infty} |x_n - y_n|^p \right)^{1/p} \tag{13.1}$$

are complete inner product spaces. The fact is that they are not even inner product spaces! More specifically, there is no inner product whose induced metric is given by (13.1). To see this, observe that, according to Theorem 13.1, any norm that comes from an inner product must satisfy the parallelogram law

$$\|\boldsymbol{x}+\boldsymbol{y}\|^2+\|\boldsymbol{x}-\boldsymbol{y}\|^2=2\|\boldsymbol{x}\|^2+2\|\boldsymbol{y}\|^2$$

But the norm in (13.1) does not satisfy this law. To see this, take $\boldsymbol{x}=(1,1,0,\dots)$ and $\boldsymbol{y}=(1,-1,0,\dots)$. Then

$$\|\boldsymbol{x}+\boldsymbol{y}\|_p=2,\ \|\boldsymbol{x}-\boldsymbol{y}\|_p=2$$

and

$$\|\boldsymbol{x}\|_p=2^{1/p},\ \|\boldsymbol{y}\|_p=2^{1/p}$$

Thus, the left side of the parallelogram law is 8 and the right side is $4\cdot 2^{2/p}$, which equals 8 if and only if $p=2$.

Just as any metric space has a completion, so does any inner product space.

Theorem 13.6 *Let V be an inner product space. Then there exists a Hilbert space H and an isometry $\tau\colon V\to H$ for which τV is dense in H. Moreover, H is unique up to isometric isomorphism.*

Proof. We know that the metric space (V,d), where d is induced by the inner product, has a unique completion (V',d'), which consists of equivalence classes of Cauchy sequences in V. If $(x_n)\in\overline{(x_n)}\in V'$ and $(y_n)\in\overline{(y_n)}\in V'$, then we set

$$\overline{(x_n)}+\overline{(y_n)}=\overline{(x_n+y_n)},\ r\overline{(x_n)}=\overline{(rx_n)}$$

and

$$\langle\overline{(x_n)},\overline{(y_n)}\rangle=\lim_{n\to\infty}\langle x_n,y_n\rangle$$

It is easy to see that since (x_n) and (y_n) are Cauchy sequences, so are (x_n+y_n) and (rx_n). In addition, these definitions are well-defined, that is, they are independent of the choice of representative from each equivalence class. For instance, if $(\widehat{x}_n)\in\overline{(x_n)}$, then

$$\lim_{n\to\infty}\|x_n-\widehat{x}_n\|=0$$

and so

$$|\langle x_n,y_n\rangle-\langle\widehat{x}_n,y_n\rangle|=|\langle x_n-\widehat{x}_n,y_n\rangle|\le\|x_n-\widehat{x}_n\|\|y_n\|\to 0$$

(The Cauchy sequence (y_n) is bounded.) Hence,

$$\langle\overline{(x_n)},\overline{(y_n)}\rangle=\lim_{n\to\infty}\langle x_n,y_n\rangle=\lim_{n\to\infty}\langle\widehat{x}_n,y_n\rangle=\langle\overline{(\widehat{x}_n)},\overline{(y_n)}\rangle$$

We leave it to the reader to show that V' is an inner product space under these operations.

Moreover, the inner product on V' induces the metric d', since

$$\begin{aligned}\langle(\overline{x_n - y_n}), (\overline{x_n - y_n})\rangle &= \lim_{n\to\infty}\langle x_n - y_n, x_n - y_n\rangle \\ &= \lim_{n\to\infty} d(x_n, y_n)^2 \\ &= d'((x_n), (y_n))^2\end{aligned}$$

Hence, the metric space isometry $\tau: V \to V'$ is an isometry of inner product spaces, since

$$\langle\tau x, \tau y\rangle = d'(\tau x, \tau y)^2 = d(x, y)^2 = \langle x, y\rangle$$

Thus, V' is a complete inner product space and τV is a dense subspace of V' that is isometrically isomorphic to V. We leave the issue of uniqueness to the reader.□

The next result concerns subspaces of inner product spaces.

Theorem 13.7

1) *Any complete subspace of an inner product space is closed.*
2) *A subspace of a Hilbert space is a Hilbert space if and only if it is closed.*
3) *Any finite-dimensional subspace of an inner product space is closed and complete.*

Proof. Parts 1) and 2) follow from Theorem 12.6. Let us prove that a finite-dimensional subspace S of an inner product space V is closed. Suppose that (x_n) is a sequence in S, $(x_n) \to x$ and $x \notin S$. Let $\mathcal{B} = \{b_1, \ldots, b_m\}$ be an orthonormal Hamel basis for S. The Fourier expansion

$$s = \sum_{i=1}^{m} \langle x, b_i\rangle b_i$$

in S has the property that $x - s \neq 0$ but

$$\langle x - s, b_j\rangle = \langle x, b_j\rangle - \langle s, b_j\rangle = 0$$

Thus, if we write $y = x - s$ and $y_n = x_n - s \in S$, the sequence (y_n), which is in S, converges to a vector y that is orthogonal to S. But this is impossible, because $y_n \perp y$ implies that

$$\|y_n - y\|^2 = \|y_n\|^2 + \|y\|^2 \geq \|y\|^2 \not\to 0$$

This proves that S is closed.

To see that any finite-dimensional subspace S of an inner product space is complete, let us embed S (as an inner product space in its own right) in its completion S'. Then S (or rather an isometric copy of S) is a finite-dimensional subspace of a complete inner product space S' and as such it is closed. However, S is dense in S' and so $S = S'$, which shows that S is complete.□

Infinite Series

Since an inner product space allows both addition of vectors and convergence of sequences, we can define the concept of infinite sums, or infinite series.

Definition Let V be an inner product space. The **nth partial sum** of the sequence (x_k) in V is

$$s_n = x_1 + \cdots + x_n$$

If the sequence (s_n) of partial sums converges to a vector $s \in V$, that is, if

$$\|s_n - s\| \to 0 \text{ as } n \to \infty$$

then we say that the series $\sum x_n$ **converges** to s and write

$$\sum_{n=1}^{\infty} x_n = s$$

□

We can also define absolute convergence.

Definition *A series $\sum x_k$ is said to be* **absolutely convergent** *if the series*

$$\sum_{n=1}^{\infty} \|x_k\|$$

converges.□

The key relationship between convergence and absolute convergence is given in the next theorem. Note that completeness is required to guarantee that absolute convergence implies convergence.

Theorem 13.8 *Let V be an inner product space. Then V is complete if and only if absolute convergence of a series implies convergence.*
Proof. Suppose that V is complete and that $\sum \|x_k\| < \infty$. Then the sequence s_n of partial sums is a Cauchy sequence, for if $n > m$, we have

$$\|s_n - s_m\| = \left\| \sum_{k=m+1}^{n} x_k \right\| \le \sum_{k=m+1}^{n} \|x_k\| \to 0$$

Hence, the sequence (s_n) converges, that is, the series $\sum x_k$ converges.

Conversely, suppose that absolute convergence implies convergence and let (x_n) be a Cauchy sequence in V. We wish to show that this sequence converges. Since (x_n) is a Cauchy sequence, for each $k > 0$, there exists an N_k

with the property that

$$i, j \geq N_k \Rightarrow \|x_i - x_j\| < \frac{1}{2^k}$$

Clearly, we can choose $N_1 < N_2 < \cdots$, in which case

$$\|x_{N_{k+1}} - x_{N_k}\| < \frac{1}{2^k}$$

and so

$$\sum_{k=1}^{\infty} \|x_{N_{k+1}} - x_{N_k}\| \leq \sum_{k=1}^{\infty} \frac{1}{2^k} < \infty$$

Thus, according to hypothesis, the series

$$\sum_{k=1}^{\infty} (x_{N_{k+1}} - x_{N_k})$$

converges. But this is a telescoping series, whose nth partial sum is

$$x_{N_{n+1}} - x_{N_1}$$

and so the subsequence (x_{N_k}) converges. Since any Cauchy sequence that has a convergent subsequence must itself converge, the sequence (x_k) converges and so V is complete.□

An Approximation Problem

Suppose that V is an inner product space and that S is a subset of V. It is of considerable interest to be able to find, for any $x \in V$, a vector in S that is *closest* to x in the metric induced by the inner product, should such a vector exist. This is the **approximation problem** for V.

Suppose that $x \in V$ and let

$$\delta = \inf_{s \in S} \|x - s\|$$

Then there is a sequence s_n for which

$$\delta_n = \|x - s_n\| \to \delta$$

as shown in Figure 13.1.

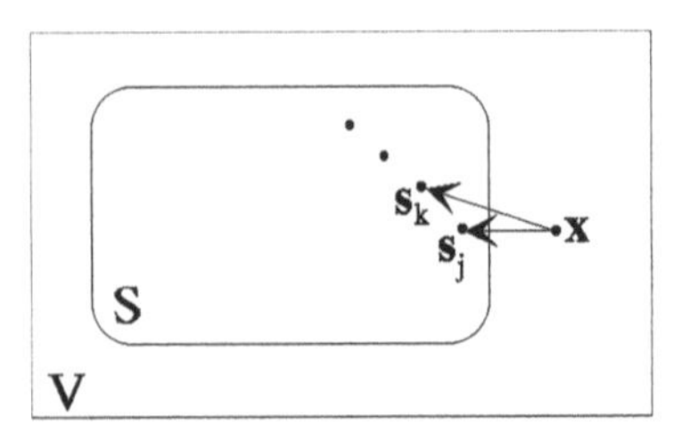

Figure 13.1

Let us see what we can learn about this sequence. First, if we let $y_k = x - s_k$, then according to the parallelogram law,

$$\|y_k + y_j\|^2 + \|y_k - y_j\|^2 = 2(\|y_k\|^2 + \|y_j\|^2)$$

or

$$\|y_k - y_j\|^2 = 2(\|y_k\|^2 + \|y_j\|^2) - 4\left\|\frac{y_k + y_j}{2}\right\|^2 \tag{13.2}$$

Now, if the set S is **convex**, that is, if

$$x, y \in S \Rightarrow rx + (1 - r)y \in S \text{ for all } 0 \le r \le 1$$

(in words, S contains the line segment between any two of its points), then $(s_k + s_j)/2 \in S$ and so

$$\left\|\frac{y_k + y_j}{2}\right\| = \left\|x - \frac{s_k + s_j}{2}\right\| \ge \delta$$

Thus, (13.2) gives

$$\|y_k - y_j\|^2 \le 2(\|y_k\|^2 + \|y_j\|^2) - 4\delta^2 \to 0$$

as $k, j \to \infty$. Hence, if S is convex, then the sequence $(y_n) = (x - s_n)$ is a Cauchy sequence and therefore so is (s_n).

If we also require that S be complete, then the Cauchy sequence (s_n) converges to a vector $\widehat{x} \in S$ and by the continuity of the norm, we must have $\|x - \widehat{x}\| = \delta$. Let us summarize and add a remark about uniqueness.

Theorem 13.9 *Let V be an inner product space and let S be a complete convex subset of V. Then for any $x \in V$, there exists a unique $\widehat{x} \in S$ for which*

$$\|x - \widehat{x}\| = \inf_{s \in S} \|x - s\|$$

The vector $\widehat{x}$ is called the **best approximation** *to x in S.*

Proof. Only the uniqueness remains to be established. Suppose that

$$\|x - \widehat{x}\| = \delta = \|x - x'\|$$

Then, by the parallelogram law,

$$\begin{aligned}\|\widehat{x} - x'\|^2 &= \|(x - x') - (x - \widehat{x})\|^2 \\ &= 2\|x - \widehat{x}\|^2 + 2\|x - x'\|^2 - \|2x - \widehat{x} - x'\|^2 \\ &= 2\|x - \widehat{x}\|^2 + 2\|x - x'\|^2 - 4\left\|x - \frac{\widehat{x} + x'}{2}\right\|^2 \\ &\leq 2\delta^2 + 2\delta^2 - 4\delta^2 = 0\end{aligned}$$

and so $\widehat{x} = x'$.□

Since any subspace S of an inner product space V is convex, Theorem 13.9 applies to complete subspaces. However, in this case, we can say more.

Theorem 13.10 *Let V be an inner product space and let S be a complete* subspace *of V. Then for any $x \in V$, the best approximation to x in S is the unique vector $x' \in S$ for which $x - x' \perp S$.*
Proof. Suppose that $x - x' \perp S$, where $x' \in S$. Then for any $s \in S$, we have $x - x' \perp s - x'$ and so

$$\|x - s\|^2 = \|x - x'\|^2 + \|x' - s\|^2 \geq \|x - x'\|^2$$

Hence $x' = \widehat{x}$ is the best approximation to x in S. Now we need only show that $x - \widehat{x} \perp S$, where $\widehat{x}$ is the best approximation to x in S. For any $s \in S$, a little computation reminiscent of completing the square gives

$$\begin{aligned}\|x - rs\|^2 &= \langle x - rs, x - rs\rangle \\ &= \|x\|^2 - \overline{r}\langle x, s\rangle - r\langle s, x\rangle + r\overline{r}\|s\|^2 \\ &= \|x\|^2 + \|s\|^2\left(r\overline{r} - \overline{r}\frac{\langle x, s\rangle}{\|s\|^2} - r\frac{\overline{\langle x, s\rangle}}{\|s\|^2}\right) \\ &= \|x\|^2 + \|s\|^2\left(r - \frac{\langle x, s\rangle}{\|s\|^2}\right)\left(\overline{r} - \frac{\overline{\langle x, s\rangle}}{\|s\|^2}\right) - \frac{|\langle x, s\rangle|^2}{\|s\|^2} \\ &= \|x\|^2 + \|s\|^2\left|r - \frac{\langle x, s\rangle}{\|s\|^2}\right|^2 - \frac{|\langle x, s\rangle|^2}{\|s\|^2}\end{aligned}$$

Now, this is smallest when

$$r = r_0 := \frac{\langle x, s\rangle}{\|s\|^2}$$

in which case

$$\|x - r_0 s\|^2 = \|x\|^2 - \frac{|\langle x, s\rangle|^2}{\|s\|^2}$$

Replacing x by $x - \widehat{x}$ gives

$$\|x - \widehat{x} - r_0 s\|^2 = \|x - \widehat{x}\|^2 - \frac{|\langle x - \widehat{x}, s\rangle|^2}{\|s\|^2}$$

But $\widehat{x}$ is the best approximation to x in S and since $\widehat{x} - r_0 s \in S$ we must have

$$\|x - \widehat{x} - r_0 s\|^2 \geq \|x - \widehat{x}\|^2$$

Hence,

$$\frac{|\langle x - \widehat{x}, s\rangle|^2}{\|s\|^2} = 0$$

or equivalently,

$$\langle x - \widehat{x}, s\rangle = 0$$

Hence, $x - \widehat{x} \perp S$.□

According to Theorem 13.9, if S is a complete subspace of an inner product space V, then for any $x \in V$, we may write

$$x = \widehat{x} + (x - \widehat{x})$$

where $\widehat{x} \in S$ and $x - \widehat{x} \in S^\perp$. Hence, $V = S + S^\perp$ and since $S \cap S^\perp = \{0\}$, we also have $V = S \odot S^\perp$. This is the projection theorem for arbitrary inner product spaces.

Theorem 13.11 (The projection theorem) *If S is a complete subspace of an inner product space V, then*

$$V = S \odot S^\perp$$

In particular, if S is a closed subspace of a Hilbert space H, then

$$H = S \odot S^\perp$$ □

Theorem 13.12 *Let S, T and T' be subspaces of an inner product space V.*
1) If $V = S \odot T$ then $T = S^\perp$.
2) If $S \odot T = S \odot T'$ then $T = T'$.

Proof. If $V = S \odot T$, then $T \subseteq S^\perp$ by definition of orthogonal direct sum. On the other hand, if $z \in S^\perp$, then $z = s + t$, for some $s \in S$ and $t \in T$. Hence,

$$0 = \langle z, s\rangle = \langle s, s\rangle + \langle t, s\rangle = \langle s, s\rangle$$

and so $s = 0$, implying that $z = t \in T$. Thus, $S^{\perp} \subseteq T$. Part 2) follows from part 1).□

Let us denote the closure of the span of a set S of vectors by $\operatorname{cspan}(S)$.

Theorem 13.13 *Let H be a Hilbert space.*
1) If A is a subset of H, then

$$\operatorname{cspan}(A) = A^{\perp\perp}$$

2) If S is a subspace of H, then

$$\operatorname{cl}(S) = S^{\perp\perp}$$

3) If K is a closed subspace of H, then

$$K = K^{\perp\perp}$$

Proof. We leave it as an exercise to show that $[\operatorname{cspan}(A)]^{\perp} = A^{\perp}$. Hence

$$H = \operatorname{cspan}(A) \odot [\operatorname{cspan}(A)]^{\perp} = \operatorname{cspan}(A) \odot A^{\perp}$$

But since $A^{\perp}$ is closed, we also have

$$H = A^{\perp} \odot A^{\perp\perp}$$

and so by Theorem 13.12, $\operatorname{cspan}(A) = A^{\perp\perp}$. The rest follows easily from part 1).□

In the exercises, we provide an example of a closed subspace K of an inner product space V for which $K \neq K^{\perp\perp}$. Hence, we cannot drop the requirement that H be a Hilbert space in Theorem 13.13.

Corollary 13.14 *If A is a* subset *of a Hilbert space H, then* $\operatorname{span}(A)$ *is dense in H if and only if $A^{\perp} = \{0\}$.*
Proof. As in the previous proof,

$$H = \operatorname{cspan}(A) \odot A^{\perp}$$

and so $A^{\perp} = \{0\}$ if and only if $H = \operatorname{cspan}(A)$.□

Hilbert Bases

We recall the following definition from Chapter 9.

Definition *A maximal orthonormal set in a Hilbert space H is called a* **Hilbert basis** *for H.*□

Zorn's lemma can be used to show that any nontrivial Hilbert space has a Hilbert basis. Again, we should mention that the concepts of Hilbert basis and Hamel basis (a maximal linearly independent set) are quite different. We will show later in this chapter that any two Hilbert bases for a Hilbert space have the same cardinality.

Since an orthonormal set $\mathcal{O}$ is maximal if and only if $\mathcal{O}^{\perp} = \{0\}$, Corollary 13.14 gives the following characterization of Hilbert bases.

Theorem 13.15 *Let $\mathcal{O}$ be an orthonormal subset of a Hilbert space H. The following are equivalent:*
1) *$\mathcal{O}$ is a Hilbert basis*
2) $\mathcal{O}^{\perp} = \{0\}$
3) *$\mathcal{O}$ is a* **total subset** *of H, that is,* $\text{cspan}(\mathcal{O}) = H$.□

Part 3) of this theorem says that a subset of a Hilbert space is a Hilbert basis if and only if it is a total orthonormal set.

Fourier Expansions

We now want to take a closer look at best approximations. Our goal is to find an explicit expression for the best approximation to any vector x from within a closed subspace S of a Hilbert space H. We will find it convenient to consider three cases, depending on whether S has finite, countably infinite, or uncountable dimension.

The Finite-Dimensional Case

Suppose that $\mathcal{O} = \{u_1, \ldots, u_n\}$ is an orthonormal set in a Hilbert space H. Recall that the Fourier expansion of any $x \in H$, with respect to $\mathcal{O}$, is given by

$$\widehat{x} = \sum_{k=1}^{n} \langle x, u_k \rangle u_k$$

where $\langle x, u_k \rangle$ is the Fourier coefficient of x with respect to u_k. Observe that

$$\langle x - \widehat{x}, u_k \rangle = \langle x, u_k \rangle - \langle \widehat{x}, u_k \rangle = 0$$

and so $x - \widehat{x} \perp \text{span}(\mathcal{O})$. Thus, according to Theorem 13.9, the Fourier expansion $\widehat{x}$ is the best approximation to x in $\text{span}(\mathcal{O})$. Moreover, since $x - \widehat{x} \perp \widehat{x}$, we have

$$\|\widehat{x}\|^2 = \|x\|^2 - \|x - \widehat{x}\|^2 \le \|x\|^2$$

and so

$$\|\widehat{x}\| \le \|x\|$$

with equality if and only if $x = \widehat{x}$, which happens if and only if $x \in \text{span}(\mathcal{O})$. Let us summarize.

Theorem 13.16 *Let $\mathcal{O} = \{u_1, \ldots, u_n\}$ be a finite orthonormal set in a Hilbert space H. For any $x \in H$, the Fourier expansion $\widehat{x}$ of x is the best approximation to x in* $\operatorname{span}(\mathcal{O})$. *We also have* **Bessel's inequality**

$$\|\widehat{x}\| \le \|x\|$$

or equivalently,

$$\sum_{k=1}^{n} |\langle x, u_k \rangle|^2 \le \|x\|^2 \tag{13.3}$$

with equality if and only if $x \in$ $\operatorname{span}(\mathcal{O})$.□

The Countably Infinite-Dimensional Case

In the countably infinite case, we will be dealing with infinite sums and so questions of convergence will arise. Thus, we begin with the following.

Theorem 13.17 *Let $\mathcal{O} = \{u_1, u_2, \ldots\}$ be a countably infinite orthonormal set in a Hilbert space H. The series*

$$\sum_{k=1}^{\infty} r_k u_k \tag{13.4}$$

converges in H if and only if the series

$$\sum_{k=1}^{\infty} |r_k|^2 \tag{13.5}$$

converges in $\mathbb{R}$. If these series converge, then they converge unconditionally (that is, any series formed by rearranging the order of the terms also converges). Finally, if the series (13.4) converges, then

$$\left\| \sum_{k=1}^{\infty} r_k u_k \right\|^2 = \sum_{k=1}^{\infty} |r_k|^2$$

Proof. Denote the partial sums of the first series by s_n and the partial sums of the second series by p_n. Then for $m \le n$

$$\|s_n - s_m\|^2 = \left\| \sum_{k=m+1}^{n} r_k u_k \right\|^2 = \sum_{k=m+1}^{n} |r_k|^2 = |p_n - p_m|$$

Hence (s_n) is a Cauchy sequence in H if and only if (p_n) is a Cauchy sequence in $\mathbb{R}$. Since both H and $\mathbb{R}$ are complete, (s_n) converges if and only if (p_n) converges.

If the series (13.5) converges, then it converges absolutely and hence unconditionally. (A real series converges unconditionally if and only if it converges absolutely.) But if (13.5) converges unconditionally, then so does (13.4). The last part of the theorem follows from the continuity of the norm.□

Now let $\mathcal{O} = \{u_1, u_2, \ldots\}$ be a countably infinite orthonormal set in H. The **Fourier expansion** of a vector $x \in H$ is defined to be the sum

$$\widehat{x} = \sum_{k=1}^{\infty} \langle x, u_k \rangle u_k \tag{13.6}$$

To see that this sum converges, observe that for any $n > 0$, (13.3) gives

$$\sum_{k=1}^{n} |\langle x, u_k \rangle|^2 \leq \|x\|^2$$

and so

$$\sum_{k=1}^{\infty} |\langle x, u_k \rangle|^2 \leq \|x\|^2$$

which shows that the series on the left converges. Hence, according to Theorem 13.17, the Fourier expansion (13.6) converges unconditionally.

Moreover, since the inner product is continuous,

$$\langle x - \widehat{x}, u_k \rangle = \langle x, u_k \rangle - \langle \widehat{x}, u_k \rangle = 0$$

and so $x - \widehat{x} \in [\text{span}(\mathcal{O})]^\perp = [\text{cspan}(\mathcal{O})]^\perp$. Hence, $\widehat{x}$ is the best approximation to x in $\text{cspan}(\mathcal{O})$. Finally, since $x - \widehat{x} \perp \widehat{x}$, we again have

$$\|\widehat{x}\|^2 = \|x\|^2 - \|x - \widehat{x}\|^2 \leq \|x\|^2$$

and so

$$\|\widehat{x}\| \leq \|x\|$$

with equality if and only if $x = \widehat{x}$, which happens if and only if $x \in \text{cspan}(\mathcal{O})$. Thus, the following analog of Theorem 13.16 holds.

Theorem 13.18 Let $\mathcal{O} = \{u_1, u_2, \ldots\}$ be a countably infinite orthonormal set in a Hilbert space H. For any $x \in H$, the Fourier expansion

$$\widehat{x} = \sum_{k=1}^{\infty} \langle x, u_k \rangle u_k$$

of x converges unconditionally and is the best approximation to x in $\text{cspan}(\mathcal{O})$. We also have **Bessel's inequality**

$$\|\widehat{x}\| \leq \|x\|$$

or equivalently,

$$\sum_{k=1}^{\infty} |\langle x, u_k \rangle|^2 \le \|x\|^2$$

with equality if and only if $x \in \operatorname{cspan}(\mathcal{O})$.□

The Arbitrary Case

To discuss the case of an arbitrary orthonormal set $\mathcal{O} = \{u_k \mid k \in K\}$, let us first define and discuss the concept of the sum of an arbitrary number of terms. (This is a bit of a digression, since we could proceed without all of the coming details – but they are interesting.)

Definition Let $\mathcal{K} = \{x_k \mid k \in K\}$ be an arbitrary family of vectors in an inner product space V. The sum

$$\sum_{k \in K} x_k$$

is said to **converge** to a vector $x \in V$ and we write

$$x = \sum_{k \in K} x_k \tag{13.7}$$

if for any $\epsilon > 0$, there exists a finite set $S \subseteq K$ for which

$$T \supset S,\ T \text{ finite} \Rightarrow \left\| \sum_{k \in T} x_k - x \right\| \le \epsilon$$ □

For those readers familiar with the language of convergence of nets, the set $\mathcal{P}_0(K)$ of all finite subsets of K is a *directed set* under inclusion (for every $A, B \in \mathcal{P}_0(K)$ there is a $C \in \mathcal{P}_0(K)$ containing A and B) and the function

$$S \to \sum_{k \in S} x_k$$

is a net in H. Convergence of (13.7) is convergence of this net. In any case, we will refer to the preceding definition as the **net definition** of convergence.

It is not hard to verify the following basic properties of net convergence for arbitrary sums.

Theorem 13.19 *Let $\mathcal{K} = \{x_k \mid k \in K\}$ be an arbitrary family of vectors in an inner product space V. If*

$$\sum_{k \in K} x_k = x \text{ and } \sum_{k \in K} y_k = y$$

then

1) **(Linearity)**

$$\sum_{k\in K}(rx_k + sy_k) = rx + sy$$

for any $r, s \in F$

2) **(Continuity)**

$$\sum_{k\in K}\langle x_k, y\rangle = \langle x, y\rangle \textit{ and } \sum_{k\in K}\langle y, x_k\rangle = \langle y, x\rangle$$

□

The next result gives a useful "Cauchy-type" description of convergence.

Theorem 13.20 *Let* $\mathcal{K} = \{x_k \mid k \in K\}$ *be an arbitrary family of vectors in an inner product space* V.

1) If the sum

$$\sum_{k\in K} x_k$$

converges, then for any $\epsilon > 0$, *there exists a finite set* $I \subseteq K$ *such that*

$$J \cap I = \emptyset,\ J \textit{ finite} \Rightarrow \left\|\sum_{k\in J} x_k\right\| \le \epsilon$$

2) If V *is a Hilbert space, then the converse of 1) also holds.*

Proof. For part 1), given $\epsilon > 0$, let $S \subseteq K$, S finite, be such that

$$T \supset S,\ T \text{ finite} \Rightarrow \left\|\sum_{k\in T} x_k - x\right\| \le \frac{\epsilon}{2}$$

If $J \cap S = \emptyset$, J finite, then

$$\begin{aligned}\left\|\sum_J x_k\right\| &= \left\|\left(\sum_J x_k + \sum_S x_k - x\right) - \left(\sum_S x_k - x\right)\right\| \\ &\le \left\|\sum_{J\cup S} x_k - x\right\| + \left\|\sum_S x_k - x\right\| \le \frac{\epsilon}{2} + \frac{\epsilon}{2} = \epsilon\end{aligned}$$

As for part 2), for each $n > 0$, let $I_n \subseteq K$ be a finite set for which

$$J \cap I_n = \emptyset,\ J \text{ finite} \Rightarrow \left\|\sum_{j\in J} x_j\right\| \le \frac{1}{n}$$

and let

$$y_n = \sum_{k\in I_n} x_k$$

Then (y_n) is a Cauchy sequence, since

$$\begin{aligned}\|y_n - y_m\| &= \left\|\sum_{I_n} x_k - \sum_{I_m} x_k\right\| = \left\|\sum_{I_n - I_m} x_k - \sum_{I_m - I_n} x_k\right\| \\ &\le \left\|\sum_{I_n - I_m} x_k\right\| + \left\|\sum_{I_m - I_n} x_k\right\| \le \frac{1}{m} + \frac{1}{n} \to 0\end{aligned}$$

Since V is assumed complete, we have $(y_n) \to y$.

Now, given $\epsilon > 0$, there exists an N such that

$$n \ge N \Rightarrow \|y_n - y\| = \left\|\sum_{I_n} x_k - y\right\| \le \frac{\epsilon}{2}$$

Setting $n = \max\{N, 2/\epsilon\}$ gives for $T \supset I_n,\ T$ finite,

$$\begin{aligned}\left\|\sum_{T} x_k - y\right\| &= \left\|\sum_{I_n} x_k - y + \sum_{T - I_n} x_k\right\| \\ &\le \left\|\sum_{I_n} x_k - y\right\| + \left\|\sum_{T - I_n} x_k\right\| \le \frac{\epsilon}{2} + \frac{1}{n} \le \epsilon\end{aligned}$$

and so $\sum_{k \in K} x_k$ converges to y.□

The following theorem tells us that convergence of an arbitrary sum implies that only countably many terms can be nonzero so, in some sense, there is no such thing as a nontrivial *uncountable* sum.

Theorem 13.21 *Let $\mathcal{K} = \{x_k \mid k \in K\}$ be an arbitrary family of vectors in an inner product space V. If the sum*

$$\sum_{k \in K} x_k$$

converges, then at most a countable number of terms x_k can be nonzero.
Proof. According to Theorem 13.20, for each $n > 0$, we can let $I_n \subseteq K$, I_n finite, be such that

$$J \cap I_n = \emptyset,\ J \text{ finite} \Rightarrow \left\|\sum_{j \in J} x_j\right\| \le \frac{1}{n}$$

Let $I = \bigcup_n I_n$. Then I is countable and

$$k \notin I \Rightarrow \{k\} \cap I_n = \emptyset \text{ for all } n \Rightarrow \|x_k\| \le \frac{1}{n} \text{ for all } n \Rightarrow x_k = 0 \qquad \square$$

Here is the analog of Theorem 13.17.

Theorem 13.22 *Let $\mathcal{O} = \{u_k \mid k \in K\}$ be an arbitrary orthonormal family of vectors in a Hilbert space H. The two series*

$$\sum_{k\in K} r_k u_k \text{ and } \sum_{k\in K} |r_k|^2$$

converge or diverge together. If these series converge, then

$$\left\| \sum_{k\in K} r_k u_k \right\|^2 = \sum_{k\in K} |r_k|^2$$

Proof. The first series converges if and only if for every $\epsilon > 0$, there exists a finite set $I \subseteq K$ such that

$$J \cap I = \emptyset,\ J \text{ finite} \Rightarrow \left\| \sum_{k\in J} r_k u_k \right\|^2 \le \epsilon^2$$

or equivalently,

$$J \cap I = \emptyset,\ J \text{ finite} \Rightarrow \sum_{k\in J} |r_k|^2 \le \epsilon^2$$

and this is precisely what it means for the second series to converge. We leave proof of the remaining statement to the reader.□

The following is a useful characterization of arbitrary sums of nonnegative real terms.

Theorem 13.23 *Let $\{r_k \mid k \in K\}$ be a collection of nonnegative real numbers. Then*

$$\sum_{k\in K} r_k = \sup_{\substack{J \text{ finite} \\ J\subseteq K}} \sum_{k\in J} r_k \tag{13.8}$$

provided that either of the preceding expressions is finite.

Proof. Suppose that

$$\sup_{\substack{J \text{ finite} \\ J\subseteq K}} \sum_{k\in J} r_k = R < \infty$$

Then, for any $\epsilon > 0$, there exists a finite set $S \subseteq K$ such that

$$R \ge \sum_{k\in S} r_k \ge R - \epsilon$$

Hence, if $T \subseteq K$ is a finite set for which $T \supset S$, then since $r_k \geq 0$,

$$R \geq \sum_{k \in T} r_k \geq \sum_{k \in S} r_k \geq R - \epsilon$$

and so

$$\left\| R - \sum_{k \in T} r_k \right\| \leq \epsilon$$

which shows that $\sum r_k$ converges to R. Finally, if the sum on the left of (13.8) converges, then the supremum on the right is finite and so (13.8) holds.□

The reader may have noticed that we have two definitions of convergence for countably infinite series: the net version and the traditional version involving the limit of partial sums. Let us write

$$\sum_{k \in \mathbb{N}^+} x_k \text{ and } \sum_{k=1}^{\infty} x_k$$

for the net version and the partial sum version, respectively. Here is the relationship between these two definitions.

Theorem 13.24 *Let H be a Hilbert space. If $x_k \in H$, then the following are equivalent:*

1) $\sum_{k \in \mathbb{N}^+} x_k$ *converges (net version) to* x

2) $\sum_{k=1}^{\infty} x_k$ *converges unconditionally to* x

Proof. Assume that 1) holds. Suppose that π is any permutation of $\mathbb{N}^+$. Given any $\epsilon > 0$, there is a finite set $S \subseteq \mathbb{N}^+$ for which

$$T \supset S,\ T \text{ finite} \Rightarrow \left\| \sum_{k \in T} x_k - x \right\| \leq \epsilon$$

Let us denote the set of integers $\{1, \ldots, n\}$ by I_n and choose a positive integer n such that $\pi(I_n) \supset S$. Then for $m \geq n$ we have

$$\pi(I_m) \supset \pi(I_n) \supset S \Rightarrow \left\| \sum_{k=1}^{m} x_{\pi(k)} - x \right\| = \left\| \sum_{k \in \pi(I_m)} x_k - x \right\| \leq \epsilon$$

and so 2) holds.

Next, assume that 2) holds, but that the series in 1) does not converge. Then there exists an $\epsilon > 0$ such that for any finite subset $I \subseteq \mathbb{N}^+$, there exists a finite subset J with $J \cap I = \emptyset$ for which

$$\left\| \sum_{k \in J} x_k \right\| > \epsilon$$

From this, we deduce the existence of a countably infinite sequence J_n of mutually disjoint finite subsets of $\mathbb{N}^+$ with the property that

$$\max(J_n) = M_n < m_{n+1} = \min(J_{n+1})$$

and

$$\left\| \sum_{k \in J_n} x_k \right\| > \epsilon$$

Now we choose any permutation $\pi\colon \mathbb{N}^+ \to \mathbb{N}^+$ with the following properties

1) $\pi([m_n, M_n]) \subseteq [m_n, M_n]$
2) if $J_n = \{j_{n,1}, \ldots, j_{n,u_n}\}$, then

$$\pi(m_n) = j_{n,1}, \pi(m_n + 1) = j_{n,2}, \ldots, \pi(m_n + u_n - 1) = j_{n,u_n}$$

The intention in property 2) is that for each n, π takes a set of consecutive integers to the integers in J_n.

For any such permutation π, we have

$$\left\| \sum_{k=m_n}^{m_n+u_n-1} x_{\pi(k)} \right\| = \left\| \sum_{k \in J_n} x_k \right\| > \epsilon$$

which shows that the sequence of partial sums of the series

$$\sum_{k=1}^{\infty} x_{\pi(k)}$$

is not Cauchy and so this series does not converge. This contradicts 2) and shows that 2) implies at least that 1) converges. But if 1) converges to $y \in H$, then since 1) implies 2) and since unconditional limits are unique, we have $y = x$. Hence, 2) implies 1).□

Now we can return to the discussion of Fourier expansions. Let $\mathcal{O} = \{u_k \mid k \in K\}$ be an arbitrary orthonormal set in a Hilbert space H. Given any $x \in H$, we may apply Theorem 13.16 to all finite subsets of $\mathcal{O}$, to deduce that

$$\sup_{\substack{J \text{ finite} \\ J \subseteq K}} \sum_{k \in J} |\langle x, u_k \rangle|^2 \leq \|x\|^2$$

and so Theorem 13.23 tells us that the sum

$$\sum_{k \in K} |\langle x, u_k \rangle|^2$$

converges. Hence, according to Theorem 13.22, the **Fourier expansion**

$$\widehat{x} = \sum_{k \in K} \langle x, u_k \rangle u_k$$

of x also converges and

$$\|\widehat{x}\|^2 = \sum_{k \in K} |\langle x, u_k \rangle|^2$$

Note that, according to Theorem 13.21, $\widehat{x}$ is a countably infinite sum of terms of the form $\langle x, u_k \rangle u_k$ and so is in $\text{cspan}(\mathcal{O})$.

The continuity of infinite sums with respect to the inner product (Theorem 13.19) implies that

$$\langle x - \widehat{x}, u_k \rangle = \langle x, u_k \rangle - \langle \widehat{x}, u_k \rangle = 0$$

and so $x - \widehat{x} \in [\text{span}(\mathcal{O})]^\perp = [\text{cspan}(\mathcal{O})]^\perp$. Hence, Theorem 3.9 tells us that $\widehat{x}$ is the best approximation to x in $\text{cspan}(\mathcal{O})$. Finally, since $x - \widehat{x} \perp \widehat{x}$, we again have

$$\|\widehat{x}\|^2 = \|x\|^2 - \|x - \widehat{x}\|^2 \leq \|x\|^2$$

and so

$$\|\widehat{x}\| \leq \|x\|$$

with equality if and only if $x = \widehat{x}$, which happens if and only if $x \in \text{cspan}(\mathcal{O})$. Thus, we arrive at the most general form of a key theorem about Hilbert spaces.

Theorem 13.25 *Let $\mathcal{O} = \{u_k \mid k \in K\}$ be an orthonormal family of vectors in a Hilbert space H. For any $x \in H$, the Fourier expansion*

$$\widehat{x} = \sum_{k \in K} \langle x, u_k \rangle u_k$$

of x converges in H and is the unique best approximation to x in $\text{cspan}(\mathcal{O})$. *Moreover, we have* **Bessel's inequality**

$$\|\widehat{x}\| \leq \|x\|$$

or equivalently,

$$\sum_{k \in K} |\langle x, u_k \rangle|^2 \le \|x\|^2$$

with equality if and only if $x \in \text{cspan}(\mathcal{O})$.□

A Characterization of Hilbert Bases

Recall from Theorem 13.15 that an orthonormal set $\mathcal{O} = \{u_k \mid k \in K\}$ in a Hilbert space H is a Hilbert basis if and only if

$$\text{cspan}(\mathcal{O}) = H$$

Theorem 13.25, then leads to the following characterization of Hilbert bases.

Theorem 13.26 *Let* $\mathcal{O} = \{u_k \mid k \in K\}$ *be an orthonormal family in a Hilbert space* H. *The following are equivalent:*

1) $\mathcal{O}$ *is a Hilbert basis (a maximal orthonormal set)*
2) $\mathcal{O}^{\perp} = \{0\}$
3) $\mathcal{O}$ *is total (that is,* $\text{cspan}(\mathcal{O}) = H$*)*
4) $x = \widehat{x}$ *for all* $x \in H$
5) *Equality holds in Bessel's inequality for all* $x \in H$, *that is,*

$$\|x\| = \|\widehat{x}\|$$

   for all $x \in H$
6) **Parseval's identity**

$$\langle x, y \rangle = \langle \widehat{x}, \widehat{y} \rangle$$

   holds for all $x, y \in H$, *that is,*

$$\langle x, y \rangle = \sum_{k \in K} \langle x, u_k \rangle \overline{\langle y, u_k \rangle}$$

Proof. Parts 1), 2) and 3) are equivalent by Theorem 13.15. Part 4) implies part 3), since $\widehat{x} \in \text{cspan}(\mathcal{O})$ and 3) implies 4) since the unique best approximation of any $x \in \text{cspan}(\mathcal{O})$ is itself and so $x = \widehat{x}$. Parts 3) and 5) are equivalent by Theorem 13.25. Parseval's identity follows from part 4) using Theorem 13.19. Finally, Parseval's identity for $y = x$ implies that equality holds in Bessel's inequality.□

Hilbert Dimension

We now wish to show that all Hilbert bases for a Hilbert space H have the same cardinality and so we can define the Hilbert dimension of H to be that cardinality.

Theorem 13.27 *All Hilbert bases for a Hilbert space H have the same cardinality. This cardinality is called the* **Hilbert dimension** *of H, which we denote by* $\mathrm{hdim}(H)$.

Proof. If H has a finite Hilbert basis, then that set is also a Hamel basis and so all finite Hilbert bases have size $\dim(H)$. Suppose next that $\mathcal{B} = \{b_k \mid k \in K\}$ and $\mathcal{C} = \{c_j \mid j \in J\}$ are infinite Hilbert bases for H. Then for each b_k, we have

$$b_k = \sum_{j \in J_k} \langle b_k, c_j \rangle c_j$$

where J_k is the countable set $\{j \mid \langle b_k, c_j \rangle \neq 0\}$. Moreover, since no c_j can be orthogonal to every b_k, we have $\bigcup_K J_k = J$. Thus, since each J_k is countable, we have

$$|J| = \left| \bigcup_{k \in K} J_k \right| \le \aleph_0 |K| = |K|$$

By symmetry, we also have $|K| \le |J|$ and so the Schröder–Bernstein theorem implies that $|J| = |K|$.□

Theorem 13.28 *Two Hilbert spaces are isometrically isomorphic if and only if they have the same Hilbert dimension.*

Proof. Suppose that $\mathrm{hdim}(H_1) = \mathrm{hdim}(H_2)$. Let $\mathcal{O}_1 = \{u_k \mid k \in K\}$ be a Hilbert basis for H_1 and $\mathcal{O}_2 = \{v_k \mid k \in K\}$ a Hilbert basis for H_2. We may define a map $\tau: H_1 \to H_2$ as follows:

$$\tau\left(\sum_{k \in K} r_k u_k\right) = \sum_{k \in K} r_k v_k$$

We leave it as an exercise to verify that τ is a bijective isometry. The converse is also left as an exercise.□

A Characterization of Hilbert Spaces

We have seen that any vector space V is isomorphic to a vector space $(F^B)_0$ of all functions from B to F that have finite support. There is a corresponding result for Hilbert spaces. Let K be any nonempty set and let

$$\ell^2(K) = \left\{ f: K \to \mathbb{C} \,\middle|\, \sum_{k \in K} |f(k)|^2 < \infty \right\}$$

The functions in $\ell^2(K)$ are referred to as **square summable functions**. (We can also define a real version of this set by replacing $\mathbb{C}$ by $\mathbb{R}$.) We define an inner product on $\ell^2(K)$ by

$$\langle f, g \rangle = \sum_{k \in K} f(k)\overline{g(k)}$$

The proof that $\ell^2(K)$ is a Hilbert space is quite similar to the proof that

$\ell^2 = \ell^2(\mathbb{N})$ is a Hilbert space and the details are left to the reader. If we define $\delta_k \in \ell^2(K)$ by

$$\delta_k(j) = \delta_{k,j} = \begin{cases} 1 & \text{if } j = k \\ 0 & \text{if } j \neq k \end{cases}$$

then the collection

$$\mathcal{O} = \{\delta_k \mid k \in K\}$$

is a Hilbert basis for $\ell^2(K)$, of cardinality $|K|$. To see this, observe that

$$\langle \delta_i, \delta_j \rangle = \sum_{k \in K} \delta_i(k)\overline{\delta_j(k)} = \delta_{i,j}$$

and so $\mathcal{O}$ is orthonormal. Moreover, if $f \in \ell^2(K)$, then $f(k) \neq 0$ for only a countable number of $k \in K$, say $\{k_1, k_2, \ldots\}$. If we define f' by

$$f' = \sum_{i=1}^{\infty} f(k_i)\delta_{k_i}$$

then $f' \in \operatorname{cspan}(\mathcal{O})$ and $f'(j) = f(j)$ for all $j \in K$, which implies that $f = f'$. This shows that $\ell^2(K) = \operatorname{cspan}(\mathcal{O})$ and so $\mathcal{O}$ is a total orthonormal set, that is, a Hilbert basis for $\ell^2(K)$.

Now let H be a Hilbert space, with Hilbert basis $\mathcal{B} = \{u_k \mid k \in K\}$. We define a map $\phi\colon H \to \ell^2(K)$ as follows. Since $\mathcal{B}$ is a Hilbert basis, any $x \in H$ has the form

$$x = \sum_{k \in K} \langle x, u_k \rangle u_k$$

Since the series on the right converges, Theorem 13.22 implies that the series

$$\sum_{k \in K} |\langle x, u_k \rangle|^2$$

converges. Hence, another application of Theorem 13.22 implies that the following series converges:

$$\phi(x) = \sum_{k \in K} \langle x, u_k \rangle \delta_k$$

It follows from Theorem 13.19 that ϕ is linear and it is not hard to see that it is also bijective. Notice that $\phi(u_k) = \delta_k$ and so ϕ takes the Hilbert basis $\mathcal{B}$ for H to the Hilbert basis $\mathcal{O}$ for $\ell^2(K)$.

Notice also that

$$\|\phi(x)\|^2 = \langle \phi(x), \phi(x) \rangle = \sum_{k \in K} |\langle x, u_k \rangle|^2 = \left\| \sum_{k \in K} \langle x, u_k \rangle u_k \right\|^2 = \|x\|^2$$

and so ϕ is an isometric isomorphism. We have proved the following theorem.

Theorem 13.29 *If H is a Hilbert space of Hilbert dimension κ and if K is any set of cardinality κ, then H is isometrically isomorphic to $\ell^2(K)$.*$\square$

The Riesz Representation Theorem

We conclude our discussion of Hilbert spaces by discussing the Riesz representation theorem. As it happens, not all linear functionals on a Hilbert space have the form "take the inner product with...," as in the finite-dimensional case. To see this, observe that if $y \in H$, then the function

$$f_y(x) = \langle x, y \rangle$$

is certainly a linear functional on H. However, it has a special property. In particular, the Cauchy–Schwarz inequality gives, for all $x \in H$,

$$|f_y(x)| = |\langle x, y \rangle| \leq \|x\| \|y\|$$

or, for all $x \neq 0$,

$$\frac{|f_y(x)|}{\|x\|} \leq \|y\|$$

Noticing that equality holds if $x = y$, we have

$$\sup_{x \neq 0} \frac{|f_y(x)|}{\|x\|} = \|y\|$$

This prompts us to make the following definition, which we do for linear transformations between Hilbert spaces (this covers the case of linear functionals).

Definition *Let $\tau: H_1 \to H_2$ be a linear transformation from H_1 to H_2. Then τ is said to be* **bounded** *if*

$$\sup_{x \neq 0} \frac{\|\tau x\|}{\|x\|} < \infty$$

If the supremum on the left is finite, we denote it by $\|\tau\|$ and call it the **norm** *of τ.*$\square$

Of course, if $f: H \to F$ is a bounded linear functional on H, then

$$\|f\| = \sup_{x \neq 0} \frac{|f(x)|}{\|x\|}$$

The set of all bounded linear functionals on a Hilbert space H is called the **continuous dual space**, or **conjugate space**, of H and denoted by H^*. Note that this differs from the algebraic dual of H, which is the set of all linear functionals on H. In the finite-dimensional case, however, since all linear functionals are bounded (exercise), the two concepts agree. (Unfortunately, there is no universal agreement on the notation for the algebraic dual versus the continuous dual. Since we will discuss only the continuous dual in this section, no confusion should arise.)

The following theorem gives some simple reformulations of the definition of norm.

Theorem 13.30 *Let $\tau: H_1 \to H_2$ be a bounded linear transformation.*

1) $\|\tau\| = \sup_{\|x\|=1} \|\tau x\|$

2) $\|\tau\| = \sup_{\|x\|\leq 1} \|\tau x\|$

3) $\|\tau\| = \inf\{c \in \mathbb{R} \mid \|\tau x\| \leq c\|x\| \textit{ for all } x \in H\}$ □

The following theorem explains the importance of bounded linear transformations.

Theorem 13.31 *Let $\tau: H_1 \to H_2$ be a linear transformation. The following are equivalent:*

1) τ *is bounded*

2) τ *is continuous at any point* $x_0 \in H$

3) τ *is continuous.*

Proof. Suppose that τ is bounded. Then

$$\|\tau x - \tau x_0\| = \|\tau(x - x_0)\| \leq \|\tau\|\|x - x_0\| \to 0$$

as $x \to x_0$. Hence, τ is continuous at x_0. Thus, 1) implies 2). If 2) holds, then for any $y \in H$, we have

$$\|\tau x - \tau y\| = \|\tau(x - y + x_0) - \tau(x_0)\| \to 0$$

as $x \to y$, since τ is continuous at x_0 and $x - y + x_0 \to x_0$ as $y \to x$. Hence, τ is continuous at any $y \in H$ and 3) holds. Finally, suppose that 3) holds. Thus, τ is continuous at 0 and so there exists a $\delta > 0$ such that

$$\|x\| \leq \delta \Rightarrow \|\tau x\| \leq 1$$

In particular,

$$\|x\| = \delta \Rightarrow \frac{\|\tau x\|}{\|x\|} \le \frac{1}{\delta}$$

and so

$$\|x\| = 1 \Rightarrow \|\delta x\| = \delta \Rightarrow \frac{\|\tau(\delta x)\|}{\|\delta x\|} \le \frac{1}{\delta} \Rightarrow \frac{\|\tau x\|}{\|x\|} \le \frac{1}{\delta}$$

Thus, τ is bounded.□

Now we can state and prove the Riesz representation theorem.

Theorem 13.32 (The Riesz representation theorem) *Let H be a Hilbert space. For any bounded linear functional f on H, there is a unique $z_0 \in H$ such that*

$$f(x) = \langle x, z_0 \rangle$$

for all $x \in H$. Moreover, $\|z_0\| = \|f\|$.

Proof. If $f = 0$, we may take $z_0 = 0$, so let us assume that $f \neq 0$. Hence, $K = \ker(f) \neq H$ and since f is continuous, K is closed. Thus

$$H = K \odot K^\perp$$

Now, the first isomorphism theorem, applied to the linear functional $f: H \to F$, implies that $H/K \approx F$ (as vector spaces). In addition, Theorem 3.5 implies that $H/K \approx K^\perp$ and so $K^\perp \approx F$. In particular, $\dim(K^\perp) = 1$.

For any $z \in K^\perp$, we have

$$x \in K \Rightarrow f(x) = 0 = \langle x, z \rangle$$

Since $\dim(K^\perp) = 1$, all we need do is find $0 \neq z \in K^\perp$ for which

$$f(z) = \langle z, z \rangle$$

for then $f(rz) = rf(z) = r\langle z, z \rangle = \langle rz, z \rangle$ for all $r \in F$, showing that $f(x) = \langle x, z \rangle$ for $x \in K$ as well.

But if $0 \neq z \in K^\perp$, then

$$z_0 = \frac{\overline{f(z)}}{\langle z, z \rangle} z$$

has this property, as can be easily checked. The fact that $\|z_0\| = \|f\|$ has already been established.□

Exercises

1. Prove that the sup metric on the metric space $C[a,b]$ of continuous functions on $[a,b]$ does not come from an inner product. Hint: let $f(t)=1$ and $g(t)=(t-\mathrm{a})/(b-\mathrm{a})$ and consider the parallelogram law.
2. Prove that any Cauchy sequence that has a convergent subsequence must itself converge.
3. Let V be an inner product space and let A and B be subsets of V. Show that
   a) $A\subseteq B\Rightarrow B^{\perp}\subseteq A^{\perp}$
   b) $A^{\perp}$ is a closed subspace of V
   c) $[\mathrm{cspan}(A)]^{\perp}=A^{\perp}$
4. Let V be an inner product space and $S\subseteq V$. Under what conditions is $S^{\perp\perp\perp}=S^{\perp}$?
5. Prove that a subspace S of a Hilbert space H is closed if and only if $S=S^{\perp\perp}$.
6. Let V be the subspace of ℓ^2 consisting of all sequences of real numbers with the property that each sequence has only a finite number of nonzero terms. Thus, V is an inner product space. Let K be the subspace of V consisting of all sequences $x=(x_n)$ in V with the property that $\Sigma x_n/n=0$. Show that K is closed, but that $K^{\perp\perp}\neq K$. Hint: For the latter, show that $K^{\perp}=\{0\}$ by considering the sequences $u=(1,\ldots,-n,\ldots)$, where the term $-n$ is in the nth coordinate position.
7. Let $\mathcal{O}=\{u_1,u_2,\ldots\}$ be an orthonormal set in H. If $x=\Sigma r_k u_k$ converges, show that

$$\|x\|^2=\sum_{k=1}^{\infty}|r_k|^2$$

8. Prove that if an infinite series

$$\sum_{k=1}^{\infty}x_k$$

   converges absolutely in a Hilbert space H, then it also converges in the sense of the "net" definition given in this section.
9. Let $\{r_k\mid k\in K\}$ be a collection of nonnegative real numbers. If the sum on the left below converges, show that

$$\sum_{k\in K}r_k=\sup_{\substack{J\text{ finite}\\ J\subseteq K}}\sum_{k\in J}r_k$$

10. Find a countably infinite sum of real numbers that converges in the sense of partial sums, but not in the sense of nets.
11. Prove that if a Hilbert space H has infinite Hilbert dimension, then no Hilbert basis for H is a Hamel basis.
12. Prove that $\ell^2(K)$ is a Hilbert space for any nonempty set K.

13. Prove that any linear transformation between finite-dimensional Hilbert spaces is bounded.
14. Prove that if $f \in H^*$, then $\ker(f)$ is a closed subspace of H.
15. Prove that a Hilbert space is separable if and only if $\operatorname{hdim}(H) \leq \aleph_0$.
16. Can a Hilbert space have countably infinite Hamel dimension?
17. What is the Hamel dimension of $\ell^2(\mathbb{N})$?
18. Let τ and σ be bounded linear operators on H. Verify the following:
    a) $\|r\tau\| = |r|\|\tau\|$
    b) $\|\tau + \sigma\| \leq \|\tau\| + \|\sigma\|$
    c) $\|\tau\sigma\| \leq \|\tau\|\|\sigma\|$
19. Use the Riesz representation theorem to show that $H^* \approx H$ for any Hilbert space H.

Chapter 14
Tensor Products

In the preceding chapters, we have seen several ways to construct new vector spaces from old ones. Two of the most important such constructions are the direct sum $U \oplus V$ and the vector space $\mathcal{L}(U,V)$ of all linear transformations from U to V. In this chapter, we consider another very important construction, known as the *tensor product*.

Universality

We begin by describing a general type of *universality* that will help motivate the definition of tensor product. Our description is strongly related to the formal notion of a *universal pair* in category theory, but we will be somewhat less formal to avoid the need to formally define categorical concepts. Accordingly, the terminology that we shall introduce is not standard, but does not contradict any standard terminology.

Referring to Figure 14.1, consider a set A and two functions f and g, with domain A.

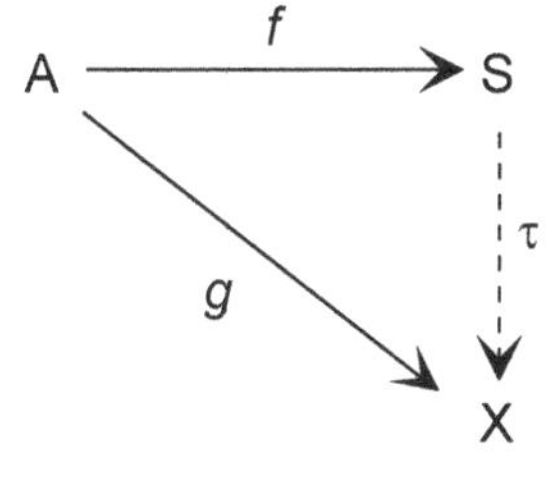

Figure 14.1

Suppose that there exists a function $\tau: S \to X$ for which this diagram *commutes*, that is,

$$g = \tau \circ f$$

This is sometimes expressed by saying that g can be **factored through** f. What does this say about the relationship between the functions f and g?

Let us think of the "information" about A contained in a function $h: A \to B$ as the way in which h *distinguishes* elements of A using *labels* from B. The relationship above implies that

$$g(a) \neq g(b) \Rightarrow f(a) \neq f(b)$$

and this can be phrased by saying that whatever ability g has to distinguish elements of A is also possessed by f. Put another way, except for labeling differences, any information about A that is contained in g is also contained in f.

If τ happens to be injective, then the *only* difference between f and g is the values of the labels. That is, the two functions have the same information about A. However, in general, τ is not required to be injective and so f may contain more information than g.

Now consider a family $\mathcal{S}$ of sets and a family

$$\mathcal{F} = \{g: A \to X \mid X \in \mathcal{S}\}$$

Assume that $S \in \mathcal{S}$ and $f: A \to S \in \mathcal{F}$. If the diagram in Figure 14.1 commutes for all $g \in \mathcal{F}$, then the information contained in every function in $\mathcal{F}$ is also contained in f. Moreover, since $f \in \mathcal{F}$, the function f cannot contain more information than is contained in the entire family and so we conclude that f contains exactly the same information as is contained in the entire family $\mathcal{F}$. In this sense, $f: A \to S$ is *universal* among all functions $g: A \to X$ in $\mathcal{F}$.

In this way, a single function $f: A \to S$, or more precisely, a single pair (S, f), can capture a mathematical concept as described by a family of functions. Some examples from linear algebra are basis for a vector space, quotient space, direct sum and bilinearity (as we will see).

Let us make a formal definition.

Definition *Referring to Figure 14.2, let A be a set and let $\mathcal{S}$ be a family of sets. Let*

$$\mathcal{F} = \{g: A \to X \mid X \in \mathcal{S}\}$$

be a family of functions, all of which have domain A and range a member of $\mathcal{S}$. Let

$$\mathcal{H} = \{\tau: X \to Y \mid X, Y \in \mathcal{S}\}$$

be a family of functions with domain and range in $\mathcal{S}$. We assume that $\mathcal{H}$ has the following structure:

1) $\mathcal{H}$ contains the identity function ι_S for each member of $\mathcal{S}$.

2) *$\mathcal{H}$ is closed under composition of functions, which is an associative operation.*
3) *For any $\tau \in \mathcal{H}$ and $f \in \mathcal{F}$, the composition $\tau \circ f$ is defined and belongs to $\mathcal{F}$.*

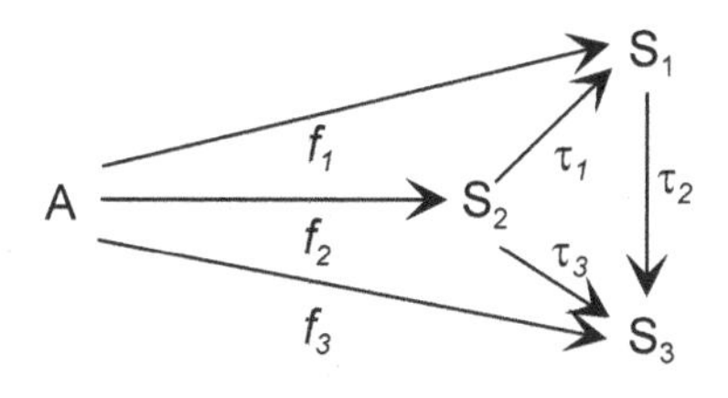

Figure 14.2

We refer to $\mathcal{H}$ as the **measuring family** *and its members as* **measuring functions**.

A pair $(S, f\colon A \to S)$, where $S \in \mathcal{S}$ and $f \in \mathcal{F}$ has the **universal property** *for the family $\mathcal{F}$* **as measured by** *$\mathcal{H}$, or is a* **universal pair** *for $(\mathcal{F}, \mathcal{H})$, if for every $g\colon A \to X$ in $\mathcal{F}$, there is a unique $\tau\colon S \to X$ in $\mathcal{H}$ for which the diagram in Figure 14.1 commutes, that is, for which*

$$g = \tau \circ f$$

or equivalently, any $g \in \mathcal{F}$ can be **factored through** *f. The unique measuring function τ is called the* **mediating morphism** *for g.*□

Note the requirement that the mediating morphism τ be unique. Universal pairs are essentially unique, as the following describes.

Theorem 14.1 *Let $(S, f\colon A \to S)$ and $(T, g\colon A \to T)$ be universal pairs for $(\mathcal{F}, \mathcal{H})$. Then there is a bijective measuring function $\mu \in \mathcal{H}$ for which $\mu S = T$. In fact, the mediating morphism of f with respect to g and the mediating morphism of g with respect to f are isomorphisms.*

Proof. With reference to Figure 14.3, there are mediating morphisms $\tau\colon S \to T$ and $\sigma\colon T \to S$ for which

$$g = \tau \circ f$$
$$f = \sigma \circ g$$

Hence,

$$g = (\tau \circ \sigma) \circ g$$
$$f = (\sigma \circ \tau) \circ f$$

However, referring to the third diagram in Figure 14.3, both $\sigma \circ \tau\colon S \to S$ and the identity map $\iota\colon S \to S$ are mediating morphisms for f and so the uniqueness of mediating morphisms implies that $\sigma \circ \tau = \iota$. Similarly $\tau \circ \sigma = \iota$ and so τ and σ are inverses of one another, making τ the desired bijection.□

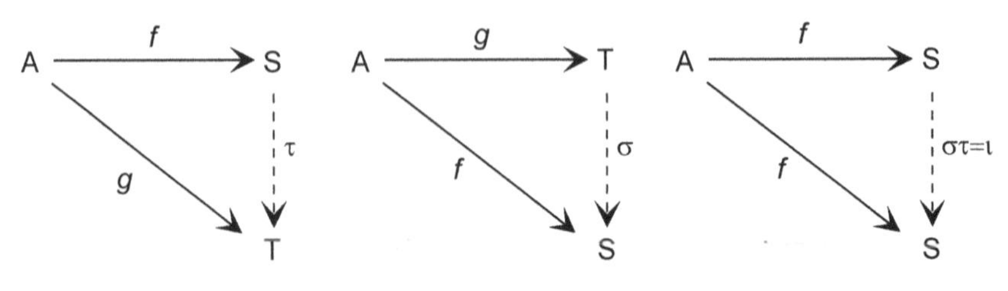

Figure 14.3

Examples of Universality

Now let us look at some examples of the universal property. Let $\mathrm{Vect}(F)$ denote the family of all vector spaces over the base field F. (We use the term *family* informally to represent what in set theory is formally referred to as a *class. A class* is a "collection" that is too large to be considered a set. For example, $\mathrm{Vect}(F)$ is a class.)

Example 14.1 (*Bases*) Let $\mathcal{B}$ be a nonempty set and let

1) $\mathcal{S} = \mathrm{Vect}(F)$
2) $\mathcal{F} =$ set functions from $\mathcal{B}$ to members of $\mathcal{F}$
3) $\mathcal{H} =$ linear transformations

If $V_{\mathcal{B}}$ is a vector space with basis $\mathcal{B}$, then the pair $(V_{\mathcal{B}}, j\colon \mathcal{B} \to V_{\mathcal{B}})$, where j is the inclusion map $jv = v$, is universal for $(\mathcal{F}, \mathcal{H})$. To see this, note that the condition that $g \in \mathcal{F}$ can be factored through j,

$$g = \tau \circ j$$

is equivalent to the statement that $\tau v = gv$ for each basis vector $v \in \mathcal{B}$. But this uniquely defines a linear transformation τ.

In fact, the universality of the pair $(V_{\mathcal{B}}, j)$ is *precisely* the statement that a linear transformation τ is uniquely determined by assigning its values arbitrarily on a basis $\mathcal{B}$, the function g doing the arbitrary assignment in this context. Note also that Theorem 14.1 implies that if $(W, k\colon \mathcal{B} \to W)$ is also universal for $(\mathcal{F}, \mathcal{H})$, then there is a bijective mediating morphism from $V_{\mathcal{B}}$ to W, that is, W and $V_{\mathcal{B}}$ are isomorphic.□

Example 14.2 (*Quotient spaces and canonical projections*) Let V be a vector space and let K be a subspace of V. Let

1) $\mathcal{S} = \mathrm{Vect}(F)$

2) $\mathcal{F}$ = linear maps with domain V, whose kernels contain K
3) $\mathcal{H}$ = linear transformations

Theorem 3.4 says precisely that the pair $(V/K, \pi: V \to V/K)$, where π is the canonical projection map, has the universal property for $\mathcal{F}$ as measured by $\mathcal{H}$.□

Example 14.3 (*Direct sums*) Let U and V be vector spaces over F. Let

1) $\mathcal{S} = \text{Vect}(F)$
2) $\mathcal{F}$ = ordered pairs $(f: U \to W, g: V \to W)$ of linear transformations
3) $\mathcal{H}$ = linear transformations

Here we have a slight variation on the definition of universal pair: In this case, $\mathcal{F}$ is a family of *pairs* of functions. For $\tau \in \mathcal{H}$ and $(f, g) \in \mathcal{F}$, we set

$$\tau \circ (f, g) = (\tau \circ f, \tau \circ g)$$

Then the pair $(U \boxplus V, (j_1, j_2): (U, V) \to U \boxplus V)$, where

$$j_1 u = (u, 0) \quad \text{and} \quad j_2 v = (0, v)$$

are called the **canonical injections**, has the universal property for $(\mathcal{F}, \mathcal{H})$. To see this, observe that for any pair $(f, g): (U, V) \to W$ in $\mathcal{F}$, the condition

$$(f, g) = \tau \circ (j_1, j_2)$$

is equivalent to

$$(f, g) = (\tau \circ j_1, \tau \circ j_2)$$

or

$$\tau(u, 0) = f(u) \quad \text{and} \quad \tau(0, v) = g(v)$$

But these conditions define a unique linear transformation $\tau: U \boxplus V \to W$.□

Thus, bases, quotient spaces and direct sums are all examples of universal pairs and it should be clear from these examples that the notion of universal property is, well, universal. In fact, it happens that the most useful definition of tensor product is through a universal property, which we now explore.

Bilinear Maps

The universality that defines tensor products rests on the notion of a bilinear map.

Definition *Let U, V and W be vector spaces over F. Let $U \times V$ be the cartesian product of U and V* as sets. *A set function*

$$f: U \times V \to W$$

is **bilinear** *if it is linear in both variables separately, that is, if*

$$f(ru + su', v) = rf(u, v) + sf(u', v)$$

and

$$f(u, rv + sv') = rf(u, v) + sf(u, v')$$

The set of all bilinear functions from $U \times V$ *to* W *is denoted by* $\hom_F(U, V; W)$. *A bilinear function* $f: U \times V \to F$ *with values in the base field* F *is called a* **bilinear form** *on* $U \times V$.□

Note that bilinearity can also be expressed in matrix language as follows: If

$$a = (a_1, \ldots, a_n) \in F^n, \quad b = (b_1, \ldots, b_m) \in F^m$$

and

$$u = (u_1, \ldots, u_n) \in U^n, \quad v = (v_1, \ldots, v_m) \in V^m$$

then $f: U \times V \to W$ is bilinear if

$$f(au^t, bv^t) = aFb^t$$

where $F = [f(u_i, v_j)]_{i,j}$.

It is important to emphasize that, in the definition of bilinear function, $U \times V$ is the *cartesian product of sets*, not the direct product of vector spaces. In other words, we do not consider any algebraic structure on $U \times V$ when defining bilinear functions, so expressions like

$$(x, y) + (z, w) \quad \text{and} \quad r(x, y)$$

are meaningless.

In fact, if V is a vector space, there are two classes of functions from $V \times V$ to W: the linear maps $\mathcal{L}(V \times V, W)$, where $V \times V = V \boxplus V$ is the direct product of vector spaces, and the bilinear maps $\hom(V, V; W)$, where $V \times V$ is just the cartesian product of sets. We leave it as an exercise to show that these two classes have only the zero map in common. In other words, the only map that is both linear and bilinear is the zero map.

We made a thorough study of bilinear forms on a finite-dimensional vector space V in Chapter 11 (although this material is not assumed here). However, bilinearity is far more important and far-reaching than its application to metric vector spaces, as the following examples show. Indeed, both multiplication and evaluation are bilinear.

Example 14.4 (Multiplication is bilinear) If A is an algebra, the product map $\mu: A \times A \to A$ defined by

$$\mu(a,b) = ab$$

is bilinear, that is, multiplication is linear in each position.□

Example 14.5 (Evaluation is bilinear) If V and W are vector spaces, then the *evaluation map* $\phi\colon \mathcal{L}(V,W) \times V \to W$ defined by

$$\phi(f,v) = fv$$

is bilinear. In particular, the evaluation map $\phi\colon V^* \times V \to F$ defined by $\phi(f,v) = fv$ is a bilinear form on $V^* \times V$.□

Example 14.6 If V and W are vector spaces, and $f \in V^*$ and $g \in W^*$, then the product map $\phi\colon V \times W \to F$ defined by

$$\phi(v,w) = f(v)g(w)$$

is bilinear. Dually, if $v \in V$ and $w \in W$, then the map $\lambda\colon V^* \times W^* \to F$ defined by

$$\lambda(f,g) = f(v)g(w)$$

is bilinear.□

It is precisely the tensor product that will allow us to generalize the previous example. In particular, if $\tau \in \mathcal{L}(U,W)$ and $\sigma \in \mathcal{L}(V,W)$, then we would like to consider a "product" map $\phi\colon U \times V \to W$ defined by

$$\phi(u,v) = \tau(u) \ ? \ \sigma(v)$$

The tensor product $\otimes$ is just the thing to replace the question mark, because it has the desired bilinearity property, as we will see. In fact, the tensor product is bilinear and nothing else, so it is *exactly* what we need!

Tensor Products

Let U and V be vector spaces. Our guide for the definition of the tensor product $U \otimes V$ will be the desire to have a universal property for bilinear functions, as measured by linearity. Referring to Figure 14.4, we want to define a vector space T and a bilinear map $t\colon U \times V \to T$ so that any bilinear map f with domain $U \times V$ can be factored through t. Intuitively speaking, t is the most "general" or "universal" bilinear map with domain $U \times V$: It is bilinear *and nothing more.*

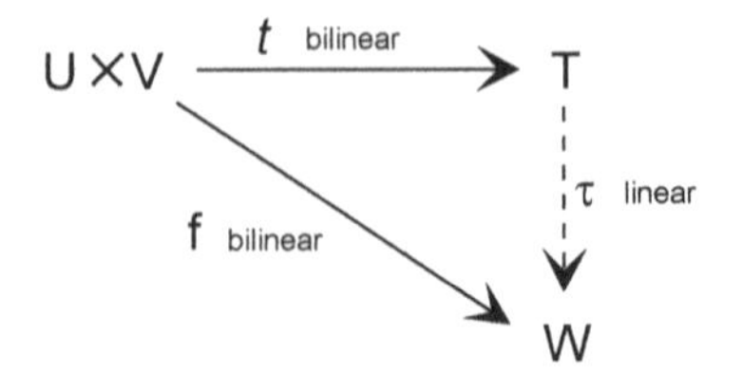

Figure 14.4

Definition *Let* $U \times V$ *be the cartesian product of two vector spaces over* F. *Let* $\mathcal{S} = \operatorname{Vect}(F)$. *Let*

$$\mathcal{F} = \bigcup_{W} \{\hom_F(U, V; W) \mid W \in \mathcal{S}\}$$

be the family of all bilinear maps from $U \times V$ *to any vector space* W. *The measuring family* $\mathcal{H}$ *is the family of all linear transformations.*

A pair $(T, t: U \times V \to T)$ *is* **universal for bilinearity** *if it is universal for* $(\mathcal{F}, \mathcal{H})$, *that is, if for every bilinear map* $f: U \times V \to W$, *there is a* unique *linear transformation* $\tau: T \to W$ *for which*

$$f = \tau \circ t$$

The map τ *is called the* **mediating morphism** *for* f.□

We can now define the tensor product via this universal property.

Definition *Let* U *and* V *be vector spaces over a field* F. *Any universal pair* $(T, t: U \times V \to T)$ *for bilinearity is called a* **tensor product** *of* U *and* V. *The vector space* T *is denoted by* $U \otimes V$ *and sometimes referred to by itself as the tensor product. The map t is called the* **tensor map** *and the elements of* $U \otimes V$ *are called* **tensors**.

It is customary to use the symbol $\otimes$ *to denote the image of any ordered pair* (u, v) *under the tensor map, that is,*

$$u \otimes v = t(u, v)$$

for any $u \in U$ *and* $v \in V$. *A tensor of the form* $u \otimes v$ *is said to be* **decomposable**, *that is, the decomposable tensors are the images under the tensor map.*□

Since universal pairs are unique up to isomorphism, we may refer to "the" tensor product of vector spaces. Note also that the tensor product $\otimes$ is not a product in the sense of a binary operation on a set. In fact, even when $V = U$, the tensor product $u \otimes u$ is not in U, but rather in $U \otimes U$.

As we will see, there are other, more constructive ways to define the tensor product. Since we have adopted the universal pair definition, the other ways to define tensor product are, for us, constructions rather than definitions. Let us examine some of these constructions.

Construction I: Intuitive but Not Coordinate Free

The universal property for bilinearity captures the essence of bilinearity and the tensor map is the most "general" bilinear function on $U \times V$. To see how this universality can be achieved in a constructive manner, let $\{e_i \mid i \in I\}$ be a basis for U and let $\{f_j \mid j \in J\}$ be a basis for V. Then a bilinear map t on $U \times V$ is uniquely determined by assigning arbitrary values to the "basis" pairs (e_i, f_j) and extending by bilinearity, that is, if $u = \sum \alpha_i e_i$ and $v = \sum \beta_j f_j$, then

$$t(u,v) = t\Big(\sum \alpha_i e_i, \sum \beta_j f_j\Big) = \sum \alpha_i \beta_j t(e_i, f_j)$$

Now, the tensor map t, being the most general bilinear map, must do this *and nothing more.* To achieve this goal, we define the tensor map t on the pairs (e_i, f_j) in such a way that the images $t(e_i, f_j)$ *do not interact*, and then extend by bilinearity.

In particular, for each ordered pair (e_i, f_j), we invent a new formal symbol, written $e_i \otimes f_j$, and define T to be the vector space with basis

$$\mathcal{D} = \{e_i \otimes f_j \mid i \in I, j \in J\}$$

The tensor map is defined by setting $t(e_i, f_j) = e_i \otimes f_j$ and extending by bilinearity. Thus,

$$t(u,v) = t\Big(\sum \alpha_i e_i, \sum \beta_j f_j\Big) = \sum \alpha_i \beta_j (e_i \otimes f_j)$$

To see that the pair (T, t) is the tensor product of U and V, if $g: U \times V \to W$ is bilinear, the universality condition $g = \tau \circ t$ is equivalent to

$$\tau(e_i \otimes f_j) = g(e_i, f_j)$$

which does indeed uniquely define a *linear* map $\tau: T \to W$. Hence, (T, t) has the universal property for bilinearity and so we can write $T = U \otimes V$ and refer to t as the tensor map.

Note that while the set $\mathcal{D} = \{e_i \otimes f_j\}$ is a basis for T (by definition), the set

$$\{u \otimes v \mid u \in U, v \in V\}$$

of decomposable tensors spans T, but is not linearly independent. This does cause some initial confusion during the learning process. For example, one cannot define a linear map on $U \otimes V$ by assigning values arbitrarily to the decomposable tensors, nor is it always easy to tell when a tensor $\sum u_i \otimes v_j$ is equal to 0. We will consider the latter issue in some detail a bit later in the chapter.

The fact that $\mathcal{D}$ is a basis for $U \otimes V$ gives the following.

Theorem 14.2 *For finite-dimensional vector spaces U and V,*

$$\dim(U \otimes V) = \dim(U) \cdot \dim(V)$$ □

Construction II: Coordinate Free

The previous construction of the tensor product is reasonably intuitive, but has the disadvantage of not being coordinate free. The following approach does not require the choice of a basis.

Let $F_{U \times V}$ be the vector space over F with basis $U \times V$. Let S be the subspace of $F_{U \times V}$ generated by all vectors of the form

$$r(u,w) + s(v,w) - (ru + sv, w) \tag{14.1}$$

and

$$r(u,v) + s(u,w) - (u, rv + sw) \tag{14.2}$$

where $r, s \in F$ and u, v and w are in the appropriate spaces. Note that these vectors are precisely what we must "identify" as the zero vector in order to enforce bilinearity. Put another way, these vectors are 0 if the ordered pairs are replaced by tensors according to our previous construction.

Accordingly, the quotient space

$$U \otimes V := \frac{F_{U \times V}}{S}$$

is also sometimes taken as the definition of the tensor product of U and V. (Strictly speaking, we should not be using the symbol $U \otimes V$ until we have shown that this is the tensor product.) The elements of $U \otimes V$ have the form

$$\left(\sum r_i(u_i, v_i)\right) + S = \sum r_i[(u_i, v_i) + S]$$

However, since $r(u,v) - (ru, v) \in S$ and $r(u,v) - (u, rv) \in S$, we can absorb the scalar in either coordinate, that is,

$$r[(u,v) + S] = (ru, v) + S = (u, rv) + S$$

and so the elements of $U \otimes V$ can be written simply as

$$\sum [(u_i, v_i) + S]$$

It is customary to denote the coset $(u,v) + S$ by $u \otimes v$, and so any element of

$U \otimes V$ has the form

$$\sum u_i \otimes v_i$$

as in the previous construction.

The tensor map $t: U \times V \to U \otimes V$ is defined by

$$t(u, v) = u \otimes v = (u, v) + S$$

This map is bilinear, since

$$\begin{aligned} t(au + bv, w) &= (ru + sv, w) + S \\ &= [r(u, w) + s(v, w)] + S \\ &= [r(u, w) + S] + [s(v, w) + S] \\ &= rt(u, w) + st(v, w) \end{aligned}$$

and similarly for the second coordinate.

We next prove that the pair $(U \otimes V, t: U \times V \to U \otimes V)$ is universal for bilinearity when $U \otimes V$ is defined as a quotient space $F_{U \times V}/S$.

Theorem 14.3 *Let U and V be vector spaces. The pair*

$$(U \otimes V, t: U \times V \to U \otimes V)$$

is the tensor product of U and V.

Proof. Consider the diagram in Figure 14.5. Here $F_{U \times V}$ is the vector space with basis $U \times V$.

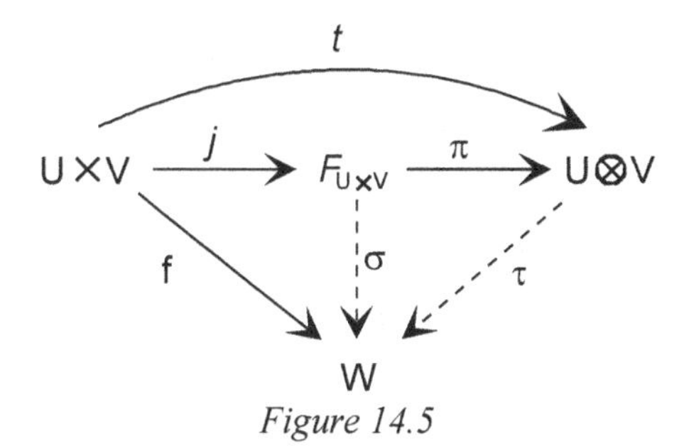

Figure 14.5

Since

$$\pi \circ j(u, v) = \pi(u, v) = (u, v) + S = u \otimes v = t(u, v)$$

we have

$$t = \pi \circ j$$

The universal property of vector spaces described in Example 14.1 implies that there is a unique linear transformation $\sigma: F_{U\times V} \to W$ for which

$$\sigma \circ j = f$$

Note that σ sends the vectors (14.1) and (14.2) that generate S to the zero vector and so $S \subseteq \ker(\sigma)$. For example,

$$\begin{aligned}\sigma[r(u,w)+s(v,w)-(ru+sv,w)] \\ &= \sigma[rj(u,w)+sj(v,w)-j(ru+sv,w)] \\ &= r\sigma j(u,w)+s\sigma j(v,w)-\sigma j(ru+sv,w) \\ &= rf(u,w)+sf(v,w)-f(ru+sv,w) \\ &= 0\end{aligned}$$

and similarly for the second coordinate. Hence, Theorem 3.4 (the universal property described in Example 14.2) implies that there exists a unique linear transformation $\tau: U \otimes V \to W$ for which

$$\tau \circ \pi = \sigma$$

Hence,

$$\tau \circ t = \tau \circ \pi \circ j = \sigma \circ j = f$$

As to uniqueness, if $\tau' \circ t = f$, then

$$\tau'[(u,v)+S] = f(u,v) = \tau[(u,v)+S]$$

and since the cosets $(u,v)+S$ generate $F_{U\times V}/S$, we conclude that $\tau' = \tau$. Thus, τ is the mediating morphism and $(U \otimes V, t)$ is universal for bilinearity.□

Let us take a moment to compare the two previous constructions. Let $\{e_i \mid i \in I\}$ and $\{f_j \mid j \in J\}$ be bases for U and V, respectively. Let (T', t') be the tensor product as constructed using these two bases and let $(T,t) = (F_{U\times V}/S, t)$ be the tensor product construction using quotient spaces.

Since both of these pairs are universal for bilinearity, Theorem 14.1 implies that the mediating morphism τ for t with respect to t', that is, the map $\tau: T' \to T$ defined by

$$\tau(e_i \otimes f_j) = (e_i, f_j) + S$$

is a vector space isomorphism. Therefore, the basis $\{(e_i \otimes f_j)\}$ of T' is sent to the set $\{(e_i, f_j) + S\}$, which is therefore a basis for T.

In other words, given any two bases $\{e_i \mid i \in I\}$ and $\{f_j \mid j \in J\}$ for U and V, respectively, the tensors $e_i \otimes f_j$ form a basis for $U \otimes V$, regardless of which construction of the tensor product we use. Therefore, we are free to think of $e_i \otimes f_j$ either as a formal symbol belonging to a basis for $U \otimes V$ or as the coset $(e_i, f_j) + S$ belonging to a basis for $U \otimes V$.

Bilinearity on $U \times V$ Equals Linearity on $U \otimes V$

The universal property for bilinearity says that to each *bilinear* function $f: U \times V \to W$, there corresponds a unique *linear* function $\tau: U \otimes V \to W$, called the mediating morphism for f. Thus, we can define the **mediating morphism map**

$$\phi: \hom(U, V; W) \to \mathcal{L}(U \otimes V, W)$$

by setting $\phi f = \tau$. In other words, ϕf is the unique linear map for which

$$(\phi f)(u \otimes v) = f(u, v)$$

Observe that ϕ is itself linear, since if $f, g \in \hom(U, V; W)$, then

$$[r\phi(f) + s\phi(g)](u \otimes v) = rf(u, v) + sg(u, v) = (rf + sg)(u, v)$$

and so $r\phi(f) + s\phi(g)$ is the mediating morphism for $rf + sg$, that is,

$$r\phi(f) + s\phi(g) = \phi(rf + sg)$$

Also, ϕ is surjective, since if $\tau: U \otimes V \to W$ is any linear map, then $f = \tau \circ t: U \times V \to W$ is bilinear and has mediating morphism τ, that is, $\phi f = \tau$. Finally, ϕ is injective, for if $\phi f = 0$, then $f = \phi f \circ t = 0$. We have established the following result.

Theorem 14.4 *Let U, V and W be vector spaces over F. Then the mediating morphism map $\phi: \hom(U, V; W) \to \mathcal{L}(U \otimes V, W)$, where ϕf is the unique linear map satisfying $f = \phi f \circ t$, is an isomorphism and so*

$$\phi: \hom(U, V; W) \approx \mathcal{L}(U \otimes V, W) \qquad \square$$

When Is a Tensor Product Zero?

Armed with the universal property of bilinearity, we can now discuss some of the basic properties of tensor products. Let us first consider the question of when a tensor $\sum u_i \otimes v_j$ is zero.

The bilinearity of the tensor product gives

$$0 \otimes v = (0 + 0) \otimes v = 0 \otimes v + 0 \otimes v$$

and so $0 \otimes v = 0$. Similarly, $u \otimes 0 = 0$. Now suppose that

$$\sum_i u_i \otimes v_i = 0$$

where we may assume that none of the vectors u_i and v_i are 0. Let $f: U \times V \to W$ be a bilinear map and let $\tau: U \otimes V \to W$ be its mediating morphism, that is, $\tau \circ t = f$. Then

$$0=\tau\Big(\sum_i u_i\otimes v_i\Big)=\sum_i(\tau\circ t)(u_i,v_i)=\sum_i f(u_i,v_i)$$

The key point is that this holds for *any* bilinear function $f:U\times V\to W$. In particular, let $\alpha\in U^*$ and $\beta\in V^*$and define f by

$$f(u,v)=\alpha(u)\beta(v)$$

which is easily seen to be bilinear. Then the previous display becomes

$$\sum_i\alpha(u_i)\beta(v_i)=0$$

If, for example, the vectors u_i are linearly independent, we can take α to be a dual vector u_k^* to get

$$0=\sum_i u_k^*(u_i)\beta(v_i)=\beta(v_k)$$

and since this holds for all linear functionals $\beta\in V^*$, it follows that $v_k=0$. We have proved the following useful result.

Theorem 14.5 *If $u_1,\dots,u_n$ are linearly independent vectors in U and $v_1,\dots,v_n$ are arbitrary vectors in V, then*

$$\sum u_i\otimes v_i=0\quad\Rightarrow\quad v_i=0\text{ for all }i$$

In particular, $u\otimes v=0$ if and only if $u=0$ or $v=0$.□

Coordinate Matrices and Rank

If $\mathcal{B}=\{u_i\mid i\in I\}$ is a basis for U and $\mathcal{C}=\{v_j\mid j\in J\}$ is a basis for V, then any vector $z\in U\otimes V$ has a unique expression as a sum

$$z=\sum_{i\in I}\sum_{j\in J}r_{i,j}(u_i\otimes v_j)$$

where only a finite number of the coefficients $r_{i,j}$ are nonzero. In fact, for a fixed $z\in U\otimes V$, we may reindex the bases so that

$$z=\sum_{i=1}^{a}\sum_{j=1}^{b}r_{i,j}(u_i\otimes v_j)$$

where none of the rows or columns of the matrix $R=(r_{i,j})$ consists only of 0's. The matrix $R=(r_{i,j})$ is called a **coordinate matrix** of z with respect to the bases $\mathcal{B}$ and $\mathcal{C}$.

Note that a coordinate matrix R is determined only up to the order of its rows and columns. We could remove this ambiguity by considering ordered bases, but this is not necessary for our discussion and adds a complication, since the bases may be infinite.

Suppose that $\mathcal{W} = \{w_i \mid i \in I\}$ and $\mathcal{X} = \{x_j \mid j \in J\}$ are also bases for U and V, respectively, and that

$$z = \sum_{i=1}^{c} \sum_{j=1}^{d} s_{i,j}(w_i \otimes x_j)$$

where $S = (s_{i,j})$ is a coordinate matrix of z with respect to these bases. We claim that the coordinate matrices R and S have the same rank, which can then be defined as the **rank** of the tensor $z \in U \otimes V$.

Each $w_1, \ldots, w_c$ is a finite linear combination of basis vectors in $\mathcal{B}$, perhaps involving some of $u_1, \ldots, u_a$ and perhaps involving other vectors in $\mathcal{B}$. We can further reindex $\mathcal{B}$ so that each w_k is a linear combination of the vectors $\mathcal{B}' = (u_1, \ldots, u_n)$, where $a \leq n$ and set

$$U_n = \operatorname{span}(u_1, \ldots, u_n)$$

Next, extend $(w_1, \ldots, w_c)$ to a basis $\mathcal{W}' = (w_1, \ldots, w_c, w_{c+1}, \ldots, w_n)$ for U_n. (Since we no longer need the rest of the basis $\mathcal{W}$, we have commandeered the symbols $w_{c+1}, \ldots, w_n$, for simplicity.) Hence

$$w_i = \sum_{h=1}^{n} a_{i,h} u_h \text{ for } i = 1, \ldots, n$$

where $A = (a_{i,h})$ is invertible of size $n \times n$.

Now repeat this process on the second coordinate. Reindex the basis $\mathcal{C}$ so that the subspace $V_m = \operatorname{span}(v_1, \ldots, v_m)$ contains $x_1, \ldots, x_d$ and extend to a basis $\mathcal{X}' = (x_1, \ldots, x_d, x_{d+1}, \ldots, x_m)$ for V_m. Then

$$x_j = \sum_{k=1}^{m} b_{j,k} v_k \text{ for } j = 1, \ldots, m$$

where $B = (b_{j,k})$ is invertible of size $m \times m$.

Next, write

$$z = \sum_{i=1}^{n} \sum_{j=1}^{m} r_{i,j}(u_i \otimes v_j)$$

by setting $r_{i,j} = 0$ for $i > a$ or $j > b$. Thus, the $n \times m$ matrix $R_1 = (r_{i,j})$ comes from R by adding $n - a$ rows of 0's to the bottom and then $m - b$ columns of 0's. In particular, R_1 and R have the same rank.

The expression for z in terms of the basis vectors $w_1, \ldots, w_c$ and $x_1, \ldots, x_d$ can also be extended using 0 coefficients to

$$z = \sum_{i=1}^{n} \sum_{j=1}^{m} s_{i,j}(w_i \otimes x_j)$$

where the $n \times m$ matrix $S_1 = (s_{i,j})$ has the same rank as S.

Now at last, we can compute. First, bilinearity gives

$$w_i \otimes x_j = \sum_{h=1}^{n} \sum_{k=1}^{m} a_{i,h} b_{j,k}(u_h \otimes v_k)$$

and so

$$\begin{aligned} z = \sum_{i=1}^{n} \sum_{j=1}^{m} s_{i,j}(w_i \otimes x_j) &= \sum_{i=1}^{n} \sum_{j=1}^{m} s_{i,j} \left(\sum_{h=1}^{n} \sum_{k=1}^{m} a_{i,h} b_{j,k}(u_h \otimes v_k) \right) \\ &= \sum_{h=1}^{n} \sum_{k=1}^{m} \left(\sum_{j=1}^{m} \sum_{i=1}^{n} (a_{i,h} s_{i,j}) b_{j,k} \right) (u_h \otimes v_k) \\ &= \sum_{h=1}^{n} \sum_{k=1}^{m} \left(\sum_{j=1}^{m} (A^t S_1)_{h,j} b_{j,k} \right) (u_h \otimes v_k) \\ &= \sum_{h=1}^{n} \sum_{k=1}^{m} (A^t S_1 B)_{h,k} (u_h \otimes v_k) \end{aligned}$$

Thus

$$\sum_{i=1}^{n} \sum_{j=1}^{m} r_{i,j}(u_i \otimes v_j) = z = \sum_{h=1}^{n} \sum_{k=1}^{m} (A^t S_1 B)_{h,k}(u_h \otimes v_k)$$

and so $R_1 = A^t S_1 B$. Since A and B are invertible, we deduce that

$$\operatorname{rk}(R) = \operatorname{rk}(R_1) = \operatorname{rk}(S_1) = \operatorname{rk}(S)$$

as desired. Moreover, in block matrix terms, we can write

$$R_1 = \begin{bmatrix} R & 0 \\ 0 & 0 \end{bmatrix}_{\text{block}} \quad \text{and} \quad S_1 = \begin{bmatrix} S & 0 \\ 0 & 0 \end{bmatrix}_{\text{block}}$$

and if we write

$$A^t = \begin{bmatrix} A^t_{a,c} & * \\ * & * \end{bmatrix}_{\text{block}} \quad \text{and} \quad B = \begin{bmatrix} B_{d,b} & * \\ * & * \end{bmatrix}_{\text{block}}$$

then $R_1 = A^t S_1 B$ implies that

$$R = A^t_{a,c} S B_{d,b}$$

We shall soon have use for the following special case. If

$$z = \sum_{i=1}^{r} u_i \otimes v_i = \sum_{i=1}^{r} w_i \otimes x_i \tag{14.3}$$

then $R = S = I_r$ and so

$$w_i = \sum_{h=1}^{r} a_{i,h} u_h \text{ for } i = 1, \ldots, r$$

and

$$x_j = \sum_{k=1}^{r} b_{j,k} v_k \text{ for } j = 1, \ldots, r$$

where if $A_{r,r} = (a_{i,h})$ and $B_{r,r} = (b_{j,k})$, then

$$I_r = A^t_{r,r} B_{r,r}$$

The Rank of a Decomposable Tensor

Recall that a tensor of the form $u \otimes v$ is said to be decomposable. If $\{u_i \mid i \in I\}$ is a basis for U and $\{v_j \mid j \in J\}$ is a basis for V, then any decomposable vector has the form

$$u \otimes v = \sum_{i,j} r_i s_j (u_i \otimes v_j)$$

Hence, the rank of a decomposable vector is 1, since the rank of a matrix whose (i,j)th entry is $r_i s_j$ is 1.

Characterizing Vectors in a Tensor Product

There are several useful representations of the tensors in $U \otimes V$.

Theorem 14.6 *Let $\{u_i \mid i \in I\}$ be a basis for U and let $\{v_j \mid j \in J\}$ be a basis for V. By an "essentially unique" sum, we mean unique up to order and presence of zero terms.*

1) Every $z \in U \otimes V$ has an essentially unique expression as a finite sum of the form

$$\sum_{i,j} r_{i,j} u_i \otimes v_j$$

where $r_{i,j} \in F$ and the tensors $u_i \otimes v_j$ are distinct.

2) *Every $z \in U \otimes V$ has an essentially unique expression as a finite sum of the form*

$$\sum_i u_i \otimes y_i$$

where $y_i \in V$ and the u_i's are distinct.

3) *Every $z \in U \otimes V$ has an essentially unique expression as a finite sum of the form*

$$\sum_i x_i \otimes v_i$$

where $x_i \in U$ and the v_i's are distinct.

4) *Every nonzero $z \in U \otimes V$ has an expression of the form*

$$z = \sum_{i=1}^{n} x_i \otimes y_i$$

where the x_i's are distinct, the y_i's are distinct and the sets $\{x_i\} \subseteq U$ and $\{y_i\} \subseteq V$ are linearly independent. As to uniqueness, n is the rank of z and so it is unique. Also, the equation

$$\sum_{i=1}^{r} x_i \otimes y_i = \sum_{i=1}^{r} w_i \otimes z_i$$

where the w_i's are distinct, the z_i's are distinct and $\{w_i\} \subseteq U$ and $\{z_i\} \subseteq V$ are linearly independent, holds if and only if there exist invertible $r \times r$ matrices $A = (a_{i,j})$ and $B = (b_{i,j})$ for which $A^t B = I$ and

$$w_i = \sum_{j=1}^{r} a_{i,j} x_j \quad and \quad z_i = \sum_{j=1}^{r} b_{i,j} y_j$$

for $i = 1, \ldots, r$.

Proof. Part 1) merely expresses the fact that $\{u_i \otimes v_j\}$ is a basis for $U \otimes V$. From part 2), we write

$$\sum_{i,j} r_{i,j} u_i \otimes v_j = \sum_i \left[u_i \otimes \sum_j r_{i,j} v_j \right] = \sum_i u_i \otimes y_i$$

Uniqueness follows from Theorem 14.5. Part 3) is proved similarly. As to part 4), we start with the expression from part 2):

$$\sum_{i=1}^{n} u_i \otimes y_i$$

where we may assume that none of the y_i's are 0. If the set $\{y_i\}$ is linearly independent, we are done. If not, then we may suppose (after reindexing if necessary) that

$$y_n = \sum_{i=1}^{n-1} r_i y_i$$

Then

$$\begin{aligned}
\sum_{i=1}^{n} u_i \otimes y_i &= \sum_{i=1}^{n-1} u_i \otimes y_i + \left(u_n \otimes \sum_{i=1}^{n-1} r_i y_i\right) \\
&= \sum_{i=1}^{n-1} u_i \otimes y_i + \sum_{i=1}^{n-1} (r_i u_n \otimes y_i) \\
&= \sum_{i=1}^{n-1} (u_i + r_i u_n) \otimes y_i
\end{aligned}$$

But the vectors $\{u_i + r_i u_n \mid 1 \le i \le n-1\}$ are linearly independent. This reduction can be repeated until the second coordinates are linearly independent. Moreover, the identity matrix I_n is a coordinate matrix for z and so $n = \text{rk}(I_n) = \text{rk}(z)$. As to uniqueness, one direction was proved earlier; see (14.3) and the other direction is left to the reader.□

The proof of Theorem 14.6 shows that if $z \neq 0$ and

$$z = \sum_{i \in I} s_i \otimes t_i$$

where $s_i \in U$ and $t_i \in V$, then if the multiset $\{s_i \mid i \in I\}$ is not linearly independent, we can rewrite z in the form

$$z = \sum_{i \in I_0} s_i \otimes t_i'$$

where $\{s_i \mid i \in I_0\}$ is linearly independent. Then we can do the same for the second coordinate to arrive so at the representation

$$z = \sum_{i=1}^{\text{rk}(x)} x_i \otimes y_i$$

where the multisets $\{x_i\}$ and $\{y_i\}$ are linearly independent sets. Therefore, $\text{rk}(x) \le |I|$ and so the rank of z is the *smallest* integer m for which z can be written as a sum of m decomposable tensors. This is often taken as the definition of the rank of a tensor.

However, we caution the reader that there is another meaning to the word rank when applied to a tensor, namely, it is the number of indices required to write the tensor. Thus, a scalar has rank 0, a vector has rank 1, the tensor z above has rank 2 and a tensor of the form

$$z = \sum_{i \in I} s_i \otimes t_i \otimes u_i$$

has rank 3.

Defining Linear Transformations on a Tensor Product

One of the simplest and most useful ways to define a linear transformation σ on the tensor product $U \otimes V$ is through the universal property, for this property says precisely that a bilinear function f on $U \times V$ gives rise to a unique (and well-defined) linear transformation on $U \otimes V$. The proof of the following theorem illustrates this well.

Theorem 14.7 *Let U and V be vector spaces. There is a unique linear transformation*

$$\theta: U^* \otimes V^* \to (U \otimes V)^*$$

defined by $\theta(f \otimes g) = f \odot g$ where

$$(f \odot g)(u \otimes v) = f(u)g(v)$$

Moreover, θ is an embedding and is an isomorphism if U and V are finite-dimensional. Thus, the tensor product $f \otimes g$ of linear functionals is (via this embedding) a linear functional on tensor products.

Proof. Informally, for fixed f and g, the function $(u, v) \to f(u)g(v)$ is bilinear in u and v and so there is a unique linear map $f \odot g$ taking $u \otimes v$ to $f(u)g(v)$. The function $(f, g) \to f \odot g$ is bilinear in f and g since

$$(rf + sg) \odot h = r(f \odot h) + s(g \odot h)$$

and so there is a unique linear map θ taking $f \otimes g$ to $f \odot g$.

More formally, for fixed f and g, the map $F_{f,g}: U \times V \to F$ defined by

$$F_{f,g}(u, v) = f(u)g(v)$$

is bilinear and so the universal property of tensor products implies that there exists a unique $f \odot g \in (U \otimes V)^*$ for which

$$(f \odot g)(u \otimes v) = f(u)g(v)$$

Next, the map $G: U^* \times V^* \to (U \otimes V)^*$ defined by

$$G(f, g) = f \odot g$$

is bilinear since, for example,

$$\begin{aligned}[(rf+sg)\odot h](u\otimes v)&=(rf+sg)(u)\cdot h(v)\\&=rf(u)h(v)+sg(u)h(v)\\&=[r(f\odot h)+s(g\odot h)](u\otimes v)\end{aligned}$$

which shows that G is linear in its first coordinate. Hence, the universal property implies that there exists a unique linear map

$$\theta: U^*\otimes V^*\to(U\otimes V)^*$$

for which

$$\theta(f\otimes g)=f\odot g$$

To see that θ is an injection, if $h\in U^*\otimes V^*$ is nonzero, then we may write h in the form

$$h=\sum_{i=1}^{n}f_i\otimes g_i$$

where the $f_i\in U^*$ are nonzero and $\{g_i\mid 1\le i\le n\}\subseteq V^*$ is linearly independent. If $\theta(h)=0$, then for any $u\in U$ and $v\in V$, we have

$$0=\theta(h)(u\otimes v)=\sum_{i=1}^{n}\theta(f_i\otimes g_i)(u\otimes v)=\sum_{i=1}^{n}f_i(u)g_i(v)$$

Hence, for each nonzero $u\in U$, the linear functional

$$\sum_{i=1}^{n}f_i(u)g_i$$

is the zero map and so the linear independence of $\{g_i\}$ implies that $f_i(u)=0$ for all i. Since u is arbitrary, it follows that $f_i=0$ for all i and so $h=0$.

Finally, in the finite-dimensional case, the map θ is a bijection since

$$\dim(U^*\otimes V^*)=\dim((U\otimes V)^*)<\infty$$

□

Combining the isomorphisms of Theorem 14.4 and Theorem 14.7, we have, for finite-dimensional vector spaces U and V,

$$U^*\otimes V^*\approx(U\otimes V)^*\approx\hom(U,V;F)$$

The Tensor Product of Linear Transformations

We wish to generalize Theorem 14.7 to arbitrary linear transformations. Let $\tau\in\mathcal{L}(U,U')$ and $\sigma\in\mathcal{L}(V,V')$. While the product $\tau(u)\sigma(v)$ does not make sense, the *tensor* product $\tau u\otimes\sigma v$ does and is bilinear in u and v, that is, the following function is bilinear:

$$f(u,v) = \tau u \otimes \sigma v$$

The same argument that we used in the proof of Theorem 14.7 will work here. Namely, the map $(u,v) \mapsto \tau u \otimes \sigma v$ from $U \times V$ to $U' \otimes V'$ is bilinear in u and v and so there is a unique linear map $(\tau \odot \sigma): U \otimes V \to U' \otimes V'$ for which

$$(\tau \odot \sigma)(u \otimes v) = \tau u \otimes \sigma v$$

The function

$$\phi: \mathcal{L}(U,U') \times \mathcal{L}(V,V') \to \mathcal{L}(U \otimes V, U' \otimes V')$$

defined by

$$\phi(\tau,\sigma) = \tau \odot \sigma$$

is bilinear, since

$$\begin{aligned}
((a\tau + b\mu) \odot \sigma)(u \otimes v) &= (a\tau + b\mu)(u) \otimes \sigma v \\
&= (a\tau u + b\mu u) \otimes \sigma v \\
&= a[\tau u \otimes \sigma v] + b[\mu u \otimes \sigma v] \\
&= a(\tau \odot \sigma)(u \otimes v) + b(\mu \odot \sigma)(u \otimes v) \\
&= (a(\tau \odot \sigma) + b(\mu \odot \sigma))(u \otimes v)
\end{aligned}$$

and similarly for the second coordinate. Hence, there is a unique linear transformation

$$\theta: \mathcal{L}(U,U') \otimes \mathcal{L}(V,V') \to \mathcal{L}(U \otimes V, U' \otimes V')$$

satisfying

$$\theta(\tau \otimes \sigma) = \tau \odot \sigma$$

that is,

$$[\theta(\tau \otimes \sigma)](u \otimes v) = \tau u \otimes \sigma v$$

To see that θ is injective, if $h \in \mathcal{L}(U,U') \otimes \mathcal{L}(V,V')$ is nonzero, then we may write

$$h = \sum_{i=1}^{n} f_i \otimes g_i$$

where the $f_i \in \mathcal{L}(U,U')$ are nonzero and the set $\{g_i\} \subseteq \mathcal{L}(V,V')$ is linearly independent. If $\theta(h) = 0$, then for all $u \in U$ and $v \in V$ we have

$$0 = \theta(h)(u \otimes v) = \sum_{i=1}^{n} \theta(f_i \otimes g_i)(u \otimes v) = \sum_{i=1}^{n} f_i(u) \otimes g_i(v)$$

Since $h \neq 0$, it follows that $f_i \neq 0$ for some i and so we may choose a $u \in U$ such that $f_i(u) \neq 0$ for some i. Moreover, we may assume, by reindexing if necessary, that the set $\{f_1(u),\ldots,f_m(u)\}$ is a maximal linearly independent subset of $\{f_1(u),\ldots,f_n(u)\}$. Hence, for each $k>m$, we have

$$f_k(u)=\sum_{i=1}^{m}\alpha_{k,i}f_i(u)$$

and so

$$\begin{aligned}
0&=\sum_{i=1}^{n}f_i(u)\otimes g_i(v)\\
&=\sum_{i=1}^{m}f_i(u)\otimes g_i(v)+\sum_{k=m+1}^{n}\left[\sum_{i=1}^{m}\alpha_{k,i}f_i(u)\right]\otimes g_k(v)\\
&=\sum_{i=1}^{m}f_i(u)\otimes g_i(v)+\sum_{k=m+1}^{n}\sum_{i=1}^{m}\alpha_{k,i}[f_i(u)\otimes g_k(v)]\\
&=\sum_{i=1}^{m}f_i(u)\otimes g_i(v)+\sum_{i=1}^{m}f_i(u)\otimes\left[\sum_{k=m+1}^{n}\alpha_{k,i}g_k(v)\right]\\
&=\sum_{i=1}^{m}f_i(u)\otimes\left[g_i(v)+\sum_{k=m+1}^{n}\alpha_{k,i}g_k(v)\right]
\end{aligned}$$

Thus, the linear independence of $\{f_1(u),\ldots,f_m(u)\}$ implies that for each $i\le m$,

$$g_i(v)+\sum_{k=m+1}^{n}\alpha_{k,i}g_k(v)=0$$

for all $v\in V$ and so

$$g_i+\sum_{k=m+1}^{n}\alpha_{k,i}g_k=0$$

But this contradicts the fact that the set $\{g_i\}$ is linearly independent. Hence, it cannot happen that $\theta(h)=0$ for $h\neq 0$ and so θ is injective.

The embedding of $\mathcal{L}(U,U')\otimes\mathcal{L}(V,V')$ into $\mathcal{L}(U\otimes V,U'\otimes V')$ means that each $\tau\otimes\sigma$ can be thought of as the linear transformation $\tau\odot\sigma$ from $U\otimes V$ to $U'\otimes V'$, defined by

$$(\tau\odot\sigma)(u\otimes v)=\tau u\otimes\sigma v$$

In fact, the notation $\tau\otimes\sigma$ is often used to denote both the tensor product of vectors (linear transformations) and the linear map $\tau\odot\sigma$, and we will do this as well. In summary, we can say that the tensor product $\tau\otimes\sigma$ of linear transformations is (up to isomorphism) a linear transformation on tensor products.

Theorem 14.8 *There is a unique linear transformation*

$$\theta: \mathcal{L}(U,U') \otimes \mathcal{L}(V,V') \to \mathcal{L}(U \otimes V, U' \otimes V')$$

defined by $\theta(\tau \otimes \sigma) = \tau \odot \sigma$ *where*

$$(\tau \odot \sigma)(u \otimes v) = \tau u \otimes \sigma v$$

Moreover, θ *is an embedding and is an isomorphism if all vector spaces are finite-dimensional. Thus, the tensor product* $\tau \otimes \sigma$ *of linear transformations is (via this embedding) a linear transformation on tensor products.* □

Let us note a few special cases of the previous theorem.

Corollary 14.9 *Let us use the symbol* $X \overset{\sim}{\hookrightarrow} Y$ *to denote the fact that there is an embedding of* X *into* Y *that is an isomorphism if* X *and* Y *are finite-dimensional.*

1) *Taking* $U' = F$ *gives*

$$U^* \otimes \mathcal{L}(V,V') \overset{\sim}{\hookrightarrow} \mathcal{L}(U \otimes V, V')$$

where

$$(f \otimes \sigma)(u \otimes v) = f(u)\sigma(v)$$

for $f \in U^*$.

2) *Taking* $U' = F$ *and* $V' = F$ *gives*

$$U^* \otimes V^* \overset{\sim}{\hookrightarrow} (U \otimes V)^*$$

where

$$(f \otimes g)(u \otimes v) = f(u)g(v)$$

3) *Taking* $V = F$ *and noting that* $\mathcal{L}(F,V') \approx V'$ *and* $U \otimes F \approx U$ *gives (letting* $W = V'$*)*

$$\mathcal{L}(U,U') \otimes W \overset{\sim}{\hookrightarrow} \mathcal{L}(U, U' \otimes W)$$

where

$$(\tau \otimes w)(u) = \tau u \otimes w$$

4) *Taking* $U' = F$ *and* $V = F$ *gives (letting* $W = V'$*)*

$$U^* \otimes W \overset{\sim}{\hookrightarrow} \mathcal{L}(U,W)$$

where

$$(f \otimes w)(u) = f(u)w$$

□

Change of Base Field

The tensor product provides a convenient way to extend the base field of a vector space that is more general than the complexification of a real vector space, discussed earlier in the book. We refer to a vector space over a field F as an **F-space** and write V_F.

Actually, there are several approaches to "upgrading" the base field of a vector space. For instance, suppose that K is an extension field of F, that is, $F \subseteq K$. If $\{b_i\}$ is a basis for V_F, then every $x \in V_F$ has the form

$$x = \sum r_i b_i$$

where $r_i \in F$. We can define a K-space V_K simply by taking all formal linear combinations of the form

$$x = \sum \alpha_i b_i$$

where $\alpha_i \in K$. Note that the dimension of V_K as a K-space is the same as the dimension of V_F as an F-space. Also, V_K is an F-space (just restrict the scalars to F) and as such, the inclusion map $j: V_F \to V_K$ sending $x \in V_F$ to $j(x) = x \in V_K$ is an F-monomorphism.

The approach described in the previous paragraph uses an arbitrarily chosen basis for V_F and is therefore not coordinate free. However, we can give a coordinate-free approach using tensor products as follows. Since K is a vector space over F, we can form the tensor product

$$W_F = K \otimes_F V_F$$

It is customary to include the subscript F on $\otimes_F$ to denote the fact that the tensor product is taken with respect to the base field F. (All relevant maps are F-bilinear and F-linear.) However, since V_F is not a K-space, the only tensor product of K and V_F that makes sense is the F-tensor product and so we will drop the subscript F.

The tensor product W_F is an F-space by definition of tensor product, but we can make it into a K-space as follows. For $\alpha \in K$, the temptation is to "absorb" the scalar α into the first coordinate,

$$\alpha(\beta \otimes v) = (\alpha\beta) \otimes v$$

but we must be certain that this is well-defined, that is,

$$\beta \otimes v = \gamma \otimes w \quad \Rightarrow \quad (\alpha\beta) \otimes v = (\alpha\gamma) \otimes w$$

But for a fixed α, the map $(\beta, v) \mapsto (\alpha\beta) \otimes v$ is bilinear and so the universal property of tensor products implies that there is a unique linear map $\beta \otimes v \mapsto (\alpha\beta) \otimes v$, which we define to be scalar multiplication by α.

To be absolutely clear, we have two distinct vector spaces: the F-space $W_F = K \otimes V_F$ defined by the tensor product and the K-space $W_K = K \otimes V_F$ with scalar multiplication by elements of K defined as absorption into the first coordinate. The spaces W_F and W_K are identical as sets and as abelian groups. It is only the "permission to multiply by" that is different. Accordingly, we can recover W_F from W_K simply by restricting scalar multiplication to scalars from F.

Thus, we can speak of "F-linear" maps τ from V_F into W_K, with the expected meaning, that is,

$$\tau(ru + sv) = r\tau u + s\tau v$$

for all scalars $r, s \in F$.

If the dimension of K as a vector space over F is d, then

$$\dim_F(W_F) = \dim_F(K \otimes V_F) = d \cdot \dim_F(V_F)$$

As to the dimension of W_K, it is not hard to see that if $\{b_i\}$ is a basis for V_F, then $\{1 \otimes b_i\}$ is a basis for W_K. Hence

$$\dim_K(W_K) = \dim_F(V_F)$$

The map $\mu: V_F \to W_F$ defined by $\mu v = 1 \otimes v$ is easily seen to be injective and F-linear and so W_F contains an isomorphic copy of V_F. We can also think of μ as mapping V_F into W_K, in which case μ is called the **K-extension map** of V_F. This map has a universal property of its own, as described in the next theorem.

Theorem 14.10 *The F-linear K-extension map $\mu: V_F \to K \otimes V_F$ has the universal property for the family of all F-linear maps from V_F into a K-space, as measured by K-linear maps. Specifically, for any F-linear map $f: V_F \to Y$, where Y is a K-space, there exists a unique K-linear map $\tau: K \otimes V_F \to Y$ for which the diagram in Figure 14.6 commutes, that is, for which*

$$\tau \circ \mu = f$$

Proof. If such a K-linear map $\tau: K \otimes V_F \to Y$ is to exist, then it must satisfy, for any $\beta \in K$,

$$\tau(\beta \otimes v) = \beta\tau(1 \otimes v) = \beta\tau\mu(v) = \beta f(v)$$

This shows that if τ exists, it is uniquely determined by f. As usual, when searching for a linear map τ on a tensor product such as $K \otimes V_F$, we look for a bilinear map. The map $g: (K \times V_F) \to Y$ defined by

$$g(\beta, v) = \beta f(v)$$

is bilinear and so there exists a unique F-linear map τ for which

$$\tau(\beta \otimes v) = \beta f(v)$$

It is easy to see that τ is also K-linear, since if $\alpha \in K$, then

$$\tau[\alpha(\beta \otimes v)] = \tau(\alpha\beta \otimes v) = \alpha\beta f(v) = \alpha\tau(\beta \otimes v)$$ □

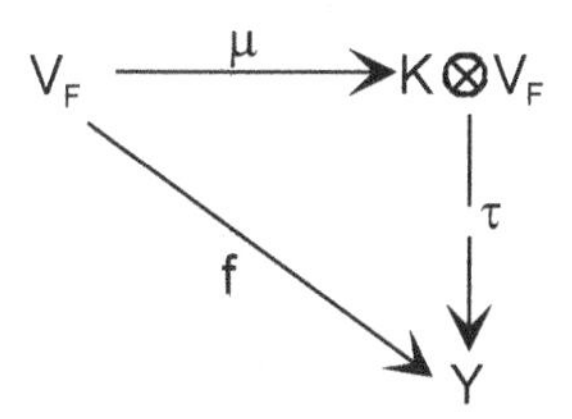

Figure 14.6

Theorem 14.10 is the key to describing how to extend an F-linear map to a K-linear map. Figure 14.7 shows an F-linear map $\tau: V \to W$ between F-spaces V and W. It also shows the K-extensions for both spaces, where $K \otimes V$ and $K \otimes W$ are K-spaces.

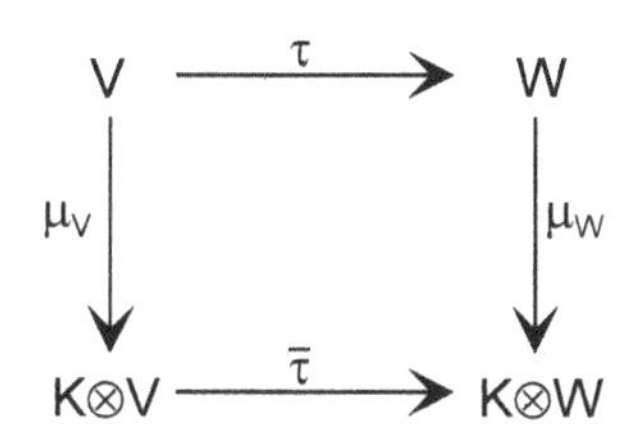

Figure 14.7

If there is a unique K-linear map $\overline{\tau}$ that makes the diagram in Figure 14.7 commute, then this would be the obvious choice for the extension of the F-linear map τ to a K-linear map.

Consider the F-linear map $\sigma = (\mu_W \circ \tau): V \to K \otimes W$ into the K-space $K \otimes W$. Theorem 14.10 implies that there is a unique K-linear map $\overline{\tau}: K \otimes V \to K \otimes W$ for which

$$\overline{\tau} \circ \mu_V = \sigma$$

that is,

$$\overline{\tau} \circ \mu_V = \mu_W \circ \tau$$

Now, $\overline{\tau}$ satisfies

$$\begin{aligned}\overline{\tau}(\beta \otimes v) &= \beta\overline{\tau}(1 \otimes v) \\ &= \beta(\overline{\tau} \circ \mu_V)(v) \\ &= \beta(\mu_W \circ \tau)(v) \\ &= \beta(1 \otimes \tau v) \\ &= \beta \otimes \tau v \\ &= (\iota_K \otimes \tau)(\beta \otimes v)\end{aligned}$$

and so $\overline{\tau} = \iota_K \otimes \tau$.

Theorem 14.11 *Let V and W be F-spaces, with K-extension maps μ_V and μ_W, respectively. (See Figure 14.7.) Then for any F-linear map $\tau: V \to W$, the map $\iota_K \otimes \tau: K \otimes V \to K \otimes W$ is the unique K-linear map that makes the diagram in Figure 14.7 commute, that is, for which*

$$\mu \circ \tau = (\iota_K \otimes \tau) \circ \nu$$ □

Multilinear Maps and Iterated Tensor Products

The tensor product operation can easily be extended to more than two vector spaces. We begin with the extension of the concept of bilinearity.

Definition *If $V_1, \ldots, V_n$ and W are vector spaces over F, a function $f: V_1 \times \cdots \times V_n \to W$ is said to be* **multilinear** *if it is linear in each coordinate separately, that is, if*

$$\begin{aligned}&f(u_1, \ldots, u_{k-1}, rv + sv', u_{k+1}, \ldots, u_n) \\ &\quad = rf(u_1, \ldots, u_{k-1}, v, u_{k+1}, \ldots, u_n) + sf(u_1, \ldots, u_{k-1}, v', u_{k+1}, \ldots, u_n)\end{aligned}$$

*for all $k = 1, \ldots, n$. A multilinear function of n variables is also referred to as an n-***linear function***. The set of all n-linear functions as defined above will be denoted by* $\hom(V_1, \ldots, V_n; W)$. *A multilinear function from $V_1 \times \cdots \times V_n$ to the base field F is called a* **multilinear form** *or n-***form***.*□

Example 14.7

1) If A is an algebra, then the product map $\mu: A \times \cdots \times A \to A$ defined by $\mu(a_1, \ldots, a_n) = a_1 \cdots a_n$ is n-linear.
2) The determinant function det: $\mathcal{M}_n \to F$ is an n-linear form on the columns of the matrices in $\mathcal{M}_n$.□

The tensor product is defined via its universal property.

Definition *As pictured in Figure 14.8, let $V_1 \times \cdots \times V_n$ be the cartesian product of vector spaces over F. A pair $(T, t: V_1 \times \cdots \times V_n \to T)$ is* **universal for multilinearity** *if for every multilinear map $f: V_1 \times \cdots \times V_n \to W$, there is a unique linear transformation $\tau: T \to W$ for which*

$$f = \tau \circ t$$

The map τ is called the **mediating morphism** *for f. If (T, t) is universal for multilinearity, then T is called the* **tensor product** *of $V_1, \ldots, V_n$ and denoted by $V_1 \otimes \cdots \otimes V_n$. The map t is called the* **tensor map.** □

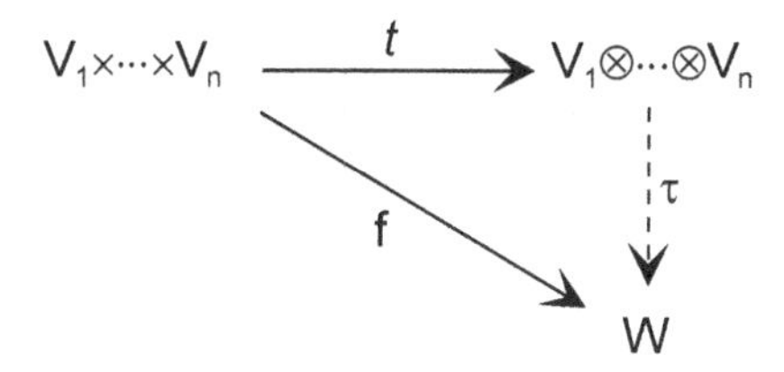

Figure 14.8

As we have seen, the tensor product is unique up to isomorphism.

The basis construction and coordinate-free construction given earlier for the tensor product of two vector spaces carry over to the multilinear case.

In particular, let $\mathcal{B}_i = \{e_{i,j} \mid j \in J_i\}$ be a basis for V_i for $i = 1, \ldots, n$. For each ordered n-tuple $(e_{1,i_1}, \ldots, e_{n,i_n})$, construct a new formal symbol $e_{1,i_1} \otimes \cdots \otimes e_{n,i_n}$ and define T to be the vector space with basis

$$\mathcal{D} = \{e_{1,i_1} \otimes \cdots \otimes e_{n,i_n} \mid i_k \in J_k\}$$

The tensor map $t: V_1 \times \cdots \times V_n \to T$ is defined by setting

$$t(e_{1,i_1}, \ldots, e_{n,i_n}) = e_{1,i_1} \otimes \cdots \otimes e_{n,i_n}$$

and extending by multilinearity. This uniquely defines a multilinear map t that is universal for multilinear functions from $V_1 \times \cdots \times V_n$.

Indeed, if $g: V_1 \times \cdots \times V_n \to W$ is multilinear, the condition $f = \tau \circ t$ is equivalent to

$$\tau(e_{1,i_1} \otimes \cdots \otimes e_{n,i_n}) = f(e_{1,i_1}, \ldots, e_{n,i_n})$$

which uniquely defines a linear map $\tau: T \to W$. Hence, (T, t) has the universal property for multilinearity.

Alternatively, we may take the coordinate-free quotient space approach as follows.

Definition *Let $V_1, \ldots, V_n$ be vector spaces over F and let $\mathcal{F}$ be the vector space with basis $V_1 \times \cdots \times V_n$. Let S be the subspace of $\mathcal{F}$ generated by all vectors of the form*

$$r(v_1,\ldots,v_{k-1},u,v_{k+1},\ldots,v_n)+s(v_1,\ldots,v_{k-1},u',v_{k+1},\ldots,v_n)$$
$$-(v_1,\ldots,v_{k-1},ru+su',v_{k+1},\ldots,v_n)$$

for $r,s\in F$, $u,u'\in V_k$ *and* $v_i\in V_i$ *for* $i\neq k$. *The quotient space* $\mathcal{F}/S$ *is the tensor product of* $V_1,\ldots,V_n$ *and the tensor map is the map*

$$t(v_1,\ldots,v_n)=(v_1,\ldots,v_n)+S \qquad \square$$

As before, we denote the coset $(v_1,\ldots,v_n)+S$ by $v_1\otimes\cdots\otimes v_n$ and so any element of $V_1\otimes\cdots\otimes V_n$ is a sum of decomposable tensors, that is,

$$\sum v_{i_1}\otimes\cdots\otimes v_{i_n}$$

where the vector space operations are linear in each variable.

Here are some of the basic properties of multiple tensor products. Proof is left to the reader.

Theorem 14.12 *The tensor product has the following properties. Note that all vector spaces are over the same field* F.

1) **(Associativity)** *There exists an isomorphism*

$$\tau:(V_1\otimes\cdots\otimes V_n)\otimes(W_1\otimes\cdots\otimes W_m)\to V_1\otimes\cdots\otimes V_n\otimes W_1\otimes\cdots\otimes W_m$$

for which

$$\tau[(v_1\otimes\cdots\otimes v_n)\otimes(w_1\otimes\cdots\otimes w_m)]=v_1\otimes\cdots\otimes v_n\otimes w_1\otimes\cdots\otimes w_m$$

In particular,

$$(U\otimes V)\otimes W\approx U\otimes(V\otimes W)\approx U\otimes V\otimes W$$

2) **(Commutativity)** *Let* π *be any permutation of the indices* $\{1,\ldots,n\}$. *Then there is an isomorphism*

$$\sigma: V_1\otimes\cdots\otimes V_n\to V_{\pi(1)}\otimes\cdots\otimes V_{\pi(n)}$$

for which

$$\sigma(v_1\otimes\cdots\otimes v_n)=v_{\pi(1)}\otimes\cdots\otimes v_{\pi(n)}$$

3) There is an isomorphism $\rho_1: F\otimes V\to V$ *for which*

$$\rho_1(r\otimes v)=rv$$

and similarly, there is an isomorphism $\rho_2: V\otimes F\to V$ *for which*

$$\rho_2(v\otimes r)=rv$$

Hence, $F\otimes V\approx V\approx V\otimes F$.$\square$

The analog of Theorem 14.4 is the following.

Theorem 14.13 *Let $V_1, \ldots, V_n$ and W be vector spaces over F. Then the mediating morphism map*

$$\phi\colon \hom(V_1, \ldots, V_n; W) \to \mathcal{L}(V_1 \otimes \cdots \otimes V_n, W)$$

defined by the fact that ϕf is the unique mediating morphism for f is an isomorphism. Thus,

$$\hom(V_1, \ldots, V_n; W) \approx \mathcal{L}(V_1 \otimes \cdots \otimes V_n, W)$$

Moreover, if all vector spaces are finite-dimensional, then

$$\dim[\hom(V_1, \ldots, V_n; W)] = \dim(W) \cdot \prod_{i=1}^{n} \dim(V_i) \qquad \square$$

Theorem 14.8 and its corollary can also be extended.

Theorem 14.14 *The linear transformation*

$$\theta\colon \mathcal{L}(U_1, U_1') \otimes \cdots \otimes \mathcal{L}(U_n, U_n') \to \mathcal{L}(U_1 \otimes \cdots \otimes U_n, U_1' \otimes \cdots \otimes U_n')$$

defined by

$$\theta(\tau_1 \otimes \cdots \otimes \tau_n)(u_1 \otimes \cdots \otimes u_n) = \tau_1 u_1 \otimes \cdots \otimes \tau_n u_n$$

is an embedding and is an isomorphism if all vector spaces are finite-dimensional. Thus, the tensor product $\tau_1 \otimes \cdots \otimes \tau_n$ of linear transformations is (via this embedding) a linear transformation on tensor products. Two important special cases of this are

$$U_1^* \otimes \cdots \otimes U_n^* \overset{\sim}{\hookrightarrow} (U_1 \otimes \cdots \otimes U_n)^*$$

where

$$(f_1 \otimes \cdots \otimes f_n)(u_1 \otimes \cdots \otimes u_n) = f_1(u_1) \cdots f_n(u_n)$$

and

$$U_1^* \otimes \cdots \otimes U_n^* \otimes V \overset{\sim}{\hookrightarrow} \mathcal{L}(U_1 \otimes \cdots \otimes U_n, V)$$

where

$$(f_1 \otimes \cdots \otimes f_n \otimes v)(u_1 \otimes \cdots \otimes u_n) = f_1(u_1) \cdots f_n(u_n) v \qquad \square$$

Tensor Spaces

Let V be a finite-dimensional vector space. For nonnegative integers p and q, the tensor product

$$T_q^p(V) = \underbrace{V \otimes \cdots \otimes V}_{p \text{ factors}} \otimes \underbrace{V^* \otimes \cdots \otimes V^*}_{q \text{ factors}} = V^{\otimes p} \otimes (V^*)^{\otimes q}$$

is called the space of **tensors of type (p, q)**, where p is the **contravariant type** and q is the **covariant type**. If $p = q = 0$, then $T_q^p(V) = F$, the base field. Here we use the notation $V^{\otimes n}$ for the n-fold tensor product of V with itself. We will also write $V^{\times n}$ for the n-fold cartesian product of V with itself.

Since $V \approx V^{**}$, we have

$$T_q^p(V) = V^{\otimes p} \otimes (V^*)^{\otimes q} \approx ((V^*)^{\otimes p} \otimes V^{\otimes q})^* \approx \hom_F((V^*)^{\times p} \times V^{\times q}, F)$$

which is the space of all multilinear functionals on

$$\underbrace{V^* \times \cdots \times V^*}_{p \text{ factors}} \times \underbrace{V \times \cdots \times V}_{q \text{ factors}}$$

In fact, tensors of type (p, q) are often defined as multilinear functionals in this way.

Note that

$$\dim(T_q^p(V)) = [\dim(V)]^{p+q}$$

Also, the associativity and commutativity of tensor products allows us to write

$$T_q^p(V) \otimes T_s^r(V) = T_{q+s}^{p+r}(V)$$

at least up to isomorphism.

Tensors of type $(p, 0)$ are called **contravariant tensors**

$$T^p(V) = T_0^p(V) = \underbrace{V \otimes \cdots \otimes V}_{p \text{ factors}}$$

and tensors of type $(0, q)$ are called **covariant tensors**

$$T_q(V) = T_q^0(V) = \underbrace{V^* \otimes \cdots \otimes V^*}_{q \text{ factors}}$$

Tensors with both contravariant and covariant indices are called **mixed tensors**.

In general, a tensor can be interpreted in a variety of ways as a multilinear map on a cartesian product, or a linear map on a tensor product. Indeed, the interpretation we mentioned above that is sometimes used as the definition is only one possibility. We simply need to decide how many of the contravariant factors and how many of the covariant factors should be "active participants" and how many should be "passive participants."

More specifically, consider a tensor of type (p,q), written

$$v_1 \otimes \cdots \otimes v_m \otimes \cdots \otimes v_p \otimes f_1 \otimes \cdots \otimes f_n \otimes \cdots \otimes f_q \in T_q^p(V)$$

where $m \le p$ and $n \le q$. Here we are choosing the first m vectors and the first n linear functionals as active participants. This determines the number of arguments of the map. In fact, we define a map from the cartesian product

$$\underbrace{V^* \times \cdots \times V^*}_{m \text{ factors}} \times \underbrace{V \times \cdots \times V}_{n \text{ factors}}$$

to the tensor product

$$\underbrace{V \otimes \cdots \otimes V}_{p-m \text{ factors}} \otimes \underbrace{V^* \otimes \cdots \otimes V^*}_{q-n \text{ factors}}$$

of the remaining factors by

$$\begin{aligned}&(v_1 \otimes \cdots \otimes v_p \otimes f_1 \otimes \cdots \otimes f_q)(h_1, \ldots, h_m, x_1, \ldots, x_n)\\ &\qquad = h_1(v_1)\cdots h_m(v_m) f_1(x_1)\cdots f_n(x_n) v_{m+1} \otimes \cdots \otimes v_p \otimes f_{n+1} \otimes \cdots \otimes f_q\end{aligned}$$

In words, the first group $v_1 \otimes \cdots \otimes v_m$ of (active) vectors interacts with the first group $h_1, \ldots, h_m$ of arguments to produce the scalar $h_1(v_1)\cdots h_m(v_m)$. The first group $f_1 \otimes \cdots \otimes f_n$ of (active) functionals interacts with the second group $x_1, \ldots, x_n$ of arguments to produce the scalar $f_1(x_1)\cdots f_n(x_n)$. The remaining (passive) vectors $v_{m+1} \otimes \cdots \otimes v_p$ and functionals $f_{n+1} \otimes \cdots \otimes f_q$ are just "copied" to the image tensor.

It is easy to see that this map is multilinear and so there is a unique linear map from the tensor product

$$\underbrace{V^* \otimes \cdots \otimes V^*}_{m \text{ factors}} \otimes \underbrace{V \otimes \cdots \otimes V}_{n \text{ factors}}$$

to the tensor product

$$\underbrace{V \otimes \cdots \otimes V}_{p-m \text{ factors}} \otimes \underbrace{V^* \otimes \cdots \otimes V^*}_{q-n \text{ factors}}$$

defined by

$$\begin{aligned}&(v_1 \odot \cdots \odot v_p \odot f_1 \odot \cdots \odot f_q)(h_1 \otimes \cdots \otimes h_m \otimes x_1 \otimes \cdots \otimes x_n)\\ &\qquad = h_1(v_1)\cdots h_m(v_m) f_1(x_1)\cdots f_n(x_n) v_{m+1} \otimes \cdots \otimes v_p \otimes f_{n+1} \otimes \cdots \otimes f_q\end{aligned}$$

Moreover, the map

$$\phi\colon V^{\otimes p} \otimes (V^*)^{\otimes q} \to \mathcal{L}((V^*)^{\otimes m} \otimes V^{\otimes n}, V^{\otimes(p-m)} \otimes (V^*)^{\otimes(q-n)})$$

defined by

$$\phi(v_1 \otimes \cdots \otimes v_p \otimes f_1 \otimes \cdots \otimes f_q) = v_1 \odot \cdots \odot v_p \odot f_1 \odot \cdots \odot f_q$$

is an isomorphism, since if $v_1 \odot \cdots \odot v_p \odot f_1 \odot \cdots \odot f_q$ is the zero map then

$$h_1(v_1)\cdots h_m(v_m)f_1(x_1)\cdots f_n(x_n)v_{m+1} \otimes \cdots \otimes v_p \otimes f_{n+1} \otimes \cdots \otimes f_q = 0$$

for all $h_i \in V^*$ and $x_i \in V$, which implies that

$$v_1 \otimes \cdots \otimes v_p \otimes f_1 \otimes \cdots \otimes f_q = 0$$

As usual, we denote the map $v_1 \odot \cdots \odot v_p \odot f_1 \odot \cdots \odot f_q$ by

$$v_1 \otimes \cdots \otimes v_p \otimes f_1 \otimes \cdots \otimes f_q$$

Theorem 11.15 *For* $0 \leq m \leq p$ *and* $0 \leq n \leq q$,

$$T_q^p(V) \approx \mathcal{L}((V^*)^{\otimes m} \otimes V^{\otimes n}, V^{\otimes(p-m)} \otimes (V^*)^{\otimes(q-n)})$$ □

When $m = p$ and $n = q$, we get

$$T_q^p(V) \approx \mathcal{L}((V^*)^{\otimes p} \otimes V^{\otimes q}, F) = ((V^*)^{\otimes p} \otimes V^{\otimes q})^*$$

as before.

Let us look at some special cases. For $q = 0$ we have

$$T_0^p(V) \approx \mathcal{L}((V^*)^{\otimes m}, V^{\otimes(p-m)})$$

where

$$(v_1 \otimes \cdots \otimes v_p)(h_1 \otimes \cdots \otimes h_m) = h_1(v_1)\cdots h_m(v_m)v_{m+1} \otimes \cdots \otimes v_p$$

When $p = q = 1$, we get for $m = 0$ and $n = 1$,

$$T_1^1(V) \approx \mathcal{L}(F \otimes V, V \otimes F) \approx \mathcal{L}(V)$$

where

$$(v \otimes f)(w) = f(w)v$$

and for $m = 1$ and $n = 0$,

$$T_1^1(V) \approx \mathcal{L}(V^* \otimes F, F \otimes V^*) \approx \mathcal{L}(V^*, V^*)$$

where

$$(v \otimes f)(h) = h(v)f$$

Finally, when $m = n = 1$, we get a multilinear form

$$(v \otimes f)(h, w) = h(v)f(w)$$

Consider also a tensor $f \otimes g$ of type $(0, 2)$. When $n = q = 2$ we get a multilinear functional $f \otimes g\colon (V \times V) \to F$ defined by

$$(f \otimes g)(v, w) = f(v)g(w)$$

This is just a bilinear form on V.

Contraction

Covariant and contravariant factors can be "combined" in the following way. Consider the map

$$h: V^{\times p} \times (V^*)^{\times q} \to T_{q-1}^{p-1}(V)$$

defined by

$$h(v_1, \ldots, v_p, f_1, \ldots, f_q) = f_1(v_1)(v_2 \otimes \cdots \otimes v_p \otimes f_1 \otimes \cdots \otimes f_q)$$

This is easily seen to be multilinear and so there is a unique linear map

$$\theta: T_q^p(V) \to T_{q-1}^{p-1}(V)$$

defined by

$$\theta(v_1 \otimes \cdots \otimes v_p \otimes f_1 \otimes \cdots \otimes f_q) = f_1(v_1)(v_2 \otimes \cdots \otimes v_p \otimes f_1 \otimes \cdots \otimes f_q)$$

This is called the **contraction** in the contravariant index 1 and covariant index 1. Of course, contraction in other indices (one contravariant and one covariant) can be defined similarly.

Example 14.8 Let $\dim(V) > 1$ and consider the tensor space $T_1^1(V)$, which is isomorphic to $\mathcal{L}(V)$ via the map

$$(v \otimes f)(w) = f(w)v$$

For a "decomposable" linear operator of the form $v \otimes f$ as defined above with $v \neq 0$ and $f \neq 0$, we have $\ker(v \otimes f) = \ker(f)$, which has codimension 1. Hence, if $f(w)(v) = (v \otimes f)(w) \neq 0$, then

$$V = \langle w \rangle \oplus \ker(f) = \langle w \rangle \oplus \mathcal{E}_0$$

where $\mathcal{E}_0$ is the eigenspace of $v \otimes f$ associated with the eigenvalue 0.

In particular, if $f(v) \neq 0$, then

$$(v \otimes f)(v) = f(v)v$$

and so v is an eigenvector for the nonzero eigenvalue $f(v)$. Hence,

$$V = \langle v \rangle \oplus \mathcal{E}_0 = \mathcal{E}_{f(v)} \oplus \mathcal{E}_0$$

and so the trace of $v \otimes f$ is

$$\operatorname{tr}(v \otimes f) = f(v) = \theta(v \otimes f)$$

where θ is the contraction map.□

The Tensor Algebra of V

Consider the contravariant tensor spaces

$$T^p(V) = T_0^p(V) = V^{\otimes p}$$

For $p = 0$ we take $T^0(V) = F$. The external direct sum

$$T(V) = \bigoplus_{p=0}^{\infty} T^p(V)$$

of these tensor spaces is a vector space with the property that

$$T^p(V) \otimes T^q(V) = T^{p+q}(V)$$

This is an example of a *graded algebra*, where $T^p(V)$ are the elements of *grade* p. The graded algebra $T(V)$ is called the **tensor algebra** over V. (We will formally define graded structures a bit later in the chapter.)

Since

$$T_q(V) = \underbrace{V^* \otimes \cdots \otimes V^*}_{q \text{ factors}} = T^q(V^*)$$

there is no need to look separately at $T_q(V)$.

Special Multilinear Maps

The following definitions describe some special types of multilinear maps.

Definition

1) *A multilinear map* $f: V^{\times n} \to W$ *is* **symmetric** *if interchanging any two coordinate positions changes nothing, that is, if*

$$f(v_1, \ldots, v_i, \ldots, v_j, \ldots, v_n) = f(v_1, \ldots, v_j, \ldots, v_i, \ldots, v_n)$$

for any $i \neq j$.

2) *A multilinear map* $f: V^{\times n} \to W$ *is* **antisymmetric** *or* **skew-symmetric** *if interchanging any two coordinate positions introduces a factor of* -1, *that is, if*

$$f(v_1, \ldots, v_i, \ldots, v_j, \ldots, v_n) = -f(v_1, \ldots, v_j, \ldots, v_i, \ldots, v_n)$$

for $i \neq j$.

3) *A multilinear map* $f: V^n \to W$ *is* **alternate** *or* **alternating** *if*

$$v_i = v_j \textit{ for some } i \neq j \quad \Rightarrow \quad f(v_1, \ldots, v_n) = 0 \qquad \square$$

As in the case of bilinear forms, we have some relationships between these concepts. In particular, if $\mathrm{char}(F) = 2$, then

$$\text{alternate} \Rightarrow \text{symmetric} \Leftrightarrow \text{skew-symmetric}$$

and if $\mathrm{char}(F) \neq 2$, then

$$\text{alternate} \Leftrightarrow \text{skew-symmetric}$$

A few remarks about permutations are in order. A **permutation** of the set $N = \{1, \ldots, n\}$ is a bijective function $\pi: N \to N$. We denote the group (under composition) of all such permutations by S_n. This is the **symmetric group** on n symbols. A **cycle** of length k is a permutation of the form $(i_1, i_2, \ldots, i_k)$, which sends i_u to i_{u+1} for $u = 1, \ldots, k-1$ and also sends i_k to i_1. All other elements of N are left fixed. Every permutation is the product (composition) of disjoint cycles.

A **transposition** is a cycle (i, j) of length 2. Every cycle (and therefore every permutation) is the product of transpositions. In general, a permutation can be expressed as a product of transpositions in many ways. However, no matter how one represents a given permutation as such a product, the number of transpositions is either always even or always odd. Therefore, we can define the **parity** of a permutation $\pi \in S_n$ to be the parity of the number of transpositions in any decomposition of π as a product of transpositions. The **sign** of a permutation is defined by

$$\mathrm{sg}(\pi) = \begin{cases} 1 & \pi \text{ has even parity} \\ -1 & \pi \text{ has odd parity} \end{cases}$$

If $\mathrm{sg}(\pi) = 1$, then π is an **even permutation** and if $\mathrm{sg}(\pi) = -1$, then π is an **odd permutation**. The sign of π is often written $(-1)^\pi$.

With these facts in mind, it is apparent that f is symmetric if and only if

$$f(v_1, \ldots, v_n) = f(v_{\pi(1)}, \ldots, v_{\pi(n)})$$

for all permutations $\pi \in S_n$ and that f is skew-symmetric if and only if

$$f(v_1, \ldots, v_n) = (-1)^\pi f(v_{\pi(1)}, \ldots, v_{\pi(n)})$$

for all permutations $\pi \in S_n$.

A word of caution is in order with respect to the notation above, which is very convenient albeit somewhat prone to confusion. It is intended that a permutation π permutes the *coordinate positions* in f, not the indices (despite appearances). Suppose, for example, that $f: \mathbb{R}^2 \times \mathbb{R}^2 \to X$ and that $\{e_1, e_2\}$ is a basis for $\mathbb{R}^2$.

If $\pi = (1\,2)$, then π applied to $f(e_1, e_1)$ gives $f(e_1, e_1)$ and not $f(e_2, e_2)$, since π permutes the two coordinate positions in $f(v_1, v_2)$.

Graded Algebras

We need to pause for a few definitions that are useful in discussing tensor algebras. An algebra A over F is said to be a **graded algebra** if as a vector space over F, A can be written in the form

$$A = \bigoplus_{i=0}^{\infty} A_i$$

for subspaces A_i of A, and where multiplication behaves nicely, that is,

$$A_i A_j \subseteq A_{i+j}$$

The elements of A_i are said to be **homogeneous of degree** i. If $a \in A$ is written

$$a = a_{i_1} + \cdots + a_{i_n}$$

for $a_{i_k} \in A_{i_k}$, $i_k \neq i_j$, then a_{i_k} is called the **homogeneous component** of a of degree i_k.

The ring of polynomials $F[x]$ provides a prime example of a graded algebra, since

$$F[x] = \bigoplus_{i=0}^{\infty} F_i[x]$$

where $F_i[x]$ is the subspace of $F[x]$ consisting of all scalar multiples of x^i.

More generally, the ring $F[x_1, \ldots, x_n]$ of polynomials in several variables is a graded algebra, since it is the direct sum of the subspaces of homogeneous polynomials of degree i. (A polynomial is **homogeneous of degree** i if each term has degree i. For example, $p = x_1 x_2^2 + x_1 x_2 x_3$ is homogeneous of degree 3.)

The Symmetric and Antisymmetric Tensor Algebras

Our discussion of symmetric and antisymmetric tensors will benefit by a discussion of a few definitions and setting a bit of notation at the outset.

Let $F_p[e_1, \ldots, e_n]$ denote the vector space of all homogeneous polynomials of degree p (together with the zero polynomial) in the *independent variables* $e_1, \ldots, e_n$. As is sometimes done in this context, we denote the product in $F_p[e_1, \ldots, e_n]$ by $\vee$, for example, writing $e_1 e_2 e_5$ as $e_1 \vee e_2 \vee e_5$. The algebra of all polynomials in $e_1, \ldots, e_n$ is denoted by $F[e_1, \ldots, e_n]$.

We will also need the counterpart of $F_p[e_1, \ldots, e_n]$ in which multiplication acts *anticommutatively*, that is, $e_i e_j = -e_j e_i$.

Definition *Let $E = (e_1, \ldots, e_n)$ be a sequence of independent variables. For $p \le n$, let $F_p^-[e_1, \ldots, e_n]$ be the vector space over F with basis*

$$\mathcal{A}_p(E) = \{e_{i_1} \cdots e_{i_p} \mid i_1 < i_2 < \cdots < i_p\}$$

consisting of all words of length p over E that are in ascending order. Let $F_0^-[e_1, \ldots, e_n] = F\epsilon$, which we identify with F by identifying ϵ with $1 \in F$. Define a product on the direct sum

$$F^-[e_1, \ldots, e_n] = \bigoplus_{p=0}^{n} F_p^-$$

as follows. First, the product $f \wedge g$ of monomials $f = x_1 \cdots x_p \in F_p^-$ and $g = y_1 \cdots y_q \in F^-$ is defined as follows:

1) *If $x_1 \cdots x_p y_1 \cdots y_q$ has a repeated factor then $f \wedge g = 1$.*
2) *Otherwise, reorder $x_1 \cdots x_p y_1 \cdots y_q$ in ascending order, say $z_1 \cdots z_{p+q}$, via the permutation σ and set*

$$f \wedge g = (-1)^\sigma z_1 \cdots z_{p+q}$$

Extend the product by distributivity to $F^-[e_1, \ldots, e_n]$. The resulting product makes $F^-[e_1, \ldots, e_n]$ into a (noncommutative) algebra over F. This product is called the **wedge product** *or* **exterior product** *on $F^-[e_1, \ldots, e_n]$.* □

For example, by definition of wedge product,

$$e_2 \wedge e_1 \wedge e_3 = -e_1 \wedge e_2 \wedge e_3$$

Let $\mathcal{B} = \{e_1, \ldots, e_n\}$ be a basis for V. It will be convenient to group the decomposable basis tensors $e_{i_1} \otimes \cdots \otimes e_{i_p}$ according to their index multiset. Specifically, for each multiset $M = \{i_1, \ldots, i_p\}$ with $1 \le i_k \le n$, let G_M be the set of all tensors

$$e_{k_1} \otimes \cdots \otimes e_{k_p}$$

where $(k_1, \ldots, k_p)$ is a permutation of $\{i_1, \ldots, i_p\}$. For example, if $M = \{2, 2, 3\}$, then

$$G_M = \{e_2 \otimes e_2 \otimes e_3, e_2 \otimes e_3 \otimes e_2, e_3 \otimes e_2 \otimes e_2\}$$

If $v \in T^p(V)$ has the form

$$v = \sum_{i_1, \ldots, i_p} \alpha_{i_1, \ldots, i_p} e_{i_1} \otimes \cdots \otimes e_{i_p}$$

where $\alpha_{i_1, \ldots, i_p} \ne 0$, then let $G_M(v)$ be the subset of G_M whose elements appear in the sum for v. For example, if

$$v = 2e_2 \otimes e_2 \otimes e_3 + 3e_2 \otimes e_3 \otimes e_2 + e_3 \otimes e_3 \otimes e_1$$

then

$$G_{\{2,2,3\}}(v) = \{e_2 \otimes e_2 \otimes e_3, e_2 \otimes e_3 \otimes e_2\}$$

Let $S_M(v)$ denote the sum of the terms of v associated with $G_M(v)$. For example,

$$S_{\{2,2,3\}}(v) = 2e_2 \otimes e_2 \otimes e_3 + 3e_2 \otimes e_3 \otimes e_2$$

Thus, v can be written in the form

$$v = \sum_M S_M(v) = \sum_M \left(\sum_{t \in G_M(v)} \alpha_t t \right)$$

where the sum is over a collection of multisets M with $S_M(v) \neq 0$. Note also that $\alpha_t \neq 0$ since $t \in G_M(v)$. Finally, let

$$u_M = e_{i_1} \otimes \cdots \otimes e_{i_p}$$

be the unique member of G_M for which $i_1 \le i_2 \le \cdots \le i_p$.

Now we can get to the business at hand.

Symmetric and Antisymmetric Tensors

Let S_p be the symmetric group on $\{1, \ldots, p\}$. For each $\sigma \in S_p$, the multilinear map $f_\sigma \colon V^{\times p} \to T^p(V)$ defined by

$$f_\sigma(x_1, \ldots, x_p) = x_{\sigma 1} \otimes \cdots \otimes x_{\sigma p}$$

determines a unique linear operator λ_σ on $T^p(V)$ for which

$$\lambda_\sigma(x_1 \otimes \cdots \otimes x_p) = x_{\sigma 1} \otimes \cdots \otimes x_{\sigma p}$$

For example, if $p = 3$ and $\sigma = (1\,2)$, then

$$\lambda_{(12)}(v_1 \otimes v_3 \otimes v_2) = v_3 \otimes v_1 \otimes v_2$$

Let $\{e_1, \ldots, e_n\}$ be a basis for V. Since λ_σ is a bijection of the basis

$$\mathcal{B} = \{e_{i_1} \otimes \cdots \otimes e_{i_p} \mid e_{i_j} \in \mathcal{B}\}$$

it follows that λ_σ is an isomorphism of $T^p(V)$. Note also that λ_σ is a permutation of each G_M, that is, the sets G_M are invariant under λ_σ.

Definition *Let V be a finite-dimensional vector space.*

1) A tensor $t \in T^p(V)$ is **symmetric** *if*

$$\lambda_\sigma t = t$$

for all permutations $\sigma \in S_p$. The set of all symmetric tensors

$$ST^p(V) = \{t \in T^p(V) \mid \lambda_\sigma t = t \text{ for all } \sigma \in S_p\}$$

is a subspace of $T^p(V)$, called the **symmetric tensor space** *of degree p over V.*

2) A tensor $t \in T^p(V)$ is **antisymmetric** *if*

$$\lambda_\sigma t = (-1)^\sigma t$$

The set of all antisymmetric tensors

$$AT^p(V) = \{t \in T^p(V) \mid \lambda_\sigma t = (-1)^\sigma t \text{ for all } \sigma \in S_p\}$$

is a subspace of $T^p(V)$, called the **antisymmetric tensor space** *or* **exterior product space** *of degree p over V.*□

We can develop the theory of symmetric and antisymmetric tensors in tandem. Accordingly, let us write (anti)symmetric to denote a tensor that is either symmetric or antisymmetrtic.

Since for any $s, t \in G_M$, there is a permutation λ_σ taking s to t, an (anti)symmetric tensor v must have $G_M(v) = G_M$ and so

$$v = \sum_M S_M(v) = \sum_M \left(\sum_{t \in G_M} \alpha_t t \right)$$

Since λ_σ is a permutation of G_M, it follows that v is symmetric if and only if

$$\lambda_\sigma(S_M(v)) = S_M(v)$$

for all $\sigma \in S_p$ and this holds if and only if the coefficients α_t of $S_M(v)$ are equal, say $\alpha_t = \alpha_M$ for all $t \in G_M$. Hence, the symmetric tensors are precisely the tensors of the form

$$v = \sum_M \left(\alpha_M \sum_{t \in G_M} t \right)$$

The tensor v is antisymmetric if and only if

$$\lambda_\sigma(S_M(v)) = (-1)^\sigma S_M(v) \tag{14.4}$$

In this case, the coefficients α_t of $S_M(v)$ differ only by sign. Before examining this more closely, we observe that M must be a set. For if M has an element k of multiplicity greater than 1, we can split G_M into two disjoint parts:

$$G_M = G'_M \cup G''_M$$

where G'_M are the tensors that have e_k in positions r and s:

$$G'_M = \{e_{i_1} \otimes \cdots \otimes \underset{\text{position } r}{e_k} \otimes \cdots \otimes \underset{\text{position } s}{e_k} \otimes \cdots \otimes e_{i_p}\}$$

Then $\lambda_{(r\,s)}$ fixes each element of G'_M and sends the elements of G''_M to other elements of G''_M. Hence, applying $\lambda_{(r\,s)}$ to the corresponding decomposition of $S_M(v)$:

$$S_M(v) = S'_M + S''_M$$

gives

$$-(S'_M + S''_M) = -S_M = \lambda_{(1\,2)}S_M = S'_M + \lambda_{(1\,2)}S''_M$$

and so $S'_M = 0$, whence $S_M(v) = 0$. Thus, M is a set.

Now, since for any $\sigma \in S_p$,

$$G_M = \{\lambda_\sigma t \mid t \in G_M\}$$

equation (14.4) implies that

$$(-1)^\sigma \sum_{t \in G_M} \alpha_t t = \lambda_\sigma \left(\sum_{t \in G_M} \alpha_t t \right) = \sum_{t \in G_M} \alpha_t \lambda_\sigma t = \sum_{t \in G_M} \alpha_{\lambda_\sigma^{-1} t} t$$

which holds if and only if $\alpha_{\lambda_\sigma^{-1} t} = (-1)^\sigma \alpha_t$, or equivalently,

$$\alpha_{\lambda_\sigma t} = (-1)^\sigma \alpha_t$$

for all $t \in G_M$ and $\sigma \in S_p$. Choosing $u = u_M = e_{i_1} \otimes \cdots \otimes e_{i_p}$, where $i_1 < \cdots < i_p$, as standard-bearer, if $\sigma_{u,t}$ denotes the permutation for which $\lambda_{\sigma_{u,t}}(u) = t$, then

$$\alpha_t = (-1)^{\sigma_{u,t}} \alpha_u$$

Thus, v is antisymmetric if and only if it has the form

$$v = \sum_M \left(\alpha_M \sum_{t \in G_M} (-1)^{\sigma_{u,t}} t \right)$$

where $\alpha_M = \alpha_u \neq 0$ and the sum is over a family of *sets*.

In summary, the symmetric tensors are

$$v = \sum_M \left(\alpha_M \sum_{t \in G_M} t \right)$$

where M is a multiset and the antisymmetric tensors are

$$v = \sum_M \left(\alpha_M \sum_{t \in G_M} (-1)^{\sigma_{u,t}} t \right)$$

where M is a set.

We can simplify these expressions considerably by representing the inside sums more succinctly. In the symmetric case, define a surjective linear map

$$\tau: T^p(V) \to F_p[e_1, \ldots, e_n]$$

by

$$\tau(e_{i_1} \otimes \cdots \otimes e_{i_p}) = e_{i_1} \vee \cdots \vee e_{i_p}$$

and extending by linearity. Since τ takes every member of G_M to the same monomial $\tau u = e_{i_1} \vee \cdots \vee e_{i_p}$, where $i_1 \le \cdots \le i_p$, we have

$$\tau v = \tau \left(\sum_M \left(\alpha_M \sum_{t \in G_M} t \right) \right) = \sum_M \alpha_M |G_M| \tau u_M$$

In the antisymmetric case, define a surjective linear map

$$\tau: T^p(V) \to F_p^-[e_1, \ldots, e_n]$$

by

$$\tau(e_{i_1} \otimes \cdots \otimes e_{i_p}) = e_{i_1} \wedge \cdots \wedge e_{i_p}$$

and extending by linearity. Since

$$\tau t = (-1)^{\sigma_{u,t}} \tau u_M$$

we have

$$\begin{aligned} \tau v &= \sum_M \left(\alpha_M \sum_{t \in G_M} (-1)^{\sigma_{u,t}} \tau t \right) \\ &= \sum_M \left(\alpha_M \sum_{t \in G_M} \tau u_M \right) \\ &= \sum_M \alpha_M |G_M| \tau u_M \end{aligned}$$

Thus, in both cases,

$$\tau v = \sum_M \alpha_M |G_M| \tau u_M$$

where $u_M = e_{i_1} \otimes \cdots \otimes e_{i_p}$ with $i_1 \le i_2 \le \cdots \le i_p$ and

$$\tau u_M = e_{i_1} \vee \cdots \vee e_{i_p} \quad \text{or} \quad \tau u_M = e_{i_1} \wedge \cdots \wedge e_{i_p}$$

depending on whether v is symmetric or antisymmetric. However, in either case, the monomials τu_M are linearly independent for distinct multisets/sets M. Therefore, if $\tau v = 0$ then $\alpha_M |G_M| = 0$ for all multisets/sets M. Hence, if $\text{char}(F) = 0$, then $\alpha_M = 0$ and so $v = 0$. This shows that the restricted maps $\tau|_{ST^p(V)}$ and $\tau|_{AT^p(V)}$ are isomorphisms.

Theorem 14.16 *Let V be a finite-dimensional vector space over a field F with* $\text{char}(F) = 0$.

1) *The symmetric tensor space $ST^p(V)$ is isomorphic to the algebra $F_p[e_1, \ldots, e_n]$ of homogeneous polynomials, via the isomorphism*

$$\tau\left(\sum \alpha_{i_1,\ldots,i_p} e_{i_1} \otimes \cdots \otimes e_{i_p}\right) = \sum \alpha_{i_1,\ldots,i_p} (e_{i_1} \vee \cdots \vee e_{i_p})$$

2) *For $p \le n$, the antisymmetric tensor space $AT^p(V)$ is isomorphic to the algebra $F_p^-[e_1, \ldots, e_n]$ of anticommutative homogeneous polynomials of degree p, via the isomorphism*

$$\tau\left(\sum \alpha_{i_1,\ldots,i_p} e_{i_1} \otimes \cdots \otimes e_{i_p}\right) = \sum \alpha_{i_1,\ldots,i_p} (e_{i_1} \wedge \cdots \wedge e_{i_p}) \qquad \square$$

The direct sum

$$ST(V) = \bigoplus_{p=0}^{\infty} ST^p(V) \approx F[e_1, \ldots, e_n]$$

is called the **symmetric tensor algebra** of V and the direct sum

$$AT(V) = \bigoplus_{p=0}^{n} AT^p(V) \approx F^-[e_1, \ldots, e_n]$$

is called the **antisymmetric tensor algebra** or the **exterior algebra** of V. These vector spaces are graded algebras, where the product is defined using the vector space isomorphisms described in the previous theorem to move the products of $F[e_1, \ldots, e_n]$ and $F^-[e_1, \ldots, e_n]$ to $ST(V)$ and $AT(V)$, respectively.

Thus, restricting the domains of the maps τ gives a nice description of the symmetric and antisymmetric tensor algebras, when $\text{char}(F) = 0$. However, there are many important fields, such as finite fields, that have nonzero characteristic. We can proceed in a different, albeit somewhat less appealing, manner that holds regardless of the characteristic of the base field. Namely, rather than restricting the domain of τ in order to get an isomorphism, we can factor out by the kernel of τ.

Consider a tensor

$$v = \sum_M S_M(v) = \sum_M \left(\sum_{t \in G_M(v)} \alpha_t t \right)$$

Since τ sends elements of different groups $G_M(v) = \{t_1, \ldots, t_k\}$ to different monomials in $F_p[e_1, \ldots, e_n]$ or $F_p^-[e_1, \ldots, e_n]$, it follows that $v \in \ker(\tau)$ if and only if $\tau(S_M(v)) = 0$ for all M, that is, if and only if

$$\alpha_{t_1} \tau t_1 + \cdots + \alpha_{t_k} \tau t_k = 0$$

In the symmetric case, τ is constant on $G_M(v)$ and so $v \in \ker(\tau)$ if and only if

$$\alpha_{t_1} + \cdots + \alpha_{t_k} = 0$$

In the antisymmetric case, $\tau t_j = (-1)^{\sigma_{1,j}} \tau t_1$ where $\lambda_{\sigma_{1,j}}(t_1) = t_j$ and so $v \in \ker(\tau)$ if and only if

$$(-1)^{\sigma_{1,1}} \alpha_{t_1} + \cdots + (-1)^{\sigma_{1,k}} \alpha_{t_k} = 0$$

In both cases, we solve for α_{t_1} and substitute into $S_M(v)$. In the symmetric case,

$$\alpha_{t_1} = -\alpha_{t_2} - \cdots - \alpha_{t_k}$$

and so

$$S_M(v) = \alpha_{t_1} t_1 + \cdots + \alpha_{t_k} t_k = \alpha_{t_2}(t_2 - t_1) + \cdots + \alpha_{t_k}(t_k - t_1)$$

In the antisymmetric case,

$$\alpha_{t_1} = -(-1)^{\sigma_{1,2}} \alpha_{t_2} - \cdots - (-1)^{\sigma_{1,k}} \alpha_{t_k}$$

and so

$$\begin{aligned} S_M(v) &= \alpha_{t_1} t_1 + \cdots + \alpha_{t_k} t_k \\ &= \alpha_{t_2}((-1)^{\sigma_{1,2}} t_2 - t_1) + \cdots + \alpha_{t_k}((-1)^{\sigma_{1,k}} t_k - t_1) \end{aligned}$$

Since $t_i \in \mathcal{B}$, it follows that $S_M(v)$ and therefore v, is in the span of tensors of the form $\lambda_\sigma(t) - t$ in the symmetric case and $(-1)^\sigma \lambda_\sigma(t) - t$ in the antisymmetric case, where $\sigma \in S_p$ and $t \in \mathcal{B}$.

Hence, in the symmetric case,

$$\ker(\tau) \subseteq I_p := \langle \lambda_\sigma(t) - t \mid t \in \mathcal{B}, \sigma \in S_p \rangle$$

and since $\tau(\lambda_\sigma(t) - t) = 0$, it follows that $\ker(\tau) = I_p$. In the antisymmetric case,

$$\ker(\tau) \subseteq I_p := \langle (-1)^\sigma \lambda_\sigma(t) - t \mid t \in \mathcal{B}, \sigma \in S_p \rangle$$

and since $\tau((-1)^\sigma \lambda_\sigma(t) - t) = 0$, it follows that $\ker(\tau) = I_p$.

We now have quotient-space characterizations of the symmetric and antisymmetric tensor spaces that do not place any restriction on the characteristic of the base field.

Theorem 14.17 *Let V be a finite-dimensional vector space over a field F.*

1) *The surjective linear map $\tau: T^p(V) \to F_p[e_1, \ldots, e_n]$ defined by*

$$\tau\left(\sum \alpha_{i_1,\ldots,i_p} e_{i_1} \otimes \cdots \otimes e_{i_p}\right) = \sum \alpha_{i_1,\ldots,i_p} e_{i_1} \vee \cdots \vee e_{i_p}$$

has kernel

$$I_p = \langle \lambda_\sigma(t) - t \mid t \in \mathcal{B}, \sigma \in S_p \rangle$$

and so

$$\frac{T^p(V)}{I_p} \approx F_p[e_1, \ldots, e_n]$$

The vector space $T^p(V)/I$ is also referred to as the **symmetric tensor space** *of degree p of V.*

2) *The surjective linear map $\tau: T^p(V) \to F_p^-[e_1, \ldots, e_n]$ defined by*

$$\tau\left(\sum \alpha_{i_1,\ldots,i_p} e_{i_1} \otimes \cdots \otimes e_{i_p}\right) = \sum \alpha_{i_1,\ldots,i_p} e_{i_1} \wedge \cdots \wedge e_{i_p}$$

has kernel

$$I_p = \langle (-1)^\sigma \lambda_\sigma(t) - t \mid t \in \mathcal{B}, \sigma \in S_p \rangle$$

and so

$$\frac{T^p(V)}{I_p} \approx F_p^-[e_1, \ldots, e_n]$$

The vector space $T^p(V)/I$ is also referred to as the **antisymmetric tensor space** *or* **exterior product space** *of degree p of V.* □

The isomorphic exterior spaces $AT^p(V)$ and $T^p(V)/I_p$ are usually denoted by $\bigwedge^p V$ and the isomorphic exterior algebras $AT(V)$ and $T(V)/I$ are usually denoted by $\bigwedge V$.

Theorem 14.18 *Let V be a vector space of dimension n.*

1) *The dimension of the symmetric tensor space $ST^p(V)$ is equal to the number of monomials of degree p in the variables $e_1, \ldots, e_n$ and this is*

$$\dim(ST^p(V_n)) = \binom{n+p-1}{p}$$

2) *The dimension of the exterior tensor space $\bigwedge^p(V)$ is equal to the number of words of length p in ascending order over the alphabet $E = \{e_1, \ldots, e_n\}$ and this is*

$$\dim(\textstyle\bigwedge^p(V_n)) = \binom{n}{p}$$

Proof. For part 1), the dimension is equal to the number of multisets of size p taken from an underlying set $\{e_1, \ldots, e_n\}$ of size n. Such multisets correspond bijectively to the solutions, in nonnegative integers, of the equation

$$x_1 + \cdots + x_n = p$$

where x_i is the multiplicity of e_i in the multiset. To count the number of solutions, invent two symbols x and $/$. Then any solution $x_i = s_i$ to the previous equation can be described by a sequence of x's and $/$'s consisting of s_1 x's followed by one $/$, followed by s_2 x's and another $/$, and so on. For example, if $p = 6$ and $n = 4$, the solution $3 + 1 + 0 + 2 = 6$ corresponds to the sequence

$$xxx/x//xx$$

Thus, the solutions correspond bijectively to sequences consisting of p x's and $n - 1$ $/$'s. To count the number of such sequences, note that such a sequence can be formed by considering $n + p - 1$ "blanks" and selecting p of these blanks for the x's. This can be done in

$$\binom{n+p-1}{p}$$

ways.□

The Universal Property

We defined tensor products through a universal property, which as we have seen is a powerful technique for determining the properties of tensor products. It is easy to show that the symmetric tensor spaces are universal for symmetric multilinear maps and the antisymmetric tensor spaces are universal for antisymmetric multilinear maps.

Theorem 14.19 *Let V be a finite-dimensional vector space with basis $\{e_1, \ldots, e_n\}$.*

1) *The pair $(F_p[x_1, \ldots, x_n], t)$, where $t\colon V^{\times p} \to F_p[x_1, \ldots, x_n]$ is the multilinear map defined by*

$$t(e_{i_1},\ldots,e_{i_p})=e_{i_1}\vee\cdots\vee e_{i_p}$$

is universal for symmetric p-linear maps with domain $V^{\times p}$; that is, for any symmetric p-linear map $f\colon V^{\times p}\to U$ where U is a vector space, there is a unique linear map $\tau\colon F_p[x_1,\ldots,x_n]\to U$ for which

$$\tau(e_{i_1}\vee\cdots\vee e_{i_p})=f(e_{i_1},\ldots,e_{i_p})$$

2) *The pair $(F_p^-[x_1,\ldots,x_n],t)$, where $t\colon V^{\times p}\to F_p^-[x_1,\ldots,x_n]$ is the multilinear map defined by*

$$t(e_{i_1},\ldots,e_{i_p})=e_{i_1}\wedge\cdots\wedge e_{i_p}$$

is universal for antisymmetric p-linear maps with domain $V^{\times p}$; that is, for any antisymmetric p-linear map $f\colon V^{\times p}\to U$ where U is a vector space, there is a unique linear map $\tau\colon F_p^-[x_1,\ldots,x_n]\to U$ for which

$$\tau(e_{i_1}\wedge\cdots\wedge e_{i_p})=f(e_{i_1},\ldots,e_{i_p})$$

Proof. For part 1), the property

$$\tau(e_{i_1}\vee\cdots\vee e_{i_p})=f(e_{i_1},\ldots,e_{i_p})$$

does indeed uniquely define a linear transformation τ, provided that it is well-defined. However,

$$e_{i_1}\vee\cdots\vee e_{i_p}=e_{j_1}\vee\cdots\vee e_{j_p}$$

if and only if the multisets $\{e_{i_1},\ldots,e_{i_p}\}$ and $\{e_{j_1},\ldots,e_{j_p}\}$ are the same, which implies that $f(e_{i_1},\ldots,e_{i_p})=f(e_{j_1},\ldots,e_{j_p})$, since f is symmetric.

For part 2), since f is antisymmetric, it is completely determined by the fact that it is alternate and by its values on the basis of ascending words $e_{i_1}\wedge\cdots\wedge e_{i_p}$. Accordingly, the condition

$$\tau(e_{i_1}\wedge\cdots\wedge e_{i_p})=f(e_{i_1},\ldots,e_{i_p})$$

uniquely defines a linear transformation τ.□

The Symmetrization Map

When $\operatorname{char}(F)=0$, we can define a linear map $S\colon T^p(V)\to ST^p(V)$, called the **symmetrization map**, by

$$St=\frac{1}{p!}\sum_{\sigma\in S_p}\lambda_\sigma t$$

Since $\lambda_\tau\lambda_\sigma=\lambda_{\tau\sigma}$, we have

$$\lambda_\tau(St) = \frac{1}{p!}\sum_{\sigma\in S_p} \lambda_\tau\lambda_\sigma t = \frac{1}{p!}\sum_{\sigma\in S_p} \lambda_{\tau\sigma} t = \frac{1}{p!}\sum_{\sigma\in S_p} \lambda_\sigma t = St$$

and so St is symmetric. The reason for the factor $1/p!$ is that if v is a symmetric tensor, then $\lambda_\sigma v = v$ and so

$$Sv = \frac{1}{p!}\sum_{\sigma\in S_p} \lambda_\sigma v = \frac{1}{p!}\sum_{\sigma\in S_p} v = v$$

that is, the symmetrization map fixes all symmetric tensors. It follows that for any tensor $t \in T^p(V)$,

$$S^2 t = St$$

Thus, S is idempotent and is therefore the projection map of $T^p(V)$ onto $\operatorname{im}(S) = ST^p(V)$.

The Determinant

The universal property for antisymmetric multilinear maps has the following corollary.

Corollary 14.20 *Let V be a vector space of dimension n over a field F. Let $E = (e_1, \ldots, e_n)$ be an ordered basis for V. Then there is at most one antisymmetric n-linear form $d\colon V^{\times n} \to F$ for which*

$$d(e_1, \ldots, e_n) = 1$$

Proof. According to the universal property for antisymmetric n-linear forms, for every antisymmetric n-linear form $f\colon V^{\times n} \to F$ satisfying $f(e_1, \ldots, e_n) = 1$, there is a unique linear map $\tau_f\colon \bigwedge^n V \to F$ for which

$$\tau_f(e_1 \wedge \cdots \wedge e_n) = f(e_1, \ldots, e_n) = 1$$

But $\bigwedge^n V$ has dimension 1 and so there is only one linear map $\sigma\colon \bigwedge^n V \to F$ with $\sigma(e_1 \wedge \cdots \wedge e_n) = 1$. Therefore, if f and g are two such forms, then $\tau_f = \sigma = \tau_g$, from which it follows that

$$f = \tau_f \circ t = \tau_g \circ t = g$$

$\square$

We now wish to construct an antisymmetric form $d\colon V^{\times n} \to F$, which is unique by the previous theorem. Let $\mathcal{B}$ be a basis for V. For any $v \in V$, write $[v]_{\mathcal{B},i}$ for the ith coordinate of the coordinate matrix $[v]_{\mathcal{B}}$. Thus,

$$v = \sum_i [v]_{\mathcal{B},i} e_i$$

For clarity, and since we will not change the basis, let us write $[v]_i$ for $[v]_{\mathcal{B},i}$.

Consider the map $d\colon V^{\times n} \to F$ defined by

$$d(v_1, \ldots, v_n) = \sum_{\sigma \in S_n} (-1)^\sigma [v_1]_{\sigma 1} \cdots [v_n]_{\sigma n}$$

Then d is multilinear since

$$\begin{aligned} d(av_1 + bu_1, \ldots, v_n) &= \sum_{\sigma \in S_n} (-1)^\sigma [av_1 + bu_1]_{\sigma 1} \cdots [v_n]_{\sigma n} \\ &= \sum_{\sigma \in S_n} (-1)^\sigma (a[v_1]_{\sigma 1} + b[u_1]_{\sigma 1}) \cdots [v_n]_{\sigma(n)} \\ &= a \sum_{\sigma \in S_n} (-1)^\sigma [v_1]_{\sigma 1} \cdots [v_n]_{\sigma n} \\ &\quad + b \sum_{\sigma \in S_n} (-1)^\sigma [u_1]_{\sigma 1} \cdots [v_n]_{\sigma n} \\ &= ad(v_1, \ldots, v_n) + bd(u_1, v_2, \ldots, v_n) \end{aligned}$$

and similarly for any coordinate position.

To see that d is alternating, and therefore antisymmetric since $\operatorname{char}(F) \neq 2$, suppose for instance that $v_1 = v_2$. For any permutation $\sigma \in S_n$, let

$$\sigma' = (\sigma 1\, \sigma 2)\sigma$$

Then $\sigma' x = \sigma x$ for $x \neq 1, 2$ and

$$\sigma' 1 = \sigma 2 \quad \text{and} \quad \sigma' 2 = \sigma 1$$

Hence, $\sigma' \neq \sigma$. Also, since $(\sigma')' = \sigma$, if the sets $\{\sigma, \sigma'\}$ and $\{\rho, \rho'\}$ intersect, then they are identical. Thus, the distinct sets $\{\sigma, \sigma'\}$ form a partition of S_n. It follows that

$$\begin{aligned} d(v_1, v_1, v_3, \ldots, v_n) &= \sum_{\sigma \in S_n} (-1)^\sigma [v_1]_{\sigma 1} [v_1]_{\sigma 2} \cdots [v_n]_{\sigma n} \\ &= \sum_{\text{pairs } \{\sigma, \sigma'\}} \left[(-1)^\sigma [v_1]_{\sigma 1} [v_1]_{\sigma 2} \cdots [v_n]_{\sigma n} + (-1)^{\sigma'} [v_1]_{\sigma' 1} [v_1]_{\sigma' 2} \cdots [v_n]_{\sigma' n} \right] \end{aligned}$$

But

$$[v_1]_{\sigma 1} [v_1]_{\sigma 2} = [v_1]_{\sigma' 1} [v_1]_{\sigma' 2}$$

and since $(-1)^\sigma = -(-1)^{\sigma'}$, the sum of the two terms involving the pair $\{\sigma, \sigma'\}$ is 0. Hence, $d(v_1, v_1, \ldots, v_n) = 0$. A similar argument holds for any coordinate pair.

Finally,

$$\begin{aligned} d(e_1,\ldots,e_n) &= \sum_{\sigma\in S_n}(-1)^\sigma[e_1]_{\sigma 1}\cdots[e_n]_{\sigma n} \\ &= \sum_{\sigma\in S_n}(-1)^\sigma \delta_{1,\sigma 1}\cdots\delta_{n,\sigma n} \\ &= 1 \end{aligned}$$

Thus, the map d is the unique antisymmetric n-linear form on $V^{\times n}$ for which $d(e_1,\ldots,e_n)=1$.

Under the ordered basis $\mathcal{E}=(e_1,\ldots,e_n)$, we can view V as the space F^n of coordinate vectors and view $V^{\times n}$ as the space $M_n(F)$ of $n\times n$ matrices, via the isomorphism

$$(v_1,\ldots,v_n)\mapsto \begin{bmatrix} [v_1]_1 & \cdots & [v_n]_1 \\ \vdots & & \vdots \\ [v_1]_n & \cdots & [v_n]_n \end{bmatrix}$$

where all coordinate matrices are with respect to $\mathcal{E}$.

With this viewpoint, d becomes an antisymmetric n-form on the columns of a matrix $A=(a_{i,j})$ given by

$$d(A)=\sum_{\sigma\in S_n}(-1)^\sigma a_{1,\sigma 1}\cdots a_{n,\sigma n}$$

This is called the **determinant** of the matrix A.

Properties of the Determinant

Let us explore some of the properties of the determinant function.

Theorem 14.21 *If $A\in M_n(F)$, then*

$$d(A)=d(A^t)$$

Proof. We have

$$\begin{aligned} d(A) &= \sum_{\sigma\in S_n}(-1)^\sigma a_{1,\sigma 1}\cdots a_{n,\sigma n} \\ &= \sum_{\sigma^{-1}\in S_n}(-1)^{\sigma^{-1}} a_{\sigma^{-1}1,1}\cdots a_{\sigma^{-1}n,n} \\ &= \sum_{\sigma\in S_n}(-1)^\sigma a_{\sigma 1,1}\cdots a_{\sigma n,n} \\ &= d(A^t) \end{aligned}$$

as desired.□

Theorem 14.22 *If $A, B \in M_n(F)$, then*

$$d(AB) = d(A)d(B)$$

Proof. Consider the map $f_A: M_n(F) \to F$ defined by

$$f_A(X) = d(AX)$$

We can consider f_A as a function on the columns of X and think of it as a composition

$$f_A: (X^{(1)}, \ldots, X^{(n)}) \overset{\alpha}{\mapsto} (AX^{(1)}, \ldots, AX^{(n)}) \overset{\beta}{\mapsto} d(AX)$$

Each step in this map is multilinear and so f_A is multilinear. It is also clear that f_A is antisymmetric and so f_A is a scalar multiple of the determinant function, say $f_A(X) = cd(X)$. Then

$$d(AX) = f_A(X) = cd(X)$$

Setting $X = I_n$ gives $d(A) = c$ and so

$$d(AX) = d(A)d(X)$$

as desired.□

Theorem 14.23 *A matrix $A \in M_n(F)$ is invertible if and only if $d(A) \neq 0$.*
Proof. If $P \in M_n(F)$ is invertible, then $PP^{-1} = I_n$ and so

$$d(P)d(P^{-1}) = 1$$

which shows that $d(P) \neq 0$ and $d(P^{-1}) = 1/d(P)$. Conversely, any matrix $A \in M_n(F)$ is equivalent to a diagonal matrix

$$A = PDQ$$

where P and Q are invertible and D is diagonal with 1's and 0's on the main diagonal. Hence,

$$d(A) = d(P)d(D)d(Q)$$

and so if $d(A) \neq 0$, then $d(D) \neq 0$, which happens if and only if $D = I_n$, whence A is invertible.□

Exercises

1. Show that if $\tau: W \to X$ is a linear map and $b: U \times V \to W$ is bilinear, then $\tau \circ b: U \times V \to X$ is bilinear.
2. Show that the only map that is both linear and n-linear (for $n \geq 2$) is the zero map.
3. Find an example of a bilinear map $\tau: V \times V \to W$ whose image $\operatorname{im}(\tau) = \{\tau(u, v) \mid u, v \in V\}$ is not a subspace of W.

4. Let $\mathcal{B} = \{u_i \mid i \in I\}$ be a basis for U and let $\mathcal{C} = \{v_j \mid j \in J\}$ be a basis for V. Show that the set

$$\mathcal{D} = \{u_i \otimes v_j \mid i \in I,\ j \in J\}$$

is a basis for $U \otimes V$ by showing that it is linearly independent and spans.
5. Prove that the following property of a pair $(W, g: U \times V \to W)$ with g bilinear characterizes the tensor product $(U \otimes V, t: U \times V \to U \otimes V)$ up to isomorphism, and thus could have been used as the definition of tensor product: For a pair $(W, g: U \times V \to W)$ with g bilinear if $\{u_i\}$ is a basis for U and $\{v_i\}$ is a basis for V, then $\{g(u_i, v_j)\}$ is a basis for W.
6. Prove that $U \otimes V \approx V \otimes U$.
7. Let X and Y be nonempty sets. Use the universal property of tensor products to prove that $\mathcal{F}_{X \times Y} \approx \mathcal{F}_X \otimes \mathcal{F}_Y$.
8. Let $u, u' \in U$ and $v, v' \in V$. Assuming that $u \otimes v \neq 0$, show that $u \otimes v = u' \otimes v'$ if and only if $u' = ru$ and $v' = r^{-1}v$, for $r \neq 0$.
9. Let $\mathcal{B} = \{b_i\}$ be a basis for U and $\mathcal{C} = \{c_i\}$ be a basis for V. Show that any function $f: \mathcal{B} \times \mathcal{C} \to W$ can be extended to a linear function $\overline{f}: U \otimes V \to W$. Deduce that the function f can be extended in a unique way to a bilinear map $\widehat{f}: U \times V \to W$. Show that all bilinear maps are obtained in this way.
10. Let S_1, S_2 be subspaces of U. Show that

$$(S_1 \otimes V) \cap (S_2 \otimes V) \approx (S_1 \cap S_2) \otimes V$$

11. Let $S \subseteq U$ and $T \subseteq V$ be subspaces of vector spaces U and V, respectively. Show that

$$(S \otimes V) \cap (U \otimes T) \approx S \otimes T$$

12. Let $S_1, S_2 \subseteq U$ and $T_1, T_2 \subseteq V$ be subspaces of U and V, respectively. Show that

$$(S_1 \otimes T_1) \cap (S_2 \otimes T_2) \approx (S_1 \cap S_2) \otimes (T_1 \otimes T_2)$$

13. Find an example of two vector spaces U and V and a nonzero vector $x \in U \otimes V$ that has at least two distinct (not including order of the terms) representations of the form

$$x = \sum_{i=1}^{n} u_i \otimes v_i$$

where the u_i's are linearly independent and so are the v_i's.
14. Let ι_X denote the identity operator on a vector space X. Prove that $\iota_V \odot \iota_W = \iota_{V \otimes W}$.

15. Suppose that $\tau_1: U \to V$, $\tau_2: V \to W$ and $\sigma_1: U' \to V_K$, $\sigma_2: V_K \to W'$. Prove that
$$(\tau_2 \circ \tau_1) \odot (\sigma_2 \circ \sigma_1) = (\tau_2 \odot \sigma_2) \circ (\tau_1 \odot \sigma_1)$$
16. Connect the two approaches to extending the base field of an F-space V to K (at least in the finite-dimensional case) by showing that $F^n \otimes_F K \approx (K)^n$.
17. Prove that in a tensor product $U \otimes U$ for which $\dim(U) \geq 2$ not all vectors have the form $u \otimes v$ for some $u, v \in U$. *Hint*: Suppose that $u, v \in U$ are linearly independent and consider $u \otimes v + v \otimes u$.
18. Prove that for the block matrix
$$M = \begin{bmatrix} A & B \\ 0 & C \end{bmatrix}_{\text{block}}$$
we have $d(M) = d(A)d(C)$.
19. Let $A, B \in M_n(F)$. Prove that if either A or B is invertible, then the matrices $A + \alpha B$ are invertible except for a finite number of α's.

The Tensor Product of Matrices

20. Let $A = (a_{i,j})$ be the matrix of a linear operator $\tau \in \mathcal{L}(V)$ with respect to the ordered basis $\mathcal{A} = (u_1, \ldots, u_n)$. Let $B = (b_{i,j})$ be the matrix of a linear operator $\sigma \in \mathcal{L}(V)$ with respect to the ordered basis $\mathcal{B} = (v_1, \ldots, v_m)$. Consider the ordered basis $\mathcal{C} = (u_i \otimes v_j)$ ordered lexicographically; that is $u_i \otimes v_j < u_\ell \otimes v_k$ if $i < \ell$ or $i = \ell$ and $j < k$. Show that the matrix of $\tau \otimes \sigma$ with respect to $\mathcal{C}$ is
$$A \otimes B = \begin{pmatrix} a_{1,1}B & a_{1,2}B & \cdots & a_{1,n}B \\ a_{2,1}B & a_{2,2}B & \cdots & a_{2,n}B \\ \vdots & \vdots & & \vdots \\ a_{n,1}B & a_{n,2}B & \cdots & a_{n,n}B \end{pmatrix}_{\text{block}}$$
This matrix is called the **tensor product**, **Kronecker product** or **direct product** of the matrix A with the matrix B.
21. Show that the tensor product is not, in general, commutative.
22. Show that the tensor product $A \otimes B$ is bilinear in both A and B.
23. Show that $A \otimes B = 0$ if and only if $A = 0$ or $B = 0$.
24. Show that
    a) $(A \otimes B)^t = A^t \otimes B^t$
    b) $(A \otimes B)^* = A^* \otimes B^*$ (when $F = \mathbb{C}$)
25. Show that if $u, v \in F^n$, then (as row vectors) $u^t v = u^t \otimes v$.
26. Suppose that $A_{m,n}, B_{p,q}, C_{n,k}$ and $D_{q,r}$ are matrices of the given sizes. Prove that
$$(A \otimes B)(C \otimes D) = (AC) \otimes (BD)$$
Discuss the case $k = r = 1$.

27. Prove that if A and B are nonsingular, then so is $A \otimes B$ and

$$(A \otimes B)^{-1} = A^{-1} \otimes B^{-1}$$

28. Prove that $\operatorname{tr}(A \otimes B) = \operatorname{tr}(A) \cdot \operatorname{tr}(B)$.
29. Suppose that F is algebraically closed. Prove that if A has eigenvalues $\lambda_1, \ldots, \lambda_n$ and B has eigenvalues $\mu_1, \ldots, \mu_m$, both lists including multiplicity, then $A \otimes B$ has eigenvalues $\{\lambda_i \mu_j \mid i \leq n, j \leq m\}$, again counting multiplicity.
30. Prove that $\det(A_{n,n} \otimes B_{m,m}) = (\det(A_{n,n}))^m (\det(B_{m,m}))^n$.

Chapter 15
Positive Solutions to Linear Systems: Convexity and Separation

It is of interest to determine conditions that guarantee the existence of *positive* solutions to homogeneous systems of linear equations

$$Ax = 0$$

where $A \in \mathcal{M}_{m,n}(\mathbb{R})$.

Definition *Let* $v = (a_1, \ldots, a_n) \in \mathbb{R}^n$.

1) v *is* **nonnegative**, *written* $v \geq 0$, *if*

$$a_i \geq 0 \textit{ for all } i$$

(The term **positive** *is also used for this property.) The set of all nonnegative vectors in* $\mathbb{R}^n$ *is the* **nonnegative orthant** *in* $\mathbb{R}^n$.

2) v *is* **strictly positive**, *written* $v > 0$, *if* v *is nonnegative but not* 0, *that is, if*

$$a_i \geq 0 \textit{ for all } i \textit{ and } a_j > 0 \textit{ for some } j$$

The set $\mathbb{R}^n_+$ *of all strictly positive vectors in* $\mathbb{R}^n$ *is the* **strictly positive orthant** *in* $\mathbb{R}^n$.

3) v *is* **strongly positive**, *written* $v \gg 0$, *if*

$$a_i > 0 \textit{ for all } i$$

The set $\mathbb{R}^n_{++}$ *of all strongly positive vectors in* $\mathbb{R}^n$ *is the* **strongly positive orthant** *in* $\mathbb{R}^n$.□

We are interested in conditions under which the system $Ax = 0$ has strictly positive or strongly positive solutions. Since the strictly and strongly positive orthants in $\mathbb{R}^n$ are not subspaces of $\mathbb{R}^n$, it is difficult to use strictly linear methods in studying this issue: we must also use geometric methods, in particular, methods of convexity.

Let us pause briefly to consider an important application of strictly positive solutions to a system $Ax = 0$. If $X = (x_1, \ldots, x_n)$ is a strictly positive solution to this system, then so is the vector

$$\Pi = \frac{1}{\Sigma x_i} X = \frac{1}{\Sigma x_i}(x_1, \ldots, x_n) = (\pi_1, \ldots, \pi_n)$$

which is a *probability distribution*, that is, $0 \leq \pi_i \leq 1$ and $\sum \pi_i = 1$. Moreover, if X is a strongly positive solution, then Π has the property that each probability is positive.

Now, the product $A\Pi$ is the expected value of the columns of A with respect to the probability distribution Π. Hence, $Ax = 0$ has a strictly (strongly) positive solution if and only if there is a strictly (strongly) positive probability distribution for which the columns of A have expected value 0. If each column of A represents the possible payoffs from a game of chance, where each row is a different possible outcome of the game, then the game is fair when the expected value of the columns is 0. Thus, $Ax = 0$ has a strictly (strongly) positive solution X if and only if the game with payoffs A and probabilities X is fair.

As another (related) example, in discrete option pricing models of mathematical finance, the absence of arbitrage opportunities in the model is equivalent to the fact that a certain vector describing the gains in a portfolio does not intersect the strictly positive orthant in $\mathbb{R}^n$. As we will see in this chapter, this is equivalent to the existence of a strongly positive solution to a homogeneous system of equations. This solution, when normalized to a probability distribution, is called a *martingale measure*.

Of course, the equation $Ax = 0$ has a strictly positive solution if and only if $\ker(A)$ contains a strictly positive vector, that is, if and only if

$$\ker(A) = \text{RowSpace}(A)^{\perp}$$

meets the strictly positive orthant in $\mathbb{R}^n$. Thus, we wish to characterize the subspaces S of $\mathbb{R}^n$ for which $S^{\perp}$ meets the strictly positive orthant in $\mathbb{R}^n$, in symbols,

$$S^{\perp} \cap \mathbb{R}_+^n \neq \emptyset$$

for these are precisely the row spaces of the matrices A for which $Ax = 0$ has a strictly positive solution. A similar statement holds for strongly positive solutions.

Looking at the real plane $\mathbb{R}^2$, we can divine the answer with a picture. A one-dimensional subspace S of $\mathbb{R}^2$ has the property that its orthogonal complement $S^{\perp}$ meets the strictly positive orthant (quadrant) in $\mathbb{R}^2$ if and only if S is the x-axis, the y-axis or a line with negative slope. For the case of the strongly positive orthant, S must have negative slope. Our task is to generalize this to $\mathbb{R}^n$.

This will lead us to the following results, which are quite intuitive in $\mathbb{R}^2$ and $\mathbb{R}^3$:

$$S^\perp \cap \mathbb{R}^n_{++} \neq \emptyset \quad \Leftrightarrow \quad S \cap \mathbb{R}^n_{+} = \emptyset \tag{15.1}$$

and

$$S^\perp \cap \mathbb{R}^n_{+} \neq \emptyset \quad \Leftrightarrow \quad S \cap \mathbb{R}^n_{++} = \emptyset \tag{15.2}$$

Let us translate these statements into the language of the matrix equation $Ax = 0$. If $S = \text{RowSpace}(A)$, then $S^\perp = \ker(A)$ and so we have

$$\ker(A) \cap \mathbb{R}^n_{++} \neq \emptyset \quad \Leftrightarrow \quad \text{RowSpace}(A) \cap \mathbb{R}^n_{+} = \emptyset$$

and

$$\ker(A) \cap \mathbb{R}^n_{+} \neq \emptyset \quad \Leftrightarrow \quad \text{RowSpace}(A) \cap \mathbb{R}^n_{++} = \emptyset$$

Now,

$$\text{RowSpace}(A) \cap \mathbb{R}^n_{+} = \{vA \mid vA > 0\}$$

and

$$\text{RowSpace}(A) \cap \mathbb{R}^n_{++} = \{vA \mid vA \gg 0\}$$

and so these statements become

$$Ax = 0 \text{ has a strongly positive solution} \quad \Leftrightarrow \quad \{vA \mid vA > 0\} = \emptyset$$

and

$$Ax = 0 \text{ has a strictly positive solution} \quad \Leftrightarrow \quad \{vA \mid vA \gg 0\} = \emptyset$$

We can rephrase these results in the form of a **theorem of the alternative**, that is, a theorem that says that exactly one of two conditions holds.

Theorem 15.1 *Let* $A \in \mathcal{M}_{m,n}(\mathbb{R})$.
1) *Exactly one of the following holds:*
   a) $Au = 0$ *for some strongly positive* $u \in \mathbb{R}^n$.
   b) $vA > 0$ *for some* $v \in \mathbb{R}^m$.
2) *Exactly one of the following holds:*
   a) $Au = 0$ *for some strictly positive* $u \in \mathbb{R}^n$.
   b) $vA \gg 0$ *for some* $v \in \mathbb{R}^m$. □

Before proving Theorem 15.1, we require some background.

Convex, Closed and Compact Sets

We shall need the following concepts.

Definition

1) *Let $x_1, \ldots, x_k \in \mathbb{R}^n$. Any linear combination of the form*

$$t_1 x_1 + \cdots + t_k x_k$$

where $0 \le t_i \le 1$ and $t_1 + \cdots + t_k = 1$ is called a **convex combination** *of the vectors $x_1, \ldots, x_k$.*

2) *A subset $X \subseteq \mathbb{R}^n$ is* **convex** *if whenever $x, y \in X$, then the line segment between x and y also lies in X, in symbols,*

$$\{tx + (1-t)y \mid 0 \le t \le 1\} \subseteq X$$

3) *A subset $X \subseteq \mathbb{R}^n$ is* **closed** *if whenever (x_n) is a convergent sequence of elements of X, then the limit is also in X.*
4) *A subset $X \subseteq \mathbb{R}^n$ is* **compact** *if it is both closed and bounded.*
5) *A subset $X \subseteq \mathbb{R}^n$ is a* **cone** *if $x \in X$ implies that $ax \in X$ for all $a \ge 0$.*□

We will also have need of the following facts from analysis.

1) A continuous function that is defined on a compact set X in $\mathbb{R}^n$ takes on maximum and minimum values at some points within the set X.
2) A subset X of $\mathbb{R}^n$ is compact if and only if every sequence in X has a subsequence that converges to a point in X.

Theorem 15.2 *Let X and Y be subsets of $\mathbb{R}^n$. Define*

$$X + Y = \{a + b \mid a \in X,\, b \in Y\}$$

1) *If X and Y are convex, then so is $X + Y$*
2) *If X is compact and Y is closed, then $X + Y$ is closed.*

Proof. For 1), let $x_0 + y_0$ and $x_1 + y_1$ be in $X + Y$. The line segment between these two points is

$$\begin{aligned} &t(x_0 + y_0) + (1-t)(x_1 + y_1) \\ &\qquad = [tx_0 + (1-t)x_1] + [ty_0 + (1-t)y_1] \in X + Y \end{aligned}$$

for $0 \le t \le 1$ and so $X + Y$ is convex.

For part 2), let $x_n + y_n$ be a convergent sequence in $X + Y$. Suppose that $x_n + y_n \to z$. We must show that $z \in X + Y$. Since x_n is a sequence in the compact set X, it has a convergent subsequence x_{n_k} whose limit x lies in X. Since $x_{n_k} + y_{n_k} \to z$ and $x_{n_k} \to x$ we can conclude that $y_{n_k} \to z - x$. Since Y is closed, it follows that $z - x \in Y$ and so $z = x + (z - x) \in X + Y$.□

Convex Hulls

We will also have use for the notion of the smallest convex set containing a given set.

Definition *The* **convex hull** *of a set $S = \{x_1, \ldots, x_k\}$ of vectors in $\mathbb{R}^n$ is the smallest convex set in $\mathbb{R}^n$ that contains S. We will denote the convex hull of S by $\mathcal{C}(S)$.*□

Here is a characterization of convex hulls.

Theorem 15.3 *Let $S = \{x_1, \ldots, x_k\}$ be a set of vectors in $\mathbb{R}^n$. Then the convex hull $\mathcal{C}(S)$ is the set Δ of all convex combinations of vectors in S, that is,*

$$\mathcal{C}(S) = \Delta := \left\{t_1x_1 + \cdots + t_kx_k \mid 0 \le t_i \le 1, \sum t_i = 1\right\}$$

Proof. Clearly, if D is a convex set that contains S, then D also contains Δ. Hence $\Delta \subseteq \mathcal{C}(S)$. To prove the reverse inclusion, we need only show that Δ is convex, since then $S \subseteq \Delta$ implies that $\mathcal{C}(S) \subseteq \Delta$. So let

$$\begin{aligned} X &= t_1x_1 + \cdots + t_kx_k \\ Y &= s_1x_1 + \cdots + s_kx_k \end{aligned}$$

be in Δ. If $a + b = 1$ and $0 \le a, b \le 1$ then

$$\begin{aligned} aX + bY &= a(t_1x_1 + \cdots + t_kx_k) + b(s_1x_1 + \cdots + s_kx_k) \\ &= (at_1 + bs_1)x_1 + \cdots + (at_k + bs_k)x_k \end{aligned}$$

But this is also a convex combination of the vectors in S, because

$$0 \le at_i + bs_i \le (a + b) \cdot \max(s_i, t_i) = \max(s_i, t_i) \le 1$$

and

$$\sum_{i=1}^{k}(at_i + bs_i) = a\sum_{i=1}^{k} t_i + b\sum_{i=1}^{k} s_i = a + b = 1$$

Thus, $aX + bY \in \Delta$.□

Theorem 15.4 *The convex hull $\mathcal{C}(S)$ of a* finite *set $S = \{x_1, \ldots, x_k\}$ of vectors in $\mathbb{R}^n$ is a compact set.*
Proof. The set

$$D = \left\{(t_1, \ldots, t_k) \mid 0 \le t_i \le 1, \sum t_i = 1\right\}$$

is closed and bounded in $\mathbb{R}^n$ and therefore compact. Define a function $f: D \to \mathbb{R}^n$ as follows: If $t = (t_1, \ldots, t_k)$, then

$$f(t) = t_1x_1 + \cdots + t_kx_k$$

To see that f is continuous, let $s = (s_1, \ldots, s_k)$ and let $M = \max(\|x_i\|)$. Given $\epsilon > 0$, if $\|s - t\| < \epsilon/kM$ then

$$|s_i - t_i| \le \|s - t\| < \frac{\epsilon}{kM}$$

and so

$$\begin{aligned}\|f(s) - f(t)\| &= \|(s_1 - t_1)x_1 + \cdots + (s_k - t_k)x_k\| \\ &\le |s_1 - t_1|\|x_1\| + \cdots + |s_k - t_k|\|x_k\| \\ &\le kM\|s - t\| \\ &= \epsilon\end{aligned}$$

Finally, since $f(D) = \mathcal{C}(S)$, it follows that $\mathcal{C}(S)$ is compact.□

Linear and Affine Hyperplanes

We next discuss hyperplanes in $\mathbb{R}^n$. A **linear hyperplane** in $\mathbb{R}^n$ is an $(n-1)$-dimensional subspace of $\mathbb{R}^n$. As such, it is the solution set of a linear equation of the form

$$a_1x_1 + \cdots + a_nx_n = 0$$

or

$$\langle N, x\rangle = 0$$

where $N = (a_1, \ldots, a_n)$ is nonzero and $x = (x_1, \ldots, x_n)$. Geometrically speaking, this is the set of all vectors in $\mathbb{R}^n$ that are perpendicular (normal) to the vector N.

An **affine hyperplane**, or just **hyperplane**, in $\mathbb{R}^n$ is a linear hyperplane that has been translated by a vector. Thus, it is the solution set to an equation of the form

$$a_1(x_1 - b_1) + \cdots + a_n(x_n - b_n) = 0$$

or equivalently,

$$\langle N, x\rangle = b$$

where $b = a_1b_1 + \cdots + a_nb_n$. We denote this hyperplane by

$$\mathcal{H}(N, b) = \{x \in \mathbb{R}^n \mid \langle N, x\rangle = b\}$$

Note that the hyperplane

$$\mathcal{H}(N, \|N\|^2) = \{x \in \mathbb{R}^n \mid \langle N, x\rangle = \|N\|^2\}$$

contains the point N, which is the point of $\mathcal{H}(N, \|N\|^2)$ closest to the origin, since Cauchy's inequality gives

$$\|N\|^2 = \langle N, x\rangle \le \|N\|\|x\|$$

and so $\|N\| \le \|x\|$ for all $x \in \mathcal{H}(N, \|N\|^2)$. Moreover, we leave it as an exercise to show that any hyperplane has the form $\mathcal{H}(N, \|N\|^2)$ for an appropriate vector N.

A hyperplane defines two **closed half-spaces**

$$\mathcal{H}_+(N,b) = \{x \in \mathbb{R}^n \mid \langle N, x\rangle \geq b\}$$
$$\mathcal{H}_-(N,b) = \{x \in \mathbb{R}^n \mid \langle N, x\rangle \leq b\}$$

and two *disjoint* **open half-spaces**

$$\mathcal{H}_+^\circ(N,b) = \{x \in \mathbb{R}^n \mid \langle N, x\rangle > b\}$$
$$\mathcal{H}_-^\circ(N,b) = \{x \in \mathbb{R}^n \mid \langle N, x\rangle < b\}$$

It is clear that

$$\mathcal{H}_+(N,b) \cap \mathcal{H}_-(N,b) = \mathcal{H}(N,b)$$

and that the sets $\mathcal{H}_+^\circ(N,b)$, $\mathcal{H}_-^\circ(N,b)$ and $\mathcal{H}(N,b)$ form a partition of $\mathbb{R}^n$.

If $N \in \mathbb{R}^n$ and $X \subseteq \mathbb{R}^n$, we let

$$\langle N, X\rangle = \{\langle N, x\rangle \mid x \in X\}$$

and write

$$\langle N, X\rangle < b$$

to denote the fact that $\langle N, x\rangle < b$ for all $x \in X$.

Definition *Two subsets X and Y of $\mathbb{R}^n$ are* **strictly separated** *by a hyperplane $\mathcal{H}(N,b)$ if X lies in one open half-space determined by $\mathcal{H}(N,b)$ and Y lies in the other open half-space; in symbols, one of the following holds:*
1) $\langle N, X\rangle < b < \langle N, Y\rangle$
2) $\langle N, Y\rangle < b < \langle N, X\rangle$ □

Note that 1) holds for N and b if and only if 2) holds for $-N$ and $-b$, and so we need only consider one of the conditions to demonstrate that two sets X and Y are *not* strictly separated. Specifically, if 1) fails for all N and b, then the condition

$$\langle -N, Y\rangle < -b < \langle -N, X\rangle$$

also fails for all N and b and so 2) also fails for all N and b, whence X and Y are not strictly separated.

Definition *Two subsets X and Y of $\mathbb{R}^n$ are* **strongly separated** *by a hyperplane $\mathcal{H}(N,b)$ if there is an $e > 0$ for which one of the following holds:*
1) $\langle N, X\rangle < b - e < b + e < \langle N, Y\rangle$
2) $\langle N, Y\rangle < b - e < b + e < \langle N, X\rangle$ □

As before, we need only consider one of the conditions to show that two sets are *not* strongly separated. Note also that if

$$\langle N, x\rangle < r \le \langle N, Y\rangle$$

for $r \in \mathbb{R}$, then $x \in \mathbb{R}^n$ and $Y \subseteq \mathbb{R}^n$ are stongly separated by the hyperplane

$$\mathcal{H}\left(N, \frac{r + \langle N, x\rangle}{2}\right)$$

Separation

Now that we have the preliminaries out of the way, we can get down to some theorems. The first is a well-known *separation theorem* that is the basis for many other separation theorems. It says that if a closed convex set $C \subseteq \mathbb{R}^n$ does not contain a vector b, then C can be *strongly* separated from b.

Theorem 15.5 *Let C be a closed convex subset of $\mathbb{R}^n$.*

1) *C contains a* unique *vector N of minimum norm, that is, there is a unique vector $N \in C$ for which*

$$\|N\| < \|x\|$$

for all $x \in C, x \ne N$.

2) *If $b \notin C$, then C lies in the closed half-space*

$$\mathcal{H}_+(N, \|N\|^2 + \langle N, b\rangle)$$

that is,

$$\langle N, C\rangle \ge \|N\|^2 + \langle N, b\rangle > \langle N, b\rangle$$

where N is the unique vector of minimum norm in the closed convex set

$$C - b = \{c - b \mid c \in C\}$$

Hence, C and b are strongly separated by the hyperplane

$$\mathcal{H}\left(N, \frac{\|N\|^2 + 2\langle N, b\rangle}{2}\right)$$

Proof. For part 1), if $0 \in C$ then this is the unique vector of minimum norm, so we may assume that $0 \notin C$. It follows that no two distinct elements of C can be negative scalar multiples of each other. For if x and rx were in C, where $r < 0$, then taking $t = -r/(1-r)$ gives

$$0 = tx + (1-t)rx \in C$$

which is false.

We first show that C contains a vector N of minimum norm. Recall that the Euclidean norm (distance) is a continuous function. Although C need not be compact, if we choose a real number s such that the closed ball

$$B_s(0) = \{z \in \mathbb{R}^n \mid \|z\| \leq s\}$$

intersects C, then that intersection $C' = C \cap B_s(0)$ is both closed and bounded and so is compact. The norm function therefore achieves its minimum on C', say at the point $N \in C' \subseteq C$. It is clear that if $\|v\| < \|N\|$ for some $v \in C$, then $v \in C'$, in contradiction to the minimality of N. Hence, N is a vector of minimum norm in C.

We establish uniqueness first for closed line segments $[u, v]$ in $\mathbb{R}^n$. If $u = rv$ where $r > 0$, then

$$\|tu + (1-t)v\| = |tr + (1-t)|\|v\|$$

is smallest when $t = 0$ for $r > 1$ and $t = 1$ for $r < 1$. Assume that u and v are not scalar multiples of each other and suppose that $x \neq y$ in $[u, v]$ have minimum norm $d > 0$. If $z = (x + y)/2$ then since x and y are also not scalar multiples of each other, the Cauchy-Schwarz inequality is strict and so

$$\begin{aligned}
\|z\|^2 &= \frac{1}{4}\|x + y\|^2 \\
&= \frac{1}{4}(\|x\|^2 + 2\langle x, y\rangle + \|y\|^2) \\
&< \frac{1}{2}(d^2 + \|x\|\|y\|) \\
&= d^2
\end{aligned}$$

which contradicts the minimality of d. Thus, $[u, v]$ has a unique point of minimum norm.

Finally, if $x \in C$ also has minimum norm, then N and x are points of minimum norm in the line segment $[N, x] \subseteq C$ and so $x = N$. Hence, C has a unique element of minimum norm.

For part 2), suppose the result is true when $0 \notin C$. Then $b \notin C$ implies that $0 \notin C - b$ and so if $N \in C - b$ has smallest norm, then

$$\langle N, C - b\rangle \geq \|N\|^2 > 0$$

Therefore,

$$\langle N, C\rangle \geq \|N\|^2 + \langle N, b\rangle > \langle N, b\rangle$$

and so C and b are strongly separated by the hyperplane

$$\mathcal{H}(N, (1/2)\|N\|^2 + \langle N, b\rangle)$$

Thus, we need only prove part 2) for $b = 0$, that is, we need only prove that

$$\langle N, C\rangle \geq \|N\|^2$$

If there is a nonzero $x \in C$ for which

$$\langle N, x\rangle < \|N\|^2$$

then $\|N\| < \|x\|$ and

$$\langle N, x\rangle = \|N\|^2 - \epsilon$$

for some $\epsilon > 0$. Then for the open line segment $f(t) = tN + (1-t)x$ with $0 < t < 1$, we have

$$\begin{aligned}\|f(t)\|^2 &= \|tN + (1-t)x\|^2 \\ &= t^2\|N\|^2 + 2t(1-t)\langle N, x\rangle + (1-t)^2\|x\|^2 \\ &< (2t - t^2)\|N\|^2 - 2t(1-t)\epsilon + (1-t)^2\|x\|^2 \\ &= (-\|N\|^2 + 2\epsilon + \|x\|^2)t^2 + 2(\|N\|^2 - \epsilon - \|x\|^2)t + \|x\|^2\end{aligned}$$

Let $p(t)$ denote the final expression above, which is a quadratic in t. It is easy to see that $p(t)$ has its minimum at the interior point of the line segment $[N, x]$ corresponding to

$$t_0 = \frac{-\|N\|^2 + \epsilon + \|x\|^2}{-\|N\|^2 + 2\epsilon + \|x\|^2} < 1$$

and so $\|f(t_0)\| < p(t_0) \leq p(1) = \|N\|$, which is a contradiction.$\square$

The next result brings us closer to our goal by replacing a single vector b with a *subspace* S disjoint from C. However, we must also require that C be bounded, and therefore compact.

Theorem 15.6 *Let C be a compact convex subset of $\mathbb{R}^n$ and let S be a subspace of $\mathbb{R}^n$ such that $C \cap S = \emptyset$. Then there exists a nonzero $N \in S^\perp$ such that*

$$\langle N, x\rangle \geq \|N\|^2 > 0$$

for all $x \in C$. Hence, the hyperplane $\mathcal{H}(N, \|N\|^2/2)$ strongly separates S and C.

Proof. Theorem 15.2 implies that the set $S + C$ is closed and convex. Furthermore, $C \cap S = \emptyset$ implies that $0 \notin S + C$ and so Theorem 15.5 implies that $S + C$ can be strongly separated from the origin. Hence, there is a nonzero $N \in \mathbb{R}^n$ such that

$$\langle N, s\rangle + \langle N, c\rangle = \langle N, s + c\rangle \geq \|N\|^2$$

for all $s \in S$ and $c \in C$. But if $\langle N, s\rangle \neq 0$ for some $s \in S$, then we can replace s by an appropriate scalar multiple of s in order to make the left side of this inequality negative, which is impossible. Hence, $\langle N, s\rangle = 0$ for all $s \in S$, that is, $N \in S^\perp$ and

$$\langle N, C\rangle \geq \|N\|^2$$

□

We can now prove (15.1) and (15.2).

Theorem 15.7 *Let S be a subspace of $\mathbb{R}^n$.*
1) $S \cap \mathbb{R}^n_+ = \emptyset$ if and only if $S^\perp \cap \mathbb{R}^n_{++} \neq \emptyset$
2) $S \cap \mathbb{R}^n_{++} = \emptyset$ if and only if $S^\perp \cap \mathbb{R}^n_+ \neq \emptyset$
Proof. In both cases, one direction is easy. It is clear that there cannot exist vectors $u \in \mathbb{R}^n_{++}$ and $v \in \mathbb{R}^n_+$ that are orthogonal. Hence, $S \cap \mathbb{R}^n_+$ and $S^\perp \cap \mathbb{R}^n_{++}$ cannot both be nonempty and so $S^\perp \cap \mathbb{R}^n_{++} \neq \emptyset$ implies $S \cap \mathbb{R}^n_+ = \emptyset$. Also, $S \cap \mathbb{R}^n_{++}$ and $S^\perp \cap \mathbb{R}^n_+$ cannot both be nonempty and so $S^\perp \cap \mathbb{R}^n_+ \neq \emptyset$ implies that $S \cap \mathbb{R}^n_{++} = \emptyset$.

For the converse in part 1), to prove that

$$S \cap \mathbb{R}^n_+ = \emptyset \quad \Rightarrow \quad S^\perp \cap \mathbb{R}^n_{++} \neq \emptyset$$

a good candidate for an element of $S^\perp \cap \mathbb{R}^n_{++}$ would be a normal to a hyperplane that separates S from a subset of $\mathbb{R}^n_+$. Note that our separation theorems do not allow us to separate S from $\mathbb{R}^n_+$, because $\mathbb{R}^n_+$ is not compact. So consider instead the convex hull Δ of the standard basis vectors $\epsilon_1, \ldots, \epsilon_n$ in $\mathbb{R}^n_+$:

$$\Delta = \{t_1\epsilon_1 + \cdots + t_n\epsilon_n \mid 0 \leq t_i \leq 1, \Sigma t_i = 1\}$$

which is compact. Moreover, $\Delta \subseteq \mathbb{R}^n_+$ implies that $\Delta \cap S = \emptyset$ and so Theorem 15.6 implies that there is a nonzero vector $N = (a_1, \ldots, a_n) \in S^\perp$ such that

$$\langle N, \delta\rangle \geq \|N\|^2$$

for all $\delta \in \Delta$. Taking $\delta = \epsilon_i$ gives

$$a_i = \langle N, \epsilon_i\rangle \geq \|N\|^2 > 0$$

and so $N \in S^\perp \cap \mathbb{R}^n_{++}$, which is therefore nonempty.

To prove part 2), again we note that there cannot exist orthogonal vectors $u \in \mathbb{R}^n_{++}$ and $v \in \mathbb{R}^n_+$ and so $S \cap \mathbb{R}^n_{++}$ and $S^\perp \cap \mathbb{R}^n_+$ cannot both be nonempty. Thus, $S^\perp \cap \mathbb{R}^n_+ \neq \emptyset$ implies that $S \cap \mathbb{R}^n_{++} = \emptyset$.

To finish the proof of part 2), we must prove that

$$S \cap \mathbb{R}^n_{++} = \emptyset \quad \Rightarrow \quad S^\perp \cap \mathbb{R}^n_+ \neq \emptyset$$

Let $\mathcal{B} = \{B_1, \ldots, B_k\}$ be a basis for S. Then $N = (a_1, \ldots, a_n) \in S^\perp$ if and only if $N \perp B_i$ for all i. In matrix terms, if

$$M = (m_{i,j}) = (B_1 \mid B_2 \mid \cdots \mid B_k)$$

has rows $R_1, \ldots, R_n$, then $N \in S^\perp$ if and only if $NM = 0$, that is,

$$a_1 R_1 + \cdots + a_n R_n = 0$$

Now, $S^\perp$ contains a strictly positive vector $N = (a_1, \ldots, a_n)$ if and only if this equation holds, where $a_i \geq 0$ for all i and $a_j > 0$ for some j. Moreover, we may assume without loss of generality that $\Sigma a_i = 1$, or equivalently, that 0 is in the convex hull $\mathcal{C}$ of the row space of M. Hence,

$$S^\perp \cap \mathbb{R}^n_+ \neq \emptyset \quad \Leftrightarrow \quad 0 \in \mathcal{C}$$

Thus, we wish to prove that

$$S \cap \mathbb{R}^n_{++} = \emptyset \quad \Rightarrow \quad 0 \in \mathcal{C}$$

or equivalently,

$$0 \notin \mathcal{C} \quad \Rightarrow \quad S \cap \mathbb{R}^n_{++} \neq \emptyset$$

Now we have something to separate. Since $\mathcal{C}$ is closed and convex, Theorem 15.5 implies that there is a nonzero vector $B = (b_1, \ldots, b_k) \in \mathbb{R}^k$ for which

$$\langle B, \mathcal{C} \rangle \geq \|B\|^2 > 0$$

Consider the vector

$$v = b_1 B_1 + \cdots + b_k B_k \in S$$

The ith coordinate of v is

$$b_1 m_{i,1} + \cdots + b_k m_{i,k} = \langle B, R_i \rangle \geq \|B\|^2 > 0$$

and so v is strongly positive. Hence, $v \in S \cap \mathbb{R}^n_{++}$, which is therefore nonempty.□

Inhomogeneous Systems

We now turn our attention to inhomogeneous systems

$$Ax = b$$

The following lemma is required.

Lemma 15.8 *Let $A \in \mathcal{M}_{m,n}(\mathbb{R})$. Then the set*

$$\mathcal{C} = \{Ay \mid y \in \mathbb{R}^n, y \geq 0\}$$

is a closed, convex cone.

Proof. We leave it as an exercise to prove that $\mathcal{C}$ is a convex cone and omit the proof that $\mathcal{C}$ is closed.□

Theorem 15.9 (Farkas's lemma) *Let $A \in \mathcal{M}_{m,n}(\mathbb{R})$ and let $b \in \mathbb{R}^m$ be nonzero. Then exactly one of the following holds:*

1) There is a strictly positive solution $u \in \mathbb{R}^n$ to the system $Ax = b$.

2) There is a vector $v \in \mathbb{R}^m$ for which $vA \leq 0$ and $\langle v, b \rangle > 0$.

Proof. Suppose first that 1) holds. If 2) also holds, then

$$(vA)u = v(Au) = \langle v, b \rangle > 0$$

However, $vA \leq 0$ and $u > 0$ imply that $(vA)u \leq 0$. This contradiction implies that 2) cannot hold.

Assume now that 1) fails to hold. By Lemma 15.8, the set

$$C = \{Ay \mid y \in \mathbb{R}^n, y \geq 0\} \subseteq \mathbb{R}^m$$

is closed and convex. The fact that 1) fails to hold is equivalent to $b \notin C$. Hence, there is a hyperplane that strongly separates b and C. All we require is that b and C be strictly separated, that is, for some $\alpha \in \mathbb{R}$ and $v \in \mathbb{R}^m$,

$$\langle v, x \rangle < \alpha < \langle v, b \rangle \text{ for all } x \in C$$

Since $0 \in C$, it follows that $\alpha > 0$ and so $\langle v, b \rangle > 0$. Also, the first inequality is equivalent to $\langle v, Ay \rangle < \alpha$, that is,

$$\langle A^t v, y \rangle < \alpha$$

for all $y \in \mathbb{R}^n, y \geq 0$. We claim that this implies that $A^t v$ cannot have any positive coordinates and thus $vA \leq 0$. For if the ith coordinate $(A^t v)_i$ is positive, then taking $y = \lambda e_i$ for $\lambda > 0$ we get

$$\lambda (A^t v)_i < \alpha$$

which does not hold for large λ. Thus, 2) holds.□

In the exercises, we ask the reader to show that the previous result cannot be improved by replacing $vA \leq 0$ in statement 2) with $vA \ll 0$.

Exercises

1. Show that any hyperplane has the form $\mathcal{H}(N, \|N\|^2)$ for an appropriate vector N.

2. If A is an $m \times n$ matrix prove that the set $\{Ax \mid x \in \mathbb{R}^n, x > 0\}$ is a convex cone in $\mathbb{R}^m$.
3. If A and B are strictly separated subsets of $\mathbb{R}^n$ and if A is finite, prove that A and B are strongly separated as well.
4. Let V be a vector space over a field F with $\operatorname{char}(F) \neq 2$. Show that a subset X of V is closed under the taking of convex combinations of any two of its points if and only if X is closed under the taking of arbitrary convex combinations, that is, for all $n \geq 1$,

$$x_1, \ldots, x_n \in X, \sum_{i=1}^{n} r_i = 1, 0 \leq r_i \leq 1 \Rightarrow \sum_{i=1}^{n} r_i x_i \in X$$

5. Explain why an $(n-1)$-dimensional subspace of $\mathbb{R}^n$ is the solution set of a linear equation of the form $a_1 x_1 + \cdots + a_n x_n = 0$.
6. Show that

$$\mathcal{H}_+(N, b) \cap \mathcal{H}_-(N, b) = \mathcal{H}(N, b)$$

and that $\mathcal{H}_+^\circ(N, b)$, $\mathcal{H}_-^\circ(N, b)$ and $\mathcal{H}(N, b)$ are pairwise disjoint and

$$\mathcal{H}_+^\circ(N, b) \cup \mathcal{H}_-^\circ(N, b) \cup \mathcal{H}(N, b) = \mathbb{R}^n$$

7. A function $T: \mathbb{R}^n \to \mathbb{R}^m$ is **affine** if it has the form $T(v) = \tau v + b$ for $b \in \mathbb{R}^m$, where $\tau \in \mathcal{L}(\mathbb{R}^n, \mathbb{R}^m)$. Prove that if $C \subseteq \mathbb{R}^n$ is convex, then so is $T(C) \subseteq \mathbb{R}^m$.
8. Find a cone in $\mathbb{R}^2$ that is not convex. Prove that a subset X of $\mathbb{R}^n$ is a convex cone if and only if $x, y \in X$ implies that $\lambda x + \mu y \in X$ for all $\lambda, \mu \geq 0$.
9. Prove that the convex hull of a set $\{x_1, \ldots, x_n\}$ in $\mathbb{R}^n$ is bounded, without using the fact that it is compact.
10. Suppose that a vector $x \in \mathbb{R}^n$ has two distinct representations as convex combinations of the vectors $v_1, \ldots, v_n$. Prove that the vectors $v_2 - v_1, \ldots, v_n - v_1$ are linearly dependent.
11. Suppose that C is a nonempty convex subset of $\mathbb{R}^n$ and that $\mathcal{H}(N, b)$ is a hyperplane disjoint from C. Prove that C lies in one of the open half-spaces determined by $\mathcal{H}(N, b)$.
12. Prove that the conclusion of Theorem 15.6 may fail if we assume only that C is closed and convex.
13. Find two nonempty convex subsets of $\mathbb{R}^2$ that are strictly separated but not strongly separated.
14. Prove that X and Y are strongly separated by $\mathcal{H}(N, b)$ if and only if

$$\langle N, x' \rangle > b \text{ for all } x' \in X_\epsilon \text{ and } \langle N, y' \rangle < b \text{ for all } y' \in Y_\epsilon$$

where $X_\epsilon = X + \epsilon B(0, 1)$ and $Y_\epsilon = Y + \epsilon B(0, 1)$ and where $B(0, 1)$ is the closed unit ball.

15. Show that Farkas's lemma cannot be improved by replacing $vA \leq 0$ in statement 2) with $vA \ll 0$. *Hint*: A nice counterexample exists for $m = 2, n = 3$.

Chapter 16
Affine Geometry

In this chapter, we will study the geometry of a finite-dimensional vector space V, along with its structure-preserving maps. *Throughout this chapter, all vector spaces are assumed to be finite-dimensional.*

Affine Geometry

The cosets of a quotient space have a special geometric name.

Definition *Let S be a subspace of a vector space V. The coset*

$$v+S=\{v+s \mid s\in S\}$$

is called a **flat** *in V with* **base** *S and* **flat representative** *v. We also refer to $v+S$ as a* **translate** *of S. The set $\mathcal{A}(V)$ of all flats in V is called the* **affine geometry** *of V. The* **dimension** $\dim(\mathcal{A}(V))$ *of $\mathcal{A}(V)$ is defined to be* $\dim(V)$.□

While a flat may have many flat representatives, it only has one base since $x+S=y+T$ implies that $x\in y+T$ and so $x+S=y+T=x+T$, whence $S=T$.

Definition *The* **dimension** *of a flat $x+S$ is* $\dim(S)$. *A flat of dimension k is called a* **k-flat**. *A 0-flat is a* **point**, *a 1-flat is a* **line** *and a 2-flat is a* **plane**. *A flat of dimension* $\dim(\mathcal{A}(V))-1$ *is called a* **hyperplane**.□

Definition *Two flats $X=x+S$ and $Y=y+T$ are said to be* **parallel** *if $S\subseteq T$ or $T\subseteq S$. This is denoted by $X \parallel Y$.*□

We will denote subspaces of V by the letters $S,T,\ldots$ and flats in V by $X,Y,\ldots$.

Here are some of the basic intersection properties of flats.

Theorem 16.1 Let S and T be subspaces of V and let $X = x + S$ and $Y = y + T$ be flats in V.

1) *The following are equivalent:*
   a) *some translate of X is in Y: $w + X \subseteq Y$ for some $w \in V$*
   b) *some translate of S is in T: $v + S \subseteq T$ for some $v \in V$*
   c) $S \subseteq T$
2) *The following are equivalent:*
   a) *X and Y are translates: $w + X = Y$ for some $w \in V$*
   b) *S and T are translates: $v + S = T$ for some $v \in V$*
   c) $S = T$
3) $X \cap Y \neq \emptyset,\ S \subseteq T \Leftrightarrow X \subseteq Y$
4) $X \cap Y \neq \emptyset,\ S = T \Leftrightarrow X = Y$
5) *If $X \parallel Y$ then $X \subseteq Y$, $Y \subseteq X$ or $X \cap Y = \emptyset$*
6) *$X \parallel Y$ if and only if some translation of one of these flats is contained in the other.*

Proof. If 1a) holds, then $-y + w + x + S \subseteq T$ and so 1b) holds. If 1b) holds, then $v \in T$ and so $S = (v + S) - v \subseteq T$ and so 1c) holds. If 1c) holds, then $y - x + X = y + S \subseteq y + T = Y$ and so 1a) holds. Part 2) is proved in a similar manner.

For part 3), $S \subseteq T$ implies that $v + X \subseteq Y$ for some $v \in V$ and so if $z \in X \cap Y$ then $v + z \in Y$ and so $v \in Y$, which implies that $X \subseteq Y$. Conversely, if $X \subseteq Y$ then part 1) implies that $S \subseteq T$. Part 4) follows similarly. We leave proof of 5) and 6) to the reader.$\square$

Affine Combinations

Let X be a nonempty subset of V. It is well known that

1) X is a subspace of V if and only if X is closed under linear combinations, or equivalently, X is closed under linear combinations of any two vectors in X.
2) The smallest subspace of V containing X is the set of all linear combinations of elements of X. In different language, the *linear hull* of X is equal to the *linear span* of X.

We wish to establish the corresponding properties of affine subspaces of V, beginning with the counterpart of a linear combination.

Definition *Let V be a vector space and let $x_i \in V$. A linear combination*

$$r_1x_1 + \cdots + r_nx_n$$

where $r_i \in F$ and $\sum r_i = 1$ is called an **affine combination** *of the vectors x_i.*$\square$

Let us refer to a nonempty subset X of V as **affine closed** if X is closed under any affine combination of vectors in X and **two-affine closed** if X is closed under affine combinations of any two vectors in X. These are not standard terms.

The **line** containing two distinct vectors $x, y \in V$ is the set

$$\overline{xy} = \{rx + (1-r)y \mid r \in F\} = y + \langle x - y \rangle$$

of all affine combinations of x and y. Thus, a subset X of V is two-affine closed if and only if X contains the line through any two of its points.

Theorem 16.2 *Let V be a vector space over a field F with* $\operatorname{char}(F) \neq 2$. *Then a subset X of V is affine closed if and only if it is two-affine closed.*
Proof. The theorem is proved by induction on the number n of terms in an affine combination. The case $n = 2$ holds by assumption. Assume the result true for affine combinations with fewer than $n \geq 3$ terms and consider the affine combination

$$z = r_1 x_1 + \cdots + r_n x_n$$

where $n \geq 3$. There are two cases to consider. If either of r_1 and r_2 is not equal to 1, say $r_1 \neq 1$, write

$$z = r_1 x_1 + (1 - r_1)\left[\frac{r_2}{1 - r_1} x_2 + \cdots + \frac{r_n}{1 - r_1} x_n\right]$$

and if $r_1 = r_2 = 1$, then since $\operatorname{char}(F) \neq 2$, we may write

$$z = 2\left[\frac{1}{2} x_1 + \frac{1}{2} x_2\right] + r_3 x_3 + \cdots + r_n x_n$$

In either case, the inductive hypothesis applies to the expression inside the square brackets and then to z.□

The requirement $\operatorname{char}(F) \neq 2$ is necessary, for if $F = \mathbb{Z}_2$, then the subset

$$X = \{(0,0), (1,0), (0,1)\}$$

of F^2 is two-affine closed but not affine closed. We can now characterize flats.

Theorem 16.3 *A nonempty subset X of a vector space V is a flat if and only if X is affine closed. Moreover, if* $\operatorname{char}(F) \neq 2$, *then X is a flat if and only if X is two-affine closed.*
Proof. Let $X = x + S$ be a flat and let $x_i = x + s_i \in X$, where $s_i \in S$. If $\Sigma r_i = 1$, then

$$\sum_i r_i x_i = \sum_i r_i (x + s_i) = x + \sum_i r_i s_i \in x + S$$

and so X is affine closed. Conversely, suppose that X is affine closed, let $x \in X$ and let $S = X - x$. If $r_i \in F$ and $s_i \in S$ then

$$r_1 s_1 + r_2 s_2 = r_1(x_1 - x) + r_2(x_2 - x) = r_1 x_1 + r_2 x_2 - (r_2 + r_1)x$$

for $x_i \in X$. Since the sum of the coefficients of x_1, x_2 and x in the last expression is 0, it follows that

$$r_1 s_1 + r_2 s_2 + x = r_1 x_1 + r_2 x_2 - (r_2 + r_1 - 1)x \in X$$

and so $r_1 s_1 + r_2 s_2 \in X - x = S$. Thus, S is a subspace of V and $X = x + S$ is a flat. The rest follows from Theorem 16.2.□

Affine Hulls

The following definition is the analog of the subspace spanned by a collection of vectors.

Definition *Let X be a nonempty set of vectors in V.*

1) *The* **affine hull** *of X, denoted by* affhull(X), *is the smallest flat containing X.*
2) *The* **affine span** *of X, denoted by* affspan(X), *is the set of all affine combinations of vectors in X.*□

Theorem 16.4 *Let X be a nonempty subset of V. Then*

$$\text{affhull}(X) = \text{affspan}(X) = x + \text{span}(X - x)$$

or equivalently, for a subspace S of V,

$$x + S = \text{affspan}(X) \quad \Leftrightarrow \quad S = \text{span}(X - x)$$

Also,

$$\dim(\text{affspan}(X)) = \dim(\text{span}(X - x))$$

Proof. Theorem 16.3 implies that $\text{affspan}(X) \subseteq \text{affhull}(X)$ and so it is sufficient to show that $A = \text{affspan}(X)$ is a flat, or equivalently, that for any $y \in A$, the set $S = A - y$ is a subspace of V. To this end, let

$$y = \sum_{i=1}^{n} r_{0,i} x_i$$

Then any two elements of S have the form $y_1 - y$ and $y_2 - y$, where

$$y_1 = \sum_{i=1}^{n} r_{1,i} x_i \quad \text{and} \quad y_2 = \sum_{i=1}^{n} r_{2,i} x_i$$

are in A. But if $s, t \in F$, then

$$\begin{aligned} z &:= s(y_1 - y) + t(y_2 - y) \\ &= \sum_{i=1}^{n} \Big(sr_{1,i} + tr_{2,i} - (s+t)r_{0,i} \Big) x_i \\ &= \sum_{i=1}^{n} \Big(sr_{1,i} + tr_{2,i} - (s+t-1)r_{0,i} \Big) x_i - y \end{aligned}$$

which is in $A - y = S$, since the last sum is an affine sum. Hence, S is a subspace of V. We leave the rest of the proof to the reader.□

The Lattice of Flats

The intersection of subspaces is a subspace, although it may be trivial. For flats, if the intersection is not empty, then it is also a flat. However, since the intersection of flats may be empty, the set $\mathcal{A}(V)$ does not form a lattice under intersection. However, we can easily fix this.

Theorem 16.5 *Let V be a vector space. The set*

$$\mathcal{A}_0(V) = \mathcal{A}(V) \cup \{\emptyset\}$$

of all flats in V, together with the empty set, is a complete lattice in which meet is intersection. In particular:

1) $\mathcal{A}_0(V)$ is closed under arbitrary intersection. In fact, if $\mathcal{F} = \{x_i + S_i \mid i \in K\}$ has nonempty intersection, then

$$\bigcap_{i \in K} \mathcal{F} = \bigcap_{i \in K} (x_i + S_i) = x + \bigcap_{i \in K} S_i$$

for some $x \in \bigcap \mathcal{F}$. In other words, the base of the intersection is the intersection of the bases.

2) The join $\bigvee \mathcal{F}$ of the family $\mathcal{F} = \{x_i + S_i \mid i \in K\}$ is the intersection of all flats containing the members of $\mathcal{F}$. Also,

$$\bigvee \mathcal{F} = \operatorname{affhull}\Big(\bigcup \mathcal{F} \Big)$$

3) If $X = x + S$ and $Y = y + T$ are flats in V, then

$$X \vee Y = x + (\langle x - y \rangle + S + T)$$

If $X \cap Y \neq \emptyset$, then

$$X \vee Y = x + (S + T)$$

Proof. For part 1), if

$$x \in \bigcap_{i \in K} (x_i + S_i)$$

then $x_i + S_i = x + S_i$ for all $i \in K$ and so

$$\bigcap_{i\in K}(x_i+S_i)=\bigcap_{i\in K}(x+S_i)=x+\bigcap_{i\in K}S_i$$

We leave proof of part 2) to the reader.

For part 3), since $x,y\in X\vee Y$, it follows that

$$X\vee Y=x+U=y+U$$

for some subspace U of V. Thus, $x-y\in U$. Also, $x+S\subseteq x+U$ implies that $S\subseteq U$ and similarly $T\subseteq U$, whence $S+T\subseteq U$ and so if $W=\langle x-y\rangle+S+T$, then $W\subseteq U$. Hence, $x+W\subseteq x+U=X\vee Y$. On the other hand,

$$X=x+S\subseteq x+W$$

and

$$Y=y+T=x-(x-y)+T\subseteq x+W$$

and so $X\vee Y\subseteq x+W$. Thus, $X\vee Y=x+W$.

If $X\cap Y\neq\emptyset$, then we may take the flat representatives for X and Y to be any element $z\in X\cap Y$, in which case part 1) gives

$$X\vee Y=z+(\langle z-z\rangle+S+T)=z+S+T$$

and since $x\in X\vee Y$, we also have $X\vee Y=x+S+T$.□

We can now describe the dimension of the join of two flats.

Theorem 16.6 *Let $X=x+S$ and $Y=y+T$ be flats in V.*
1) If $X\cap Y\neq\emptyset$, then

$$\dim(X\vee Y)=\dim(S+T)=\dim(X)+\dim(Y)-\dim(X\cap Y)$$

2) If $X\cap Y=\emptyset$, then

$$\dim(X\vee Y)=\dim(S+T)+1$$

Proof. We have seen that if $X\cap Y\neq\emptyset$, then

$$X\vee Y=x+S+T$$

and so

$$\dim(X\vee Y)=\dim(S+T)$$

On the other hand, if $X\cap Y=\emptyset$, then

$$X\vee Y=x+(\langle x-y\rangle+S+T)$$

and since $\dim(\langle x-y\rangle)=1$, we get

$$\dim(X \vee Y) = \dim(S+T) + 1$$

Finally, we have

$$\dim(S+T) = \dim(S) + \dim(T) - \dim(S \cap T)$$

and Theorem 16.5 implies that

$$\dim(X \cap Y) = \dim(S \cap T) \qquad \square$$

Affine Independence

We now discuss the affine counterpart of linear independence.

Theorem 16.7 *Let X be a nonempty set of vectors in V. The following are equivalent:*

1) *For all $x \in X$, the set*

$$(X - x) \setminus \{0\}$$

is linearly independent.

2) *For all $x \in X$,*

$$x \notin \text{affhull}(X \setminus \{x\})$$

3) *For any vectors $x_i \in X$,*

$$\sum_i r_i x_i = 0, \sum_i r_i = 0 \quad \Rightarrow \quad r_i = 0 \textit{ for all } i$$

4) *For affine combinations of vectors in X,*

$$\sum_i r_i x_i = \sum_i s_i x_i \quad \Rightarrow \quad r_i = s_i \textit{ for all } i$$

5) *When $X = \{x_1, \ldots, x_n\}$ is finite,*

$$\dim(\text{affhull}(X)) = n - 1$$

A set X of vectors satisfying any (any hence all) of these conditions is said to be **affinely independent**.

Proof. If 1) holds but there is an affine combination equal to x,

$$x = \sum_{i=1}^{n} r_i x_i$$

where $x_i \neq x$ for all i, then

$$\sum_{i=1}^{n} r_i(x_i - x) = 0$$

Since r_i is nonzero for some i, this contradicts 1). Hence, 1) implies 2). Suppose that 2) holds and

$$\sum r_i x_i = 0$$

where $\sum r_i = 0$. If some r_i, say r_1, is nonzero then

$$x_1 = -\sum_{i>1} (r_i/r_1)x_i \in \text{affhull}(X \setminus \{x_1\})$$

which contradicts 2) and so $r_i = 0$ for all i. Hence, 2) implies 3).

If 3) holds and the affine combinations satisfy

$$\sum r_i x_i = \sum s_i x_i$$

then

$$\sum (r_i - s_i)x_i = 0$$

and since $\sum(r_i - s_i) = 1 - 1 = 0$, it follows that $r_i = s_i$ for all i. Hence, 4) holds. Thus, it is clear that 3) and 4) are equivalent. If 3) holds and

$$\sum r_i(x_i - x) = 0$$

for $x \neq x_i$, then

$$\sum r_i x_i - \left(\sum r_i\right)x = 0$$

and so 3) implies that $r_i = 0$ for all i.

Finally, suppose that $X = \{x_1, \ldots, x_n\}$. Since

$$\dim(\text{affhull}(X)) = \dim(\langle X - x_i \rangle)$$

it follows that 5) holds if and only if $(X - x_i) \setminus \{0\}$, which has size $n - 1$, is linearly independent.□

Affinely independent sets enjoy some of the basic properties of linearly independent sets. For example, a nonempty subset of an affinely independent set is affinely independent. Also, any nonempty set X contains an affinely independent set.

Since the affine hull $H = \text{affhull}(X)$ of an affinely independent set X is not the affine hull of any proper subset of X, we deduce that X is a minimal affine spanning set of its affine hull.

Affine Bases and Barycentric Coordinates

We have seen that a set X is affinely independent if and only if the set

$$X_x = (X - x) \setminus \{0\}$$

is linearly independent. We have also seen that for a subsapce S of V,

$$x + S = \text{affspan}(X) \quad \Leftrightarrow \quad S = \text{span}(X_x)$$

Therefore, if by analogy, we define a subset $\mathcal{B}$ of a flat $A = x + S$ to be an **affine basis** for A if $\mathcal{B}$ is affinely independent and affspan$(\mathcal{B}) = A$, then $\mathcal{B}$ is an affine basis for $x + S$ if and only if $\mathcal{B}_x$ is a basis for S.

Theorem 16.8 *Let $A = x + S$ be a flat of dimension n. Let $\mathcal{B} = (x_1, \ldots, x_n)$ be an ordered basis for S and let $(\mathcal{B} + x) \cup \{x\} = (x_1 + x, \ldots, x_n + x, x)$ be an ordered affine basis for A. Then every $v \in A$ has a unique expression as an affine combination*

$$v = r_1 x_1 + \cdots + r_n x_n + r_{n+1} x$$

The coefficients r_i are called the **barycentric coordinates** *of v with respect to the ordered affine basis $\mathcal{B} + x$.*□

For example, in $\mathbb{R}^3$, a plane is a flat of the form $A = x + \langle v_1, v_2 \rangle$ where $\mathcal{B} = (v_1, v_2)$ is an ordered basis for a two-dimensional subspace of $\mathbb{R}^3$. Then

$$(\mathcal{B} + x) \cup \{x\} = (v_1 + x, v_2 + x, x) = (p_1, p_2, p_3)$$

are barycentric coordinates for the plane, that is, any $v \in A$ has the form

$$r_1 p_1 + r_2 p_2 + r_3 p_3$$

where $r_1 + r_2 + r_3 = 1$.

Affine Transformations

Now let us discuss some properties of maps that preserve affine structure.

Definition *A function $f: V \to V$ that preserves affine combinations, that is, for which*

$$\sum_i r_i = 1 \Rightarrow f\left(\sum_i r_i x_i\right) = \sum_i r_i f(x_i)$$

is called an **affine transformation** (*or* **affine map**, *or* **affinity**).□

We should mention that some authors require that f be bijective in order to be an affine map. The following theorem is the analog of Theorem 16.2.

Theorem 16.9 *If* $\mathrm{char}(F) \neq 2$, *then a function* $f: V \to V$ *is an affine transformation if and only if it preserves affine combinations of every pair of its points, that is, if and only if*

$$f(rx + (1-r)y) = rf(x) + (1-r)f(y) \qquad \square$$

Thus, if $\mathrm{char}(F) \neq 2$, then a map f is an affine transformation if and only if it sends the line through x and y to the line through $f(x)$ and $f(y)$. It is clear that linear transformations are affine transformations. So are the following maps.

Definition *Let* $v \in V$. *The affine map* $T_v: V \to V$ *defined by*

$$T_v(x) = x + v$$

for all $x \in V$, *is called* **translation** *by* v.$\square$

It is not hard to see that any composition $T_v \circ \tau$, where $\tau \in \mathcal{L}(V)$, is affine. Conversely, any affine map must have this form.

Theorem 16.10 *A function* $f: V \to V$ *is an affine transformation if and only if it is a linear operator followed by a translation,*

$$f = T_v \circ \tau$$

where $v \in V$ *and* $\tau \in \mathcal{L}(V)$.

Proof. We leave proof that $T_v \circ \tau$ is an affine transformation to the reader. Let f be an affine map and suppose that $f0 = -z$. Then $T_z \circ f0 = 0$. Moreover, letting $\tau = T_z \circ f$, we have

$$\begin{aligned}
\tau(ru + sv) &= f(ru + sv) + z \\
&= f(ru + sv + (1 - r - s)0) + z \\
&= rfu + sfv - (1 - r - s)z + z \\
&= r\tau u + s\tau v
\end{aligned}$$

and so τ is linear.$\square$

Corollary 16.11

1) *The composition of two affine transformations is an affine transformation.*
2) *An affine transformation* $f = T_v \circ \tau$ *is bijective if and only if* τ *is bijective.*
3) *The set* $\mathrm{aff}(V)$ *of all bijective affine transformations on* V *is a group under composition of maps, called the* **affine group** *of* V.$\square$

Let us make a few group-theoretic remarks about $\mathrm{aff}(V)$. The set $\mathrm{trans}(V)$ of all translations of V is a subgroup of $\mathrm{aff}(V)$. We can define a function $\phi: \mathrm{aff}(V) \to \mathcal{L}(V)$ by

$$\phi(T_v \circ \tau) = \tau$$

It is not hard to see that ϕ is a well-defined group homomorphism from aff(V) onto $\mathcal{L}(V)$, with kernel trans(V). Hence, trans(V) is a normal subgroup of aff(V) and

$$\frac{\text{aff}(V)}{\text{trans}(V)} \approx \mathcal{L}(V)$$

Projective Geometry

If $\dim(V) = 2$, the join (affine hull) of any two distinct points in V is a line. On the other hand, it is not the case that the intersection of any two lines is a point, since the lines may be parallel. Thus, there is a certain asymmetry between the concepts of points and lines in V. This asymmetry can be removed by constructing the *projective plane*. Our plan here is to very briefly describe one possible construction of projective geometries of all dimensions.

By way of motivation, let us consider Figure 16.1.

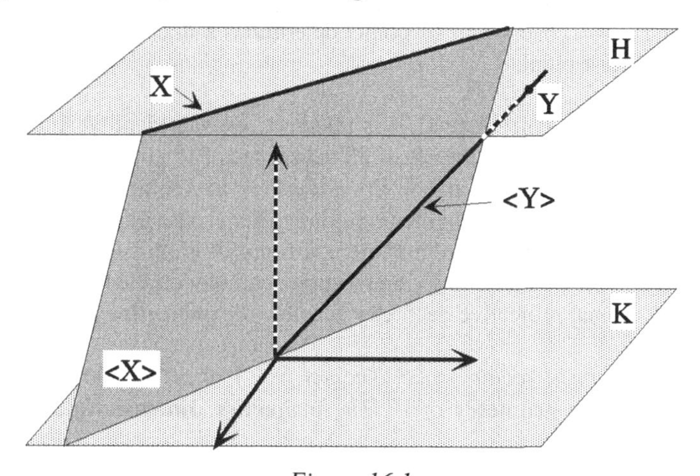

Figure 16.1

Note that H is a hyperplane in a 3-dimensional vector space V and that $0 \notin H$. Now, the set $\mathcal{A}(H)$ of all flats of V that lie in H is an affine geometry of dimension 2. (According to our definition of affine geometry, H must be a vector space in order to define $\mathcal{A}(H)$. However, we hereby extend the definition of affine geometry to include the collection of all flats contained in a flat of V.)

Figure 16.1 shows a one-dimensional flat X and its linear span $\langle X \rangle$, as well as a zero-dimensional flat Y and its span $\langle Y \rangle$. Note that, for any flat X in H, we have

$$\dim(\langle X \rangle) = \dim(X) + 1$$

Note also that if L_1 and L_2 are any two distinct lines in H, the corresponding planes $\langle L_1 \rangle$ and $\langle L_2 \rangle$ have the property that their intersection is a line through the origin, *even if the lines are parallel.* We are now ready to define projective geometries.

Let V be a vector space of any dimension and let H be a hyperplane in V not containing the origin. To each flat X in H, we associate the subspace $\langle X \rangle$ of V generated by X. Thus, the linear span function $P: \mathcal{A}(H) \to \mathcal{S}(V)$ maps affine subspaces X of H to subspaces $\langle X \rangle$ of V. The span function is not surjective: Its image is the set of all subspaces that are *not* contained in the base subspace K of the flat H.

The linear span function is one-to-one and its inverse is intersection with H,

$$P^{-1}U = U \cap H$$

for any subspace U not contained in K.

The affine geometry $\mathcal{A}(H)$ is, as we have remarked, somewhat incomplete. In the case $\dim(H) = 2$, every pair of points determines a line but not every pair of lines determines a point.

Now, since the linear span function P is injective, we can identify $\mathcal{A}(H)$ with its image $P(\mathcal{A}(H))$, which is the set of all subspaces of V not contained in the base subspace K. This view of $\mathcal{A}(H)$ allows us to "complete" $\mathcal{A}(H)$ by including the base subspace K. In the three-dimensional case of Figure 16.1, the base plane, in effect, adds a projective line at infinity. With this inclusion, every pair of lines intersects, parallel lines intersecting at a point on the line at infinity. This two-dimensional projective geometry is called the **projective plane**.

Definition *Let V be a vector space. The set $\mathcal{S}(V)$ of all subspaces of V is called the* **projective geometry** *of V. The* **projective dimension** $\operatorname{pdim}(S)$ *of $S \in \mathcal{S}(V)$ is defined as*

$$\operatorname{pdim}(S) = \dim(S) - 1$$

The **projective dimension** *of $\mathcal{P}(V)$ is defined to be* $\operatorname{pdim}(V) = \dim(V) - 1$. *A subspace of projective dimension* 0 *is called a* **projective point** *and a subspace of projective dimension* 1 *is called a* **projective line.** □

Thus, referring to Figure 16.1, a projective point is a line through the origin and, provided that it is not contained in the base plane K, it meets H in an affine point. Similarly, a projective line is a plane through the origin and, provided that it is not K, it will meet H in an affine line. In short,

$$\begin{aligned} \operatorname{span}(\text{affine point}) &= \text{line through the origin} = \text{projective point} \\ \operatorname{span}(\text{affine line}) &= \text{plane through the origin} = \text{projective line} \end{aligned}$$

The linear span function has the following properties.

Theorem 16.12 *The linear span function $P: \mathcal{A}(H) \to \mathcal{S}(V)$ from the affine geometry $\mathcal{A}(H)$ to the projective geometry $\mathcal{S}(V)$ defined by $PX = \langle X \rangle$ satisfies the following properties:*

1) *The linear span function is injective, with inverse given by*

$$P^{-1}U = U \cap H$$

for all subspaces U not contained in the base subspace K of H.

2) *The image of the span function is the set of all subspaces of V that are not contained in the base subspace K of H.*
3) *$X \subseteq Y$ if and only if $\langle X \rangle \subseteq \langle Y \rangle$*
4) *If X_i are flats in H with nonempty intersection, then*

$$\operatorname{span}\left(\bigcap_{i \in K} X_i\right) = \bigcap_{i \in K} \operatorname{span}(X_i)$$

5) *For any collection of flats in H,*

$$\operatorname{span}\left(\bigvee_{i \in K} X_i\right) = \bigoplus_{i \in K} \operatorname{span}(X_i)$$

6) *The linear span function preserves dimension, in the sense that*

$$\operatorname{pdim}(\operatorname{span}(X)) = \dim(X)$$

7) *$X \parallel Y$ if and only if one of $\langle X \rangle \cap K$ and $\langle Y \rangle \cap K$ is contained in the other.*

Proof. To prove part 1), let $x + S$ be a flat in H. Then $x \in H$ and so $H = x + K$, which implies that $S \subseteq K$. Note also that $\langle x + S \rangle = \langle x \rangle + S$ and

$$z \in \langle x + S \rangle \cap H = (\langle x \rangle + S) \cap (x + K) \Rightarrow z = rx + s = x + k$$

for some $s \in S$, $k \in K$ and $r \in F$. This implies that $(1 - r)x \in K$, which implies that either $x \in K$ or $r = 1$. But $x \in H$ implies $x \notin K$ and so $r = 1$, which implies that $z = x + s \in x + S$. In other words,

$$\langle x + S \rangle \cap H \subseteq x + S$$

Since the reverse inclusion is clear, we have

$$\langle x + S \rangle \cap H = x + S$$

This establishes 1).

To prove 2), let U be a subspace of V that is not contained in K. We wish to show that U is in the image of the linear span function. Note first that since $U \not\subseteq K$ and $\dim(K) = \dim(V) - 1$, we have $U + K = V$ and so

$$\dim(U \cap K) = \dim(U) + \dim(K) - \dim(U + K) = \dim(U) - 1$$

Now let $0 \neq x \in U - K$. Then

$$\begin{aligned} x \notin K &\Rightarrow \langle x \rangle + K = V \\ &\Rightarrow rx + k \in H \text{ for some } 0 \neq r \in F,\ k \in K \\ &\Rightarrow rx \in H \end{aligned}$$

Thus, $rx \in U \cap H$ for some $0 \neq r \in F$. Hence, the flat $rx + (U \cap K)$ lies in H and

$$\dim(rx + (U \cap K)) = \dim(U \cap K) = \dim(U) - 1$$

which implies that $\operatorname{span}(rx + (U \cap K)) = \langle rx \rangle + (U \cap K)$ lies in U and has the same dimension as U. In other words,

$$\operatorname{span}(rx + (U \cap K)) = \langle rx \rangle + (U \cap K) = U$$

We leave proof of the remaining parts of the theorem as exercises.□

Exercises

1. Show that if $x_1, \ldots, x_n \in V$, then the set $S = \{\Sigma r_i x_i \mid \Sigma r_i = 0\}$ is a subspace of V.
2. Prove that if $X \subseteq V$ is nonempty then
$$\operatorname{affhull}(X) = x + \langle X - x \rangle$$
3. Prove that the set $X = \{(0,0),\ (1,0),\ (0,1)\}$ in $(\mathbb{Z}_2)^2$ is closed under the formation of lines, but not affine hulls.
4. Prove that a flat contains the origin 0 if and only if it is a subspace.
5. Prove that a flat X is a subspace if and only if for some $x \in X$ we have $rx \in X$ for some $1 \neq r \in F$.
6. Show that the join of a collection $\mathcal{C} = \{x_i + S_i \mid i \in K\}$ of flats in V is the intersection of all flats that contain all flats in $\mathcal{C}$.
7. Is the collection of all flats in V a lattice under set inclusion? If not, how can you "fix" this?
8. Suppose that $X = x + S$ and $Y = y + T$. Prove that if $\dim(X) = \dim(Y)$ and $X \parallel Y$, then $S = T$.
9. Suppose that $X = x + S$ and $Y = y + T$ are disjoint hyperplanes in V. Show that $S = T$.
10. (The parallel postulate) Let X be a flat in V and $v \notin X$. Show that there is exactly one flat containing v, parallel to X and having the same dimension as X.
11. a) Find an example to show that the join $X \vee Y$ of two flats may not be the set of all lines connecting all points in the union of these flats.
    b) Show that if X and Y are flats with $X \cap Y \neq \emptyset$, then $X \vee Y$ is the union of all lines $\overline{xy}$ where $x \in X$ and $y \in Y$.
12. Show that if $X \parallel Y$ and $X \cap Y = \emptyset$, then
$$\dim(X \vee Y) = \max\{\dim(X), \dim(Y)\} + 1$$

13. Let $\dim(V) = 2$. Prove the following:
    a) The join of any two distinct points is a line.
    b) The intersection of any two nonparallel lines is a point.
14. Let $\dim(V) = 3$. Prove the following:
    a) The join of any two distinct points is a line.
    b) The intersection of any two nonparallel planes is a line.
    c) The join of any two lines whose intersection is a point is a plane.
    d) The intersection of two coplanar nonparallel lines is a point.
    e) The join of any two distinct parallel lines is a plane.
    f) The join of a line and a point not on that line is a plane.
    g) The intersection of a plane and a line not on that plane is a point.
15. Prove that $f: V \to V$ is a surjective affine transformation if and only if $f = \tau \circ T_w$ for some $w \in V$ and $\tau \in \mathcal{L}(V)$.
16. Verify the group-theoretic remarks about the group homomorphism $\phi: \mathrm{aff}(V) \to \mathcal{L}(V)$ and the subgroup $\mathrm{trans}(V)$ of $\mathrm{aff}(V)$.

Chapter 17
Singular Values and the Moore–Penrose Inverse

Singular Values

Let U and V be finite-dimensional inner product spaces over $\mathbb{C}$ or $\mathbb{R}$ and let $\tau \in \mathcal{L}(U,V)$. The spectral theorem applied to $\tau^*\tau$ can be of considerable help in understanding the relationship between τ and its adjoint τ^*. This relationship is shown in Figure 17.1. Note that U and V can be decomposed into direct sums

$$U = A \oplus B \quad \text{and} \quad V = C \oplus D$$

in such a manner that $\tau\colon A \to C$ and $\tau^*\colon C \to A$ act symmetrically in the sense that

$$\tau\colon u_i \mapsto s_i v_i \quad \text{and} \quad \tau^*\colon v_i \mapsto s_i u_i$$

Also, both τ and τ^* are zero on B and D, respectively.

We begin by noting that $\tau^*\tau \in \mathcal{L}(U)$ is a positive Hermitian operator. Hence, if $r = \mathrm{rk}(\tau) = \mathrm{rk}(\tau^*\tau)$, then U has an ordered orthonormal basis

$$\mathcal{B} = (u_1, \ldots, u_r, u_{r+1}, \ldots, u_n)$$

of eigenvectors for $\tau^*\tau$, where the corresponding eigenvalues can be arranged so that

$$\lambda_1 \ge \cdots \ge \lambda_r > 0 = \lambda_{r+1} = \cdots = \lambda_n$$

The set $(u_{r+1}, \ldots, u_n)$ is an ordered orthonormal basis for $\ker(\tau^*\tau) = \ker(\tau)$ and so $(u_1, \ldots, u_r)$ is an ordered orthonormal basis for $\ker(\tau)^\perp = \mathrm{im}(\tau^*)$.

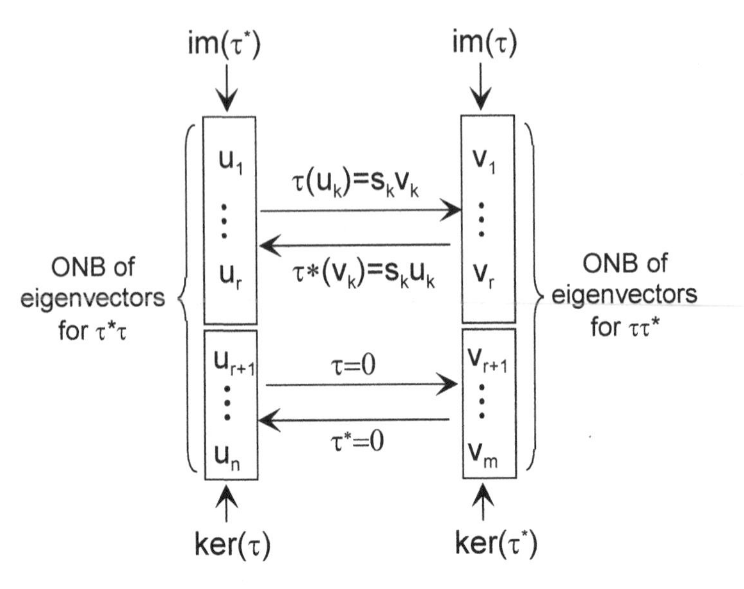

Figure 17.1

For $i = 1, \ldots, r$, the positive numbers $s_i = \sqrt{\lambda_i}$ are called the **singular values** of τ. If we set $s_i = 0$ for $i > r$, then

$$\tau^* \tau u_i = s_i^2 u_i$$

for $i = 1, \ldots, n$. We can achieve some "symmetry" here between τ and τ^* by setting $v_i = (1/s_i)\tau u_i$ for each $i \leq r$, giving

$$\tau u_i = \begin{cases} s_i v_i & i \leq r \\ 0 & i > r \end{cases}$$

and

$$\tau^* v_i = \begin{cases} s_i u_i & i \leq r \\ 0 & i > r \end{cases}$$

The vectors $v_1, \ldots, v_r$ are orthonormal, since if $i, j \leq r$, then

$$\langle v_i, v_j \rangle = \frac{1}{s_i s_j} \langle \tau u_i, \tau u_j \rangle = \frac{1}{s_i s_j} \langle \tau^* \tau u_i, u_j \rangle = \frac{s_i}{s_j} \langle u_i, u_j \rangle = \delta_{i,j}$$

Hence, $(v_1, \ldots, v_r)$ is an orthonormal basis for $\operatorname{im}(\tau) = \ker(\tau^*)^\perp$, which can be extended to an orthonormal basis $\mathcal{C} = (v_1, \ldots, v_m)$ for V, the extension $(v_{r+1}, \ldots, v_m)$ being an orthonormal basis for $\ker(\tau^*)$. Moreover, since

$$\tau \tau^* v_i = s_i \tau u_i = s_i^2 v_i$$

the vectors $v_1, \ldots, v_r$ are eigenvectors for $\tau\tau^*$ with the same eigenvalues $\lambda_i = s_i^2$ as for $\tau^*\tau$. This completes the picture in Figure 17.1.

Theorem 17.1 *Let U and V be finite-dimensional inner product spaces over $\mathbb{C}$ or $\mathbb{R}$ and let $\tau \in \mathcal{L}(U,V)$ have rank r. Then there are ordered orthonormal bases $\mathcal{B}$ and $\mathcal{C}$ for U and V, respectively, for which*

$$\mathcal{B} = (\underbrace{u_1, \ldots, u_r}_{\textit{ONB for } \operatorname{im}(\tau^*)}, \underbrace{u_{r+1}, \ldots, u_n}_{\textit{ONB for } \ker(\tau)})$$

and

$$\mathcal{C} = (\underbrace{v_1, \ldots, v_r}_{\textit{ONB for } \operatorname{im}(\tau)}, \underbrace{v_{r+1}, \ldots, v_m}_{\textit{ONB for } \ker(\tau^*)})$$

Moreover, for $1 \le k \le r$,

$$\begin{aligned} \tau u_i &= s_i v_i \\ \tau^* v_i &= s_i u_i \end{aligned}$$

where $s_i > 0$ are called the **singular values** *of τ, defined by*

$$\tau^* \tau u_i = s_i^2 u_i, \, s_i > 0$$

for $i \le r$. The vectors $u_1, \ldots, u_r$ are called the **right singular vectors** *for τ and the vectors $v_1, \ldots, v_r$ are called the* **left singular vectors** *for τ.*□

The matrix version of the previous discussion leads to the well-known **singular-value decomposition** of a matrix. Let $A \in \mathcal{M}_{m,n}(F)$ and let $\mathcal{B} = (u_1, \ldots, u_n)$ and $\mathcal{C} = (v_1, \ldots, v_m)$ be the orthonormal bases from U and V, respectively, in Theorem 17.1, for the operator τ_A. Then

$$[\tau]_{\mathcal{B},\mathcal{C}} = \Sigma = \operatorname{diag}(s_1, s_2, \ldots, s_r, 0, \ldots, 0)$$

A change of orthonormal bases from the standard bases to $\mathcal{C}$ and $\mathcal{D}$ gives

$$A = [\tau_A]_{\mathcal{E}_n,\mathcal{E}_m} = M_{\mathcal{C},\mathcal{E}_m}[\tau_A]_{\mathcal{B},\mathcal{C}} M_{\mathcal{E}_n,\mathcal{B}} = P\Sigma Q^*$$

where $P = M_{\mathcal{C},\mathcal{E}_m}$ and $Q = M_{\mathcal{B},\mathcal{E}_n}$ are unitary/orthogonal. This is the singular-value decomposition of A.

As to uniqueness, if $A = P\Sigma Q^*$, where P and Q are unitary and Σ is diagonal, with diagonal entries λ_i, then

$$A^*A = (P\Sigma Q^*)^* P\Sigma Q^* = Q\Sigma^*\Sigma Q^*$$

and since $\Sigma^*\Sigma = \operatorname{diag}(\lambda_1^2, \ldots, \lambda_n^2)$, it follows that the λ_i^2's are eigenvalues of A^*A, that is, they are the squares of the singular values along with a sufficient number of 0's. Hence, Σ is uniquely determined by A, up to the order of the diagonal elements.

We state without proof the following uniqueness facts and refer the reader to [48] for details. If $n \le m$ and if the eigenvalues λ_i are distinct, then P is uniquely determined up to multiplication on the right by a diagonal matrix of the form $D = \operatorname{diag}(z_1, \ldots, z_m)$ with $|z_i| = 1$. If $n < m$, then Q is never uniquely determined. If $m = n = r$, then for any given P there is a unique Q. Thus, we see that, in general, the singular-value decomposition is not unique.

The Moore–Penrose Generalized Inverse

Singular values lead to a generalization of the inverse of an operator that applies to all linear transformations. The setup is the same as in Figure 17.1. Referring to that figure, we are prompted to define a linear transformation $\tau^+: V \to U$ by

$$\tau^+ v_i = \begin{cases} \frac{1}{s_i} u_i & \text{for } i \le r \\ 0 & \text{for } i > r \end{cases}$$

since then

$$\begin{aligned} (\tau^+\tau)|_{\langle u_1,\ldots,u_r\rangle} &= \iota \\ (\tau^+\tau)|_{\langle u_{r+1},\ldots,u_n\rangle} &= 0 \end{aligned}$$

and

$$\begin{aligned} (\tau\tau^+)|_{\langle v_1,\ldots,v_r\rangle} &= \iota \\ (\tau\tau^+)|_{\langle v_{r+1},\ldots,v_m\rangle} &= 0 \end{aligned}$$

Hence, if $n = m = r$, then $\tau^+ = \tau^{-1}$. The transformation τ^+ is called the **Moore–Penrose generalized inverse** or **Moore–Penrose pseudoinverse** of τ. We abbreviate this as MP inverse.

Note that the composition $\tau^+\tau$ is the identity on the largest possible subspace of U on which any composition of the form $\sigma\tau$ could be the identity, namely, the orthogonal complement of the kernel of τ. A similar statement holds for the composition $\tau\tau^+$. Hence, τ^+ is as "close" to an inverse for τ as is possible.

We have said that if τ is invertible, then $\tau^+ = \tau^{-1}$. More is true: If τ is injective, then $\tau^+\tau = \iota$ and so τ^+ is a left inverse for τ. Also, if τ is surjective, then τ^+ is a right inverse for τ. Hence the MP inverse τ^+ generalizes the one-sided inverses as well.

Here is a characterization of the MP inverse.

Theorem 17.2 *Let $\tau \in \mathcal{L}(U, V)$. The MP inverse τ^+ of τ is completely characterized by the following four properties:*
1) $\tau\tau^+\tau = \tau$
2) $\tau^+\tau\tau^+ = \tau^+$
3) $\tau\tau^+$ *is Hermitian*
4) $\tau^+\tau$ *is Hermitian*

Proof. We leave it to the reader to show that τ^+ does indeed satisfy conditions 1)–4) and prove only the uniqueness. Suppose that ρ and σ satisfy 1)–4) when substituted for τ^+. Then

$$\begin{aligned}\rho &= \rho\tau\rho \\ &= (\rho\tau)^*\rho \\ &= \tau^*\rho^*\rho \\ &= (\tau\sigma\tau)^*\rho^*\rho \\ &= \tau^*\sigma^*\tau^*\rho^*\rho \\ &= (\sigma\tau)^*\tau^*\rho^*\rho \\ &= \sigma\tau\tau^*\rho^*\rho \\ &= \sigma\tau\rho\tau\rho \\ &= \sigma\tau\rho\end{aligned}$$

and

$$\begin{aligned}\sigma &= \sigma\tau\sigma \\ &= \sigma(\tau\sigma)^* \\ &= \sigma\sigma^*\tau^* \\ &= \sigma\sigma^*(\tau\rho\tau)^* \\ &= \sigma\sigma^*\tau^*\rho^*\tau^* \\ &= \sigma\sigma^*\tau^*(\tau\rho)^* \\ &= \sigma\sigma^*\tau^*\tau\rho \\ &= \sigma\tau\sigma\tau\rho \\ &= \sigma\tau\rho\end{aligned}$$

which shows that $\rho = \sigma$.□

The MP inverse can also be defined for matrices. In particular, if $A \in M_{m,n}(F)$, then the matrix operator τ_A has an MP inverse τ_A^+. Since this is a linear transformation from F^n to F^m, it is just multiplication by a matrix $\tau_A^+ = \tau_B$. This matrix B is the **MP inverse** for A and is denoted by A^+.

Since $\tau_A^+ = \tau_{A^+}$ and $\tau_{AB} = \tau_A\tau_B$, the matrix version of Theorem 17.2 implies that A^+ is completely characterized by the four conditions

1) $AA^+A = A$
2) $A^+AA^+ = A^+$
3) AA^+ is Hermitian
4) A^+A is Hermitian

Moreover, if

$$A = U_1\Sigma U_2^*$$

is the singular-value decomposition of A, then

$$A^+ = U_2 \Sigma' U_1^*$$

where Σ' is obtained from Σ by replacing all nonzero entries by their multiplicative inverses. This follows from the characterization above and also from the fact that for $i \le r$,

$$U_2 \Sigma' U_1^* v_i = U_2 \Sigma' e_i = s_i^{-1} U_2 e_i = s_i^{-1} u_i$$

and for $i > r$,

$$U_2 \Sigma' U_1^* v_i = U_2 \Sigma' e_i = 0$$

Least Squares Approximation

Let us now discuss the most important use of the MP inverse. Consider the system of linear equations

$$Ax = v$$

where $A \in M_{m,n}(F)$. (As usual, $F = \mathbb{C}$ or $F = \mathbb{R}$.) This system has a solution if and only if $v \in \operatorname{im}(\tau_A)$. If the system has no solution, then it is of considerable practical importance to be able to solve the system

$$Ax = \widehat{v}$$

where $\widehat{v}$ is the unique vector in $\operatorname{im}(\tau_A)$ that is closest to v, as measured by the unitary (or Euclidean) distance. This problem is called the **linear least squares** problem. Any solution to the system $Ax = \widehat{v}$ is called a **least squares solution** to the system $Ax = v$. Put another way, a least squares solution to $Ax = v$ is a vector x for which $\|Ax - v\|$ is minimized.

Suppose that w and z are least squares solutions to $Ax = v$. Then

$$Aw = \widehat{v} = Az$$

and so $w - z \in \ker(A)$. (We will write A for τ_A.) Thus, if w is a particular least squares solution, then the set of all least squares solutions is $w + \ker(A)$. Among all solutions, the most interesting is the solution of minimum norm. Note that if there is a least squares solution w that lies in $\ker(A)^\perp$, then for any $z \in \ker(A)$, we have

$$\|w + z\|^2 = \|w\|^2 + \|z\|^2 \ge \|w\|^2$$

and so w will be the unique least squares solution of minimum norm.

Before proceeding, we recall (Theorem 9.14) that if S is a subspace of a finite-dimensional inner product space V, then the best approximation to a vector $v \in V$ from within S is the unique vector $\widehat{v} \in S$ for which $v - \widehat{v} \perp S$. Now we can see how the MP inverse comes into play.

Theorem 17.3 *Let $A \in M_{m,n}(F)$. Among the least squares solutions to the system*

$$Ax = \widehat{v}$$

there is a unique solution of minimum norm, given by A^+v, where A^+ is the MP inverse of A.

Proof. A vector w is a least squares solution if and only if $Aw = \widehat{v}$. Using the characterization of the best approximation $\widehat{v}$, we see that w is a solution to $Aw = \widehat{v}$ if and only if

$$Aw - v \perp \operatorname{im}(A)$$

Since $\operatorname{im}(A)^\perp = \ker(A^*)$ this is equivalent to

$$A^*(Aw - v) = 0$$

or

$$A^*Aw = A^*v$$

This system of equations is called the **normal equations** for $Ax = v$. Its solutions are precisely the least squares solutions to the system $Ax = v$.

To see that $w = A^+v$ is a least squares solution, recall that, in the notation of Figure 17.1,

$$AA^+v_i = \begin{cases} v_i & i \le r \\ 0 & i > r \end{cases}$$

and so

$$A^*A(A^+v_i) = \begin{cases} A^*v_i & i \le r \\ 0 & i > r \end{cases} = A^*v_i$$

and since $\mathcal{C} = (v_1, \ldots, v_m)$ is a basis for V, we conclude that A^+v satisfies the normal equations. Finally, since $A^+v \in \ker(A)^\perp$, we deduce by the preceding remarks that A^+v is the unique least squares solution of minimum norm.□

Exercises

1. Let $\tau \in \mathcal{L}(U)$. Show that the singular values of τ^* are the same as those of τ.
2. Find the singular values and the singular value decomposition of the matrix

$$A = \begin{bmatrix} 3 & 1 \\ 6 & 2 \end{bmatrix}$$

   Find A^+.
3. Find the singular values and the singular value decomposition of the matrix

$$A = \begin{bmatrix} 1 & 2 & 0 \\ 2 & 0 & 2 \end{bmatrix}$$

Find A^+. *Hint*: Is it better to work with A^*A or AA^*?

4. Let $X = (x_1\ x_2\ \cdots\ x_m)^t$ be a column matrix over $\mathbb{C}$. Find a singular-value decomposition of X.
5. Let $A \in M_{m,n}(F)$ and let $B \in M_{m+n,m+n}(F)$ be the square matrix

$$B = \begin{bmatrix} 0 & A \\ A^* & 0 \end{bmatrix}_{\text{block}}$$

Show that, counting multiplicity, the nonzero eigenvalues of B are precisely the singular values of A together with their negatives. *Hint*: Let $A = U_1 \Sigma U_2^*$ be a singular-value decomposition of A and try factoring B into a product USU^* where U is unitary. Do not read the following second hint unless you get stuck. *Second Hint*: Verify the block factorization

$$B = \begin{bmatrix} 0 & U_1 \\ U_2 & 0 \end{bmatrix} \begin{bmatrix} 0 & \Sigma^* \\ \Sigma & 0 \end{bmatrix} \begin{bmatrix} 0 & U_2^* \\ U_1^* & 0 \end{bmatrix}$$

What are the eigenvalues of the middle factor on the right? (Try $\epsilon_1 + \epsilon_{n+1}$ and $\epsilon_1 - \epsilon_{n+1}$.)

6. Use the results of the previous exercise to show that a matrix $A \in M_{m,n}(F)$, its adjoint A^*, its transpose A^t and its conjugate $\overline{A}$ all have the same singular values. Show also that if U and U' are unitary, then A and UAU' have the same singular values.
7. Let $A \in M_n(F)$ be nonsingular. Show that the following procedure produces a singular-value decomposition $A = U_1 \Sigma U_2^*$ of A.
   a) Write $A = UDU^*$ where $D = \operatorname{diag}(\lambda_1, \ldots, \lambda_n)$ and the λ_i's are positive and the columns of U form an orthonormal basis of eigenvectors for A. (We never said that this was a practical procedure.)
   b) Let $\Sigma = \operatorname{diag}(\lambda_1^{1/2}, \ldots, \lambda_n^{1/2})$ where the square roots are nonnegative. Also let $U_1 = U$ and $U_2 = A^* U \Sigma^{-1}$.
8. If $A = (a_{i,j})$ is an $n \times m$ matrix, then the **Frobenius norm** of A is

$$\|A\|_F = \left(\sum_{i,j} a_{i,j}^2 \right)^{1/2}$$

Show that $\|A\|_F^2 = \sum s_i^2$ is the sum of the squares of the singular values of A.

Chapter 18
An Introduction to Algebras

Motivation

We have spent considerable time studying the structure of a linear operator $\tau \in \mathcal{L}_F(V)$ on a finite-dimensional vector space V over a field F. In our studies, we defined the $F[x]$-module V_τ and used the decomposition theorems for modules over a principal ideal domain to dissect this module. We concentrated on an individual operator τ, rather than the entire vector space $\mathcal{L}_F(V)$. In fact, we have made relatively little use of the fact that $\mathcal{L}_F(V)$ is an algebra under composition. In this chapter, we give a brief introduction to the theory of algebras, of which $\mathcal{L}_F(V)$ is the most general, in the sense of Theorem 18.2 below.

Associative Algebras

An algebra is a combination of a ring and a vector space, with an axiom that links the ring product with scalar multiplication.

Definition *An* **(associative) algebra** *A over a field F, or an* ***F*-algebra**, *is a nonempty set A, together with three operations, called* **addition** *(denoted by $+$),* **multiplication** *(denoted by juxtaposition) and* **scalar multiplication** *(also denoted by juxtaposition), for which the following properties hold:*

1) *A is a vector space over F under addition and scalar multiplication.*
2) *A is a ring with identity under addition and multiplication.*
3) *If $r \in F$ and $a, b \in A$, then*

$$r(ab) = (ra)b = a(rb)$$

An algebra is **finite-dimensional** *if it is finite-dimensional as a vector space. An algebra is* **commutative** *if A is a commutative ring. An element $a \in A$ is* **invertible** *if there is $b \in A$ for which $ab = ba = 1$.*□

Our definition requires that A have a multiplicative identity. Such algebras are called **unital algebras**. Algebras without unit are also of great importance, but we will not study them here. Also, in this chapter, we will assume that all algebras are associative. Nonassociative algebras, such as Lie algebras and Jordan algebras, are important as well.

The Center of an Algebra

Definition *The* **center** *of an F-algebra A is the set*

$$Z(A) = \{a \in A \mid ax = xa \text{ for all } x \in A\}$$

of all elements of A that commute with every element of A.□

The center of an algebra is never trivial since it contains a copy of F:

$$\{r1 \mid r \in F\} \subseteq Z(A)$$

Definition *An F-algebra A is* **central** *if its center is as small as possible, that is, if*

$$Z(A) = \{r1 \mid r \in F\}$$ □

From a Vector Space to an Algebra

If V is a vector space over a field F and if $\mathcal{B} = \{b_i \mid i \in I\}$ is a basis for V, then it is natural to wonder whether we can form an F-algebra simply by defining a product for the basis elements and then using the distributive laws to extend the product to V. In particular, we choose a set of constants $\alpha_k^{i,j}$ with the property that for each pair (i, j), only finitely many of the $\alpha_k^{i,j}$ are nonzero. Then we set

$$b_i b_j = \sum_k \alpha_k^{i,j} b_k$$

and make multiplication bilinear, that is,

$$\left(\sum_{i=1}^n r_i b_i\right) b_k = \sum_{i=1}^n r_i b_i b_k$$

$$b_k \left(\sum_{i=1}^n r_i b_i\right) = \sum_{i=1}^n r_i b_k b_i$$

and

$$r\left(\sum_{i=1}^n r_i b_i\right) = \sum_{i=1}^n r r_i b_i$$

for $r \in F$. It is easy to see that this does define a nonunital associative algebra A provided that

$$(b_i b_j)b_k = b_i(b_j b_k)$$

for all $i, j, k \in I$ and that A is commutative if and only if

$$b_i b_j = b_j b_i$$

for all $i, j \in I$. The constants $\alpha_k^{i,j}$ are called the **structure constants** for the algebra A. To get a unital algebra, we can take for a given $i \in I$, the structure constants to be

$$\alpha_k^{i,j} = \delta_{k,j} = \alpha_k^{j,i}$$

in which case b_i is the multiplicative identity. (An alternative is to adjoin a new element to the basis and define its structure constants in this way.)

Examples

The following examples will make it clear why algebras are important.

Example 18.1 If $F \leq E$ are fields, then E is a vector space over F. This vector space structure, along with the ring structure of E, is an algebra over F.□

Example 18.2 The ring $F[x]$ of polynomials is an algebra over F.□

Example 18.3 The ring $\mathcal{M}_n(F)$ of all $n \times n$ matrices over a field F is an algebra over F, where scalar multiplication is defined by

$$M = (a_{i,j}), \quad r \in F \quad \Rightarrow \quad rM = (ra_{i,j})$$

□

Example 18.4 The set $\mathcal{L}_F(V)$ of all linear operators on a vector space V over a field F is an F-algebra, where addition is addition of functions, multiplication is composition of functions and scalar multiplication is given by

$$(r\sigma)(v) = r[\sigma v]$$

The identity map $\iota \in \mathcal{L}_F(V)$ is the multiplicative identity and the zero map $0 \in \mathcal{L}_F(V)$ is the additive identity. This algebra is also denoted by $\text{End}_F(V)$, since the linear operators on V are also called endomorphisms of V.□

Example 18.5 If G is a group and F is a field, then we can form a vector space $F[G]$ over F by taking all formal F-linear combinations of elements of G and treating G as a basis for $F[G]$. This vector space can be made into an F-algebra where the structure constants are determined by the group product, that is, if $g_i g_j = g_u$, then $\alpha_k^{i,j} = \delta_{k,u}$. The group identity $g_1 = 1$ is the algebra identity since $g_1 g_j = g_j$ and so $\alpha_k^{1,j} = \delta_{k,j}$ and similarly, $\alpha_k^{j,1} = \delta_{k,j}$.

The resulting associative algebra $F[G]$ is called the **group algebra** over F. Specifically, the elements of $F[G]$ have the form

$$x = r_1 g_1 + \cdots + r_n g_n$$

where $r_i \in F$ and $g_i \in G$. If

$$y = s_1 h_1 + \cdots + s_m h_m$$

then we can include additional terms with 0 coefficients and reindex if necessary so that we may assume that $m = n$ and $g_i = h_i$ for all i. Then the sum in $F[G]$ is given by

$$\left(\sum_{i=1}^{n} r_i g_i\right) + \left(\sum_{i=1}^{n} s_i g_i\right) = \sum_{i=1}^{n} (r_i + s_i) g_i$$

Also, the product is given by

$$\left(\sum_{i=1}^{n} r_i g_i\right)\left(\sum_{j=1}^{m} s_i h_i\right) = \sum_{i,j} r_i s_j g_i h_j$$

and the scalar product is

$$s\left(\sum_{i=1}^{n} r_i g_i\right) = \sum_{i=1}^{n} s r_i g_i$$ □

The Usual Suspects

Algebras have substructures and structure-preserving maps, as do groups, rings and other algebraic structures.

Subalgebras

Definition *Let A be an F-algebra. A* **subalgebra** *of A is a subset B of A that is a subring of A (with the same identity as A) and a subspace of A.*□

The intersection of subalgebras is a subalgebra and so the family of all subalgebras of A is a complete lattice, where meet is intersection and the join of a family $\mathcal{F}$ of subalgebras is the intersection of all subalgebras of A that contain the members of $\mathcal{F}$.

The **subalgebra generated** by a nonempty subset X of an algebra A is the smallest subalgebra of A that contains X and is easily seen to be the set of all linear combinations of finite products of elements of X, that is, the subspace spanned by the products of finite subsets of elements of X:

$$\langle X \rangle_{\text{alg}} = \langle x_1 \cdots x_n \mid x_i \in X \rangle$$

Alternatively, $\langle X \rangle_{\text{alg}}$ is the set of all polynomials in the variables in X. In particular, the algebra generated by a single element $x \in A$ is the set of all polynomials in x over F.

Ideals and Quotients

In defining the notion of an ideal of an algebra A, we must consider the fact that A may be noncommutative.

Definition *A* **(two-sided) ideal** *of an associative algebra A is a nonempty subset I of A that is closed under addition and subtraction, that is,*

$$a, b \in I \quad \Rightarrow \quad a+b, a-b \in I$$

and also left and right multiplication by elements of A, that is,

$$k \in I, \quad a, b \in A \quad \Rightarrow \quad akb \in I$$ □

The **ideal generated** by a nonempty subset X of A is the smallest ideal containing X and is equal to

$$\langle X \rangle_{\text{ideal}} = \left\{ \sum_{i=1}^{n} a_i x_i b_i \mid x_i \in X, a_i, b_i \in A \right\}$$

Definition *An algebra A is* **simple** *if*

1) *The product in A is not trivial, that is, $ab \neq 0$ for at least one pair of elements $a, b \in A$*
2) *A has no proper nonzero ideals.* □

Definition *If I is an ideal in A, then the* **quotient algebra** *is the quotient ring/quotient space*

$$A/I = \{a + I \mid a \in A\}$$

with operations

$$\begin{aligned} (a+I)+(b+I) &= (a+b)+I \\ (a+I)(b+I) &= ab+I \\ r(a+I) &= ra+I \end{aligned}$$

where $a, b \in A$ and $r \in F$. These operations make A/I an F-algebra. □

Homomorphisms

Definition *If A and B are F-algebras a map $\sigma: A \to B$ is an* **algebra homomorphism** *if it is a ring homomorphism as well as a linear transformation, that is,*

$$\sigma(a+a') = \sigma a + \sigma a', \quad \sigma(aa') = (\sigma a)(\sigma a'), \quad \sigma 1 = 1$$

and

$$r(\sigma a) = \sigma(ra)$$

for $r \in F$. □

The usual terms monomorphism, epimorphism, isomorphism, embedding, endomorphism and automorphism apply to algebras with the analogous meaning as for vector spaces and modules.

Example 18.6 Let V be an n-dimensional vector space over F. Fix an ordered basis $\mathcal{B}$ for V. Consider the map $\mu: \mathcal{L}(V) \to \mathcal{M}_n(F)$ defined by

$$\mu(\sigma) = [\sigma]_{\mathcal{B}}$$

where $[\sigma]_{\mathcal{B}}$ is the matrix representation of σ with respect to the ordered basis $\mathcal{B}$. This map is a vector space isomorphism and since

$$[\tau\sigma]_{\mathcal{B}} = [\tau]_{\mathcal{B}}[\sigma]_{\mathcal{B}}$$

it is also an algebra isomorphism.□

Another View of Algebras

If A is an algebra over F, then A contains a copy of F. Specifically, we define a function $\lambda: F \to A$ by

$$\lambda r = r1$$

for all $r \in F$, where $1 \in A$ is the multiplicative identity. The elements $r1$ are in the center of A, since for any $a \in A$,

$$(r1)a = r(1a) = ra$$

and

$$a(r1) = r(a1) = ra$$

Thus, $\lambda: F \to Z(A)$. To see that λ is a ring homomorphism, we have

$$\begin{aligned}
&\lambda(1_F) = 1_F \cdot 1 = 1\\
&\lambda(r+s) = (r+s)1 = r1 + s1 = \lambda(r) + \lambda(s)\\
&\lambda(rs) = (rs)1 = r(s1) = r\lambda(s) = r(1 \cdot \lambda(s)) = (r \cdot 1)\lambda(s) = \lambda(r)\lambda(s)
\end{aligned}$$

Moreover, if $r1 = 0$ and $r \neq 0$, then

$$0 = r^{-1}(r1) = 1_F \cdot 1 = 1$$

and so provided that $0 \neq 1$ in A, we have $r = 0$. Thus, λ is an embedding.

Theorem 18.1

1) *If A is an associative algebra over F and if $0 \neq 1$ in A, then the map $\lambda: F \to Z(A)$ defined by*

$$\lambda r = r1$$

is an embedding of the field F into the center $Z(A)$ of the ring A. Thus, F can be embedded as a subring of $Z(A)$.

2) *Conversely, if* R *is a ring with identity and if* $F \subseteq Z(R)$ *is a field, then* R *is an* F*-algebra with scalar multiplication defined by the product in* R.□

One interesting consequence of this theorem is that a ring R whose center does not contain a field is not an algebra over *any* field F. This happens, for example, with the ring $\mathbb{Z}_6$.

The Regular Representation of an Algebra

An algebra homomorphism $\sigma: A \to \mathcal{L}_F(V)$ is called a **representation** of the algebra A in $\mathcal{L}_F(V)$. A representation σ is **faithful** if it is injective, that is, if σ is an embedding. In this case, A is isomorphic to a subalgebra of $\mathcal{L}_F(V)$.

Actually, the endomorphism algebras $\mathcal{L}_F(V)$ are the most general algebras possible, in the sense that any algebra A has a faithful representation in some endomorphism algebra.

Theorem 18.2 *Any associative* F*-algebra* A *is isomorphic to a subalgebra of the endomorphism algebra* $\mathcal{L}_F(A)$*. In fact, if* μ_a *is the left multiplication map defined by*

$$\mu_a x = ax$$

then the map $\mu: A \to \mathcal{L}_f(A)$ *is an algebra embedding, called the* **left regular representation** *of* A.□

When $\dim(A) = n < \infty$, we can select an ordered basis $\mathcal{B}$ for A and represent the elements of $\mathcal{L}_F(A)$ by matrices. This gives an embedding of A into the matrix algebra $\mathcal{M}_n(F)$, called the **left regular matrix representation** of A with respect to the ordered basis $\mathcal{B}$.

Example 18.7 Let $G = \{1, \alpha, \ldots, \alpha^{n-1}\}$ be a finite cyclic group. Let

$$\mathcal{B} = (1, \alpha, \ldots, \alpha^{n-1})$$

be an ordered basis for the group algebra $F[G]$. The multiplication map μ_k that is multiplcation by α^k is a shifting of $\mathcal{B}$ (with wraparound) and so the matrix representation of μ_k is the matrix whose columns are obtained from the identity matrix by shifting k columns to the right (with wrap around). For example,

$$[\mu_1]_{\mathcal{B}} = \begin{bmatrix} 0 & 0 & \cdots & 0 & 1 \\ 1 & 0 & \cdots & 0 & 0 \\ 0 & 1 & \ddots & 0 & 0 \\ \vdots & \vdots & \ddots & \vdots & \vdots \\ 0 & 0 & \cdots & 1 & 0 \end{bmatrix}$$

These matrices are called **circulant matrices**.□

Since the endomorphism algebras $\mathcal{L}_F(V)$ are of obvious importance, let us examine them a bit more closely.

Theorem 18.3 *Let V be a vector space over a field F.*

1) *The algebra $\mathcal{L}_F(V)$ has center*
$$Z = \{r1 \mid r \in F\}$$
and so $\mathcal{L}_F(V)$ is central.
2) *The set I of all elements of $\mathcal{L}_F(V)$ that have finite rank is an ideal of $\mathcal{L}_F(V)$ and is contained in all other ideals of $\mathcal{L}_F(V)$.*
3) *$\mathcal{L}_F(V)$ is simple if and only if V is finite-dimensional.*

Proof. We leave the proof of parts 1) and 3) as exercises. For part 2), we leave it to the reader to show that I is an ideal of $\mathcal{L}_F(V)$. Let J be a nonzero ideal of $\mathcal{L}_F(V)$. Let $f \in \mathcal{L}_F(V)$ have rank 1. Then there is a basis $\mathcal{B} = \mathcal{B}_1 \cup \mathcal{B}_2$ (a disjoint union) and a nonzero $w \in V$ for which $\mathcal{B}_1$ is a finite set, $f(\mathcal{B}_2) = \{0\}$ and $f(b) = r_b w$ for all $b \in \mathcal{B}$. Thus, f is a linear combination over F of endomorphisms f_b defined by
$$f_b(b) = w, \quad f_b(\mathcal{B} \setminus \{b\}) = \{0\}$$
Hence, we need only show that $f_b \in J$.

If $\sigma \in J$ is nonzero, then there is an $e \in \mathcal{B}$ for which $\sigma e = u \neq 0$. If $\tau \in \mathcal{L}_F(V)$ is defined by
$$\tau b = e, \quad \tau(\mathcal{B} \setminus \{b\}) = \{0\}$$
and $\lambda \in \mathcal{L}_F(V)$ is defined by
$$\lambda u = w, \quad \lambda(\mathcal{B} \setminus \{u\}) = \{0\}$$
then
$$\lambda\sigma\tau(b) = w, \quad \lambda\sigma\tau(\mathcal{B} \setminus \{b\}) = \{0\}$$
and so $f_b = \lambda\sigma\tau \in J$.□

Annihilators and Minimal Polynomials

If A is an F-algebra and $a \in A$, then it may happen that a satisfies a nonzero polynomial $p(x) \in F[x]$. This always happens, in particular, if A is finite-dimensional, since in this case the powers
$$1, a, a^2, \ldots$$
must be linearly dependent and so there is a nonzero polynomial in a that is equal to 0.

Definition *Let A be an F-algebra. An element $a \in A$ is* **algebraic** *if there is a nonzero polynomial $p(x) \in F[x]$ for which $p(a) = 0$. If a is algebraic, the* *monic polynomial $m_a(x)$ of smallest degree that is satisfied by a is called the* **minimal polynomial** *of a.* □

If $a \in A$ is algebraic over F, then the subalgebra generated by a over F is

$$F[a] = \{p(a) \mid p(x) \in F[x], \deg(p) < \deg(m_a)\}$$

and this is isomorphic to the quotient algebra

$$F[a] \approx \frac{F[x]}{\langle m_a(x) \rangle}$$

where $\langle m_a(x) \rangle$ is the ideal generated by the minimal polynomial of a. We leave the details of this as an exercise.

The minimal polynomial can be used to tell when an element is invertible.

Theorem 18.4

1) *The minimal polynomial $m_a(x)$ of $a \in A$ generates the* **annihilator** *of a, that is, the ideal*

$$\operatorname{ann}(a) = \{f(x) \in F[x] \mid f(a) = 0\}$$

of all polynomials that annihilate a.

2) *The element $a \in A$ is invertible if and only if $m_a(x)$ has nonzero constant term.*

Proof. We prove only the second statement. If a is invertible but

$$m_a(x) = xp(x)$$

then $0 = m_a(a) = ap(a)$. Multiplying by a^{-1} gives $p(a) = 0$, which contradicts the minimality of $\deg(m_a(x))$. Conversely, if

$$m_a(x) = \alpha_0 + \alpha_1 x + \cdots + \alpha_{n-1}x^{n-1} + x^n$$

where $\alpha_0 \neq 0$, then

$$0 = \alpha_0 + \alpha_1 a + \cdots + \alpha_{n-1}a^{n-1} + a^n$$

and so

$$\frac{-1}{\alpha_0}(\alpha_1 + \alpha_2 a + \cdots + \alpha_{n-1}a^{n-2} + a^{n-1})a = 1$$

and so

$$a^{-1} = \frac{-1}{\alpha_0}(\alpha_1 + \alpha_2 a + \cdots + \alpha_{n-1}a^{n-2} + a^{n-1})$$

□

Theorem 18.5 *If A is a finite-dimensional F-algebra, then every element of A is algebraic. There are infinite-dimensional algebras in which all elements are algebraic.*

Proof. The first statement has been proved. To prove the second, let us consider the complex field $\mathbb{C}$ as a $\mathbb{Q}$-algebra. The set A of algebraic elements of $\mathbb{C}$ is a field, known as the field of **algebraic numbers**. These are the complex numbers that are roots of some nonzero polynomial with rational (or integral) coefficients.

To see this, if $a \in A$, then the subalgebra $\mathbb{Q}[a]$ is finite-dimensional. Also, $\mathbb{Q}[a]$ is a field. To prove this, first note that since $\mathbb{C}$ is a field, the minimal polynomial of any nonzero $a \in A$ is irreducible, for if $m_a(x) = p(x)q(x)$, then $0 = p(a)q(a)$ and so one of $p(a)$ and $q(a)$ is 0, which implies that $p(x) = m_a(x)$ or $q(x) = m_a(x)$. Since $m_a(x)$ is irreducible, it has nonzero constant term and so the inverse of a is a polynomial in a, that is, $a^{-1} \in \mathbb{Q}[a]$. Of course, $\mathbb{Q}[a]$ is closed under addition and multiplication and so $\mathbb{Q}[a]$ is a subfield of $\mathbb{C}$.

Thus, $\mathbb{C}$ is an algebra over $\mathbb{Q}[a]$. By similar reasoning, if $b \in A$, then the minimal polynomial of b over $\mathbb{Q}[a]$ is irreducible and so $b^{-1} \in \mathbb{Q}[a][b]$. Since $\mathbb{Q}[a][b] = \mathbb{Q}[a,b]$ is the set of all polynomials in the "variables" a and b, it is closed under addition and multiplication as well. Hence, $\mathbb{Q}[a,b]$ is a finite-dimensional algebra over $\mathbb{Q}[a]$, as well as a subfield of $\mathbb{C}$. Now,

$$\dim_{\mathbb{Q}}(\mathbb{Q}[a,b]) = \dim_{\mathbb{Q}[a]}(\mathbb{Q}[a,b]) \cdot \dim_{\mathbb{Q}}(\mathbb{Q}[a])$$

and so $\mathbb{Q}[a,b]$ is finite-dimensional over $\mathbb{Q}$. Hence, the elements of $\mathbb{Q}[a,b]$ are algebraic over $\mathbb{Q}$, that is, $\mathbb{Q}[a,b] \subseteq A$. But $\mathbb{Q}[a,b]$ contains $a^{-1}, a+b, a-b$ and ab and so A is a field.

We claim that A is not finite-dimensional over $\mathbb{Q}$. This follows from the fact that for every prime p, the polynomial $x^n - p$ is irreducible over $\mathbb{Q}$ (by Eisenstein's criterion). Hence, if a is a complex root of $x^n - p$, then a has minimal polynomial $x^n - p$ over $\mathbb{Q}$ and so the dimension of $\mathbb{Q}[a]$ over $\mathbb{Q}$ is n. Hence, A cannot be finite-dimensional.□

The Spectrum of an Element

Let A be an algebra. A nonzero element $a \in A$ is a **left zero divisor** if $ab = 0$ for some $b \neq 0$ and a **right zero divisor** if $ca = 0$ for some $c \neq 0$. In the exercises, we ask the reader to show that an *algebraic* element is a left zero divisor if and only if it is a right zero divisor.

Theorem 18.6 *Let A be a algebra. An algebraic element $a \in A$ is invertible if and only if it is not a zero divisor.*

Proof. If a is invertible and $ab = 0$, then multiplying by a^{-1} gives $b = 0$. Conversely, suppose that a is not invertible but $ab = 0$ implies $b = 0$. Then

$m_a(x) = xp(x)$ for some nonzero polynomial $p(x)$ and so $0 = ap(a)$, which implies that $p(a) = 0$, a contradiction to the minimality of $m_a(x)$.□

We have seen that the eigenvalues of a linear operator τ on a finite-dimensional vector space are the roots of the minimal polynomial of τ, or equivalently, the scalars r for which $\tau - r\iota$ is not invertible. By analogy, we can define the eigenvalues of an element a of an algebra A.

Theorem 18.7 *Let A be an algebra and let $a \in A$ be algebraic. An element $r \in F$ is a root of the minimal polynomial $m_a(x)$ if and only if $a - r1$ is not invertible in A.*

Proof. If $a - r1$ is not invertible, then

$$m_{a-r1}(x) = xp(x)$$

and since $m_a(x + r1)$ is satisfied by $a - r1$, it follows that

$$xp(x) = m_{a-r1}(x) \mid m_a(x + r1)$$

Hence, $(x - r1) \mid m_a(x)$. Alternatively, if $a - r1$ is not invertible, then there is a nonzero $b \in A$ such that $(a - r1)b = 0$, that is, $ab = rb$. Hence, for any polynomial $p(x)$ we have $p(a)b = p(r)b$. Setting $p(x) = m_a(x)$ gives $m_a(r) = 0$.

Conversely, if $m_a(r) = 0$, then $m_a(x) = (x - r1)p(x)$ and so $0 = (a - r1)p(a)$, which shows that $a - r1$ is a zero divisor and therefore not invertible.□

Definition *Let A be an F-algebra and let $a \in A$ be algebraic. The roots of the minimal polynomial of a are called the* **eigenvalues** *of a. The set of all eigenvalues of a*

$$\operatorname{Spec}(a) = \{r \in F \mid m_a(r) = 0\}$$

is called the **spectrum** *of a.*□

Note that $a \in A$ is invertible if and only if $0 \notin \operatorname{Spec}(a)$.

Theorem 18.8 (The spectral mapping theorem) *Let A be an algebra over an algebraically closed field F. Let $a \in A$ and let $p(x) \in F[x]$. Then*

$$\operatorname{Spec}(p(a)) = p(\operatorname{Spec}(a)) = \{p(r) \mid r \in \operatorname{Spec}(a)\}$$

Proof. We leave it as an exercise to show that $p(\operatorname{Spec}(a)) \subseteq \operatorname{Spec}(p(a))$. For the reverse inclusion, let $r \in \operatorname{Spec}(p(a))$ and suppose that

$$p(x) - r = (x - r_1)^{e_1} \cdots (x - r_n)^{e_n}$$

Then

$$p(a) - r1 = (a - r_1 1)^{e_1} \cdots (a - r_n 1)^{e_n}$$

and since the left-hand side is not invertible, neither is one of the factors $a - r_k 1$, whence $r_k \in \text{Spec}(a)$. But

$$p(r_k) - r = 0$$

and so $r = p(r_k) \in p(\text{Spec}(a))$. Hence, $\text{Spec}(p(a)) \subseteq p(\text{Spec}(a))$.□

Theorem 18.9 *Let A be an algebra over an algebraically closed field F. If $a, b \in A$, then*

$$\text{Spec}(ab) = \text{Spec}(ba)$$

Proof. If $0 \neq r \notin \text{Spec}(ba)$, then $ba - r1$ is invertible and a simple computation gives

$$(ab - r1)[a(ba - r1)^{-1}b - 1] = r$$

and so $ab - r1$ is invertible and $r \notin \text{Spec}(ab)$. If $0 \notin \text{Spec}(ba)$, then ba is invertible. We leave it as an exercise to show that this implies that ab is also invertible and so $0 \notin \text{Spec}(ab)$. Thus, $\text{Spec}(ab) \subseteq \text{Spec}(ba)$ and by symmetry, equality must hold.□

Division Algebras

Some important associative algebras A have the property that all nonzero elements are invertible and yet A is not a field since it is not commutative.

Definition *An associative algebra D over a field F is a* **division algebra** *if every nonzero element has a multiplicative inverse.*□

Our goal in this section is to classify all finite-dimensional division algebras over the real field $\mathbb{R}$, over any algebraically closed field F and over any finite field. The classification of finite-dimensional division algebras over the rational field $\mathbb{Q}$ is quite complicated and we will not treat it here.

The Quaternions

Perhaps the most famous noncommutative division algebra is the following. Define a real vector space $\mathbb{H}$ with basis

$$\mathcal{B} = \{1, i, j, k\}$$

To make $\mathbb{H}$ into an F-algebra, define the product of basis vectors as follows:

1) $1x = x1 = x$ for all $x \in \mathcal{B}$
2) $i^2 = j^2 = k^2 = -1$
3) $ij = k, jk = i, ki = j$
4) $ji = -k, kj = -i, ik = -j$

Note that 3) can be stated as follows: The product of two consecutive elements i, j, k is the next element (with wraparound). Also, 4) says that $yx = -xy$ for $x, y \in \{i, j, k\}$. This product is extended to all of $\mathbb{H}$ by distributivity.

We leave it to the reader to verify that $\mathbb{H}$ is a division algebra, called Hamilton's **quaternions**, after their discoverer William Rowan Hamilton (1805-1865). (Readers familiar with group theory will recognize the quaternion group $Q = \{\pm 1, \pm i, \pm j, \pm k\}$.) The quaternions have applications in geometry, computer science and physics.

Finite-Dimensional Division Algebras over an Algebraically Closed Field

It happens that there are no interesting finite-dimensional division algebras over an algebraically closed field.

Theorem 18.10 *If D is a finite-dimensional division algebra over an algebraically closed field F then $D = F$.*
Proof. Let $a \in D$ have minimal polynomial $m_a(x)$. Since a division algebra has no zero divisors, $m_a(x)$ must be irreducible over F and so must be linear. Hence, $m_a(x) = x - r$ and so $a = r \in F$.$\square$

Finite-Dimensional Division Algebras over a Finite Field

The finite-dimensional division algebras over a finite field are also easily described: they are all commutative and so are finite fields. The proof, however, is a bit more challenging. To understand the proof, we need two facts: the class equation and some information about the complex roots of unity. So let us briefly describe what we need.

The Class Equation

Those who have studied group theory have no doubt encountered the famous class equation. Let G be a finite group. Each $a \in G$ can be thought of as a permutation σ_a of G defined by

$$\sigma_a x = axa^{-1}$$

for all $x \in G$. The set of all conjugates axa^{-1} of x is denoted by $\{x\}^G$ and so

$$\{x\}^G = \{\sigma_a x \mid a \in G\}$$

This set is also called a **conjugacy class** in G. Now, the following are equivalent:

$$\begin{aligned} \sigma_a x &= \sigma_b x \\ axa^{-1} &= bxb^{-1} \\ b^{-1}axab^{-1} &= x \\ b^{-1}a &\in C_G(x) \end{aligned}$$

where

$$C_G(x) = \{g \in G \mid gx = xg\}$$

is the **centralizer** of x. But $b^{-1}a \in C_G(x)$ if and only if a and b are in the same coset of $C_G(x)$. Thus, there is a one-to-one correspondence between the conjugates of x and the cosets of $C_G(x)$. Hence,

$$\left|\{x\}^G\right| = (G : C_G(x))$$

Since the distinct conjugacy classes form a partition of G (because conjugacy is an equivalence relation), we have

$$|G| = \sum_{x \in S} \left|\{x\}^G\right| = \sum_{x \in S} (G : C_G(x))$$

where S is a set consisting of exactly one element from each conjugacy class $\{x\}^G$. Note that a conjugacy class $\{x\}^G$ has size 1 if and only if $axa^{-1} = x$ for all $a \in G$, that is, $xa = ax$ for all $a \in G$ and these are precisely the elements in the center $Z(G)$ of G. Hence, the previous equation can be written in the form

$$|G| = |Z(G)| + \sum_{x \in S'} (G : C_G(x))$$

where S' is a set consisting of exactly one element from each conjugacy class $\{x\}^G$ of size greater than 1. This is the **class equation** for G.

The Complex Roots of Unity

If n is a positive integer, then the complex nth **roots of unity** are the complex solutions to the equation

$$x^n - 1 = 0$$

The set U_n of complex nth roots of unity is a cyclic group of order n. To see this, note first that U_n is an abelian group since $a, b \in U_n$ implies that $ab \in U_n$ and $a^{-1} \in U_n$. Also, since $x^n - 1$ has no multiple roots, U_n has order n.

Now, in any finite abelian group G, if m is the maximum order of all elements of G, then $g^m = 1$ for all $g \in G$. Thus, if no element of U_n has order n, then $m < n$ and every $g \in G$ satisfies the equation $x^m - 1 = 0$, which has fewer than n solutions. This contradiction implies that some element of U_n must have order n and so U_n is cyclic.

The elements of U_n that generate U_n, that is, the elements of order n are called the **primitive** nth roots of unity. We denote the set of primitive nth roots of unity by Ω_n. Hence, if $a \in \Omega_n$, then

$$\Omega_n = \{a^k \mid (n,k) = 1\}$$

has size $\phi(n)$, where ϕ is the Euler phi function. (The value $\phi(k)$ is defined to be the number of positive integers less than or equal to k and relatively prime to k.)

The nth **cyclotomic polynomial** is defined by

$$Q_n(x) = \prod_{w \in \Omega_n} (x - w)$$

Thus,

$$\deg(Q_n(x)) = \phi(n)$$

Since every nth root of unity is a primitive dth root of unity for some $d \mid n$ and since every primitive dth root of unity for $d \mid n$ is also an nth root of unity, we deduce that

$$U_n = \bigcup_{d \mid n} \Omega_d$$

where the union is a disjoint one. It follows that

$$x^n - 1 = \prod_{d \mid n} Q_d(x)$$

Finally, we show that $Q_n(x)$ is monic and has integer coefficients by induction on n. It is clear from the definition that $Q_n(x)$ is monic. Since $Q_1(x) = x - 1$, the result is true for $n = 1$. If p is a prime, then all nonidentity pth roots of unity are primitive and so

$$Q_p(x) = \frac{x^p - 1}{x - 1} = x^{p-1} + x^{p-2} + \cdots + x + 1$$

and the result holds for $n = p$. Assume the result holds for all proper divisors of n. Then

$$x^n - 1 = Q_n(x) \prod_{\substack{d \mid n \\ d < n}} Q_d(x) = Q_n(x) R(x)$$

By the induction hypothesis, $R(x)$ has integer coefficients and it follows that $Q_n(x)$ must also have integer coefficients.

Wedderburn's Theorem

Now we can prove Wedderburn's theorem.

Theorem 18.11 (Wedderburn's theorem) *If D is a finite division algebra, then D is a field.*

Proof. We must show that D is commutative. Let $G = D^*$ be the multiplicative group of all nonzero elements of D. The class equation is

$$|D^*| = |Z(D^*)| + \sum (D^* : C(\beta))$$

where the sum is taken over one representative β from each conjugacy class of size greater than 1. If we assume for the purposes of contradiction that D is not commutative, that is, that $Z(D^*) \neq D^*$, then the sum on the far right is not an empty sum and so $|C^*(\beta)| < |D^*|$ for some $\beta \in D^*$.

The sets $Z(D)$ and $C(\beta)$ are subalgebras of D and, in fact, $Z(D)$ is a commutative division algebra; that is, a field. Let $|Z(D)| = z \geq 2$. Since $Z(D) \subseteq C(\beta)$, we may view $C(\beta)$ and D as vector spaces over $Z(D)$ and so

$$|C(\beta)| = z^{k(\beta)} \quad \text{and} \quad |D| = z^n$$

for integers $1 \leq k(\beta) < n$. The class equation now gives

$$z^n - 1 = z - 1 + \sum_{k(\beta)} \frac{z^n - 1}{z^{k(\beta)} - 1}$$

and since $z^{k(\beta)} - 1 \mid z^n - 1$, it follows that $k(\beta) \mid n$.

If $Q_n(x)$ is the nth cyclotomic polynomial, then $Q_n(z)$ divides $z^n - 1$. But $Q_n(z)$ also divides each summand on the far right above, since its roots are not roots of $z^{k(\beta)} - 1$. It follows that $Q_n(z) \mid z - 1$. On the other hand,

$$Q_n(z) = \prod_{\omega \in \Omega_n} (z - \omega)$$

and since $\omega \in \Omega_n$ implies that $|z - \omega| > z - 1$, we have a contradiction. Hence $Z(D^*) = D^*$ and D is commutative, that is, D is a field.□

Finite-Dimensional Real Division Algebras

We now consider the finite-dimensional division algebras over the real field $\mathbb{R}$. In 1877, Frobenius proved that there are only three such division algebras.

Theorem 18.12 (Frobenius, 1877) *If D is a finite-dimensional division algbera over $\mathbb{R}$, then*

$$D = \mathbb{R}, \quad D = \mathbb{C} \quad \textit{or} \quad D = \mathbb{H}$$

Proof. Note first that the minimal polynomial $m_d(x)$ of any $d \in D$ is either linear, in which case $d \in \mathbb{R}$ or irreducible quadratic $m_d(x) = x^2 + rx + s$ with $r^2 - 4s < 0$. Completing the square gives

$$0 = m_d(a) = a^2 + ra + s = \left(a + \frac{1}{2}r\right)^2 + \frac{1}{4}(4s - r^2)$$

Hence, any $a \in D$ has the form

$$a = \left(a + \frac{1}{2}r\right) - \frac{1}{2}r = \alpha + t$$

where $t \in \mathbb{R}$ and either $\alpha = 0$ or $\alpha^2 < 0$. Hence, $\alpha^2 \in \mathbb{R}$ but $\alpha \notin \mathbb{R}$. Thus, every element of D is the sum of an element of $\mathbb{R}$ and an element of the set

$$D' = \{\alpha \in D \mid \alpha^2 \leq 0\}$$

that is, as sets:

$$D = \mathbb{R} + D'$$

Also, $\mathbb{R} \cap D' = \{0\}$. To see that D' is a subspace of D, let $u, v \in D'$. We wish to show that $u + v \in D'$. If $v = ru$ for some $r \in \mathbb{R}$, then $u + v = (1 + r)u \in D'$. So assume that u and v are linearly independent. Then u and v are nonzero and so also nonreal.

Now, $u + v$ and $u - v$ cannot both be real, since then u and v would be real. We have seen that

$$u + v = r + \alpha$$

and

$$u - v = s + \beta$$

where $r, s \in \mathbb{R}$, at least one of α or β is nonzero and $\alpha^2, \beta^2 \leq 0$. Then

$$(u + v)^2 + (u - v)^2 = (r + \alpha)^2 + (s + \beta)^2$$

and so

$$2u^2 + 2v^2 = r^2 + 2r\alpha + \alpha^2 + s^2 + 2s\beta + \beta^2$$

Collecting the real part on one side gives

$$2r\alpha + 2s\beta = 2u^2 + 2v^2 - (r^2 + \alpha^2 + s^2 + \beta^2)$$

Now, if we knew that α, β and 1 were linearly independent over $\mathbb{R}$ we could conclude that $r = s = 0$ and so

$$(u + v)^2 = \alpha^2 \leq 0 \quad \text{and} \quad (u - v)^2 = \beta^2 \leq 0$$

which shows that $u + v$ and $u - v$ are in D'.

To see that $\{\alpha, \beta, 1\}$ is linearly independent, it is equivalent to show that $\{u, v, 1\}$ is linearly independent. But if

$$v = au + b$$

for $a, b \in \mathbb{R}$, then

$$v^2 = a^2u^2 + 2abu + b^2$$

and since $u \notin \mathbb{R}$, it follows that $ab = 0$ and so $a = 0$ or $b = 0$. But $a \neq 0$ since $v \notin \mathbb{R}$ and $b \neq 0$ since $\{u, v\}$ are linearly independent.

Thus, D' is a subspace of D and

$$D = \mathbb{R} \oplus D'$$

We now look at D', which is a real vector space. If $D' = \{0\}$, then $D = \mathbb{R}$ and we are done, so assume otherwise. If $a \in D'$ is nonzero, then $a^2 = -r^2$ where $r \in \mathbb{R}$. Hence, $i = ar^{-1} \in D'$ satisfies $i^2 = -1$. If

$$D' = \mathbb{R}i = \{ri \mid r \in \mathbb{R}\}$$

then $D = \mathbb{R} \oplus \mathbb{R}i = \mathbb{C}$ and we are done. If not, then $\mathbb{R}i$ is a proper subspace of D'.

In the quaternion field, there is an element j for which $ij + ji = 0$. So we seek a $j \in D' \setminus \mathbb{R}i$ with this property. To this end, define a bilinear form on D' by

$$\langle u, v \rangle = -(uv + vu)$$

Then it is easy to see that this form is a real inner product on D' (positive definite, symmetric and bilinear). Hence, if $\mathbb{R}i$ is a proper subspace of D', then

$$D' = \mathbb{R}i \odot S$$

where $\odot$ denotes the orthogonal direct sum. If $u \in S$ is nonzero, then $u^2 = -r^2$ for $r \in \mathbb{R}$ and so if $j = ur^{-1}$, then

$$j^2 = -1 \quad \text{and} \quad ji + ij = 0$$

Now, $\mathbb{R}j$ is a subspace of S and so

$$D' = \mathbb{R}i \odot \mathbb{R}j \odot T$$

Setting $k = ij$, we have

$$-\langle i, k \rangle = ik + ki = iij + iji = 0$$

and

$$-\langle j, k \rangle = jk + kj = jij + ijj = 0$$

and so $k \in T$ and we can write

$$D' = \mathbb{R}i \odot \mathbb{R}j \odot \mathbb{R}k \odot U$$

Now, if $U \neq \{0\}$, then there is a $u \in U$ for which $u^2 = -1$ and

$$\begin{aligned} ui &= -iu \\ uj &= -ju \\ uk &= -ku \end{aligned}$$

The third equation is $u(ij) = -(ij)u$ and so

$$u(ij) = -(ij)u = iuj = -uij$$

whence $uij = 0$, which is false. Hence, $U = \{0\}$ and

$$D = \mathbb{R} \oplus \mathbb{R}i \oplus \mathbb{R}j \oplus \mathbb{R}k = \mathbb{H}$$

This completes the proof.□

Exercises

1. Prove that the subalgebra generated by a nonempty subset X of an algebra A is the subspace spanned by the products of finite subsets of elements of X:
$$\langle X \rangle_{\text{alg}} = \langle x_1 \cdots x_n \mid x_i \in X \rangle$$
2. Verify that the group algebra $F[G]$ is indeed an associative algebra over F.
3. Show that the kernel of an algebra homomorphism is an ideal.
4. Let A be a finite-dimensional algebra over F and let B be a subalgebra. Show that if $b \in B$ is invertible, then $b^{-1} \in B$.
5. If A is an algebra and $S \subseteq A$ is nonempty, define the **centralizer** $C_A(S)$ of S to be the set of elements of A that commute with all elements of S. Prove that $C_A(S)$ is a subalgebra of A.
6. Show that $\mathbb{Z}_6$ is not an algebra over any field.
7. Let $A = F[a]$ be the algebra generated over F by a single algebraic element a. Show that A is isomorphic to the quotient algebra $F[x]/\langle f(x) \rangle$, where $\langle f(x) \rangle$ is the ideal generated by $f(x) \in F[x]$. What can you say about $f(x)$? What is the dimension of A? What happens if a is not algebraic?
8. Let $G = \{1 = a_0, \ldots, a_n\}$ be a finite group. For $x \in F[G]$ of the form
$$x = r_1 a_1 + \cdots + r_n a_n$$
let $T(x) = r_1 + \cdots + r_n$. Prove that $T: F[G] \to F$ is an algebra homomorphism, where F is an algebra over itself.
9. Prove the **first isomorphism theorem** of algebras: A homomorphism $\sigma: A \to B$ of F-algebras induces an isomorphism $\overline{\sigma}: A/\ker(\sigma) \approx \operatorname{im}(\sigma)$ defined by $\overline{\sigma}(a\ker(\sigma)) = \sigma a$.
10. Prove that the quaternion field is an F-algebra and a field. *Hint*: For
$$x = r_0 + r_1 i + r_2 j + r_3 k \neq 0$$

($r_0 = r_0 1$) consider

$$\overline{x} = r_0 - r_1 i - r_2 j - r_3 k$$

11. Describe the left regular representation of the quaternions using the ordered basis $\mathcal{B} = (1, i, j, k)$.
12. Let S_n be the group of permutations (bijective functions) of the ordered set $X = (x_1, \ldots, x_n)$, under composition. Verify the following statements. Each $\sigma \in S_n$ defines a linear isomorphism τ_σ on the vector space V with basis X over a field F. This defines an algebra homomorphism $f\colon F[S_n] \to \mathcal{L}_F(V)$ with the property that $f(\sigma) = \tau_\sigma$. What does the matrix representation of a $\sigma \in S_n$ look like? Is the representation f faithful?
13. Show that the center of the algebra $\mathcal{L}_F(V)$ is

$$Z = \{r1 \mid r \in F\}$$

14. Show that $\mathcal{L}_F(V)$ is simple if and only if $\dim(V) < \infty$.
15. Prove that for $n \geq 3$, the matrix algebras $\mathcal{M}_n(F)$ are central and simple.
16. An element $a \in A$ is **left-invertible** if there is a $b \in A$ for which $ba = 1$, in which case b is called a **left inverse** of a. Similarly, $a \in A$ is **right-invertible** if there is a $b \in A$ for which $ab = 1$, in which case b is called a **right inverse** of a. Left and right inverses are called **one-sided inverses** and an ordinary inverse is called a **two-sided inverse**. Let $a \in A$ be algebraic over F.
    a) Prove that $ab = 0$ for some $b \neq 0$ if and only if $ca = 0$ for some $c \neq 0$. Does c necessarily equal b?
    b) Prove that if a has a one-sided inverse b, then b is a two-sided inverse. Does this hold if a is not algebraic? *Hint*: Consider the algebra $A = \mathcal{L}_F(F[x])$.
    c) Let $a, b \in A$ be algebraic. Show that ab is invertible if and only if a and b are invertible, in which case ba is also invertible.

Chapter 19
The Umbral Calculus

In this chapter, we give a brief introduction to an area called the *umbral calculus*. This is a linear-algebraic theory used to study certain types of polynomial functions that play an important role in applied mathematics. We give only a brief introduction to the subject, emphasizing the algebraic aspects rather than the applications. For more on the umbral calculus, may we suggest *The Umbral Calculus*, by Roman [1984]?

One bit of notation: The **lower factorial numbers** are defined by

$$(n)_k = n(n-1)\cdots(n-k+1)$$

Formal Power Series

We begin with a few remarks concerning formal power series. Let $\mathcal{F}$ denote the algebra of formal power series in the variable t, with complex coefficients. Thus, $\mathcal{F}$ is the set of all formal sums of the form

$$f(t) = \sum_{k=0}^{\infty} a_k t^k \tag{19.1}$$

where $a_k \in \mathbb{C}$ (the complex numbers). Addition and multiplication are purely formal:

$$\sum_{k=0}^{\infty} a_k t^k + \sum_{k=0}^{\infty} b_k t^k = \sum_{k=0}^{\infty} (a_k + b_k) t^k$$

and

$$\left(\sum_{k=0}^{\infty} a_k t^k\right)\left(\sum_{k=0}^{\infty} b_k t^k\right) = \sum_{k=0}^{\infty}\left(\sum_{j=0}^{k} a_j b_{k-j}\right) t^k$$

The **order** $o(f)$ of f is the smallest exponent of t that appears with a nonzero coefficient. The order of the zero series is defined to be $+\infty$. Note that a series

f has a multiplicative inverse, denoted by f^{-1}, if and only if $o(f) = 0$. We leave it to the reader to show that

$$o(fg) = o(f) + o(g)$$

and

$$o(f+g) \geq \min\{o(f), o(g)\}$$

If f_k is a sequence in $\mathcal{F}$ with $o(f_k) \to \infty$ as $k \to 0$, then for any series

$$g(t) = \sum_{k=0}^{\infty} b_k t^k$$

we may substitute f_k for t^k to get the series

$$h(t) = \sum_{k=0}^{\infty} b_k f_k(t)$$

which is well-defined since the coefficient of each power of t is a finite sum. In particular, if $o(f) \geq 1$, then $o(f^k) \to \infty$ and so the **composition**

$$(g \circ f)(t) = g(f(t)) = \sum_{k=0}^{\infty} b_k f^k(t)$$

is well-defined. It is easy to see that $o(g \circ f) = o(g)o(f)$.

If $o(f) = 1$, then f has a compositional inverse, denoted by $\overline{f}$ and satisfying

$$(f \circ \overline{f})(t) = (\overline{f} \circ f)(t) = t$$

A series f with $o(f) = 1$ is called a **delta series.**

The sequence of powers f^k of a delta series f forms a **pseudobasis** for $\mathcal{F}$, in the sense that for any $g \in \mathcal{F}$, there exists a unique sequence of constants a_k for which

$$g(t) = \sum_{k=0}^{\infty} a_k f^k(t)$$

Finally, we note that the formal derivative of the series (19.1) is given by

$$\partial_t f(t) = f'(t) = \sum_{k=1}^{\infty} k a_k t^{k-1}$$

The operator ∂_t is a derivation, that is,

$$\partial_t(fg) = \partial_t(f)g + f\partial_t(g)$$

The Umbral Algebra

Let $\mathcal{P} = \mathbb{C}[x]$ denote the algebra of polynomials in a single variable x over the complex field. One of the starting points of the umbral calculus is the fact that any formal power series in $\mathcal{F}$ can play three different roles: as a formal power series, as a linear functional on $\mathcal{P}$ and as a linear operator on $\mathcal{P}$. Let us first explore the connection between formal power series and linear functionals.

Let $\mathcal{P}^*$ denote the vector space of all linear functionals on $\mathcal{P}$. Note that $\mathcal{P}^*$ is the algebraic dual space of $\mathcal{P}$, as defined in Chapter 2. It will be convenient to denote the action of $L \in \mathcal{P}^*$ on $p(x) \in \mathcal{P}$ by

$$\langle L \mid p(x) \rangle$$

(This is the "bra-ket" notation of Paul Dirac.) The vector space operations on $\mathcal{P}^*$ then take the form

$$\langle L + M \mid p(x) \rangle = \langle L \mid p(x) \rangle + \langle M \mid p(x) \rangle$$

and

$$\langle rL \mid p(x) \rangle = r\langle L \mid p(x) \rangle,\ r \in \mathbb{C}$$

Note also that since any linear functional on $\mathcal{P}$ is uniquely determined by its values on a basis for $\mathcal{P}$, the functional $L \in \mathcal{P}^*$ is uniquely determined by the values $\langle L \mid x^n \rangle$ for $n \geq 0$.

Now, any formal series in $\mathcal{F}$ can be written in the form

$$f(t) = \sum_{k=0}^{\infty} \frac{a_k}{k!} t^k$$

and we can use this to define a linear functional $f(t)$ by setting

$$\langle f(t) \mid x^n \rangle = a_n$$

for $n \geq 0$. In other words, the linear functional $f(t)$ is defined by

$$f(t) = \sum_{k=0}^{\infty} \frac{\langle f(t) \mid x^k \rangle}{k!} t^k$$

where the expression $f(t)$ on the left is just a formal power series. Note in particular that

$$\langle t^k \mid x^n \rangle = n!\delta_{n,k}$$

where $\delta_{n,k}$ is the Kronecker delta function. This implies that

$$\langle t^k \mid p(x) \rangle = p^{(k)}(0)$$

and so t^k is the functional "kth derivative at 0." Also, t^0 is evaluation at 0.

As it happens, any linear functional L on $\mathcal{P}$ has the form $f(t)$. To see this, we simply note that if

$$f_L(t) = \sum_{k=0}^{\infty} \frac{\langle L \mid x^k \rangle}{k!} t^k$$

then

$$\langle f_L(t) \mid x^n \rangle = \langle L \mid x^n \rangle$$

for all $n \geq 0$ and so as linear functionals, $L = f_L(t)$.

Thus, we can define a map $\phi: \mathcal{P}^* \to \mathcal{F}$ by $\phi(L) = f_L(t)$.

Theorem 19.1 *The map* $\phi: \mathcal{P}^* \to \mathcal{F}$ *defined by* $\phi(L) = f_L(t)$ *is a vector space isomorphism from* $\mathcal{P}^*$ *onto* $\mathcal{F}$.

Proof. To see that ϕ is injective, note that

$$f_L(t) = f_M(t) \Rightarrow \langle L \mid x^n \rangle = \langle M \mid x^n \rangle \text{ for all } n \geq 0 \Rightarrow L = M$$

Moreover, the map ϕ is surjective, since for any $f \in \mathcal{F}$, the linear functional $L = f(t)$ has the property that $\phi(L) = f_L(t) = f(t)$. Finally,

$$\begin{aligned} \phi(rL + sM) &= \sum_{k=0}^{\infty} \frac{\langle rL + sM \mid x^k \rangle}{k!} t^k \\ &= r\sum_{k=0}^{\infty} \frac{\langle L \mid x^k \rangle}{k!} t^k + s\sum_{k=0}^{\infty} \frac{\langle M \mid x^k \rangle}{k!} t^k \\ &= r\phi(L) + s\phi(M) \end{aligned}$$

□

From now on, we shall identify the vector space $\mathcal{P}^*$ with the vector space $\mathcal{F}$, using the isomorphism $\phi: \mathcal{P}^* \to \mathcal{F}$. Thus, we think of linear functionals on $\mathcal{P}$ simply as formal power series. The advantage of this approach is that $\mathcal{F}$ is more than just a vector space—it is an algebra. Hence, we have automatically defined a multiplication of linear functionals, namely, the product of formal power series. The algebra $\mathcal{F}$, when thought of as both the algebra of formal power series and the algebra of linear functionals on $\mathcal{P}$, is called the **umbral algebra**.

Let us consider an example.

Example 19.1 For $a \in \mathbb{C}$, the **evaluation functional** $\epsilon_a \in \mathcal{P}^*$ is defined by

$$\langle \epsilon_a \mid p(x) \rangle = p(a)$$

In particular, $\langle \epsilon_a \mid x^n \rangle = a^n$ and so the formal power series representation for this functional is

$$f_{\epsilon_a}(t) = \sum_{k=0}^{\infty} \frac{\langle \epsilon_a \mid x^k \rangle}{k!} t^k = \sum_{k=0}^{\infty} \frac{a^k}{k!} t^k = e^{at}$$

which is the exponential series. If e^{bt} is evaluation at b, then

$$e^{at} e^{bt} = e^{(a+b)t}$$

and so the product of evaluation at a and evaluation at b is evaluation at $a + b$.$\square$

When we are thinking of a delta series $f \in \mathcal{F}$ as a linear functional, we refer to it as a **delta functional**. Similarly, an invertible series $f \in \mathcal{F}$ is referred to as an **invertible functional**. Here are some simple consequences of the development so far.

Theorem 19.2

1) *For any $f \in \mathcal{F}$,*

$$f(t) = \sum_{k=0}^{\infty} \frac{\langle f(t) \mid x^k \rangle}{k!} t^k$$

2) *For any $p \in \mathcal{P}$,*

$$p(x) = \sum_{k \geq 0} \frac{\langle t^k \mid p(x) \rangle}{k!} x^k$$

3) *For any $f, g \in \mathcal{F}$,*

$$\langle f(t)g(t) \mid x^n \rangle = \sum_{k=0}^{n} \binom{n}{k} \langle f(t) \mid x^k \rangle \langle g(t) \mid x^{n-k} \rangle$$

4) $o(f(t)) > \deg p(x) \Rightarrow \langle f(t) \mid p(x) \rangle = 0$
5) *If $o(f_k) = k$ for all $k \geq 0$, then*

$$\left\langle \sum_{k=0}^{\infty} a_k f_k(t) \middle| p(x) \right\rangle = \sum_{k \geq 0} a_k \langle f_k(t) \mid p(x) \rangle$$

where the sum on the right is a finite one.

6) *If $o(f_k) = k$ for all $k \geq 0$, then*

$$\langle f_k(t) \mid p(x) \rangle = \langle f_k(t) \mid q(x) \rangle \text{ for all } k \geq 0 \Rightarrow p(x) = q(x)$$

7) *If $\deg p_k(x) = k$ for all $k \geq 0$, then*

$$\langle f(t) \mid p_k(x) \rangle = \langle g(t) \mid p_k(x) \rangle \text{ for all } k \geq 0 \Rightarrow f(t) = g(t)$$

Proof. We prove only part 3). Let

$$f(t) = \sum_{k=0}^{\infty} \frac{a_k}{k!} t^k \text{ and } g(t) = \sum_{j=0}^{\infty} \frac{b_j}{j!} t^j$$

Then

$$f(t)g(t) = \sum_{m=0}^{\infty} \left(\frac{1}{m!} \sum_{k=0}^{m} \binom{m}{k} a_k b_{m-k} \right) t^m$$

and applying both sides of this (as linear functionals) to x^n gives

$$\langle f(t)g(t) \mid x^n \rangle = \sum_{k=0}^{n} \binom{n}{k} a_k b_{n-k}$$

The result now follows from the fact that part 1) implies $a_k = \langle f(t) \mid x^k \rangle$ and $b_{n-k} = \langle g(t) \mid x^{n-k} \rangle$.□

We can now present our first "umbral" result.

Theorem 19.3 For any $f(t) \in \mathcal{F}$ and $p(x) \in \mathcal{P}$,

$$\langle f(t) \mid xp(x) \rangle = \langle \partial_t f(t) \mid p(x) \rangle$$

Proof. By linearity, we need only establish this for $p(x) = x^n$. But if

$$f(t) = \sum_{k=0}^{\infty} \frac{a_k}{k!} t^k$$

then

$$\begin{aligned} \langle \partial_t f(t) \mid x^n \rangle &= \left\langle \sum_{k=1}^{\infty} \frac{a_k}{(k-1)!} t^{k-1} \middle| x^n \right\rangle \\ &= \sum_{k=1}^{\infty} \frac{a_k}{(k-1)!} \delta_{k-1,n} \\ &= a_{n+1} \\ &= \langle f(t) \mid x^{n+1} \rangle \end{aligned}$$

□

Let us consider a few examples of important linear functionals and their power series representations.

Example 19.2

1) We have already encountered the **evaluation functional** e^{at}, satisfying

$$\langle e^{at} \mid p(x) \rangle = p(a)$$

2) The **forward difference functional** is the delta functional $e^{at} - 1$, satisfying

$$\langle e^{at} - 1 \mid p(x) \rangle = p(a) - p(0)$$

3) The **Abel functional** is the delta functional te^{at}, satisfying

$$\langle te^{at} \mid p(x) \rangle = p'(a)$$

4) The invertible functional $(1-t)^{-1}$ satisfies

$$\langle (1-t)^{-1} \mid p(x) \rangle = \int_0^\infty p(u)e^{-u}\, du$$

as can be seen by setting $p(x) = x^n$ and expanding the expression $(1-t)^{-1}$.

5) To determine the linear functional f satisfying

$$\langle f(t) \mid p(x) \rangle = \int_0^a p(u)\, du$$

we observe that

$$f(t) = \sum_{k=0}^{\infty} \frac{\langle f(t) \mid x^k \rangle}{k!} t^k = \sum_{k=0}^{\infty} \frac{a^{k+1}}{(k+1)!} t^k = \frac{eat^{at} - 1}{t}$$

The inverse $t/(e^{at} - 1)$ of this functional is associated with the Bernoulli polynomials, which play a very important role in mathematics and its applications. In fact, the numbers

$$B_n = \left\langle \frac{t}{e^{at} - 1} \middle| x^n \right\rangle$$

are known as the **Bernoulli numbers.**□

Formal Power Series as Linear Operators

We now turn to the connection between formal power series and linear operators on $\mathcal{P}$. Let us denote the kth derivative operator on $\mathcal{P}$ by t^k. Thus,

$$t^k p(x) = p^{(k)}(x)$$

We can then extend this to formal series in t,

$$f(t) = \sum_{k=0}^{\infty} \frac{a_k}{k!} t^k \tag{19.2}$$

by defining the linear operator $f(t)\colon \mathcal{P} \to \mathcal{P}$ by

$$f(t)p(x) = \sum_{k=0}^{\infty} \frac{a_k}{k!}[t^k p(x)] = \sum_{k\geq 0} \frac{a_k}{k!} p^{(k)}(x)$$

the latter sum being a finite one. Note in particular that

$$f(t)x^n = \sum_{k=0}^{n} \binom{n}{k} a_k x^{n-k} \tag{19.3}$$

With this definition, we see that each formal power series $f \in \mathcal{F}$ plays three roles in the umbral calculus, namely, as a formal power series, as a linear functional and as a linear operator. The two notations $\langle f(t) \mid p(x)\rangle$ and $f(t)p(x)$ will make it clear whether we are thinking of f as a functional or as an operator.

It is important to note that $f = g$ in $\mathcal{F}$ if and only if $f = g$ as linear functionals, which holds if and only if $f = g$ as linear operators. It is also worth noting that

$$[f(t)g(t)]p(x) = f(t)[g(t)p(x)]$$

and so we may write $f(t)g(t)p(x)$ without ambiguity. In addition,

$$f(t)g(t)p(x) = g(t)f(t)p(x)$$

for all $f, g \in \mathcal{F}$ and $p \in \mathcal{P}$.

When we are thinking of a delta series f as an operator, we call it a **delta operator**. The following theorem describes the key relationship between linear functionals and linear operators of the form $f(t)$.

Theorem 19.4 *If $f, g \in \mathcal{F}$, then*

$$\langle f(t)g(t) \mid p(x)\rangle = \langle f(t) \mid g(t)p(x)\rangle$$

for all polynomials $p(x) \in \mathcal{P}$.

Proof. If f has the form (19.2), then by (19.3),

$$\langle t^0 \mid f(t)x^n\rangle = \left\langle t^0 \middle| \sum_{k=0}^{n} \binom{n}{k} a_k x^{n-k} \right\rangle = a_n = \langle f(t) \mid x^n\rangle \tag{19.4}$$

By linearity, this holds for x^n replaced by any polynomial $p(x)$. Hence, applying this to the product fg gives

$$\begin{aligned}\langle f(t)g(t) \mid p(x)\rangle &= \langle t^0 \mid f(t)g(t)p(x)\rangle \\ &= \langle t^0 \mid f(t)[g(t)p(x)]\rangle = \langle f(t) \mid g(t)p(x)\rangle\end{aligned}$$

□

Equation (19.4) shows that applying the linear functional $f(t)$ is equivalent to applying the operator $f(t)$ and then following by evaluation at $x = 0$.

Here are the operator versions of the functionals in Example 19.2.

Example 19.3

1) The operator e^{at} satisfies

$$e^{at}x^n = \sum_{k=0}^{\infty} \frac{a^k}{k!} t^k x^n = \sum_{k=0}^{n} \binom{n}{k} a^k x^{n-k} = (x+a)^n$$

and so

$$e^{at}p(x) = p(x+a)$$

for all $p \in \mathcal{P}$. Thus e^{at} is a **translation operator**.

2) The **forward difference operator** is the delta operator $e^{at} - 1$, where

$$(e^{at} - 1)p(x) = p(x+a) - p(a)$$

3) The **Abel operator** is the delta operator te^{at}, where

$$te^{at}p(x) = p'(x+a)$$

4) The invertible operator $(1-t)^{-1}$ satisfies

$$(1-t)^{-1}p(x) = \int_0^{\infty} p(x+u)e^{-u}du$$

5) The operator $(e^{at} - 1)/t$ is easily seen to satisfy

$$\frac{e^{at} - 1}{t} p(x) = \int_x^{x+a} p(u)\, du$$ □

We have seen that all linear functionals on $\mathcal{P}$ have the form $f(t)$, for $f \in \mathcal{F}$. However, not all linear operators on $\mathcal{P}$ have this form. To see this, observe that

$$\deg\,[f(t)p(x)] \le \deg p(x)$$

but the linear operator $\phi: \mathcal{P} \to \mathcal{P}$ defined by $\phi(p(x)) = xp(x)$ does not have this property.

Let us characterize the linear operators of the form $f(t)$. First, we need a lemma.

Lemma 19.5 *If T is a linear operator on P and $Tf(t) = f(t)T$ for some delta series $f(t)$, then* $\deg(Tp(x)) \le \deg(p(x))$.

Proof. For any $m \ge 0$,

$$\deg(Tx^m) - 1 = \deg(f(t)Tx^m) = \deg(Tf(t)x^m)$$

and so

$$\deg(Tx^m) = \deg(Tf(t)x^m) + 1$$

Since $\deg(f(t)x^m) = m - 1$ we have the basis for an induction. When $m = 0$ we get $\deg(T1) = 1$. Assume that the result is true for $m - 1$. Then

$$\deg(Tx^m) = \deg(Tf(t)x^m) + 1 \le m - 1 + 1 = m$$

□

Theorem 19.6 *The following are equivalent for a linear operator* $T: \mathcal{P} \to \mathcal{P}$.

1) T *has the form* $f(t)$, *that is, there exists an* $f \in \mathcal{F}$ *for which* $T = f(t)$, *as linear operators.*
2) T *commutes with the derivative operator, that is,* $Tt = tT$.
3) T *commutes with any delta operator* $g(t)$, *that is,* $Th(t) = h(t)T$.
4) T *commutes with any translation operator, that is,* $Te^{at} = e^{at}T$.

Proof. It is clear that 1) implies 2). For the converse, let

$$g(t) = \sum_{k=0}^{\infty} \frac{\langle t^0 \mid Tx^k \rangle}{k!} t^k$$

Then

$$\langle g(t) \mid x^n \rangle = \langle t^0 \mid Tx^k \rangle$$

Now, since T commutes with t, we have

$$\begin{aligned}
\langle t^n \mid Tx^k \rangle &= \langle t^0 \mid t^n Tx^k \rangle \\
&= \langle t^0 \mid Tt^n x^k \rangle \\
&= (k)_n \langle t^0 \mid Tx^{k-n} \rangle \\
&= (k)_n \langle t^0 \mid g(t)x^{k-n} \rangle \\
&= \langle t^n \mid g(t)x^k \rangle
\end{aligned}$$

and since this holds for all n and k we get $T = g(t)$. We leave the rest of the proof as an exercise.□

Sheffer Sequences

We can now define the principal object of study in the umbral calculus. When referring to a sequence $s_n(x)$ in $\mathcal{P}$, we shall always assume that $\deg s_n(x) = n$ for all $n \ge 0$.

Theorem 19.7 *Let* f *be a delta series, let* g *be an invertible series and consider the geometric sequence*

$$g, gf, gf^2, gf^3, \ldots$$

in $\mathcal{F}$. *Then there is a unique sequence* $s_n(x)$ *in* $\mathcal{P}$ *satisfying the* **orthogonality conditions**

$$\langle g(t)f^k(t) \mid s_n(x)\rangle = n!\delta_{n,k} \tag{19.5}$$

for all $n, k \geq 0$.

Proof. The uniqueness follows from Theorem 19.2. For the existence, if we set

$$s_n(x) = \sum_{j=0}^{n} a_{n,j}x_j$$

and

$$g(t)f^k(t) = \sum_{i=k}^{\infty} b_{k,i}t^i$$

where $b_{k,k} \neq 0$, then (19.5) is

$$\begin{aligned} n!\delta_{n,k} &= \left\langle \sum_{i=k}^{\infty} b_{k,i}t^i \middle| \sum_{j=0}^{n} a_{n,j}x_j \right\rangle \\ &= \sum_{i=k}^{\infty}\sum_{j=0}^{n} b_{k,i}a_{n,j}\langle t^i \middle| x_j\rangle \\ &= \sum_{i=k}^{n} b_{k,i}a_{n,i}i! \end{aligned}$$

Taking $k = n$ we get

$$a_{n,n} = \frac{1}{b_{n,n}}$$

For $k = n - 1$ we have

$$0 = b_{n-1,n-1}a_{n,n-1}(n-1)! + b_{n-1,n}a_{n,n}n!$$

and using the fact that $a_{n,n} = 1/b_{n,n}$ we can solve this for $a_{n,n-1}$. By successively taking $k = n, n-1, n-2, \ldots$ we can solve the resulting equations for the coefficients $a_{n,k}$ of the sequence $s_n(x)$.□

Definition *The sequence* $s_n(x)$ *in (19.5) is called the* **Sheffer sequence** *for the ordered pair* $(g(t), f(t))$. *We shorten this by saying that* $s_n(x)$ *is* **Sheffer for** $(g(t), f(t))$.□

Two special types of Sheffer sequences deserve explicit mention.

Definition *The Sheffer sequence for a pair of the form* $(1, f(t))$ *is called the* **associated sequence** *for* $f(t)$. *The Sheffer sequence for a pair of the form* $(g(t), t)$ *is called the* **Appell sequence** *for* $g(t)$.□

Note that the sequence $s_n(x)$ is Sheffer for $(g(t), f(t))$ if and only if

$$\langle g(t)f^k(t) \mid s_n(x)\rangle = n!\delta_{n,k}$$

which is equivalent to

$$\langle f^k(t) \mid g(t)s_n(x)\rangle = n!\delta_{n,k}$$

which, in turn, is equivalent to saying that the sequence $p_n(x) = g(t)s_n(x)$ is the associated sequence for $f(t)$.

Theorem 19.8 *The sequence $s_n(x)$ is Sheffer for $(g(t), f(t))$ if and only if the sequence $p_n(x) = g(t)s_n(x)$ is the associated sequence for $f(t)$.*□

Before considering examples, we wish to describe several characterizations of Sheffer sequences. First, we require a key result.

Theorem 19.9 (The expansion theorems) *Let $s_n(x)$ be Sheffer for $(g(t), f(t))$.*
1) For any $h \in \mathcal{F}$,

$$h(t) = \sum_{k=0}^{\infty} \frac{\langle h(t) \mid s_k(x)\rangle}{k!} g(t)f^k(t)$$

2) For any $p \in \mathcal{P}$,

$$p(x) = \sum_{k\geq 0} \frac{\langle g(t)f^k(t) \mid p(x)\rangle}{k!} s_k(x)$$

Proof. Part 1) follows from Theorem 19.2, since

$$\begin{aligned}\left\langle \sum_{k=0}^{\infty} \frac{\langle h(t) \mid s_k(x)\rangle}{k!} g(t)f^k(t) \middle| s_n(x) \right\rangle &= \sum_{k=0}^{\infty} \frac{\langle h(t) \mid s_k(x)\rangle}{k!} n!\delta_{n,k} \\ &= \langle h(t) \mid s_n(x)\rangle\end{aligned}$$

Part 2) follows in a similar way from Theorem 19.2.□

We can now begin our characterization of Sheffer sequences, starting with the generating function. The idea of a generating function is quite simple. If $r_n(x)$ is a sequence of polynomials, we may define a formal power series of the form

$$g(t, x) = \sum_{k=0}^{\infty} \frac{r_k(x)}{k!} t^k$$

This is referred to as the **(exponential) generating function** for the sequence $r_n(x)$. (The term exponential refers to the presence of $k!$ in this series. When this is not present, we have an ordinary generating function.) Since the series is a formal one, knowing $g(t, x)$ is equivalent (in theory, if not always in practice) to knowing the polynomials $r_n(x)$. Moreover, a knowledge of the generating function of a sequence of polynomials can often lead to a deeper understanding of the sequence itself, that might not be otherwise easily accessible. For this reason, generating functions are studied quite extensively.

For the proofs of the following characterizations, we refer the reader to Roman [1984].

Theorem 19.10 (Generating function)

1) *The sequence $p_n(x)$ is the associated sequence for a delta series $f(t)$ if and only if*

$$e^{y\overline{f}(t)} = \sum_{k=0}^{\infty} \frac{p_k(y)}{k!} t^k$$

where $\overline{f}(t)$ is the compositional inverse of $f(t)$.

2) *The sequence $s_n(x)$ is Sheffer for $(g(t), f(t))$ if and only if*

$$\frac{1}{g(\overline{f}(t))} e^{y\overline{f}(t)} = \sum_{k=0}^{\infty} \frac{s_k(y)}{k!} t^k$$

The sum on the right is called the **generating function** *of $s_n(x)$.*

Proof. Part 1) is a special case of part 2). For part 2), the expression above is equivalent to

$$\frac{1}{g(t)} e^{yt} = \sum_{k=0}^{\infty} \frac{s_k(y)}{k!} f^k(t)$$

which is equivalent to

$$e^{yt} = \sum_{k=0}^{\infty} \frac{s_k(y)}{k!} g(t) f^k(t)$$

But if $s_n(x)$ is Sheffer for $(f(t), g(t))$, then this is just the expansion theorem for e^{yt}. Conversely, this expression implies that

$$s_n(y) = \langle e^{yt} \mid s_n(x) \rangle = \sum_{k=0}^{\infty} \frac{s_k(y)}{k!} \langle g(t) f^{\,k}(t) \mid s_n(x) \rangle$$

and so $\langle g(t) f^{\,k}(t) \mid s_n(x) \rangle = n!\delta_{n,k}$, which says that $s_n(x)$ is Sheffer for (f, g).$\square$

We can now give a representation for Sheffer sequences.

Theorem 19.11 (Conjugate representation)

1) *A sequence $p_n(x)$ is the associated sequence for $f(t)$ if and only if*

$$p_n(x) = \sum_{k=0}^{n} \frac{1}{k!} \langle \overline{f}(t)^k \mid x^n \rangle x^k$$

2) *A sequence $s_n(x)$ is Sheffer for $(g(t), f(t))$ if and only if*

$$s_n(x) = \sum_{k=0}^{n} \frac{1}{k!} \langle g(\overline{f}(t))^{-1} \overline{f}(t)^k \mid x^n \rangle x^k$$

Proof. We need only prove part 2). We know that $s_n(x)$ is Sheffer for $(g(t), f(t))$ if and only if

$$\frac{1}{g(\overline{f}(t))} e^{y\overline{f}(t)} = \sum_{k=0}^{\infty} \frac{s_k(y)}{k!} t^k$$

But this is equivalent to

$$\left\langle \frac{1}{g(\overline{f}(t))} e^{y\overline{f}(t)} \mid x^n \right\rangle = \left\langle \sum_{k=0}^{\infty} \frac{s_k(y)}{k!} t^k \middle| x^n \right\rangle = s_n(y)$$

Expanding the exponential on the left gives

$$\sum_{k=0}^{\infty} \frac{\langle g(\overline{f}(t))^{-1} \overline{f}(t)^k \mid x^n \rangle}{k!} y^k = \left\langle \sum_{k=0}^{\infty} \frac{s_k(y)}{k!} t^k \middle| x^n \right\rangle = s_n(y)$$

Replacing y by x gives the result.□

Sheffer sequences can also be characterized by means of linear operators.

Theorem 19.12 (Operator characterization)

1) *A sequence $p_n(x)$ is the associated sequence for $f(t)$ if and only if*
   a) $p_n(0) = \delta_{n,0}$
   b) $f(t)p_n(x) = np_{n-1}(x)$ *for* $n \geq 0$
2) *A sequence $s_n(x)$ is Sheffer for $(g(t), f(t))$ for some invertible series $g(t)$ if and only if*

$$f(t)s_n(x) = ns_{n-1}(x)$$

for all $n \geq 0$.

Proof. For part 1), if $p_n(x)$ is associated with $f(t)$, then

$$p_n(0) = \langle e^{0t} \mid p_n(x) \rangle = \langle f(t)^0 \mid p_n(x) \rangle = 0!\delta_{n,0}$$

and

$$\begin{aligned}\langle f(t)^k \mid f(t)p_n(x)\rangle &= \langle f(t)^{k+1} \mid p_n(x)\rangle \\ &= n!\delta_{n,k+1} \\ &= n(n-1)!\delta_{n-1,k} \\ &= n\langle f(t)^k \mid p_{n-1}(x)\rangle\end{aligned}$$

and since this holds for all $k \geq 0$ we get 1b). Conversely, if 1a) and 1b) hold, then

$$\begin{aligned}\langle f(t)^k \mid p_n(x)\rangle &= \langle t^0 \mid f(t)^k p_n(x)\rangle \\ &= (n)_k p_{n-k}(0) \\ &= (n)_k \delta_{n-k,0} \\ &= n!\delta_{n,k}\end{aligned}$$

and so $p_n(x)$ is the associated sequence for $f(t)$.

As for part 2), if $s_n(x)$ is Sheffer for $(g(t), f(t))$, then

$$\begin{aligned}\langle g(t)f(t)^k \mid f(t)s_n(x)\rangle &= \langle g(t)f(t)^{k+1} \mid s_n(x)\rangle \\ &= n!\delta_{n,k+1} \\ &= n(n-1)!\delta_{n-1,k} \\ &= n\langle g(t)f(t)^k \mid s_{n-1}(x)\rangle\end{aligned}$$

and so $f(t)s_n(x) = ns_{n-1}(x)$, as desired. Conversely, suppose that

$$f(t)s_n(x) = ns_{n-1}(x)$$

and let $p_n(x)$ be the associated sequence for $f(t)$. Let T be the invertible linear operator on V defined by

$$Ts_n(x) = p_n(x)$$

Then

$$Tf(t)s_n(x) = nTs_{n-1}(x) = np_{n-1}(x) = f(t)p_n(x) = f(t)Ts_n(x)$$

and so Lemma 19.5 implies that $T = g(t)$ for some invertible series $g(t)$. Then

$$\begin{aligned}\langle g(t)f(t)^k \mid s_n(x)\rangle &= \langle f(t)^k \mid g(t)s_n(x)\rangle \\ &= \langle t^0 \mid f(t)^k p_n(x)\rangle \\ &= (n)_k p_{n-k}(0) \\ &= (n)_k \delta_{n-k,0} \\ &= n!\delta_{n,k}\end{aligned}$$

and so $s_n(x)$ is Sheffer for $(g(t), f(t))$.□

We next give a formula for the action of a linear operator $h(t)$ on a Sheffer sequence.

Theorem 19.13 *Let $s_n(x)$ be a Sheffer sequence for $(g(t), f(t))$ and let $p_n(x)$ be associated with $f(t)$. Then for any $h(t)$ we have*

$$h(t)s_n(x) = \sum_{k=0}^{n} \binom{n}{k} \langle h(t) \mid s_k(x) \rangle p_{n-k}(x)$$

Proof. By the expansion theorem

$$h(t) = \sum_{k=0}^{\infty} \frac{\langle h(t) \mid s_k(x) \rangle}{k!} g(t) f^k(t)$$

we have

$$\begin{aligned} h(t)s_n(x) &= \sum_{k=0}^{\infty} \frac{\langle h(t) \mid s_k(x) \rangle}{k!} g(t) f^k(t) s_n(x) \\ &= \sum_{k=0}^{\infty} \frac{\langle h(t) \mid s_k(x) \rangle}{k!} (n)_k p_{n-k}(x) \end{aligned}$$

which is the desired formula.□

Theorem 19.14

1) **(The binomial identity)** *A sequence $p_n(x)$ is the associated sequence for a delta series $f(t)$ if and only if it is of* **binomial type**, *that is, if and only if it satisfies the identity*

$$p_n(x+y) = \sum_{k=0}^{n} \binom{n}{k} p_k(y) p_{n-k}(x)$$

for all $y \in \mathbb{C}$.

2) **(The Sheffer identity)** *A sequence $s_n(x)$ is Sheffer for $(g(t), f(t))$ for some invertible $g(t)$ if and only if*

$$s_n(x+y) = \sum_{k=0}^{n} \binom{n}{k} p_k(y) s_{n-k}(x)$$

for all $y \in \mathbb{C}$, where $p_n(x)$ is the associated sequence for $f(t)$.

Proof. To prove part 1), if $p_n(x)$ is an associated sequence, then taking $h(t) = e^{yt}$ in Theorem 19.13 gives the binomial identity. Conversely, suppose that the sequence $p_n(x)$ is of binomial type. We will use the operator characterization to show that $p_n(x)$ is an associated sequence. Taking $x = y = 0$ we have for $n = 0$,

$$p_0(0) = p_0(0)p_0(0)$$

and so $p_0(0) = 1$. Also,

$$p_1(0) = p_0(0)p_1(0) + p_1(0)p_0(0) = 2p_1(0)$$

and so $p_1(0) = 0$. Assuming that $p_i(0) = 0$ for $i = 1, \ldots, m-1$ we have

$$p_m(0) = p_0(0)p_m(0) + p_m(0)p_0(0) = 2p_m(0)$$

and so $p_m(0) = 0$. Thus, $p_n(0) = \delta_{n,0}$.

Next, define a linear functional $f(t)$ by

$$\langle f(t) \mid p_n(x) \rangle = \delta_{n,1}$$

Since $\langle f(t) \mid 1 \rangle = \langle f(t) \mid p_0(x) \rangle = 0$ and $\langle f(t) \mid p_1(x) \rangle = 1 \neq 0$ we deduce that $f(t)$ is a delta series. Now, the binomial identity gives

$$\begin{aligned}\langle f(t) \mid e^{yt} p_n(x) \rangle &= \sum_{k=0}^{n} \binom{n}{k} p_k(y) \langle f(t) \mid p_{n-k}(x) \rangle \\ &= \sum_{k=0}^{n} \binom{n}{k} p_k(y) \delta_{n-k,1} \\ &= n p_{n-1}(y)\end{aligned}$$

and so

$$\langle e^{yt} \mid f(t) p_n(x) \rangle = \langle e^{yt} \mid n p_{n-1}(x) \rangle$$

and since this holds for all y, we get $f(t)p_n(x) = np_{n-1}(x)$. Thus, $p_n(x)$ is the associated sequence for $f(t)$.

For part 2), if $s_n(x)$ is a Sheffer sequence, then taking $h(t) = e^{yt}$ in Theorem 19.13 gives the Sheffer identity. Conversely, suppose that the Sheffer identity holds, where $p_n(x)$ is the associated sequence for $f(t)$. It suffices to show that $g(t)s_n(x) = p_n(x)$ for some invertible $g(t)$. Define a linear operator T by

$$Ts_n(x) = p_n(x)$$

Then

$$e^{yt} T s_n(x) = e^{yt} p_n(x) = p_n(x+y)$$

and by the Sheffer identity,

$$Te^{yt} s_n(x) = \sum_{k=0}^{n} \binom{n}{k} p_k(y) T s_{n-k}(x) = \sum_{k=0}^{n} \binom{n}{k} p_k(y) p_{n-k}(x)$$

and the two are equal by part 1). Hence, T commutes with e^{yt} and is therefore of the form $g(t)$, as desired.□

Examples of Sheffer Sequences

We can now give some examples of Sheffer sequences. While it is often a relatively straightforward matter to verify that a given sequence is Sheffer for a given pair $(g(t), f(t))$, it is quite another matter to find the Sheffer sequence for a given pair. The umbral calculus provides two formulas for this purpose, one of which is direct, but requires the usually very difficult computation of the series $(f(t)/t)^{-n}$. The other is a recurrence relation that expresses each $s_n(x)$ in terms of previous terms in the Sheffer sequence. Unfortunately, space does not permit us to discuss these formulas in detail. However, we will discuss the recurrence formula for associated sequences later in this chapter.

Example 19.4 The sequence $p_n(x) = x^n$ is the associated sequence for the delta series $f(t) = t$. The generating function for this sequence is

$$e^{yt} = \sum_{k=0}^{\infty} \frac{y^k}{k!} t^k$$

and the binomial identity is the well-known binomial formula

$$(x+y)^n = \sum_{k=0}^{n} \binom{n}{k} x^k y^{n-k}$$

Example 19.5 The **lower factorial polynomials**

$$(x)_n = x(x-1)\cdots(x-n+1)$$

form the associated sequence for the forward difference functional

$$f(t) = e^t - 1$$

discussed in Example 19.2. To see this, we simply compute, using Theorem 19.12. Since $(0)_0$ is defined to be 1, we have $(0)_n = \delta_{n,0}$. Also,

$$\begin{aligned}(e^t - 1)(x)_n &= (x+1)_n - (x)_n \\ &= [(x+1)x(x-1)\cdots(x-n+2)] - [x(x-1)\cdots(x-n+1)] \\ &= x(x-1)\cdots(x-n+2)[(x+1) - (x-n+1)] \\ &= nx(x-1)\cdots(x-n+2) \\ &= n(x)_{n-1}\end{aligned}$$

The generating function for the lower factorial polynomials is

$$e^{y\log(1+t)} = \sum_{k=0}^{\infty} \frac{(y)_k}{k!} t^k$$

which can be rewritten in the more familiar form

$$(1+t)^y = \sum_{k=0}^{\infty} \binom{y}{k} t^k$$

Of course, this is a formal identity, so there is no need to make any restrictions on t. The binomial identity in this case is

$$(x+y)_n = \sum_{k=0}^{n} \binom{n}{k} (x)_k (y)_{n-k}$$

which can also be written in the form

$$\binom{x+y}{n} = \sum_{k=0}^{n} \binom{x}{k} \binom{y}{n-k}$$

This is known as the **Vandermonde convolution formula**.

Example 19.6 The **Abel polynomials**

$$A_n(x;a) = x(x-an)^{n-1}$$

form the associated sequence for the **Abel functional**

$$f(t) = te^{at}$$

also discussed in Example 19.2. We leave verification of this to the reader. The generating function for the Abel polynomials is

$$e^{y\bar{f}(t)} = \sum_{k=0}^{\infty} \frac{y(y-ak)^{k-1}}{k!} t^k$$

Taking the formal derivative of this with respect to y gives

$$\bar{f}(t)e^{y\bar{f}(t)} = \sum_{k=0}^{\infty} \frac{k(y-a)(y-ak)^{k-1}}{k!} t^k$$

which, for $y=0$, gives a formula for the compositional inverse of the series $f(t) = te^{at}$,

$$\bar{f}(t) = \sum_{k=1}^{\infty} \frac{(-a)^k k^{k-1}}{(k-1)!} t^k$$

Example 19.7 The famous **Hermite polynomials** $H_n(x)$ form the Appell sequence for the invertible functional

$$g(t) = e^{t^2/2}$$

We ask the reader to show that $s_n(x)$ is the Appell sequence for $g(t)$ if and only if $s_n(x) = g(t)^{-1}x^n$. Using this fact, we get

$$H_n(x) = e^{-t^2/2}x^n = \sum_{k\geq 0}(-\frac{1}{2})^k \frac{(n)_{2k}}{k!} x^{n-k}$$

The generating function for the Hermite polynomials is

$$e^{yt-t^2/2} = \sum_{k=0}^{\infty} \frac{H_k(y)}{k!} t^k$$

and the Sheffer identity is

$$H_n(x+y) = \sum_{k=0}^{n} \binom{n}{k} H_k(x) y^{n-k}$$

We should remark that the Hermite polynomials, as defined in the literature, often differ from our definition by a multiplicative constant.□

Example 19.8 The well-known and important **Laguerre polynomials** $L_n^{(\alpha)}(x)$ of order α form the Sheffer sequence for the pair

$$g(t) = (1-t)^{-\alpha-1}, \ f(t) = \frac{t}{t-1}$$

It is possible to show (although we will not do so here) that

$$L_n^{(\alpha)}(x) = \sum_{k=0}^{n} \frac{n!}{k!} \binom{\alpha+n}{n-k} (-x)^k$$

The generating function of the Laguerre polynomials is

$$\frac{1}{(1-t)^{\alpha+1}} e^{yt/(t-1)} = \sum_{k=0}^{\infty} \frac{L_k^{(\alpha)}(x)}{k!} t^k$$

As with the Hermite polynomials, some definitions of the Laguerre polynomials differ by a multiplicative constant.□

We presume that the few examples we have given here indicate that the umbral calculus applies to a significant range of important polynomial sequences. In Roman [1984], we discuss approximately 30 different sequences of polynomials that are (or are closely related to) Sheffer sequences.

Umbral Operators and Umbral Shifts

We have now established the basic framework of the umbral calculus. As we have seen, the umbral algebra plays three roles: as the algebra of formal power series in a single variable, as the algebra of all linear functionals on $\mathcal{P}$ and as the algebra of all linear operators on $\mathcal{P}$ that commute with the derivative operator. Moreover, since $\mathcal{F}$ is an algebra, we can consider geometric sequences

$$g,\ gf, gf^2, gf^3, \ldots$$

in $\mathcal{F}$, where $o(g) = 0$ and $o(f) = 1$. We have seen by example that the orthogonality conditions

$$\langle g(t) f^k(t) \mid s_n(x) \rangle = n! \delta_{n,k}$$

define important families of polynomial sequences.

While the machinery that we have developed so far does unify a number of topics from the classical study of polynomial sequences (for example, special cases of the expansion theorem include Taylor's expansion, the Euler–MacLaurin formula and Boole's summation formula), it does not provide much new insight into their study. Our plan now is to take a brief look at some of the deeper results in the umbral calculus, which center on the interplay between operators on $\mathcal{P}$ and their adjoints, which are operators on the umbral algebra $\mathcal{F} = \mathcal{P}^*$.

We begin by defining two important operators on $\mathcal{P}$ associated with each Sheffer sequence.

Definition *Let $s_n(x)$ be Sheffer for $(g(t), f(t))$. The linear operator $\lambda_{g,f}: \mathcal{P} \to \mathcal{P}$ defined by*

$$\lambda_{g,f}(x^n) = s_n(x)$$

is called the **Sheffer operator** *for the pair $(g(t), f(t))$, or for the sequence $s_n(x)$. If $p_n(x)$ is the associated sequence for $f(t)$, the Sheffer operator*

$$\lambda_f(x^n) = p_n(x)$$

is called the **umbral operator** *for $f(t)$, or for $p_n(x)$.*$\square$

Definition *Let $s_n(x)$ be Sheffer for $(g(t), f(t))$. The linear operator $\theta_{g,f}: \mathcal{P} \to \mathcal{P}$ defined by*

$$\theta_{g,f}[s_n(x)] = s_{n+1}(x)$$

is called the **Sheffer shift** *for the pair $(g(t), f(t))$, or for the sequence $s_n(x)$. If $p_n(x)$ is the associated sequence for $f(t)$, the Sheffer operator*

$$\theta_f[p_n(x)] = p_{n+1}(x)$$

is called the **umbral shift** *for $f(t)$, or for $p_n(x)$.*$\square$

It is clear that each Sheffer sequence uniquely determines a Sheffer operator and vice versa. Hence, knowing the Sheffer operator of a sequence is equivalent to knowing the sequence.

Continuous Operators on the Umbral Algebra

It is clearly desirable that a linear operator T on the umbral algebra $\mathcal{F}$ pass under infinite sums, that is, that

$$T\left(\sum_{k=0}^{\infty} a_k f_k(t)\right) = \sum_{k=0}^{\infty} a_k T[f_k(t)] \tag{19.6}$$

whenever the sum on the left is defined, which is precisely when $o(f_k(t)) \to \infty$ as $k \to \infty$. Not all operators on $\mathcal{F}$ have this property, which leads to the following definition.

Definition *A linear operator T on the umbral algebra $\mathcal{F}$ is* **continuous** *if it satisfies* (19.6).□

The term continuous can be justified by defining a topology on $\mathcal{F}$. However, since no additional topological concepts will be needed, we will not do so here. Note that in order for (19.6) to make sense, we must have $o(T[f_k(t)]) \to \infty$. It turns out that this condition is also sufficient.

Theorem 19.15 A linear operator T on $\mathcal{F}$ is continuous if and only if

$$o(f_k) \to \infty \Rightarrow o(T(f_k)) \to \infty \tag{19.7}$$

Proof. The necessity is clear. Suppose that (19.7) holds and that $o(f_k) \to \infty$. For any $m \geq 0$, we have

$$\left\langle T\sum_{k=0}^{\infty} a_k f_k(t) \middle| x^n \right\rangle = \left\langle T\sum_{k=0}^{m} a_k f_k(t) \middle| x^n \right\rangle + \left\langle T\sum_{k>m} a_k f_k(t) \middle| x^n \right\rangle \tag{19.8}$$

Since

$$o\left(\sum_{k>m} a_k f_k(t)\right) \to \infty$$

(19.7) implies that we may choose m large enough that

$$o\left(T\sum_{k>m} a_k f_k(t)\right) > n$$

and

$$o(T[f_k(t)]) > n \text{ for } k > m$$

Hence, (19.8) gives

$$\begin{aligned}\left\langle T\sum_{k=0}^{\infty} a_k f_k(t) \middle| x^n \right\rangle &= \left\langle T\sum_{k=0}^{m} a_k f_k(t) \middle| x^n \right\rangle \\ &= \left\langle \sum_{k=0}^{m} a_k T[f_k(t)] \middle| x^n \right\rangle \\ &= \left\langle \sum_{k=0}^{\infty} a_k T[f_k(t)] \middle| x^n \right\rangle\end{aligned}$$

which implies the desired result.□

Operator Adjoints

If $\tau: \mathcal{P} \to \mathcal{P}$ is a linear operator on $\mathcal{P}$, then its (operator) adjoint $\tau^\times$ is an operator on $\mathcal{P}^* = \mathcal{F}$ defined by

$$\tau^\times[h(t)] = h(t) \circ \tau$$

In the symbolism of the umbral calculus, this is

$$\langle \tau^\times h(t) \mid p(x) \rangle = \langle h(t) \mid \tau p(x) \rangle$$

(We have reduced the number of parentheses used to aid clarity.)

Let us recall the basic properties of the adjoint from Chapter 3.

Theorem 19.16 *For $\tau, \sigma \in \mathcal{L}(\mathcal{P})$,*
1) $(\tau + \sigma)^\times = \tau^\times + \sigma^\times$
2) $(r\tau)^\times = r\tau^\times$ *for any $r \in \mathbb{C}$*
3) $(\tau\sigma)^\times = \sigma^\times \tau^\times$
4) $(\tau^{-1})^\times = (\tau^\times)^{-1}$ *for any invertible $\tau \in \mathcal{L}(\mathcal{P})$* □

Thus, the map $\phi: \mathcal{L}(\mathcal{P}) \to \mathcal{L}(\mathcal{F})$ that sends $\tau: \mathcal{P} \to \mathcal{P}$ to its adjoint $\tau^\times: \mathcal{F} \to \mathcal{F}$ is a linear transformation from $\mathcal{L}(\mathcal{P})$ to $\mathcal{L}(\mathcal{F})$. Moreover, since $\tau^\times = 0$ implies that $\langle h(t) \mid \tau p(x) \rangle = 0$ for all $h(t) \in \mathcal{F}$ and $p(x) \in \mathcal{P}$, which in turn implies that $\tau = 0$, we deduce that ϕ is injective. The next theorem describes the range of ϕ.

Theorem 19.17 *A linear operator $T \in \mathcal{L}(\mathcal{F})$ is the adjoint of a linear operator $\mathcal{L} \in \mathcal{L}(\mathcal{P})$ if and only if T is continuous.*
Proof. First, suppose that $T = \tau^\times$ for some $\tau \in \mathcal{L}(\mathcal{P})$ and let $o(f_k(t)) \to \infty$. If $n \geq 0$, then for all $0 \leq i \leq n$ we have

$$\langle \tau^\times f_k(t) \mid x^i \rangle = \langle f_k(t) \mid \tau x^i \rangle$$

and so it is only necessary to take k large enough that $o(f_k(t)) > \deg \tau(x^i)$ for all $0 \leq i \leq n$, whence

$$\langle \tau^\times f_k(t) \mid x^i \rangle = 0$$

for all $0 \le i \le n$ and so $o(\tau^\times f_k(t)) > n$. Thus, $o(\tau^\times f_k(t)) \to \infty$ and $\tau^\times$ is continuous.

For the converse, assume that T is continuous. If T did have the form $\tau^\times$, then

$$\langle Tt^k \mid x^n \rangle = \langle \tau^\times t^k \mid x^n \rangle = \langle t^k \mid \tau x^n \rangle$$

and since

$$\tau x^n = \sum_{k \ge 0} \frac{\langle t^k \mid \tau x^n \rangle}{k!} x^k$$

we are prompted to *define* τ by

$$\tau x^n = \sum_{k \ge 0} \frac{\langle Tt^k \mid x^n \rangle}{k!} x^k$$

This makes sense since $o(Tt^k) \to \infty$ as $k \to \infty$ and so the sum on the right is a finite sum. Then

$$\langle \tau^\times t^m \mid x^n \rangle = \langle t^m \mid \tau x^n \rangle = \sum_{k \ge 0} \frac{\langle Tt^k \mid x^n \rangle}{k!} \langle t^m \mid x^k \rangle = \langle Tt^m \mid x^n \rangle$$

which implies that $Tt^m = \tau^\times t^m$ for all $m \ge 0$. Finally, since T and $\tau^\times$ are both continuous, we have $T = \tau^\times$.$\square$

Umbral Operators and Automorphisms of the Umbral Algebra

Figure 19.1 shows the map ϕ, which is an isomorphism from the vector space $\mathcal{L}(\mathcal{P})$ onto the space of all continuous linear operators on $\mathcal{F}$. We are interested in determining the images under this isomorphism of the set of umbral operators and the set of umbral shifts, as pictured in Figure 19.1.

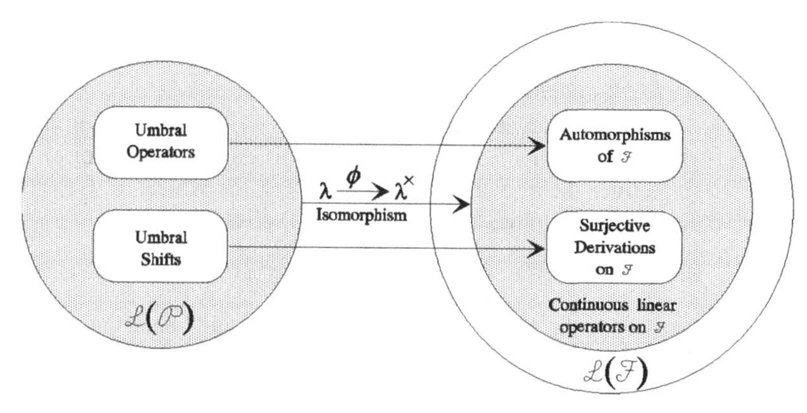

Figure 19.1

Let us begin with umbral operators. Suppose that λ_f is the umbral operator for the associated sequence $p_n(x)$, with delta series $f(t) \in \mathcal{F}$. Then

$$\langle \lambda_f^\times f(t)^k \mid x^n \rangle = \langle f(t)^k \mid \lambda_f x^n \rangle = \langle f(t)^k \mid p_n(x) \rangle = n!\delta_{n,k} = \langle t^k \mid x^n \rangle$$

for all k and n. Hence, $\lambda_f^\times f(t)^k = t^k$ and the continuity of $\lambda_f^\times$ implies that

$$\lambda_f^\times t^k = \overline{f}(t)^k$$

More generally, for any $h(t) \in \mathcal{F}$,

$$\lambda_f^\times h(t) = h(\overline{f}(t)) \tag{19.9}$$

In words, $\lambda_f^\times$ is composition by $\overline{f}(t)$.

From (19.9), we deduce that $\lambda_f^\times$ is a vector space isomorphism and that

$$\lambda_f^\times [g(t)h(t)] = g(\overline{f}(t))h(\overline{f}(t)) = \lambda_f^\times g(t)\lambda_f^\times h(t)$$

Hence, $\lambda_f^\times$ is an automorphism of the umbral algebra $\mathcal{F}$. It is a pleasant fact that this characterizes umbral operators. The first step in the proof of this is the following, whose proof is left as an exercise.

Theorem 19.18 *If T is an automorphism of the umbral algebra, then T preserves order, that is, $o(Tf(t)) = o(f(t))$. In particular, T is continuous.* □

Theorem 19.19 *A linear operator λ on $\mathcal{P}$ is an umbral operator if and only if its adjoint is an automorphism of the umbral algebra $\mathcal{F}$. Moreover, if λ_f is an umbral operator, then*

$$\lambda_f^\times h(t) = h(\overline{f}(t))$$

for all $h(t) \in \mathcal{F}$. *In particular,* $\lambda_f^\times f(t) = t$.

Proof. We have already shown that the adjoint of λ_f is an automorphism satisfying (19.9). For the converse, suppose that $\lambda^\times$ is an automorphism of $\mathcal{F}$. Since $\lambda^\times$ is surjective, there is a unique series $f(t)$ for which $\lambda^\times f(t) = t$. Moreover, Theorem 19.18 implies that $f(t)$ is a delta series. Thus,

$$n!\delta_{n,k} = \langle t^k \mid x^n \rangle = \langle \lambda^\times f(t)^k \mid x^n \rangle = \langle f(t)^k \mid \lambda x^n \rangle$$

which shows that λx^n is the associated sequence for $f(t)$ and hence that λ is an umbral operator.□

Theorem 19.19 allows us to fill in one of the boxes on the right side of Figure 19.1. Let us see how we might use Theorem 19.19 to advantage in the study of associated sequences.

We have seen that the isomorphism $\lambda \mapsto \lambda^\times$ maps the set $\mathcal{U}$ of umbral operators on $\mathcal{P}$ onto the set $\operatorname{aut}(\mathcal{F})$ of automorphisms of $\mathcal{F} = \mathcal{P}^*$. But $\operatorname{aut}(\mathcal{F})$ is a group under composition. So if

$$\lambda_f\colon x^n \to p_n(x) \text{ and } \lambda_g\colon x^n \to q_n(x)$$

are umbral operators, then since

$$(\lambda_g \circ \lambda_f)^\times = \lambda_f^\times \circ \lambda_g^\times$$

is an automorphism of $\mathcal{F}$, it follows that the composition $\lambda_g \circ \lambda_f$ is an umbral operator. In fact, since

$$(\lambda_g \circ \lambda_f)^\times f(g(t)) = \lambda_f^\times \circ \lambda_g^\times f(g(t)) = \lambda_f^\times f(t) = t$$

we deduce that $\lambda_g \circ \lambda_f = \lambda_{f \circ g}$. Also, since

$$\lambda_{\overline{f}} \circ \lambda_f = \lambda_{f \circ \overline{f}} = \lambda_t = \iota$$

we have $\lambda_f^{-1} = \lambda_{\overline{f}}$.

Thus, the set $\mathcal{U}$ of umbral operators is a group under composition with

$$\lambda_g \circ \lambda_f = \lambda_{f \circ g}$$

and

$$\lambda_f^{-1} = \lambda_{\overline{f}}$$

Let us see how this plays out with respect to associated sequences. If the associated sequence for $f(t)$ is

$$p_n(x) = \sum_{k=0}^{n} p_{n,k} x^k$$

then $\lambda_f: x^n \to p_n(x)$ and so $\lambda_{f \circ g} = \lambda_g \circ \lambda_f$ is the umbral operator for the associated sequence

$$(\lambda_g \circ \lambda_f) x^n = \lambda_g p_n(x) = \sum_{k=0}^{n} p_{n,k} \lambda_g x^k = \sum_{k=0}^{n} p_{n,k} q_k(x)$$

This sequence, denoted by

$$p_n(\boldsymbol{q}(x)) = \sum_{k=0}^{n} p_{n,k} q_k(x) \tag{19.10}$$

is called the **umbral composition** of $p_n(x)$ with $q_n(x)$. The umbral operator $\lambda_{\overline{f}} = \lambda_f^{-1}$ is the umbral operator for the associated sequence $r_n(x) = \Sigma r_{n,k} x^k$ where

$$\lambda_f^{-1} x^n = r_n(x)$$

and so

$$x^n = \sum_{k=0}^{n} r_{n,k} p_k(x)$$

Let us summarize.

Theorem 19.20

1) *The set $\mathcal{U}$ of umbral operators on $\mathcal{P}$ is a group under composition, with*

$$\lambda_g \circ \lambda_f = \lambda_{f \circ g} \quad \textit{and} \quad \lambda_f^{-1} = \lambda_{\overline{f}}$$

2) *The set of associated sequences forms a group under umbral composition*

$$p_n(\boldsymbol{q}(x)) = \sum_{k=0}^{n} p_{n,k} q_k(x)$$

In particular, the umbral composition $p_n(\boldsymbol{q}(x))$ is the associated sequence for the composition $f \circ g$, that is,

$$\lambda_{f \circ g}: x^n \to p_n(\boldsymbol{q}(x))$$

The identity is the sequence x^n and the inverse of $p_n(x)$ is the associated sequence for the compositional inverse $\overline{f}(t)$.

3) *Let* $\lambda_f \in \mathcal{U}$ *and* $g(t) \in \mathcal{F}$. *Then as operators,*

$$\lambda_f g(t) = \lambda_f^{-1} g(t) \lambda_f$$

4) *Let* $\lambda_f \in \mathcal{U}$ *and* $g(t) \in \mathcal{F}$. *Then*

$$\lambda_f g(\overline{f}(t)) = g(t)\lambda_f$$

Proof. We prove 3) as follows. For any $h(t) \in \mathcal{F}$ and $p(x) \in P$,

$$\begin{aligned} \langle h(t) \mid \lambda^\times g(t)p(x)\rangle &= \langle [\lambda^\times g(t)]h(t) \mid p(x)\rangle \\ &= \langle \lambda^\times [g(t)(\lambda^{-1})^\times h(t)] \mid p(x)\rangle \\ &= \langle g(t)(\lambda^{-1})^\times h(t) \mid \lambda p(x)\rangle \\ &= \langle (\lambda^{-1})^\times h(t) \mid g(t)\lambda p(x)\rangle \\ &= \langle h(t) \mid \lambda^{-1} g(t)\lambda p(x)\rangle \end{aligned}$$

which gives the desired result. Part 4) follows immediately from part 3) since λ_f is composition by $\overline{f}$.□

Sheffer Operators

If $s_n(x)$ is Sheffer for (g, f), then the linear operator $\lambda_{g,f}$ defined by

$$\lambda_{g,f}(x^n) = s_n(x)$$

is called a **Sheffer operator**. Sheffer operators are closely related to umbral operators, since if $p_n(x)$ is associated with $f(t)$, then

$$s_n(x) = g^{-1}(t)p_n(x) = g^{-1}(t)\lambda_f x^n$$

and so

$$\lambda_{g,f} = g^{-1}(t)\lambda_f$$

It follows that the Sheffer operators form a group with composition

$$\begin{aligned} \lambda_{g,f} \circ \lambda_{h,k} &= g^{-1}(t)\lambda_f h^{-1}(t)\lambda_k \\ &= g^{-1}(t)h^{-1}(f(t))\lambda_f\lambda_k \\ &= [g(t)h(f(t))]^{-1}\lambda_{k\circ f} \\ &= \lambda_{g\cdot(h\circ f),k\circ f} \end{aligned}$$

and inverse

$$\lambda_{g,f}^{-1} = \lambda_{g^{-1}(\overline{f}),\overline{f}}$$

From this, we deduce that the umbral composition of Sheffer sequences is a Sheffer sequence. In particular, if $s_n(x)$ is Sheffer for (g, f) and $t_n(x) = \Sigma t_{n,k}x^k$ is Sheffer for (h, k), then

$$\begin{aligned}\lambda_{g,f}\circ\lambda_{h,k}(x^n)&=\sum_{k=0}^{n}t_{n,k}\lambda_{g,f}x^k\\&=\sum_{k=0}^{n}t_{n,k}s_k(x)\\&=t_n(\boldsymbol{s}(x))\end{aligned}$$

is Sheffer for $(g\cdot(h\circ f),k\circ f)$.

Umbral Shifts and Derivations of the Umbral Algebra

We have seen that an operator on $\mathcal{P}$ is an umbral operator if and only if its adjoint is an automorphism of $\mathcal{F}$. Now suppose that $\theta_f\in\mathcal{L}(\mathcal{P})$ is the umbral shift for the associated sequence $p_n(x)$, associated with the delta series $f(t)\in\mathcal{F}$. Then

$$\begin{aligned}\langle\theta_f^\times f(t)^k\mid p_n(x)\rangle&=\langle f(t)^k\mid\theta_f p_n(x)\rangle\\&=\langle f(t)^k\mid p_{n+1}(x)\rangle\\&=(n+1)!\delta_{n+1,k}\\&=k(k-1)!\delta_{n,k-1}\\&=\langle kf(t)^{k-1}\mid p_n(x)\rangle\end{aligned}$$

and so

$$\theta_f^\times f(t)^k=kf(t)^{k-1}\tag{19.11}$$

This implies that

$$\theta_f^\times[f(t)^kf(t)^j]=\theta_f^\times[f(t)^k]f(t)^j+f(t)^k\theta_f^\times[f(t)^j]\tag{19.12}$$

and further, by continuity, that

$$\theta_f^\times[g(t)h(t)]=[\theta_f^\times g(t)]h(t)+g(t)[\theta_f^\times g(t)]\tag{19.13}$$

Let us pause for a definition.

Definition *Let $\mathcal{A}$ be an algebra. A linear operator ∂ on $\mathcal{A}$ is a* **derivation** *if*

$$\partial(ab)=(\partial a)b+a\partial b$$

for all $a,b\in\mathcal{A}$.□

Thus, we have shown that the adjoint of an umbral shift is a derivation of the umbral algebra $\mathcal{F}$. Moreover, the expansion theorem and (19.11) show that $\theta_f^\times$ is surjective. This characterizes umbral shifts. First we need a preliminary result on surjective derivations.

Theorem 19.21 *Let ∂ be a surjective derivation on the umbral algebra $\mathcal{F}$. Then $\partial c = 0$ for any* constant $c \in \mathcal{F}$ *and $o(\partial f(t)) = o(f(t)) - 1$, if $o(f(t)) \geq 1$. In particular, ∂ is continuous.*

Proof. We begin by noting that

$$\partial 1 = \partial 1^2 = \partial 1 + \partial 1 = 2\partial 1$$

and so $\partial c = c\partial 1 = 0$ for all constants $c \in \mathcal{F}$. Since ∂ is surjective, there must exist an $h(t) \in \mathcal{F}$ for which

$$\partial h(t) = 1$$

Writing $h(t) = h_0 + th_1(t)$, we have

$$1 = \partial[h_0 + th_1(t)] = (\partial t)h_1(t) + t\partial h_1(t)$$

which implies that $o(\partial t) = 0$. Finally, if $o(h(t)) = k \geq 1$, then $h(t) = t^k h_1(t)$, where $o(h_1(t)) = 0$ and so

$$o[\partial h(t)] = o[\partial t^k h_1(t)] = o[t^k \partial h(t) + kt^{k-1}h_1(t)\partial t] = k - 1 \qquad \square$$

Theorem 19.22 *A linear operator θ on $\mathcal{P}$ is an umbral shift if and only if its adjoint is a surjective derivation of the umbral algebra $\mathcal{F}$. Moreover, if θ_f is an umbral shift, then $\theta_f^\times = \partial_f$ is derivation with respect to $f(t)$, that is,*

$$\theta_f^\times f(t)^k = kf(t)^{k-1}$$

for all $k \geq 0$. In particular, $\theta_f^\times f(t) = 1$.

Proof. We have already seen that $\theta_f^\times$ is derivation with respect to $f(t)$. For the converse, suppose that $\theta^\times$ is a surjective derivation. Theorem 19.21 implies that there is a delta functional $f(t)$ such that $\theta^\times f(t) = 1$. If $p_n(x)$ is the associated sequence for $f(t)$, then

$$\begin{aligned}
\langle f(t)^k \mid \theta p_n(x)\rangle &= \langle \theta^\times f(t)^k \mid p_n(x)\rangle \\
&= \langle kf(t)^{k-1}\theta^\times f(t) \mid p_n(x)\rangle \\
&= \langle kf(t)^{k-1} \mid p_n(x)\rangle \\
&= (n+1)!\delta_{n+1,k} \\
&= \langle f(t)^k \mid p_{n+1}(x)\rangle
\end{aligned}$$

Hence, $\theta p_n(x) = p_{n+1}(x)$, that is, $\theta = \theta_f$ is the umbral shift for $p_n(x)$.$\square$

We have seen that the fact that the set of all automorphisms on $\mathcal{F}$ is a group under composition shows that the set of all associated sequences is a group under umbral composition. The set of all surjective derivations on $\mathcal{F}$ does not form a group. However, we do have the chain rule for derivations!

Theorem 19.23 (The chain rule) *Let ∂_f and ∂_g be surjective derivations on $\mathcal{F}$. Then*

$$\partial_g = (\partial_g f(t))\partial_f$$

Proof. This follows from

$$\partial_g f(t)^k = kf(t)^{k-1}\partial_g f(t) = (\partial_g f(t))\partial_f f(t)^k$$

and so continuity implies the result.□

The chain rule leads to the following umbral result.

Theorem 19.24 *If θ_f and θ_g are umbral shifts, then*

$$\theta_g = \theta_f \circ \partial_g f(t)$$

Proof. Taking adjoints in the chain rule gives

$$\theta_g = \theta_f \circ (\partial_g f(t))^{\times} = \theta_f \circ \partial_g f(t)$$ □

We leave it as an exercise to show that $\partial_g f(t) = [\partial_f g(t)]^{-1}$. Now, by taking $g(t) = t$ in Theorem 19.24 and observing that $\theta_t x^n = x^{n+1}$ and so θ_t is multiplication by x, we get

$$\theta_f = x\partial_f t = x[\partial_t f(t)]^{-1} = x[f'(t)]^{-1}$$

Applying this to the associated sequence $p_n(x)$ for $f(t)$ gives the following important recurrence relation for $p_n(x)$.

Theorem 19.25 (The recurrence formula) *Let $p_n(x)$ be the associated sequence for $f(t)$. Then*

1) $p_{n+1}(x) = x[f'(t)]^{-1}p_n(x)$

2) $p_{n+1}(x) = x\lambda_f[\overline{f}(t)]'x^n$

Proof. The first part is proved. As to the second, using Theorem 19.20 we have

$$\begin{aligned} p_{n+1}(x) &= x[f'(t)]^{-1}p_n(x) \\ &= x[f'(t)]^{-1}\lambda_f x^n \\ &= x\lambda_f[f'(\overline{f}(t))]^{-1}x^n \\ &= x\lambda_f[\overline{f}(t)]'x^n \end{aligned}$$ □

Example 19.9 The recurrence relation can be used to find the associated sequence for the forward difference functional $f(t) = e^t - 1$. Since $f'(t) = e^t$, the recurrence relation is

$$p_{n+1}(x) = xe^{-t}p_n(x) = xp_n(x-1)$$

Using the fact that $p_0(x) = 1$, we have

$$p_1(x) = x,\ p_2(x) = x(x-1),\ p_3(x) = x(x-1)(x-2)$$

and so on, leading easily to the lower factorial polynomials

$$p_n(x) = x(x-1)\cdots(x-n+1) = (x)_n$$

□

Example 19.10 Consider the delta functional

$$f(t) = \log(1+t)$$

Since $\overline{f}(t) = e^t - 1$ is the forward difference functional, Theorem 19.20 implies that the associated sequence $\phi_n(x)$ for $f(t)$ is the inverse, under umbral composition, of the lower factorial polynomials. Thus, if we write

$$\phi_n(x) = \sum_{k=0}^{n} S(n,k)x^k$$

then

$$x^n = \sum_{k=0}^{n} S(n,k)(x)_k$$

The coefficients $S(n,k)$ in this equation are known as the **Stirling numbers of the second kind** and have great combinatorial significance. In fact, $S(n,k)$ is the number of partitions of a set of size n into k blocks. The polynomials $\phi_n(x)$ are called the **exponential polynomials**.

The recurrence relation for the exponential polynomials is

$$\phi_{n+1}(x) = x(1+t)\phi_n(x) = x(\phi_n(x) + \phi_n'(x))$$

Equating coefficients of x^k on both sides of this gives the well-known formula for the Stirling numbers

$$S(n+1,k) = S(n,k-1) + kS(n,k)$$

Many other properties of the Stirling numbers can be derived by umbral means.□

Now we have the analog of part 3) of Theorem 19.20.

Theorem 19.26 *Let θ_f be an umbral shift. Then*

$$\theta_f^{\times} g(t) = g(t)\theta_f - \theta_f g(t)$$

Proof. We have

$$\begin{aligned}\langle f^k(t) \mid \theta_f^\times g(t)p_n(x)\rangle &= \langle [\theta_f^\times g(t)]f^k(t) \mid p_n(x)\rangle \\ &= \langle \theta_f^\times [g(t)f^k(t)] - g(t)\theta_f^\times f^k(t) \mid p_n(x)\rangle \\ &= \langle \theta_f^\times [g(t)f^k(t)] \mid p_n(x)\rangle - \langle kg(t)f^{k-1}(t) \mid p_n(x)\rangle \\ &= \langle f^k(t) \mid g(t)\theta_f p_n(x)\rangle - \langle kf^{k-1}(t) \mid g(t)p_n(x)\rangle \\ &= \langle f^k(t) \mid g(t)\theta_f p_n(x)\rangle - \langle \theta_f^\times f^k(t) \mid g(t)p_n(x)\rangle \\ &= \langle f^k(t) \mid g(t)\theta_f p_n(x)\rangle - \langle f^k(t) \mid \theta_f g(t)p_n(x)\rangle\end{aligned}$$

from which the result follows.□

If $f(t) = t$, then θ_f is multiplication by x and $\theta_f^\times$ is the derivative with respect to t and so the previous result becomes

$$g'(t) = g(t)x - xg(t)$$

as operators on $\mathcal{P}$. The right side of this is called the **Pincherle derivative** of the operator $g(t)$. (See [104].)

Sheffer Shifts

Recall that the linear map

$$\theta_{g,f}[s_n(x)] = s_{n+1}(x)$$

where $s_n(x)$ is Sheffer for $(g(t), f(t))$ is called a Sheffer shift. If $p_n(x)$ is associated with $f(t)$, then $g(t)s_n(x) = p_n(x)$ and so

$$g^{-1}(t)p_{n+1}(x) = \theta_{g,f}[g^{-1}(t)p_n(x)]$$

and so

$$\theta_{g,f} = g^{-1}(t)\theta_f g(t)$$

From Theorem 19.26, the recurrence formula and the chain rule, we have

$$\begin{aligned}\theta_{g,f} &= g^{-1}(t)\theta_f g(t) \\ &= g^{-1}(t)[g(t)\theta_f - \theta_f^\times g(t)] \\ &= \theta_f - g^{-1}(t)\partial_f g(t) \\ &= \theta_f - g^{-1}(t)\partial_f g(t) \\ &= \theta_f - g^{-1}(t)\partial_f t\partial_t g(t) \\ &= x[f'(t)]^{-1} - g^{-1}(t)[f'(t)]^{-1}g'(t) \\ &= \left[x - \frac{g'(t)}{g(t)}\right]\frac{1}{f'(t)}\end{aligned}$$

We have proved the following.

Theorem 19.27 *Let $\theta_{g,f}$ be a Sheffer shift. Then*

1) $\theta_{g,f} = \left[x - \frac{g'(t)}{g(t)}\right]\frac{1}{f'(t)}$
2) $s_{n+1}(x) = \left[x - \frac{g'(t)}{g(t)}\right]\frac{1}{f'(t)} s_n(x)$ $\square$

The Transfer Formulas

We conclude with a pair of formulas for the computation of associated sequences.

Theorem 19.28 (The transfer formulas) *Let $p_n(x)$ be the associated sequence for $f(t)$. Then*

1) $p_n(x) = f'(t)\left(\frac{f(t)}{t}\right)^{-n-1} x^n$
2) $p_n(x) = x\left(\frac{f(t)}{t}\right)^{-n} x^{n-1}$

Proof. First we show that 1) and 2) are equivalent. Write $g(t) = f(t)/t$. Then

$$\begin{aligned} f'(t)g(t)^{-n-1}x^n &= [tg(t)]'g(t)^{-n-1}x^n \\ &= g(t)^{-n}x^n + tg'(t)g(t)^{-n-1}x^n \\ &= g(t)^{-n}x^n + ng'(t)g(t)^{-n-1}x^{n-1} \\ &= g(t)^{-n}x^n + [g(t)^{-n}]'x^{n-1} \\ &= g(t)^{-n}x^n - [g(t)^{-n}x - xg(t)^{-n}]x^{n-1} \\ &= xg(t)^{-n}x^{n-1} \end{aligned}$$

To prove 1), we verify the operation conditions for an associated sequence for the sequence $q_n(x) = f'(t)g(t)^{-n-1}x^n$. First, when $n \geq 1$ the fourth equality above gives

$$\begin{aligned} \langle t^0 \mid q_n(x)\rangle &= \langle t^0 \mid f'(t)g(t)^{-n-1}x^n\rangle \\ &= \langle t^0 \mid g(t)^{-n}x^n - [g(t)^{-n}]'x^{n-1}\rangle \\ &= \langle g(t)^{-n} \mid x^n\rangle - \langle [g(t)^{-n}]' \mid x^{n-1}\rangle \\ &= \langle g(t)^{-n} \mid x^n\rangle - \langle g(t)^{-n} \mid x^n\rangle \\ &= 0 \end{aligned}$$

If $n = 0$, then $\langle t^0 \mid q_n(x)\rangle = 1$, and so in general, we have $\langle t^0 \mid q_n(x)\rangle = \delta_{n,0}$ as required.

For the second required condition,

$$\begin{aligned} f(t)q_n(x) &= f(t)f'(t)g(t)^{-n-1}x^n \\ &= tg(t)f'(t)g(t)^{-n-1}x^n \\ &= nf'(t)g(t)^{-n-1}x^{n-1} \\ &= nq_{n-1}(x) \end{aligned}$$

Thus, $q_n(x)$ is the associated sequence for $f(t)$.$\square$

A Final Remark

Unfortunately, space does not permit a detailed discussion of examples of Sheffer sequences nor the application of the umbral calculus to various classical problems. In [105], one can find a discussion of the following polynomial sequences:

The lower factorial polynomials and Stirling numbers
The exponential polynomials and Dobinski's formula
The Gould polynomials
The central factorial polynomials
The Abel polynomials
The Mittag-Leffler polynomials
The Bessel polynomials
The Bell polynomials
The Hermite polynomials
The Bernoulli polynomials and the Euler–MacLaurin expansion
The Euler polynomials
The Laguerre polynomials
The Bernoulli polynomials of the second kind
The Poisson–Charlier polynomials
The actuarial polynomials
The Meixner polynomials of the first and second kinds
The Pidduck polynomials
The Narumi polynomials
The Boole polynomials
The Peters polynomials
The squared Hermite polynomials
The Stirling polynomials
The Mahler polynomials
The Mott polynomials

and more. In [105], we also find a discussion of how the umbral calculus can be used to approach the following types of problems:

The connection constants problem
Duplication formulas
The Lagrange inversion formula
Cross sequences
Steffensen sequences
Operational formulas
Inverse relations
Sheffer sequence solutions to recurrence relations
Binomial convolution

Finally, it is possible to generalize the classical umbral calculus that we have described in this chapter to provide a context for studying polynomial sequences such as those of the names Gegenbauer, Chebyshev and Jacobi. Also, there is a q-version of the umbral calculus that involves the **q-binomial coefficients** (also known as the **Gaussian coefficients**)

$$\binom{n}{k}_q = \frac{(1-q)\cdots(1-q^n)}{(1-q)\cdots(1-q^k)(1-q)\cdots(1-q^{n-k})}$$

in place of the binomial coefficients. There is also a logarithmic version of the umbral calculus, which studies the *harmonic logarithms* and sequences of *logarithmic type*. For more on these topics, please see [103], [106] and [107].

Exercises

1. Prove that $o(fg) = o(f) + o(g)$, for any $f, g \in \mathcal{F}$.
2. Prove that $o(f+g) \geq \min\{o(f), o(g)\}$, for any $f, g \in \mathcal{F}$.
3. Show that any delta series has a compositional inverse.
4. Show that for any delta series f, the sequence f^k is a pseudobasis.
5. Prove that ∂_t is a derivation.
6. Show that $f \in \mathcal{F}$ is a delta functional if and only if $\langle f \mid 1 \rangle = 0$ and $\langle f \mid x \rangle \neq 0$.
7. Show that $f \in \mathcal{F}$ is invertible if and only if $\langle f \mid 1 \rangle \neq 0$.
8. Show that $\langle f(at) \mid p(x) \rangle = \langle f(t) \mid p(ax) \rangle$ for any $a \in \mathbb{C}$, $f \in \mathcal{F}$ and $p \in \mathcal{P}$.
9. Show that $\langle te^{at} \mid p(x) \rangle = p'(a)$ for any polynomial $p(x) \in \mathcal{P}$.
10. Show that $f = g$ in $\mathcal{F}$ if and only if $f = g$ as linear functionals, which holds if and only if $f = g$ as linear operators.
11. Prove that if $s_n(x)$ is Sheffer for $(g(t), f(t))$, then $f(t)s_n(x) = ns_{n-1}(x)$. *Hint*: Apply the functionals $g(t)f^k(t)$ to both sides.
12. Verify that the Abel polynomials form the associated sequence for the Abel functional.
13. Show that a sequence $s_n(x)$ is the Appell sequence for $g(t)$ if and only if $s_n(x) = g(t)^{-1}x^n$.
14. If f is a delta series, show that the adjoint $\lambda_f^\times$ of the umbral operator λ_f is a vector space isomorphism of $\mathcal{F}$.
15. Prove that if T is an automorphism of the umbral algebra, then T preserves order, that is, $o(Tf(t)) = o(f(t))$. In particular, T is continuous.
16. Show that an umbral operator maps associated sequences to associated sequences.
17. Let $p_n(x)$ and $q_n(x)$ be associated sequences. Define a linear operator α by $\alpha\colon p_n(x) \to q_n(x)$. Show that α is an umbral operator.
18. Prove that if ∂_f and ∂_g are surjective derivations on $\mathcal{F}$, then $\partial_g f(t) = [\partial_f g(t)]^{-1}$.

References

General References

[1] Jacobson, N., *Basic Algebra I*, second edition, W.H. Freeman, 1985.
[2] Snapper, E. and Troyer, R., *Metric Affine Geometry*, Dover Publications, 1971.

General Linear Algebra

[3] Akivis, M., Goldberg, V., *An Introduction To Linear Algebra and Tensors*, Dover, 1977.
[4] Blyth, T., Robertson, E., *Further Linear Algebra*, Springer, 2002.
[5] Brualdi, R., Friedland, S., Klee, V., *Combinatorial and Graph-Theoretical Problems in Linear Algebra*, Springer, 1993.
[6] Curtis, M., Place, P., *Abstract Linear Algebra*, Springer, 1990.
[7] Fuhrmann, P., *A Polynomial Approach to Linear Algebra*, Springer, 1996.
[8] Gel'fand, I. M., *Lectures On Linear Algebra*, Dover, 1989.
[9] Greub, W., *Linear Algebra*, Springer, 1995.
[10] Halmos, P. R., *Linear Algebra Problem Book*, Mathematical Association of America, 1995.
[11] Halmos, P. R., *Finite-Dimensional Vector Spaces*, Springer, 1974.
[12] Hamilton, A. G., *Linear Algebra*, Cambridge University Press, 1990.
[13] Jacobson, N., *Lectures in Abstract Algebra II: Linear Algebra*, Springer, 1953.
[14] Jänich, K., *Linear Algebra*, Springer, 1994.
[15] Kaplansky, I., *Linear Algebra and Geometry: A Second Course*, Dover, 2003.
[16] Kaye, R., Wilson, R., *Linear Algebra*, Oxford University Press, 1998.
[17] Kostrikin, A. and Manin, Y., *Linear Algebra and Geometry*, Gordon and Breach Science Publishers, 1997.
[18] Lax, P., *Linear Algebra*, John Wiley, 1996.
[19] Lewis, J. G., *Proceedings of the 5th SIAM Conference On Applied Linear Algebra*, SIAM, 1994.

[20] Marcus, M., Minc, H., *Introduction to Linear Algebra*, Dover, 1988.
[21] Mirsky, L., *An Introduction to Linear Algebra*, Dover, 1990.
[22] Nef, W., *Linear Algebra*, Dover, 1988.
[23] Nering, E. D., *Linear Algebra and Matrix Theory*, John Wiley, 1976.
[24] Pettofrezzo, A., *Matrices and Transformations*, Dover, 1978.
[25] Porter, G., Hill, D., *Interactive Linear Algebra: A Laboratory Course Using Mathcad*, Springer, 1996.
[26] Schneider, H., Barker, G., *Matrices and Linear Algebra*, Dover, 1989.
[27] Schwartz, J., *Introduction to Matrices and Vectors*, Dover, 2001.
[28] Shapiro, H., A survey of canonical forms and invariants for unitary similarity, *Linear Algebra and Its Applications* 147:101–167 (1991).
[29] Shilov, G., *Linear Algebra*, Dover, 1977.
[30] Wilkinson, J., *The Algebraic Eigenvalue Problem*, Oxford University Press, 1988.

Matrix Theory

[31] Antosik, P., Swartz, C., *Matrix Methods in Analysis*, Springer, 1985.
[32] Bapat, R., Raghavan, T., *Nonnegative Matrices and Applications*, Cambridge University Press, 1997.
[33] Barnett, S., *Matrices*, Oxford University Press, 1990.
[34] Bellman, R., *Introduction to Matrix Analysis*, SIAM, 1997.
[35] Berman, A., Plemmons, R., *Non-negative Matrices in the Mathematical Sciences*, SIAM, 1994.
[36] Bhatia, R., *Matrix Analysis*, Springer, 1996.
[37] Bowers, J., *Matrices and Quadratic Forms*, Oxford University Press, 2000.
[38] Boyd, S., El Ghaoui, L., Feron, E.; Balakrishnan, V., *Linear Matrix Inequalities in System and Control Theory*, SIAM, 1994.
[39] Chatelin, F., *Eigenvalues of Matrices*, John Wiley, 1993.
[40] Ghaoui, L., *Advances in Linear Matrix Inequality Methods in Control*, SIAM, 1999.
[41] Coleman, T., Van Loan, C., *Handbook for Matrix Computations*, SIAM, 1988.
[42] Duff, I., Erisman, A., Reid, J., *Direct Methods for Sparse Matrices*, Oxford University Press, 1989.
[43] Eves, H., *Elementary Matrix Theory*, Dover, 1966.
[44] Franklin, J., *Matrix Theory*, Dover, 2000.
[45] Gantmacher, F.R., *Matrix Theory I*, American Mathematical Society, 2000.
[46] Gantmacher, F.R., *Matrix Theory II*, American Mathematical Society, 2000.
[47] Gohberg, I., Lancaster, P., Rodman, L., *Invariant Subspaces of Matrices with Applications*, John Wiley, 1986.
[48] Horn, R. and Johnson, C., *Matrix Analysis*, Cambridge University Press, 1985.

[49] Horn, R. and Johnson, C., *Topics in Matrix Analysis*, Cambridge University Press, 1991.
[50] Jennings, A., McKeown, J. J., *Matrix Computation*, John Wiley, 1992.
[51] Joshi, A. W., *Matrices and Tensors in Physics*, John Wiley, 1995.
[52] Laub, A., *Matrix Analysis for Scientists and Engineers*, SIAM, 2004.
[53] Lütkepohl, H., *Handbook of Matrices*, John Wiley, 1996.
[54] Marcus, M., Minc, H., *A Survey of Matrix Theory and Matrix Inequalities*, Dover, 1964.
[55] Meyer, C., *Matrix Analysis and Applied Linear Algebra*, SIAM, 2000.
[56] Muir, T., *A Treatise on the Theory of Determinants*, Dover, 2003.
[57] Perlis, S., *Theory of Matrices*, Dover, 1991.
[58] Serre, D., *Matrices: Theory and Applications*, Springer, 2002.
[59] Stewart, G., *Matrix Algorithms*, SIAM, 1998.
[60] Stewart, G., *Matrix Algorithms Volume II: Eigensystems*, SIAM, 2001.
[61] Watkins, D., *Fundamentals of Matrix Computations*, John Wiley, 1991.

Multilinear Algebra

[62] Marcus, M., *Finite Dimensional Multilinear Algebra, Part I*, Marcel Dekker, 1971.
[63] Marcus, M., *Finite Dimensional Multilinear Algebra, Part II*, Marcel Dekker, 1975.
[64] Merris, R., *Multilinear Algebra*, Gordon & Breach, 1997.
[65] Northcott, D. G., *Multilinear Algebra*, Cambridge University Press, 1984.

Applied and Numerical Linear Algebra

[66] Anderson, E., *LAPACK User's Guide*, SIAM, 1995.
[67] Axelsson, O., *Iterative Solution Methods*, Cambridge University Press, 1994.
[68] Bai, Z., *Templates for the Solution of Algebraic Eigenvalue Problems: A Practical Guide*, SIAM, 2000.
[69] Banchoff, T., Wermer, J., *Linear Algebra Through Geometry*, Springer, 1992.
[70] Blackford, L., *ScaLAPACK User's Guide*, SIAM, 1997.
[71] Ciarlet, P. G., *Introduction to Numerical Linear Algebra and Optimization*, Cambridge University Press, 1989.
[72] Campbell, S., Meyer, C., *Generalized Inverses of Linear Transformations*, Dover, 1991.
[73] Datta, B., Johnson, C., Kaashoek, M., Plemmons, R., Sontag, E., *Linear Algebra in Signals, Systems and Control*, SIAM, 1988.
[74] Demmel, J., *Applied Numerical Linear Algebra*, SIAM, 1997.
[75] Dongarra, J., Bunch, J. R., Moler, C. B., Stewart, G. W., *Linpack Users' Guide*, SIAM, 1979.
[76] Dongarra, J., *Numerical Linear Algebra for High-Performance Computers*, SIAM, 1998.

[77] Dongarra, J., *Templates for the Solution of Linear Systems: Building Blocks For Iterative Methods*, SIAM, 1993.
[78] Faddeeva, V. N., *Computational Methods of Linear Algebra*, Dover,
[79] Frazier, M., *An Introduction to Wavelets Through Linear Algebra*, Springer, 1999.
[80] George, A., Gilbert, J., Liu, J., *Graph Theory and Sparse Matrix Computation*, Springer, 1993.
[81] Golub, G., Van Dooren, P., *Numerical Linear Algebra, Digital Signal Processing and Parallel Algorithms*, Springer, 1991.
[82] Granville S., *Computational Methods of Linear Algebra*, 2nd edition, John Wiley, 2005.
[83] Greenbaum, A., *Iterative Methods for Solving Linear Systems*, SIAM, 1997.
[84] Gustafson, K., Rao, D., *Numerical Range: The Field of Values of Linear Operators and Matrices*, Springer, 1996.
[85] Hackbusch, W., *Iterative Solution of Large Sparse Systems of Equations*, Springer, 1993.
[86] Jacob, B., *Linear Functions and Matrix Theory*, Springer, 1995.
[87] Kuijper, M., *First-Order Representations of Linear Systems*, Birkhäuser, 1994.
[88] Meyer, C., Plemmons, R., *Linear Algebra, Markov Chains, and Queueing Models*, Springer, 1993.
[89] Neumaier, A., *Interval Methods for Systems of Equations*, Cambridge University Press, 1991.
[90] Nevanlinna, O., *Convergence of Iterations for Linear Equations*, Birkhäuser, 1993.
[91] Olshevsky, V., *Fast Algorithms for Structured Matrices: Theory and Applications*, SIAM, 2003.
[92] Plemmon, R.J., Gallivan, K.A., Sameh, A.H., *Parallel Algorithms for Matrix Computations*, SIAM, 1990.
[93] Rao, K. N., *Linear Algebra and Group Theory for Physicists*, John Wiley, 1996.
[94] Reichel, L., Ruttan, A., Varga, R., *Numerical Linear Algebra*, Walter de Gruyter, 1993.
[95] Saad, Y., *Iterative Methods for Sparse Linear Systems*, SIAM, 2003.
[96] Scharlau, W., *Quadratic and Hermitian Forms*, Springer, 1985.
[97] Snapper, E., Troyer, R., *Metric Affine Geometry*, Dover,
[98] Spedicato, E., *Computer Algorithms for Solving Linear Algebraic Equations*, Springer, 1991.
[99] Trefethen, L., Bau, D., *Numerical Linear Algebra*, SIAM, 1997.
[100] Van Dooren, P., Wyman, B., *Linear Algebra for Control Theory*, Springer, 1994.
[101] Vorst, H., *Iterative Krylov Methods for Large Linear Systems*, Cambridge University Press, 2003.
[102] Young, D., *Iterative Solution of Large Linear Systems*, Dover, 2003.

The Umbral Calculus

[103] Loeb, D. and Rota, G.-C., Formal Power Series of Logarithmic Type, *Advances in Mathematics*, Vol. 75, No. 1, (May 1989) 1–118.
[104] Pincherle, S. "Operatori lineari e coefficienti di fattoriali." *Alti Accad. Naz. Lincei, Rend. Cl. Fis. Mat. Nat.* (6) 18, 417–519, 1933.
[105] Roman, S., *The Umbral Calculus*, Pure and Applied Mathematics vol. 111, Academic Press, 1984.
[106] Roman, S., The logarithmic binomial formula, *American Mathematical Monthly* 99 (1992) 641–648.
[107] Roman, S., The harmonic logarithms and the binomial formula, *Journal of Combinatorial Theory*, series A, 63 (1992) 143–163.

Index of Symbols

$C[p(x)]$: the companion matrix of $p(x)$
$c_\tau(x)$: characteristic polynomial of τ
crk(A): column rank of A
cs(A): column space of A
diag$(A_1, \ldots, A_n)$: a block diagonal matrix with A_i's on the block diagonal
ElemDiv(τ): the multiset of elementary divisors
InvFact(τ): the multiset of invariant factors of τ
$\mathcal{J}(\lambda_i, e_{i,j})$: Jordan block
$m_\tau(x)$: minimal polynomial of τ
null(τ): the nullity of τ
π_S: canonical projection modulo S
R_f: Riesz vector for $f \in V^*$
rk(τ): the rank of τ
rrk(A): row rank of A
rs(A): row space of A
$\rho_{A,B}$: projection onto A along B
τ_A: the multiplication by A operator
supp(f): the support of a function
V_τ: the F-vector space/$F[x]$-module where $p(x)v = p(\tau)v$
$V^{\mathbb{C}}$: the complexification of V
$:=$: assignment, for example, $u := \langle S \rangle$ means that u stands for $\langle S \rangle$
$\leq$: subspace or submodule
$<$: proper subspace or proper submodule
$\langle S \rangle$: subspace/ideal spanned by S
$\langle\langle S \rangle\rangle$: submodule spanned by S
$\overset{\sim}{\hookrightarrow}$: an embedding that is an isomorphism when all is finite-dimensional.
$\sim$: similarity of matrices or operators, associate in a ring.
$\times$: cartesian product
$\odot$: orthogonal direct sum
$\boxtimes$: external direct product
$\boxplus$: external direct sum
$\oplus$: internal direct sum
$\bowtie$:$x \bowtie y$ means $\langle x, y \rangle = \langle y, x \rangle$

Index

Graduate Texts in Mathematics

(continued from page ii)

76 Iitaka. Algebraic Geometry.
77 Hecke. Lectures on the Theory of Algebraic Numbers.
78 Burris/Sankappanavar. A Course in Universal Algebra.
79 Walters. An Introduction to Ergodic Theory.
80 Robinson. A Course in the Theory of Groups. 2nd ed.
81 Forster. Lectures on Riemann Surfaces.
82 Bott/Tu. Differential Forms in Algebraic Topology.
83 Washington. Introduction to Cyclotomic Fields. 2nd ed.
84 Ireland/Rosen. A Classical Introduction to Modern Number Theory. 2nd ed.
85 Edwards. Fourier Series. Vol. II. 2nd ed.
86 van Lint. Introduction to Coding Theory. 2nd ed.
87 Brown. Cohomology of Groups.
88 Pierce. Associative Algebras.
89 Lang. Introduction to Algebraic and Abelian Functions. 2nd ed.
90 Brøndsted. An Introduction to Convex Polytopes.
91 Beardon. On the Geometry of Discrete Groups.
92 Diestel. Sequences and Series in Banach Spaces.
93 Dubrovin/Fomenko/Novikov. Modern Geometry—Methods and Applications. Part I. 2nd ed.
94 Warner. Foundations of Differentiable Manifolds and Lie Groups.
95 Shiryaev. Probability. 2nd ed.
96 Conway. A Course in Functional Analysis. 2nd ed.
97 Koblitz. Introduction to Elliptic Curves and Modular Forms. 2nd ed.
98 Bröcker/tom Dieck. Representations of Compact Lie Groups.
99 Grove/Benson. Finite Reflection Groups. 2nd ed.
100 Berg/Christensen/Ressel. Harmonic Analysis on Semigroups: Theory of Positive Definite and Related Functions.
101 Edwards. Galois Theory.
102 Varadarajan. Lie Groups, Lie Algebras and Their Representations.
103 Lang. Complex Analysis. 3rd ed.
104 Dubrovin/Fomenko/Novikov. Modern Geometry—Methods and Applications. Part II.
105 Lang. $SL_2(\mathrm{R})$.
106 Silverman. The Arithmetic of Elliptic Curves.
107 Olver. Applications of Lie Groups to Differential Equations. 2nd ed.
108 Range. Holomorphic Functions and Integral Representations in Several Complex Variables.
109 Lehto. Univalent Functions and Teichmüller Spaces.
110 Lang. Algebraic Number Theory.
111 Husemöller. Elliptic Curves. 2nd ed.
112 Lang. Elliptic Functions.
113 Karatzas/Shreve. Brownian Motion and Stochastic Calculus. 2nd ed.
114 Koblitz. A Course in Number Theory and Cryptography. 2nd ed.
115 Berger/Gostiaux. Differential Geometry: Manifolds, Curves, and Surfaces.
116 Kelley/Srinivasan. Measure and Integral. Vol. I.
117 J.-P. Serre. Algebraic Groups and Class Fields.
118 Pedersen. Analysis Now.
119 Rotman. An Introduction to Algebraic Topology.
120 Ziemer. Weakly Differentiable Functions: Sobolev Spaces and Functions of Bounded Variation.
121 Lang. Cyclotomic Fields I and II. Combined 2nd ed.
122 Remmert. Theory of Complex Functions. *Readings in Mathematics*
123 Ebbinghaus/Hermes et al. Numbers. *Readings in Mathematics*
124 Dubrovin/Fomenko/Novikov. Modern Geometry—Methods and Applications Part III.
125 Berenstein/Gay. Complex Variables: An Introduction.
126 Borel. Linear Algebraic Groups. 2nd ed.
127 Massey. A Basic Course in Algebraic Topology.
128 Rauch. Partial Differential Equations.
129 Fulton/Harris. Representation Theory: A First Course. *Readings in Mathematics*
130 Dodson/Poston. Tensor Geometry.
131 Lam. A First Course in Noncommutative Rings. 2nd ed.
132 Beardon. Iteration of Rational Functions.
133 Harris. Algebraic Geometry: A First Course.
134 Roman. Coding and Information Theory.
135 Roman. Advanced Linear Algebra. 3rd ed.
136 Adkins/Weintraub. Algebra: An Approach via Module Theory.
137 Axler/Bourdon/Ramey. Harmonic Function Theory. 2nd ed.

138 Cohen. A Course in Computational Algebraic Number Theory.
139 Bredon. Topology and Geometry.
140 Aubin. Optima and Equilibria. An Introduction to Nonlinear Analysis.
141 Becker/Weispfenning/Kredel. Gröbner Bases. A Computational Approach to Commutative Algebra.
142 Lang. Real and Functional Analysis. 3rd ed.
143 Doob. Measure Theory.
144 Dennis/Farb. Noncommutative Algebra.
145 Vick. Homology Theory. An Introduction to Algebraic Topology. 2nd ed.
146 Bridges. Computability: A Mathematical Sketchbook.
147 Rosenberg. Algebraic *K*-Theory and Its Applications.
148 Rotman. An Introduction to the Theory of Groups. 4th ed.
149 Ratcliffe. Foundations of Hyperbolic Manifolds. 2nd ed.
150 Eisenbud. Commutative Algebra with a View Toward Algebraic Geometry.
151 Silverman. Advanced Topics in the Arithmetic of Elliptic Curves.
152 Ziegler. Lectures on Polytopes.
153 Fulton. Algebraic Topology: A First Course.
154 Brown/Pearcy. An Introduction to Analysis.
155 Kassel. Quantum Groups.
156 Kechris. Classical Descriptive Set Theory.
157 Malliavin. Integration and Probability.
158 Roman. Field Theory.
159 Conway. Functions of One Complex Variable II.
160 Lang. Differential and Riemannian Manifolds.
161 Borwein/Erdélyi. Polynomials and Polynomial Inequalities.
162 Alperin/Bell. Groups and Representations.
163 Dixon/Mortimer. Permutation Groups.
164 Nathanson. Additive Number Theory: The Classical Bases.
165 Nathanson. Additive Number Theory: Inverse Problems and the Geometry of Sumsets.
166 Sharpe. Differential Geometry: Cartan's Generalization of Klein's Erlangen Program.
167 Morandi. Field and Galois Theory.
168 Ewald. Combinatorial Convexity and Algebraic Geometry.
169 Bhatia. Matrix Analysis.
170 Bredon. Sheaf Theory. 2nd ed.
171 Petersen. Riemannian Geometry. 2nd ed.
172 Remmert. Classical Topics in Complex Function Theory.
173 Diestel. Graph Theory. 2nd ed.
174 Bridges. Foundations of Real and Abstract Analysis.
175 Lickorish. An Introduction to Knot Theory.
176 Lee. Riemannian Manifolds.
177 Newman. Analytic Number Theory.
178 Clarke/Ledyaev/Stern/Wolenski. Nonsmooth Analysis and Control Theory.
179 Douglas. Banach Algebra Techniques in Operator Theory. 2nd ed.
180 Srivastava. A Course on Borel Sets.
181 Kress. Numerical Analysis.
182 Walter. Ordinary Differential Equations.
183 Megginson. An Introduction to Banach Space Theory.
184 Bollobas. Modern Graph Theory.
185 Cox/Little/O'Shea. Using Algebraic Geometry. 2nd ed.
186 Ramakrishnan/Valenza. Fourier Analysis on Number Fields.
187 Harris/Morrison. Moduli of Curves.
188 Goldblatt. Lectures on the Hyperreals: An Introduction to Nonstandard Analysis.
189 Lam. Lectures on Modules and Rings.
190 Esmonde/Murty. Problems in Algebraic Number Theory. 2nd ed.
191 Lang. Fundamentals of Differential Geometry.
192 Hirsch/Lacombe. Elements of Functional Analysis.
193 Cohen. Advanced Topics in Computational Number Theory.
194 Engel/Nagel. One-Parameter Semigroups for Linear Evolution Equations.
195 Nathanson. Elementary Methods in Number Theory.
196 Osborne. Basic Homological Algebra.
197 Eisenbud/Harris. The Geometry of Schemes.
198 Robert. A Course in p-adic Analysis.
199 Hedenmalm/Korenblum/Zhu. Theory of Bergman Spaces.
200 Bao/Chern/Shen. An Introduction to Riemann–Finsler Geometry.
201 Hindry/Silverman. Diophantine Geometry: An Introduction.
202 Lee. Introduction to Topological Manifolds.
203 Sagan. The Symmetric Group: Representations, Combinatorial Algorithms, and Symmetric Functions.
204 Escofier. Galois Theory.
205 Félix/Halperin/Thomas. Rational Homotopy Theory. 3rd ed.
206 Murty. Problems in Analytic Number Theory. *Readings in Mathematics*
207 Godsil/Royle. Algebraic Graph Theory.
208 Cheney. Analysis for Applied Mathematics.
209 Arveson. A Short Course on Spectral Theory.
210 Rosen. Number Theory in Function Fields.
211 Lang. Algebra. Revised 3rd ed.
212 Matoušek. Lectures on Discrete Geometry.
213 Fritzsche/Grauert. From Holomorphic Functions to Complex Manifolds.

214 Jost. Partial Differential Equations. 2nd ed.
215 Goldschmidt. Algebraic Functions and Projective Curves.
216 D. Serre. Matrices: Theory and Applications.
217 Marker. Model Theory: An Introduction.
218 Lee. Introduction to Smooth Manifolds.
219 Maclachlan/Reid. The Arithmetic of Hyperbolic 3-Manifolds.
220 Nestruev. Smooth Manifolds and Observables.
221 Grünbaum. Convex Polytopes. 2nd ed.
222 Hall. Lie Groups, Lie Algebras, and Representations: An Elementary Introduction.
223 Vretblad. Fourier Analysis and Its Applications.
224 Walschap. Metric Structures in Differential Geometry.
225 Bump. Lie Groups.
226 Zhu. Spaces of Holomorphic Functions in the Unit Ball.
227 Miller/Sturmfels. Combinatorial Commutative Algebra.
228 Diamond/Shurman. A First Course in Modular Forms.
229 Eisenbud. The Geometry of Syzygies.
230 Stroock. An Introduction to Markov Processes.
231 Björner/Brenti. Combinatorics of Coxeter Groups.
232 Everest/Ward. An Introduction to Number Theory.
233 Albiac/Kalton. Topics in Banach Space Theory.
234 Jorgenson. Analysis and Probability.
235 Sepanski. Compact Lie Groups.
236 Garnett. Bounded Analytic Functions.
237 Martínez-Avendaño/Rosenthal. An Introduction to Operators on the Hardy-Hilbert Space.
238 Aigner, A Course in Enumeration.
239 Cohen, Number Theory, Vol. I.
240 Cohen, Number Theory, Vol. II.
241 Silverman. The Arithmetic of Dynamical Systems.
242 Grillet. Abstract Algebra. 2nd ed.
243 Geoghegan. Topological Methods in Group Theory.
244 Bondy/Murty. Graph Theory.
245 Gilman/Kra/Rodriguez. Complex Analysis.
246 Kaniuth. A Course in Commutative Banach Algebras.